GEOMECHANICS 93

PROCEEDINGS OF THE INTERNATIONAL CONFERENCE GEOMECHANICS 93
HRADEC / OSTRAVA / CZECH REPUBLIC / 28 - 30 SEPTEMBER 1993

Geomechanics 93

Strata Mechanics / Numerical Methods / Water Jet Cutting / Mechanical Rock Disintegration

Edited by
ZIKMUND RAKOWSKI
Institute of Geonics of Academy of Sciences of Czech Republic, Ostrava, Czech Republic

A.A. BALKEMA / ROTTERDAM / BROOKFIELD / 1994

The texts of the various papers in this volume were set individually by typists under the supervision of each of the authors concerned.

Published by
A.A. Balkema, P.O. Box 1675, 3000 BR Rotterdam, Netherlands
A.A. Balkema Publishers, Old Post Road, Brookfield, VT 05036, USA

ISBN 90 5410 354 X

Printed in the Netherlands

Geomechanics 93, Rakowski (ed.) © 1994 Balkema, Rotterdam, ISBN 90 5410 354 X

Table of contents

Strata mechanics

Technical notes

Numerical methods
Main lectures

Numerical methods
Lectures

Numerical methods
Technical notes

Water jet cutting
Main lectures

Water jet cutting
Lectures

Mechanical rock disintegration

Technical notes

Geomechanics 93, Rakowski (ed.) © 1994 Balkema, Rotterdam, ISBN 90 5410 354 X

About GEONICS

Z. Rakowski

The process commonly called *going underground* has been very well observed recently around the world. In short it means rather rapidly developing transference of numbers of even highly civilized human activities to the underground space. Geosphere, having been traditionally used as source of energy and raw materials, has now become the space where any anthropogenic activity could be placed and realized. In this way, apart from mining, we can also find various facilities for industrial transport, waste deponies, sport and/or even cultural and social activities underground. Main reasons for it could be found in:

– Generally lower accessibility and growing prices of the land;
– specific properties of rock materials which could be effectively utilized (like thermoconductivity, permeability, strength, etc.);
– the possibility of separating some dangerous anthropogenic activities from the important components of the environment such as hydrosphere, atmosphere, etc.;
– comparatively large scale of geosphere, etc.

Observing the processes connected with *going underground* or accompanying it one could see some similarities in the approaches to the rock environment with the same ones to the atmosphere or hydrosphere some decades ago. Neglecting some fundamental principles in the past led us to the recent problems with the environment in global scale. To avoid such a development in case of geosphere it seems to be just the right time now for initiating certain changes in the approach to this problem. Most generally the most important requirements regarding anthropogenic activities in geosphere could be defined as follows:

a) As great adaptability of anthropogenic activities to the original nature of the geosphere as possible.

b) As small environmental impact on the rock mass itself and eco-systems on the surface as possible.

Fulfilling the above mentioned requirements will be influenced by several important factors like:

– The level of knowledge of natural properties and processes in the geosphere;
– a certain degree of similarity of the character of anthropogenic processes to the same one of the nature of geosphere;
– a certain degree of utilization of primary energy being stored in the rock mass for underground human activities.

Such a formulation leads to the following definition of scientific and research areas:

– Properties of the rock mass and its state;
– natural and induced processes in geosphere;
– simulation (modelling) of the induced processes;
– environmental impact of these processes.

Such an approach requires inclusion of many disciplines and so it forms typical interdisciplinarity by involving the disciplines both from GEOsciences and different ones characterizing anthropogeNIC activities (e.g. techNICS, physICS). Here the composed word GEONICS is well described as the unification of the above mentioned interdisciplinarity of numbers of disciplines aimed to a common goal which should be highly sophisticated and environmentally fully sensitive utilizing of underground space. The parallel to BIONICS which has been respected as a very modern scientific discipline for many years is not accidental.

GEOMECHANICS should also be one of the fields forming GEONICS. The proceedings of the International Conference GEOMECHANICS 93 contain many papers which are more or less based on such an approach.

The conference Geomechanics 93 worked in four sections: strata mechanics, numerical methods, water jet cutting and mechanical rock disintegration.

The strata mechanics section was generally dominated by papers dealing with rock mass modification problems, and rock bolting and anchoring formed an essential part of them. Great attention was paid traditionally to rock bursts and ways and methods of defining rock and rock mass properties. Non-linear geomechanical problems, modelling of jointed rock mass and advanced numerical techniques formed the main part of the second section. Engineering applications and case histories were also presented. Up-to-date progress in water jet cutting techniques and technology in rock and rock-like materials is visible from the papers of the third section. Traditional ways of disintegration of rocks were accomplished by interesting non-traditional approaches in section four.

The 13th Plenary Session of the International Bureau of Strata Mechanics of World Mining Congress was held at the same place after the conference.

The very good discussions in the working sessions and generally a very good atmosphere during the conference allowed the statement that GEOMECHANICS 93 was a very successful event which enabled scientists and experts from 16 countries of three continents to share new scientific knowledge and professional experience. The published proceedings could be useful material for wide international communities.

Geomechanics 93, Rakowski (ed.) © 1994 Balkema, Rotterdam, ISBN 90 5410 354 X

Organization

CONFERENCE CHAIRMAN
Z. Rakowski, Academy of Sciences of the Czech Republic, Institute of Geonics, Ostrava

SECTIONS
1. Strata mechanics – Chairman: Z. Rakowski
2. Numerical methods – Chairman: R. Blaheta
3. Water jet cutting – Chairmen: J. Vašek and M. Mazurkiewicz
4. Mechanical rock disintegration – Chairman: F. Sekula

INTERNATIONAL PROGRAMME COMMITTEE

R. Blaheta (2) (Czech Republic)
V.A. Brenner (4) (Russia)
N.S. Bulytchev (1) (Russia)
M. Doležalová (2) (Czech Republic)
Z. Dostál (2) (Czech Republic)
R.J. Fowell (3) (UK)
O. Del Greco (1) (Italy)
A. Kidybiński (1) (Poland)
A. Wahab Khair (1) (USA)
R. Kobayashi (3) (Japan)
P. Konečný (Czech Republic)
H. Louis (3) (Germany)
M. Mazurkiewicz (3) (USA)
Z. Mróz (2) (Poland)
Z. Rakowski (1) (Czech Republic)
F. Sekula (4) (Slovak Republic)
T.R. Stacey (1) (South Africa)
D. Summers (3) (USA)
J. Vašek (3) (Czech Republic)
M.M. Vijay (3) (Canada)
F.L. Wilke (4) (Germany)
T. Yahiro (3) (Japan)

Geomechanics 93, Rakowski (ed.) © 1994 Balkema, Rotterdam, ISBN 90 5410 354 X

Acknowledgements

The conference Geomechanics 93 was kindly sponsored by the International Bureau of Strata Mechanics of World Mining Congress and the International Society of Water Jet Technology. On behalf of the Organizing Committee I would like to acknowledge the activities of Prof. Kidybinski (President of the IBSM) and Dr M.M. Vijay (President of the ISWJT) in this matter. The Institute of Geonics of the Czech Academy of Sciences in Ostrava provided all the necessary secretarial and administrative assistance.

Z. Rakowski

Strata mechanics
Main lectures

Geomechanics 93, Rakowski (ed.) © 1994 Balkema, Rotterdam, ISBN 90 5410 354 X

Towards a methodology for mechanics of underground structures

N. S. Bulytchev
Tula State Technical University, Russia

ABSTRACT: Choice of adequate rock mass models both for stability estimation of an underground openings and for design and analysis of support for permanent mine workings and tunnel linings is subject of this paper. Problems of the linear or nonlinear models, external and internal plausibility of the ones, analytical or numerical approach, a required precision of rock mass chracteristics determination are discussed

1 INTRODUCTION

Mechanics of underground structures is the theory of underground structures, theory of analysis of working supports and tunnel linings. It has risen as a branch of rock mechanics in which the focus of attention has shifted from rock massif to structures.

The basic principle in mechanics of underground structures is that of joint deformation (interaction) of the rock massif and the lining. In accordance with this principle, the basic design lay-out for all types of linings and other supporting elements subject to any kinds of loads is the contact interaction of the lining and the rock massif being deformed, in which the lining and massif are elements of a common deformable "lining-rock" system.

The principle mentioned above is realised in different methods of investigation such as: phisical (object) modelling, for instance the photoelastic one, the method of equivalent materials, method of centrifugal modelling, and method of nuverical simulation, for example the finite elements method. However, all these methods are not theoretical ones, they are rather modelling arts. They may be included in the mechanics of underground structures as means of development of the theory.

The aim of the underground structures mechanics is both to gain new understanding (knowledge) and elaborate practical methods of lining analysis.

Design and computation of real underground objects may be fulfiled by several methods at the same time. And that is the most preferabl taking into account the decrease of errors. As to the underground structures mechanics its fundamental methods are the analytic solutions of corresponding problems of deformable body mechanics. It is not a borrowing of finished solutions. The underground structures mechanics formulates new problems, the solution and methods of which are often not known.

In this way the underground structures mechanics is becaming a branch of deformable body mechanics closely connected with rock mtchanics.

2 CHOICE OF ROCK MASS MODEL

There is an opinion nowadays, that the better a model is themore elements it contains, namely a nonlinear model is better than a lintar one, a three axial model is better than aplane one, a heterogeneous model is better than a homogeneous one and so on. There are the following reasons for such opinion: evident successes of numerical (computer) modelling and the ease of access to computer packages, that provides the designer with a number of computer programs. Starfield and Cundal (1988) say "nowadays every body builds models".

The ability to include geological details in the design of a model proves to be correct for modelling a concrete underground object wyich is provided by the necessary input-data. According to the theory we suggest a better a model is a more simpl one. Following G. Polya one can distinquish an internal (intrinsic, inherent) and external (outward) plausibility of a model. The external plausibility of a model is characterised by including mre and more details. We agree with Stafield and Candell that at best these efforts are waste of time, at worst they are counter-productive, concealing the wood for the trees. One can lose intellectual control of that model instead of obtaining effect.

Another thing is inherent plausibility of the model that is able to give maximal information and understanding of research problem.

The most useful model of the rock mass in the underground structures mechanics is the elastic model. Elastycity is the fundamental property of all natural bodies. These words of Academician A. Krylov can in full measure apply to the rock mass. Elastic model, or accurately linear deformable medium, has got only two chracteristics, namely the E_o modulus of total deformation and the ν_o Poisson's ratio. It makes easy experiments of receiving thease input-data for designing.

The linear deformable model allows to obtain rigorous analytical solutions of the problems of underground structures mechanics both the direct and inverse ones. Note that analytical solution is always better then a numerical model.

The base of the application of the linear deformable model is the u_L displacement being small, that is shwn in Figure 1. It is a joint displacementof the lining and the rock mass after the lining had been installed in contact with the surrounding rock massif.

Application of the linearly deformable medium as a model of rock mass allows the

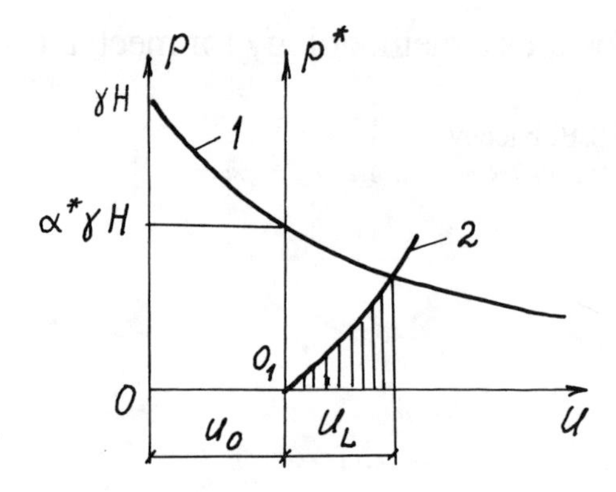

Fig. 1 Ground support interaction diagram 1 and curve 2 characterised rigidity of lining

initial gravity or tectonic stress state of the rock massif to be taken into account

It is obviously very important for underground structure designing.

3 CHOICE OF DESIGN SCHEME

Design scheme must be adequate to the underground object being simulated and at the same time the design scheme is a simplification of reality rather than an imitation of one.

Nowadays only plane design schemes are applied. It may be the elastic infinite plane with any configuration supported opening, as shown in Figure 2, when H >> b,

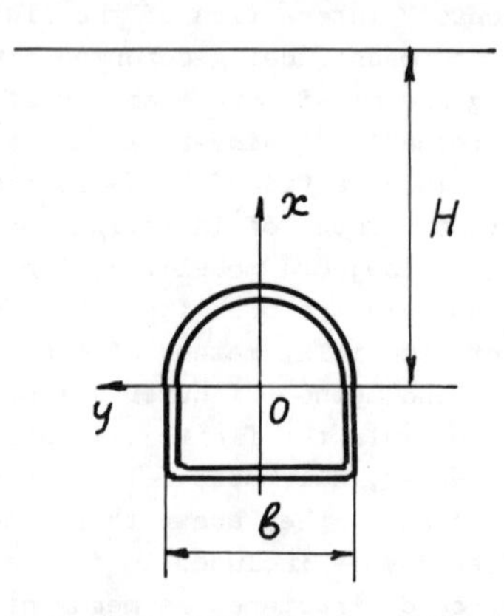

Fig. 2 Designe scheme of arbitrary cross section tunnel lining

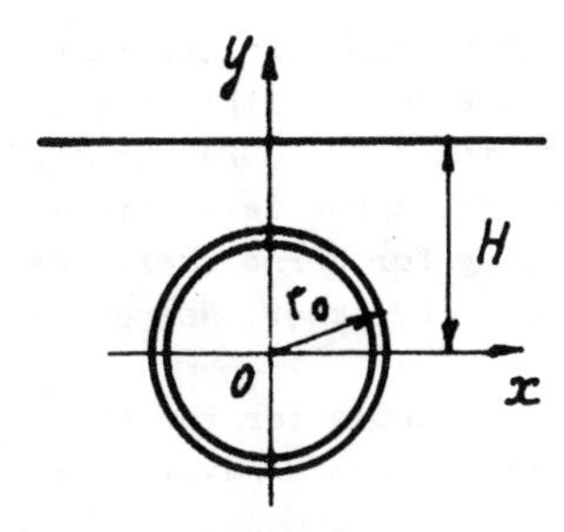

Fig. 3 Designe scheme of shallow embedding tunnel linings of circular cross section

or semiplane for shallow embedded tunnel linings (Figure 3), when H and r are commensurable values.

The elastic plane simulates rock massif. It may be isonropic or anisotropic, homogeneous (usually) or heterogeneous (in a particular case). It may have the property of the linear creep according to real rock mass properties.

The designe scheme may be not only continuous, as shown in Figures 2 and 3, but discrete, as shown in Figure 4, where the rock massif is simulated by a set of coupled elastic bearings (rods), and the concentrated forces are in accordance with the stresses being removed, that models the formation of the opening. That design scheme is applied to analysis of assembled lining from prefabricated elements, espesially those have articulated joints, when analytical solution is lacking (Bulychev, 1982, Bulychev and other, 1988).

4 OVERCOMING THREE DIMENTIONAL STRESS STATE NEAR FACE OF WORKING

Transition from a three dimentional stress strain state of a rock massif near the face of a working is realised by the α^* factor, that may be understended from scheme shown in Figure 5 and may be expressed by formula:

$$\alpha^* = (u_\infty - u_o) / u_\infty \qquad (1)$$

where u_o is initial displacement of the rock mass before a lining is installed, u_∞ is complete rock displacement.

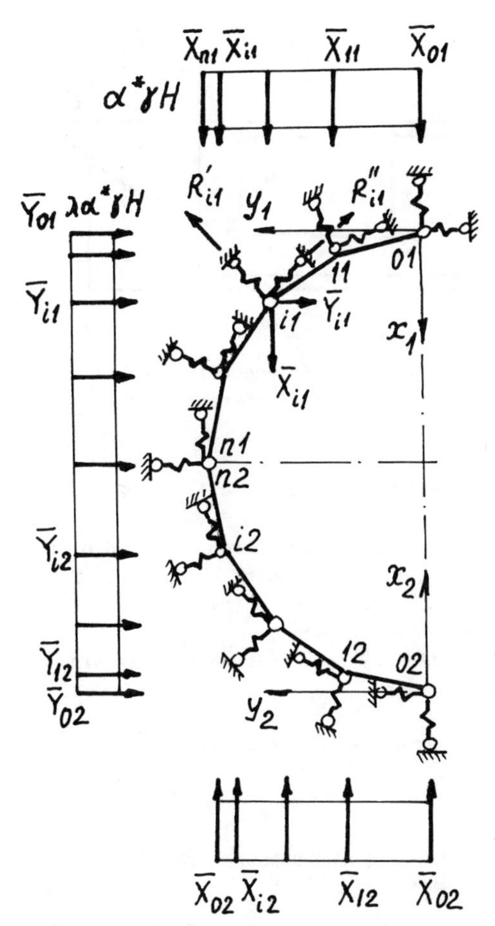

Fig. 4 Design scheme of prefabricated lining

The α^* factor is applied in design schemes of an underground structures as multiplier to the initial stress state components ($\alpha^* \gamma H$). This corresponds to traçsfer the p axis in Figure 1 to posicion p* and alimination this way the u_o initial displacement from the designe schemes.

The α^* factor may be determined from an one-dementional analysis of an elasto-plastic problem, as shown in Figure 1, from a numtrical modelling or full-scale measurement. For example there are interesting results of the numerical experiments in the paper by Baudendistel (1979)In particularly there are in that paper following data:

$1/r_o$	0	0.25	0.50	1.0	2.0	3.0

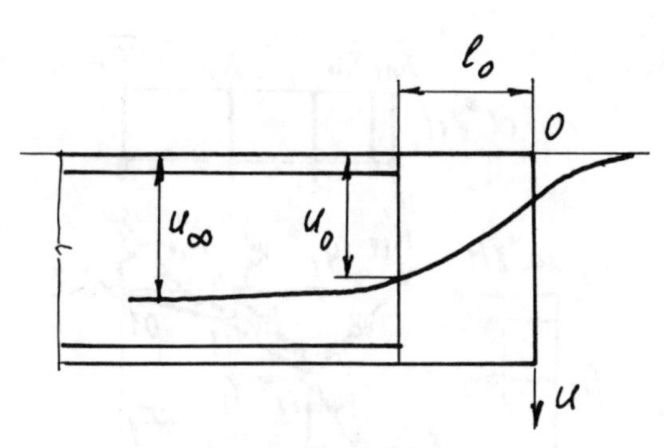

Fig. 5 Radial displacement of opening surface near face of opening

f_A=α* 0.72 0.41 0.23 0.11 0.02 0

where l is a distance from the face of an opening, r_o is radius of an opening, f_A is coefficient of loading (by Baudendistel, it is equal to α*).
That data may be expressed by following correlation:

$$\alpha^* = 0.64 \exp (-1.75\ l_o / r_o) \quad (2)$$

5 PROGRESS OF UNDERGROUND STRUCTURES MECHANICS

Mathematical models, algorythms, methods of designing and computation programs, based on above described principles, have been developed for the followingstructures types: a monolythic concrete lining of excavations of an arbitrary cross-section shape symmetric about the vertical axis (Fotieva, 1974), ashotcrete support following roughness of the excavation cross-section counter with the amplitude of the roughness sametimes exceeding the support thickness (Bulychev, Fotieva & Streltsov,1986: 141-180, Fotieva & Sammal, 1991, 1992), a frame metall support (Fotieva Petrenko & Sammal, 1988), every kinds of multy-layer lining, such as: steel-concrete cast iron - concrete and others including shft linings thinking by drilling (Bulychev,1982, 1989, Bulychev, Fotieva & Streltsov, 1986: 32-81), a rock bolting support withallowanse for interaction of the rock bolt system with the rock mass and that of the bolts within the system (Bulychev & Stepanjn, 1992). Optimal design methodfor tyree layer steel - concrete lining for bored shafts has to be mentioned also (Bulychev, Nechaev & others 1986).

Computation methods for the above mentioned support kinds have been developed taking into account complicating factors, such as existence of a zone of rock strengthening around the excavation (Fotieva, Petrenko & Sammal, 1988), as well as computation method for a shaft support where there is a zone of antifiltration grouting (Bulychev, Fotieva & Streltsov, 1988: 223-285), computation metyods for a shotcrete lining (Fotieva & Sammal,1987, 1991, 1992), computation methods for various kinds of supports combined with anchors (Fotieva, Sammal, Klimov & Kireeva, 1988).

Support computation methods including those for a multy-layerlining have been developed for parallel excavations influencing one another with allowance for the sequence of excavation work and support installation (Fotieva & Antsiferov, 1988).

Support computation is performed for gravity initial stress field, tectonic initial stress field with an arbitrary direction of the principal stresses in the plane of excavation cross-section, underground water pressure including filtration through the support and grouting

Rock excavation support analysis for earthquake seismic effects should be specially mentioned, since they are of great importance for tunnels and mining workings in seismically active regions. The peculiarity of this kind of analysis is impossibility to predict the direction of the propagation of longitudinal and transverse waves and their combination on reaching a rock excavation. The methods of analysis are based on analytical determination of the most unfavourable stressed state in tach structures section with various combinations of the actions both of longitudinal and transverse seismic waves of different character and of any direction in the excavation cross-section plane.

The methods of support analysis enumerated show a high level of agreement with practical results. We can say that at present there are no data contradicting these analysis methods and mathematical models. This for example, holds true for experimental-analytical methods of lining analysis based on back analysis of lining-rock interaction on the basis of field measurement (Bulychev, Fotieva & Streltsov 1986: 133-140, Bulychev & Savin, 1988).

The mathematical models of support-rock intraction enable one to determine the required accuracy of rock mass deformation chracteristics (Bulychev, Fotieva & Kalinin 1986).

Onthe basis of the mathematical models one can determine the domain of application of standart rock excavation supports,which is expressed as components of the initial stress field, or immediately in the form of the limiting depth of support applications and the support strength safety fector under various conditions, whis makes the work of designers much easier (Bulychev, Fotieva & Streltsov, 1986, Fotieva, Kazakevich & Sammal, 1986, Fotieva & Sammal, 1991, 1992).

The above mentioned mathematical models and computation methods for supports are widely used in Russia and abroad for design and construction shafts,horisontal permanent openings of coal and ore mines, traffic and utility tunnels, underground structures of hydro-power and pumped storage stations.

REFERENCES

Baudendistel, M. 1979. Zum Entwurf von Tunneln mit grossem Ausbruchquerschnitt. Rock Mechanics. 8: 75-100.

Bylychev, I. N. 1982. Design procedure for uncclosed and prefabricated support structures for permanent rock excavations based on the scheme of support contact interaction with the rock mass. In Mech. of Underground Structure. Tula. 36-42.

Bulychev, N. S. 1982. Mechanics of underground structures. Moskow. NeБra.

Bulychev, N. S. 1989. Mechanics of underground structure in instans and problems. Moskow. Nedra.

Bulychev, N. S. & Fotieva, N. N. 1988. Mathematical modelling and design of linings of hydraulic tunnels. In M. Romana (ed.) Rock Mech. and Power Plants, p. 269-276. Rottrdam, Balkema.

Bulychev, N. S.,Fotieva, N. N. & Kalinin, N. B. 1986. Influence of the accuracy of rock mass characteristics determination on designe of tunnel linings. Gidrotechnicheskoe stritelstvo. 11: 38-39

Bulychev, N. S.,Fotieva, N. N.,Rozenvasser G. V. & Shmrin, Yu. E. 1988. Design of prefabricated linings of collector tunnels with allowancefor their contact interaction with the rock mass. Osnovan. fundamenty i mechanika gruntov. 5: 18-20.

Bulychev, N. S., Nechaev, V. I., Naumov, S. A. & Shvetsov V. V. 1986. In A. Kidybinski & M. Kwasnewski (ed.) Mining System Adjusted to High Rock Pressure Conditions, p.145-150. Rotterdam, Balkema.

Bulychev, N. S. & Stepanjan, M. N. 1992. Rock bolting calculation at static and stismic loadings.In P. K. Kaiser & D. R. McCreath (ed.) Rock Support in Mining and Underground Construction, p. 699-703. Rotterdam, Balkema.

Fotieva, N. N. 1980. Design of support of underground structurts in stismically active regions. Moskow. Ntdra.

Fotieva, N. N. & Antsiferov, S. V. 1988. Design of multy-layer linings of a complexof parallel circular tunnels for earthquake seismic effects. In Mech. of Undergr. Structures,p.30-38. Tula.

Fotieva, N. N., Kazakevich, E. V. & Sammal,A. S. 1986. Determination of the sphere of application of a light support using shotcrete.Shachtnoe stroitelstvo. 4: 9-11.

Fotieva, N. N. & Sammal, A. S. 1987. Design of a shotcrete support of underground structures with allowance for a layer of monolythic rock (strengthened by concrete). Fiziko-technicheskie problemy razrabotki poleznych iskopaemych. 2: 3-8.

Fotieva, N. N. & Sammal, A. S. 1991. Determining the maximum permissible depth in which working anchoring with shotcrete lining in combination with anchors is possible. In G. Beer, J. R. Booker & J.

P. Carter (ed.) Computer Methods and Advances in Geomechanics, p 1961-1066. Rotterdam, Balkema.

Fotieva, N. N. & Sammal, A. S. 1992. Determining the border of the field applying composite lining of shafts from shotcrete in combination with anchors. InP. K. Kaiser & D. R. McCreath (ed.) Rock Support in Mining and Underground Construction, p. 157-164. Rotterdam, Balkema.

Fotieva, N. N., Petrenko, A. K. & Sammal, A. S. 1988. Approximate computation of a frame metal support in a strengthened rock mass. In Numerical Methods of Stabimity Assessment of Underground Structures, p. 8-11. Apatity (in russ.)

Polya, G. 1954. Mathematics and plausible reasoning. Princeton Univ. Press.

Starfield, A. M. & Cundall, P. A. 1988. Towards a methodology for rock mechanics modelling. Int. J. of Rock Mech. Min. Sci. & Geomech.,25, 3: 99-106.

Geomechanics 93, Rakowski (ed.) © 1994 Balkema, Rotterdam, ISBN 90 5410 354 X

Behaviour of composite rock specimens under uniaxial compressive tests

Otello Del Greco
Dipartimento di Georisorse e Territorio, Politecnico di Torino, Italy

ABSTRACT: This study concerns the behaviour of laboratory specimens composed of disks of different rocks in uniaxial compressive tests. The aim of the research has been to understand the behaviour of structures, such as stratified rock pillars in underground excavations or masonries in ancient monuments made up of different stonework materials. The tests showed a reduction of the disk specimens' strength, compared to the continuous specimens' strength, for two different reasons: the same disk structure and the contact between the materials with different mechanical characteristics. Finally, in such cases the failure mechanism of the disk specimens is different from that of each single rock component.

1 INTRODUCTION

This study was started in 1986. In that period the purposes of the study were different from those that later developed. In 1986 the Geotechnical group of the "Dipartimento di Georisorse e Territorio" of the Technical University of Turin was encharged with the analysing of the static conditions of the XIV century columns of the Orvieto Cathedral. The main walls and columns of this church are made up of an alternation of two different rock blocks: travertine and "basaltina" (volcanic silicate) (fig.1). Several new cracks had formed in that period, in particular, in the columns in the axial direction.

In the first study, uniaxial compressive tests were carried out using specimens made up of alternated disks of the two rocks. These specimens were not intended to represent physical models of the column but they were considered as a way of analysing a possible interaction between two rocks with different physical and mechanical characteristics.

The results of these first tests led to a follow up study using new

Fig.1 - The nave of the Orvieto Cathedral.

lithotypes especially chosen on the basis of a wide range of various deformability and strngth features. Different materials were placed between the rock disks to simulate the filling material which is often present in real stuctures.

This study presents other possible applications such as the evaluation of the strength of natural stratified sedimentary rock structures.

Tab A - Rock properties

Q = compressive strength, Es = secant Young modulus, vs = secant Poisson's ratio, T = indirect tensile strength.

Rocks	Q(MPa)	Es(MPa)	vs	T(MPa)
Travertine	88-90	67000	0.32	7.2-9.6
"Basaltina"	100-105	40000	0.28	7.3-12.0
"Perlato di Sicilia"	130-170	82000	0.31	3.5-6.6
"Finale" stone	12-14	14000	0.21	2.8-4.7
"Montorfano" granite	140-160	33000	0.16	4.7-7.9
"Lasa" marble	86-92	61000	0.25	5.1-7.2
"Vicenza" stone	32-34	14000	0.19	1.9-3.9

2 ROCK CHARACTERISTICS AND TEST PROCEDURES

The rocks that were used for the tests were chosen in such a way as to cover a wide resistence and deformability range, with the common condition of being substantially isotropic in order to not introduce an additional factor of the mechanical anisotropy which would have made the test programme more complex and the results more difficult to interpretate.

The following lithotypes were therefore used: Travertine (sedimentary limestone of a chemical origin); "Basaltina" (tephritic leucite; volcanic silicate); "Perlato di Sicilia" (sedimentary limestone very compact, used as an ornamental stone); "Finale stone" (sedimentary limestone, of an organic origin, used as a construction material); "Montorfano" granite; "Lasa marble" (crystalline limestone of a metamorphic origin); "Vicenza stone" (sedimentary limestone of an organic origin, used as a construction material).

Classical tests, such as uniaxial compressive tests were carried out for each of these lithotypes in order to characterize the rock materials. The tests were carried out, according to the method suggested by the International Society for Rock Mechanics on cylindrical 54 mm diameter samples. The obtained results, reported in Table A, make up the values to which the successive test results refer.

Specimens were then made up of 4 disks of the same rocks separated by different types of contacts (free contact, glued surfaces, interposition of a rubber disk) in order to evaluate the influence of the single discontinuity with different shear strengths.

Finally, tests were carried out on specimens made up of 4 disks of two alternated materials (fig.2), again varying the types of contact between the disks. These contacts simulated the effect of a possible bond between the stratifications, or of a very deformable filling characterized by poor resistence, thus obtaining useful qualitative information to interpretate the behaviour of composite materials.

In order to understand the stress-strain behaviour and the failure mechanism of the various set up, one should consider that the press plates influence the boundary conditions of the tests. In the case of a discontinuous

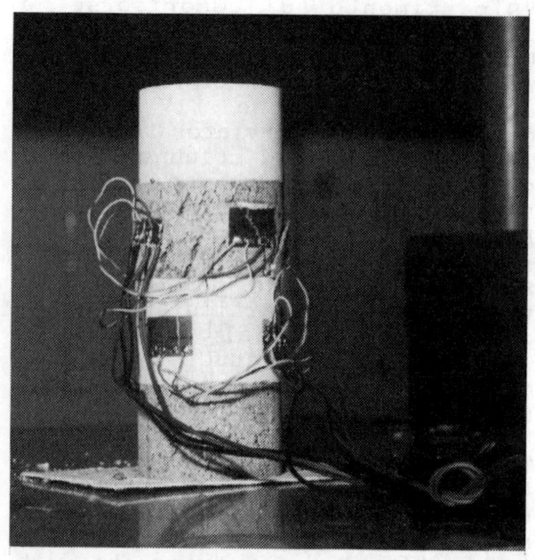

Fig.2 - Specimen composed by disks of two rocks ("Perlato di Sicilia" and "Basaltina"), equipped with electrical strain gauges.

specimen this effect is less induced in the two central disks, and this depends on the friction characteristics of the contact surfaces between the disks.

For these reasons, many composite specimens were instrumented with electrical strain gauges on the two central disks in order to record both vertical and horizontal strains.

3 ANALYTICAL ASPECTS

Composite structures, such as the studied specimens, have basically anisotropic characteristics. Their resistence is therefore greatly influenced by the direction of the applied loads and by the interaction of the different materials in contact.

In a mixed specimen subjected to uniaxial load condition, lateral deformability difference between two materials determines a shear state of stress at contact and, consequently, radial stresses within the rock disks. This fact determines a complex state of stress even under uniaxial load conditions.

The shear stress magnitude is directly influenced by the contact features: cohesive strength (as in the case of rock bridges or very rough surfaces) or only friction contacts (with high or low friction angle due to surface roughness or clay filling behaviour).

A simplified analytical approach can give a quantitative interpretation of the experimental results. A cylindrical specimen made up of disks of different thicknesses under a uniaxial compressive load is therefore considered (fig.3).

In these conditions the following equilibrium equation, on an axial direction, is valid:

$$\sigma_{y_a} = \sigma_{y_b} = \sigma \qquad (1)$$

while in the radial direction one obtains the following equilibrium condition , in the hypothesis that the stresses are uniformally distributed in the two materials in contact:

$$\sigma_{r_a} \cdot H + \sigma_{r_b} \cdot h = 0 \qquad (2)$$

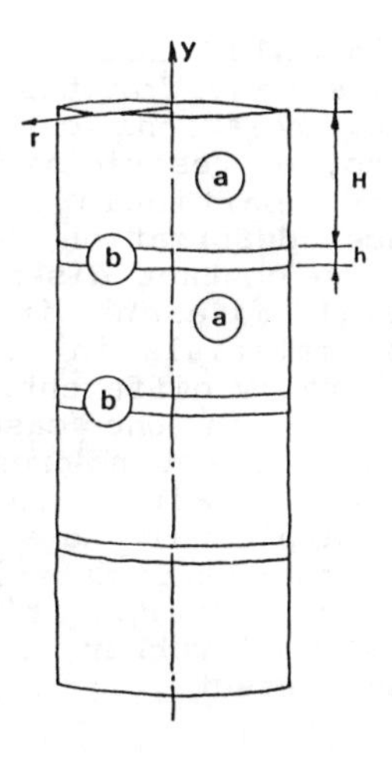

Fig.3 - Schematic representation of the studied specimens.

If congruent radial strains are hypothesised, one obtains the following relation:

$$\varepsilon_a = \varepsilon_b$$

$$\frac{\sigma_{r_a} - \nu_a \cdot \sigma}{E_a} = \frac{\sigma_{r_b} - \nu_b \cdot \sigma}{E_b} \qquad (3)$$

and by joining (3) with relation (2) one obtains:

$$\left.\begin{aligned} \sigma_{r_a} &= \frac{\sigma\,(\nu_a - \nu_b \cdot E_a/E_b)}{1 + E_a/E_b \cdot H/h} \\ \sigma_{r_b} &= \frac{\sigma\,(\nu_b - \nu_a \cdot E_b/E_a)}{1 + E_b/E_a \cdot h/H} \end{aligned}\right\} \qquad (4)$$

These expressions determine the same stress value with opposite signs for adjacent rock disks with different elastic constants but with the same heights. The tensile radial stresses in one of the two materials heavily influences the global behaviour and the failure mechanism as the tensile rock strength is much lower (1/15 - 1/8) than the compressive strength.

The contact areas between the disks must therefore be considered as weakness zones as the the tensile fractures can start there and spread even if the loads are lower than the compressive strength of the rock components.

One should note that the hypothesis made for the obtained analytical relations can, in some situations, be restrictive. This is true, in particular, in the congruence deformation hypothesis in the free contact disks case, as the radial movement in the two different materials in contact can be considerably different.

Instead, in the case of the glued disks one can presume that the radial strains in the two materials are equal and also in the tests of the same material disks either with interpositioning of thin sheets of rubber or glue, or with free contact.

4 EXPERIMENTAL OBSERVATIONS

4.1 General remarks

By carrying out the compressive tests the failure behaviour of the continuous specimens and those with disks, was macroscopically observed, showing some notable differences.

The experimental data are given in Table A and B where they are given as mean values. By analysing this data some general observations can be made:

a) the compressive strength of continuous specimens is higher in comparison to the specimens made up of four disks of the same material; b) the compressive strength of mixed specimens is lower than the minimun strength of the two materials; c) the values of the elasticity modulus of the single disk are generally higher than those of the standard specimen; d) Poisson's ratio is higher in the discontinuous specimens; e) the failure mechanism depends on the combination of the materials, in the sense that a material breaks down following different paths according to its coupling with the other materials; f) the type of contact between the disks alters the values of the mechanical parameters and this contact has a certain influence on stress distribution.

There are three main features to emphasize: the first concerns the strength of the specimens and the failure mechanism according to the

Tab B - Experimental values of tests on disk specimens: TR = Travertine; BA = "Basaltina"; FI = "Finale" stone; PE = "Perlato di Sicilia"; MO = "Montorfano" granite; LA = "Lasa" marble; VI = "Vicenza" stone

Rock	Specimen	Q (MPa)	Es (GPa)		ν_s	
TR	continuous	89	67		.32	
TR	free contact	78	57		.48	
BA	continuous	102	40		.28	
BA	free contact	88	41		.42	
FI	continuous	13	14		.21	
FI	free contact	11.3	17		.36	
PS	continuous	150	82		.31	
PS	free contact	101.7	161		.77	
PS	glued	71.7	180		.51	
PS	rubber	85	119		.56	
			TR	BA	TR	BA
TR-BA	free contact	62.8	74.5	52	.49	.35
TR-BA	glued	61.5	102	66	.64	.49
TR-BA	rubber	32.8	64.1	36.2	.33	.36
			PS	BA	PS	BA
PS-BA	free contact	72.5	128.7	37.9	.52	.37
PS-BA	glued	44.9	190.2	61.2	.70	.70
PS-BA	rubber	45.2	148.3	44.5	.54	.34
			PS	FI	PS	FI
PS-FI	free contact	13.0	67.9	26.9	.39	.32
PS-FI	glued	10.1	121.4	49.7	.62	.39
PS-FI	rubber	14.3	89.4	12.5	.46	.24
MO	continuous	160	33.0		.16	
MO	free contact	118	50.0		.32	
MO	glued	125	51.0		.33	
MO	rubber	67	55.0		.23	
LA	continuous	92	61.0		.25	
LA	free contact	84	77.0		.35	
LA	glued	52	52.0		.38	
LA	rubber	74	75.0		.39	
VI	continuous	34	14.0		.18	
VI	free contact	21	15.5		.21	
VI	glued	24	17.7		.29	
VI	rubber	21	16.0		.40	
			MO	LA	MO	LA
MO-LA	free contact	70.6	38.0	73.0	.18	.42
MO-LA	glued	62.0	24.0	68.0	.20	.34
MO-LA	rubber	72.6	36.7	91.6	.26	.42
			MO	VI	MO	VI
MO-VI	free contact	26.6	20.1	17.5	.14	.39
MO-VI	glued	31.4	26.3	37.6	.30	.86
MO-VI	rubber	22.1	27.4	16.2	.25	.65
			LA	VI	LA	VI
LA-VI	free contact	31.0	59.0	22.3	.32	.47
LA-VI	glued	32.0	60.8	15.0	.30	.34
LA-VI	rubber	29.9	105.6	19.1	.39	.39

type of contact surface between the disks; the second involves the stress distribution in the specimens and the consequent failure of the materials with different couplings; the last feature concerns strength value

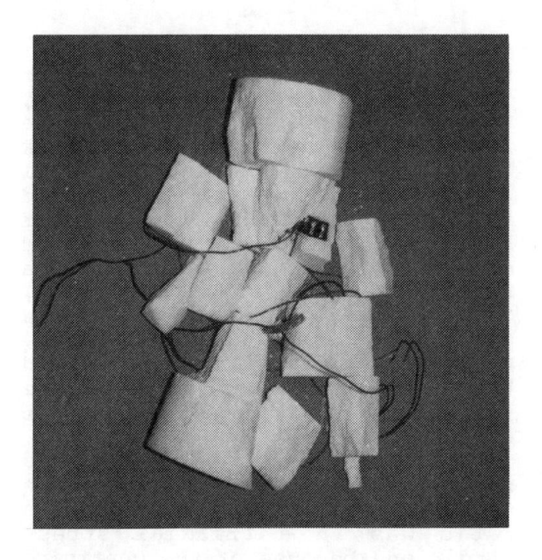

Fig.4 - Specimen composed by disks of one type rock (Lasa Marble) on free contact, after failure.

Fig.5 - Specimen composed by disks of one type rock (Montorfano Granite) with rubber contact, after failure.

changes when coupling the disks.

Failure occurred in the tested materials, in continuous specimens, with the creation of inclined surfaces with respect to the loading axe, or with wedge shaped splinters. It is probable that in this standard situation there was a failure when the strength envelope, for example the Mohr-Coulomb failure criterion, was reached by the path.

4.2 Influence of the type of disks contact surface.

The specimens made up of diks of the same rock material with free contacts again showed inclined surface failures (fig.4); in the more resistant rock materials, subvertical fractures also sometimes occurred.

When a thin rubber element or glue were placed between the rock disks of the same material, the behaviour of the specimens resulted to be changed, since radial stress components were transmitted trough the elements, because of the different transverse deformability of the glue or rubber, causing radial movements.

Joining with rubber usually produced tensile stresses in the rock disks which break with subvertical surfaces without any trace of reciprocal slipping between the broken elements (fig.5). The glued disk tests are those which gave the most scattered results. This is probably due to the fact that the glue's effectiveness can change from rock to rock and also because the uniformity of the action of the glue on the surfaces in contact is less controllable. In general however, the glue produce continuity between the disks and the failure results with inclined surfaces in a way similar to the continuous samples of a standard type.

Experimental data reported in fig.6 show that the ratio between the compressive strength of continuous specimens and specimens made up of free disks of a single type of rock is constant ($\simeq 1.2$) for all the tested rocks. Different kinds of contacts determine a constant value in the strength ratio even though with a larger deviation than the free disks specimens.

4.3 Behaviour of mixed rock specimens

When coupling different materials one could verify different failure

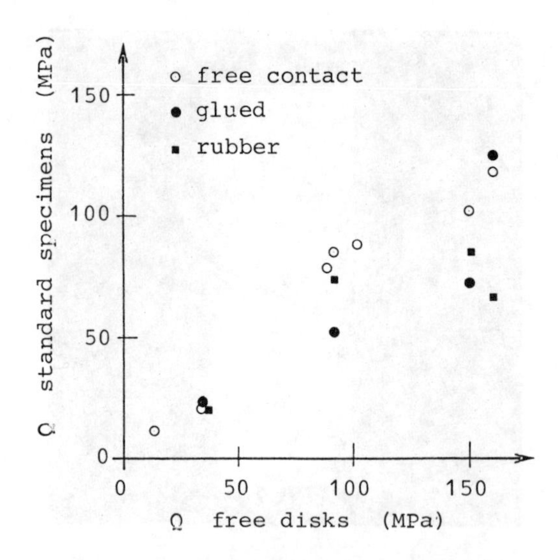

Fig.6 - Standard specimen compressive strength vs disk specimens (one rock type) compressive strength.

mechanisms connected with the compressive strength, the Poisson's ratio and the modulus of elasticity of the two rocks. Therefore, when coupling two materials with different strengths, the failure in the weaker material occurred on inclined planes or cone shaped envelopes. The occurrence of pulverised rock due to shearing slip was often combined with these surfaces. In the more resistant material the failure occurred as vertical, clear cut fractures, without slip powder. It is probable that the more resistant rock induces a less confining lateral stress on the weaker rock, caused by the different values of Poisson's ratio of the coupled materials. This fact involves a related radial stress component, determining failures with inclined surfaces or cone shape elements, while in the material characterised by the lower values of Poisson's ratio a tensile strength is evident.

The failure process was progressive. At the joint surface the rock with a higher Poisson's ratio was in a triaxial stress condition because of the confining effect due to the material at a lower Poisson's ratio, which follows a tensile path. The first cracks in this disk determines the loss of confining in the more expanding material, with a consequent change to a uniaxial stress condition and the reaching of the strength envelope.

In a group of three rock types, the term of intermediate strength changes its failure mechanism according to its relative strength value if coupled with the other two rock materials. Lasa Marble, coupled with Montorfano Granite, in

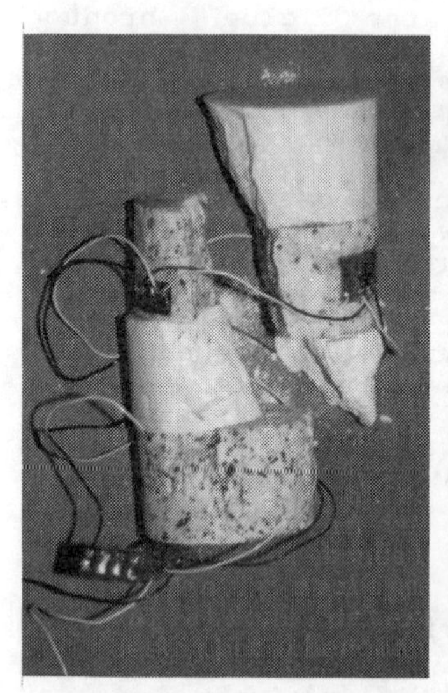

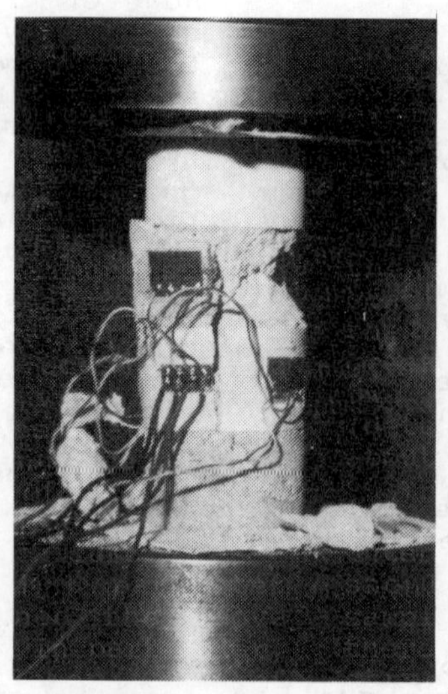

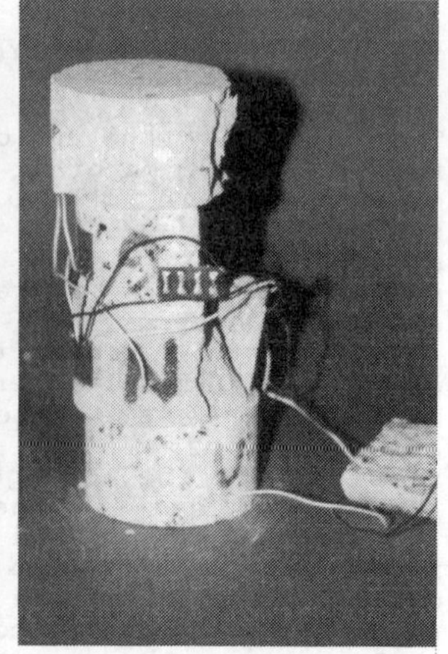

Fig.7 -Different failure mechanisms in the various tested rock couplings: a) Granite - Marble; b) Marble - Vicenza Stone; c) Granite - Vicenza Stone.

fact, brokes with inclined surfaces and cone shaoed elements, if coupled with Vicenza Stone it brokes with clear cut vertical divisions, due to tensile stresses (fig.7). The type of failure in each material is similar both in the central disks and in the external disks, although these disks also suffered the effect of the contact with the testing press plates. A common factor was the continuity in the spreading of fractures through the joint surfaces.

The combinations Perlato - Basaltina (fig.2) and Travertine - Basaltina showed a similar behaviour; the tensile stress caused a failure in Perlato and in Travertine. The joining of Perlato and Finale Stone determined, in some cases, only a wedge shaped failure in Finale Stone without failure in the other rock. This fact was due to the high difference in resistance and deformability, so that the occurrence of the weaker rock material was more important than the coupling of different rock types.

Joining with glue does not considerably modify the described failure mechanism, and sometimes determines the presence of inclined failure surfaces in the more resistant material.

Joining with rubber in mixed specimens usually produced tensile stresses in both rocks. The inclined surfaces occurred in the external disk of the weaker material in contact with the press plate, while in the central disks there were tensile cracks in both rock types (fig.8). An examination of the broken samples showed that the propagation of failure started vertically in the weaker material of the internal disks. At complete failure it was possible to recognize that the weak disk in contact with the press plate failed with a cone shaped surface (due to the triaxial effect of the plate), while those in the middle broke as a result of tensile failure.

The mechanical parameter variations determined the above described phenomena. In particular, the material with lower value of Poisson's ratio underwent a tensile failure induced by the other

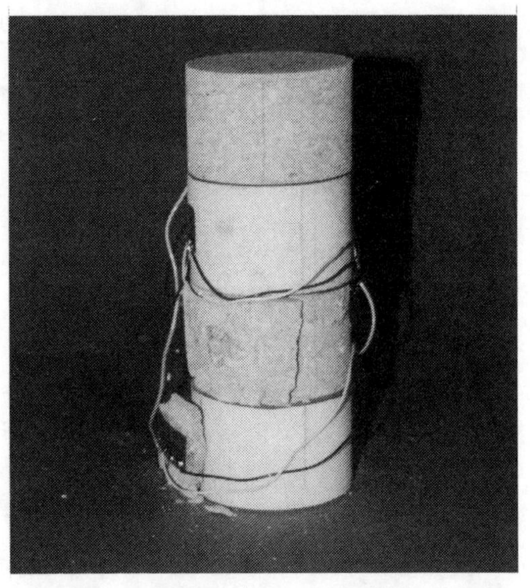

Fig.8 - Composite specimen (Lasa marble and Vicenza Stone) with rubber contact, after failure.

material with the exception of rubber. Finally, it is worth noting the behaviour of Vicenza and Finale Stone. These natural materials have a large porosity (22-27%) with a large gap in bulk density (2100 kg/m3) and grain density (2660 kg/m3 for the Vicenza Stone). Such materials, in compressive tests, showed a compaction according to the reduction of voids, and a consequently hardening linked to an increase of the elasticity modulus. At high stress levels, the lateral expansion was more evident, with a consequent increase of the Poisson's ratio.

Figure 9 shows that the difference of resistance between the mixed disk specimen and the continuous specimens (referring to the lower compressive strength) is greatest when the ratio between the two coupled rock stiffnesses (determined with standard tests) is about 0.5-0.6.

In order to understand the experimental data, another interpretation way is possible: one should consider that a standard sample is slimmer than a disk, and therefore shows a lower compressive strength. Actually, the superimposition of four disks involves a loss in the compressive

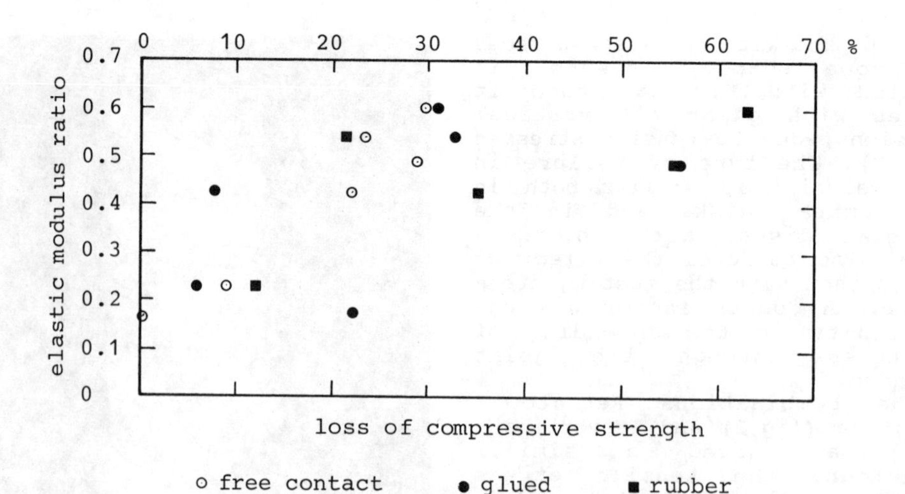

Fig.9 - Elastic modulus ratio of rocks vs loss of compressive strength of composite specimens.

strength, because the continuous specimen gives an effect of lateral confining from the bases to the central portion due to the transmission of a stress with a radial component. This confining effect was feeble in the discontinuous specimen, so that it made up for the influence of the stumpy shape of each single disk.

Further development of this study is in progress with the tests conducted on other lithitypes, including some natural anisotropic rocks.

REFERENCES

Bathe K.J., 1982. Finite Element Procedure in Engineering analysis. Prentice Hall, New York.

Del Piero G., 1983. Le costruzioni in muratura. Collana di Ing. Strutturale n.2, CISM.

Horino F., 1968. Effect of planes of weakness on uniaxial compressive strength of model mine pillars, Report of Invest. 7155, Bureau of Mines, US Dept. of Interior.

Desai C.S., Siriwardane H.J.,1984. Constitutive laws for engineering materials. Prentice Hall.

Vutukuri V.S., Lama R.D., Saluja S.S.,1974. Handbook on mechanical properties of rocks. Trans Tech Publ. vol.1.

Del Greco O., Ferreo A.M., Peila D., 19891. Behaviour of laboratory specimens composed by different rocks. ISRM Int. Congr. on Rock Mech., Aachen.

Geomechanics 93, Rakowski (ed.) © 1994 Balkema, Rotterdam, ISBN 90 5410 354 X

Bolted rock mass behavior

A.Wahab Khair
West Virginia University, Morgantown, W.Va., USA

ABSTRACT: This paper starts with a brief chronicle of rock reinforcement by bolts. It traces the development and theory behind the different types of roof bolting principles and their mechanisms. Two case studies are reviewed and the effectiveness of each bolting principle in regard to geotechnical data are analyzed. Results of the above case studies demonstrate that no particular roof reinforcement system is a cure for all geologic and mining conditions. Hence roof reinforcements may require combined effects of the mechanisms offered by more than one rock reinforcement system.

1 INTRODUCTION

Since roof control is one of the most troublesome problems confronting mining engineers. Roof bolts as a media of roof control in underground mines is no longer an innovation. The present understanding of theory and practice of bolting was largely developed during the period of 1950-60 (Panek and McCormick, 1973).

The general practices involved in roof-bolting are: suspension of friable strata by anchoring bolts in relatively more stable overlying strata (suspension effect), clamping of various bedded planes to form a beam (friction effect), keying effect in order to inhibit movement along persistent fractures, and developing of compression in roof strata (truss bolting) for stress redistribution and load transformation within the roof strata (Fairhurst and Singh 1974; Panek 1957; Panek 1964; Neal et al. 1976).

1.1 Effect of Roof Bolting System on Roof Strata

When an underground excavation is made, the in-situ stresses are disturbed, resulting in deformation of rock/coal mass surrounding the opening. The magnitude of these stress changes and strata movements depend on the initial state of in-situ stresses, mechanical and geological properties of rock/coal strata and excavation geometry. In a typical underground coal mine the roof rock often consists of laminate strata which is deformed during or shortly after excavation is made. These changes of stress field and strata movements are idealized in Figure 1 (Stefanko 1980). The design concept in arriving at the proper reinforcement system for laminated roof strata should be based on either inhibiting strata separation and movement along the bedding planes or reducing the magnitude of the induced stress below the strength of strata. In some instances the application of both principles may require roof strata stability of an opening.

Suspension Effect: Basically there are two types of reinforcement generated by the suspension effect, namely, a) simple suspension and b) beam suspension.

In simple suspension, the principal weight of the immediate roof rock, assumed to be loose rock, is supported by the upper strata which can take an additional load and the bolts are anchored at that horizon. The suspension load on the bolt, p, is given by the following equations (7):

$$p = \frac{wtBL}{(n_1 + 1)\ (n_2 + 1)} \tag{1.1}$$

where w is the unit weight of rock, L is the roof span, B is the length of the opening, t is the thickness of the roof layer, n_1 is the number of rows of bolts, and n_2 is the number of bolts per row. The effect on the roof strata will be equivalent to the reaction load p distributed over the

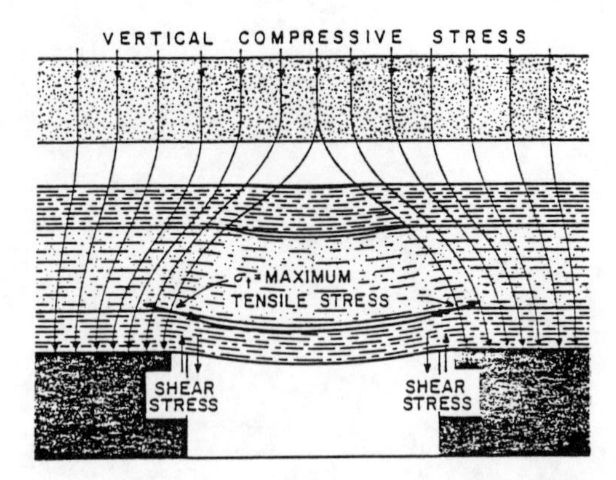

Figure 1. Critical stresses in bedded strata (6).

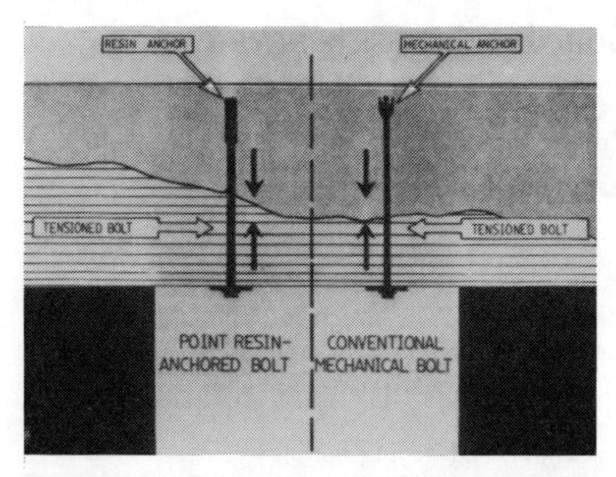

Figure 2. Comparison of mechanical and point resin-anchored bolts (8).

area of the plate, A_p. The distributed load, S_d, can be calculated as:

$$S_d = \frac{P}{A_p} \tag{1.2}$$

The beam suspension principle is usually implied in horizontally bedded strata, which is most common in underground coal mines. These strata are either unbonded or the bonding strength between the layers are relatively low in comparison to the tensile strength of the massive or typical thick sedimentary rocks. After an excavation is made, the immediate roof strata usually detaches from the overlying strata immediately or shortly after a period of time (Obert and Duvall 1967). The suspension effect under this condition is two-fold: a) to transfer part of the weight of the weak strata to the competent strata through roof bolting and b) to induce compressive stress normal to the bedding plane of roof strata. This normal compressive stress assists the immediate roof strata to maintain bonded conditions with the overlying strata and avoid detachment. Figure 2 (Karabin and Hoch 1979) employs the two functions of the suspension effect. The suspension effect by roof bolting in a horizontally bedded strata depends mainly on the relative flexural rigidities of all bolted roof strata (Peng 1986). Furthermore, the bending stress and deflection are inversely proportional to the thickness and its square, respectively. The maximum bending stress and deflection in the thick bed, bolted strata, is much less than the thin, unbolted strata hence the suspended beam strata becomes more stable than the unbolted condition.

Friction Effect: The friction reinforcement principle is usually employed when there is no competent strata within the distance of the bolt length above the roof line. Under this condition roof bolts are used to bond multiple layers of immediate roof rock into a monolitic beam which is inherently stronger or subjected to a lesser bending stress and deflection. The basic concept in friction reinforcement is to increase shear resistance between the layers and avoid a slip on the interface between the thin layers. The stability of the bolted strata increases with the thickness of the bolted strata and increasing friction resistance. The first parameter is related to the bolt length while the second parameter is affected by the spacing between the bolts and tension applied to the bolt. The relationship between the decrease in maximum bending strain, ΔE_f, due to the friction effect and the maximum bending strain of the unbolted strata, E_{nfs}, is expressed by the following equation (Peng 1986):

$$\frac{\Delta E_f}{E_{nfs}} = -0.375\,\mu\,(bL)^{0.5}\,[\frac{NP(\ell/t_{avg}-1)}{W_{avg}}]^{0.33} \tag{1.3}$$

where μ is the coefficient of friction between the bedding planes, b is the spacing between adjacent rows of bolts, L is the roof span, N is the number of bolts per row, p is the bolt tension, ℓ is the bolt length, t_{avg} is the average thickness of the bolted roof, and W_{avg} is the average unit weight of the bolted rocks. The reinforcement factor, RF, due to friction effect, is then defined (Panek 1973) as:

$$RF = \frac{1}{1 + (\Delta E_f/E_{nfs})} \tag{1.4}$$

The above equations are used, within the assumed limitations, to evaluate the effectiveness of the friction effect of roof bolting. Figure 3 (Karabin and Hoch 1979) reflects reinforcement principle of friction effect.

Keying Effect: The principle of the keying effect is based on the development of resistance force in the joint interface. The design concept must determine the proper reinforcement system of the roof bolts across these planes of weakness in order to prevent or reduce movement. Figure 4a-b exhibits the keystoning and keying bolting systems. The roof bolting system in Figure 4a is commonly used in hard ground, making use of the natural strength of the arch at lower levels thus maintaining pre-existing stresses surrounding rock and limiting roof movement. While installation exhibited in Figure 4b exemplifies the keying effect for fracture strata. The stability condition for keystoning/keying effect requires application of sufficient compressive stress on the strata to inhibit strata movement along the fractured surface.

Roof Truss Effect: Effect of trusses over the roof behavior and how they react when the roof starts to work on them is largely unknown. One widely accepted theory is that the tension members cause a resultant force to be exerted on the mine roof at the hole collars, giving an uplift force and putting the rock into compression in the shaded areas as shown in Figure 5 (Neal et al. 1976). As a result, the tensile stress in the midspan of the roof is reduced and neutral axis of the strata moves upward. As tension in the roof truss increases, the tensile stress continues to decrease until the neutral axis of the lowest roof stratum is eliminated and the midspan is completely in compression. Therefore, a roof truss provides support by creating a compressive zone at the midspan of the opening as opposed to a tensile stress area in an unsupported roof. Since rock is much weaker in tension than in compression, the roof is much safer in compression. Khair studied the performance of the truss bolts and made the following conclusions (Khair 1983):

1. Physical and analytical studies of the truss bolted roof indicated that proper tensioning of the truss bolt reduces the deformation (sagging) of the roof. Less sag indicates less tensile stress develops at the upper portion of the layers above the ribs. Furthermore, upward thrust developed by truss induces compression forces along the span of the entry thus increasing the shearing resistance of roof rock along the fracture planes.

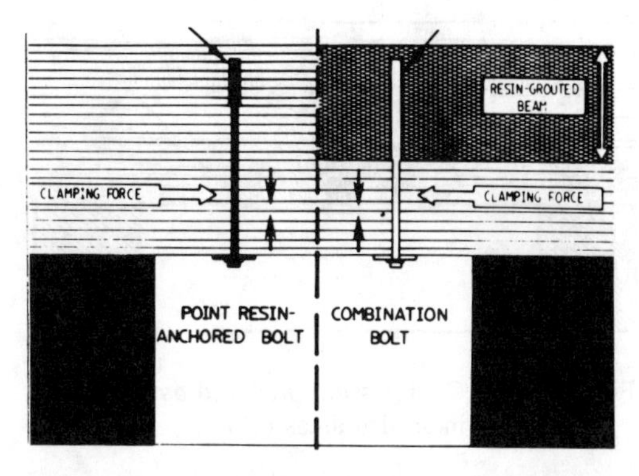

Figure 3. Bi-modal reinforcing characteristics of combination bolting system (8).

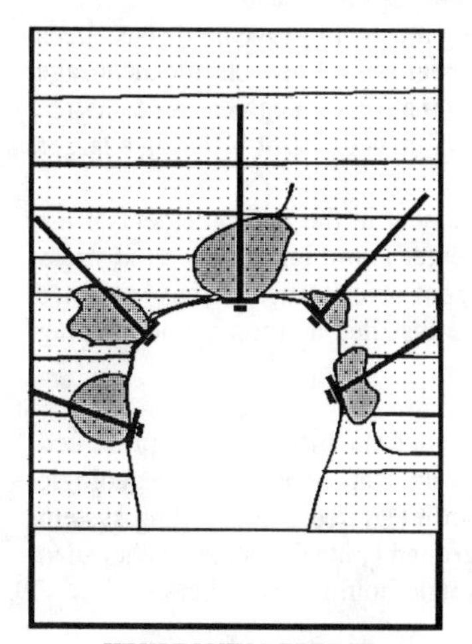

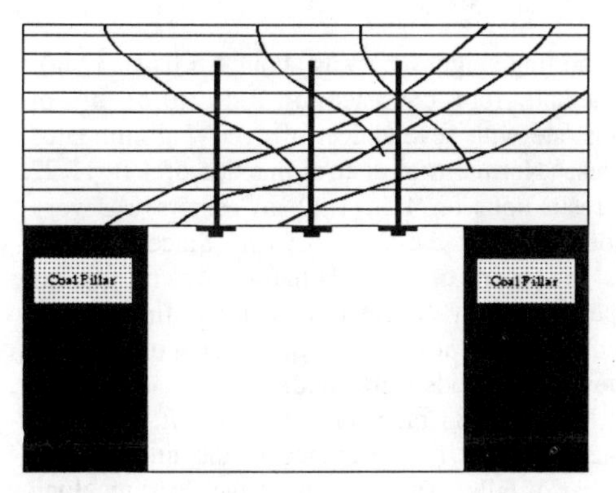

Figure 4. Keystoning (a) and keying (b) effects.

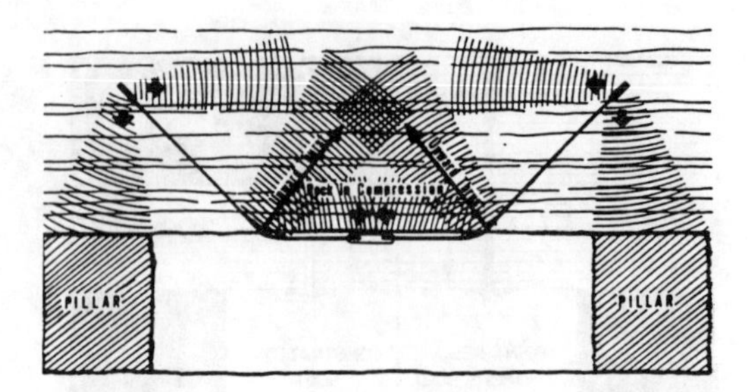

Figure 5. Compression produced by roof truss in shaded areas (15).

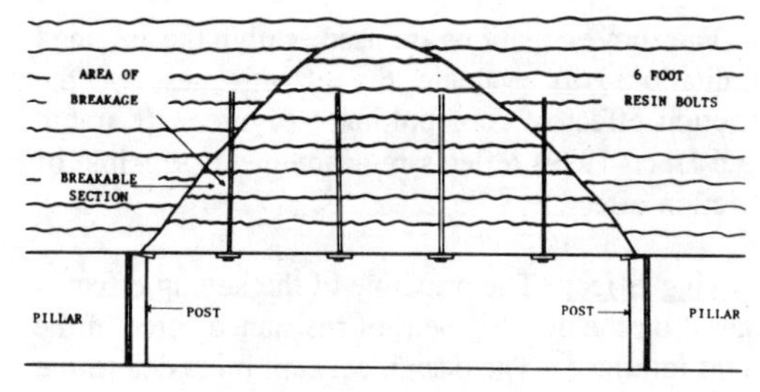

Figure 6. Typical roof fall using 6-ft. resin bolts on 4-ft. centers. (1 ft. = 0.305 m)

2. As the thickness of the immediate roof reduces, the effect of the truss is minimized to the immediate surroundings of the entry.

3. Since not much improvement has been observed for 60 degree inclined chord anchorage when compared to that of 45 degrees, a 45 degree anchorage is preferable because of covering a larger span.

4. The improvement in the reinforcement of the roof span in the center is minimized once the stress in the immediate roof layer turned to tensile, hence the truss should be installed as soon as the opening is made.

Case Studies: Two case histories are presented below to demonstrate that under adverse geologic conditions, only an integrated strata bolting system can alleviate the ground control problem rather than application of a single bolting principle.

2 CASE I

The mining height here was about 4.5 ft. (1.37 m). Immediate roof consisted of 2 ft. (0.61 m) of sandstone with several feet of overlying laminated shales. Normal roof control practice of 4 ft. (1.22 m) resin bolts on 4 ft (1.22 m) centers was very inadequate near the shaft airway. Since this area was the heart of the ventilation system, it was imperative to try different methods and find the one which would provide adequate support. The following methods were tried:

1. 6 ft. (1.83 m) resin bolts on 4 ft. (1.22 m) centers showed no difference in the number and manner of falls. This was due to the shearing along the rib until the rib side bolts failed which then induced the failure of center bolts resulting in a fall of 6.5 ft. by 8 ft. (1.98 m x 2.44 m) high arch as shown in Figure 6.

2. 6 ft. (1.83 m) resin bolts on 4 ft. (1.22 m) centers with 6 in. x 16 in. x 18 ft. (15.24 cm x 40.64 cm x 5.49 m) headers were installed on 8 ft. (2.44 m) centers with 6 in. x 6 in. (15.24 cm x 15.24 cm) rib posts on 4 ft. (1.22 m) centers and doubled under each header was employed. Under this system several rib side areas sheared and the roof started sagging. Headings and intersections were cribbed by 60 lb. (27.22 kg) and 85 lb. (38.56 kg) mine rails but the rails were severely bowing. This caused a great inconvenience in transportation. Extreme downward pressure sheared off the washer around the head of some rib-bolts and allowed the roof bolt plate and header to drop away. This effectively eliminated the support that the bolt had been providing.

3. 6 ft. (1.83 m) combination bolts with 3 ft. (0.92 m) resin were tried but to no avail (Figure 7). The torque build-up was so great, that the bolts were embedded into the roof.

4. Passive resin truss system with 6 ft. (1.83 m) resin bolts and T-5 channel on 5 ft. (1.53 m) centers provided the necessary support. Comparison of this with a full truss system consisting of 6 ft. (1.83 m) resin bolts on 5 ft. (1.53 m) centers indicated that the full truss system was more effective in reducing roof sag. Hence, the full truss system was employed to drive approximately 2000 ft. (610 m) of headings to intersect the shaft (Figure 8).

3 CASE II

The mine is located within the Allegheny mountain section of the Appalachian Plateau Province. The local geologic structures includes folds, faults, and joints. Overburden depth varies from 183-244 m and topographic relief is less than 152 m. The mining height is approximately 2 m and the entries

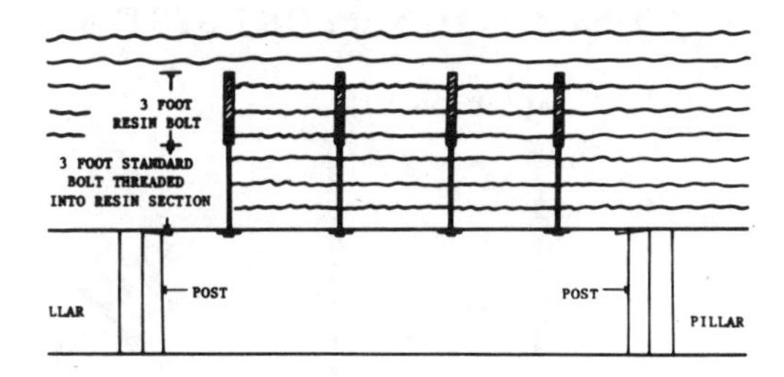

Figure 7. Point anchor, 6-ft. combination bolts with 3 ft. resin. (1 ft. = 0.305 m)

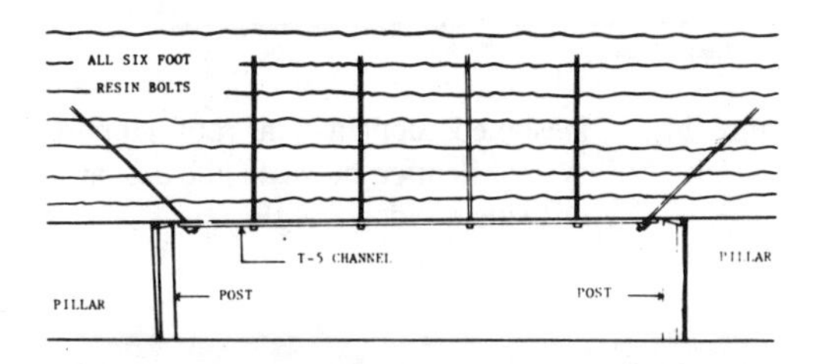

Figure 8. Resin truss, 6-ft. resin bolts on 4-ft. centers with 6-in. x 16-in. x 19 ft. headers. (1 ft. = 0.305 m) and 1 in. = 2.54 cm)

are 5.5 m wide. Roof strata consists of laminated shale with poor bonding strength. In-situ stresses were much higher than normally expected values. Due to adverse geologic conditions and tectonic stresses, cutter roof developed at the corner of the entry (Figure 9). Further lateral strata movement increased bending stress on the strata at the opposite corner and resulted in caving of the roof strata. The strata movements are simulated in the schematic of Figure 10. Typical roof strata reinforcement included roof bolting systems supplemented with beam support system. The roof bolting systems consisted of four 6 ft. (1.83 m) resin grouted bolts on 4 ft. (1.22 m) center supplemented with two 12 ft. (3.66 m) combination bolt on 6 ft. (1.83 m) center (Figure 11). Despite dense roof reinforcement, roof caving often occurred. Experiments were carried out, combining friction, suspension, and trussing effects. A primary roof bolt system consisting of four vertical and two inclined bolts were designed according to the configuration of Figure 12 and implemented in the mine (Figure 13). The design concept was: a) to increase bolting density near the corners where the strata is still intact thus increasing the friction

Figure 9. Cutter roof at the corner of the entry due to adverse geologic conditions and tectonic stress.

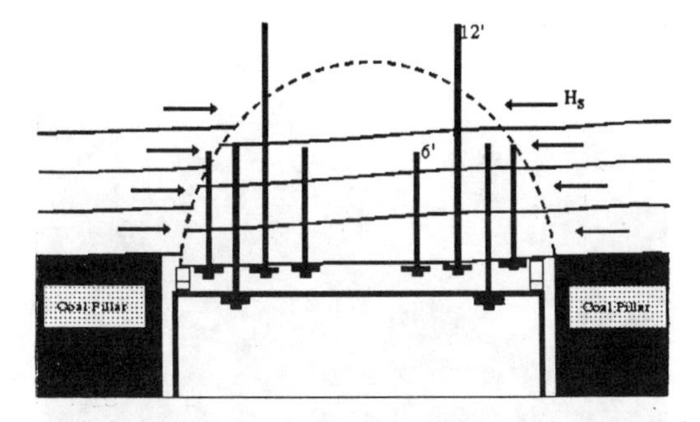

Figure 10. Simulation of strata movements.

Figure 11. Typical roof reinforcement system practiced in this mine.

resistance of the strata at the interface, b) to build a thicker beam strata in order to reduce the bending stresses and deformation in the strata, and c) transferring downward weight of the roof strata to

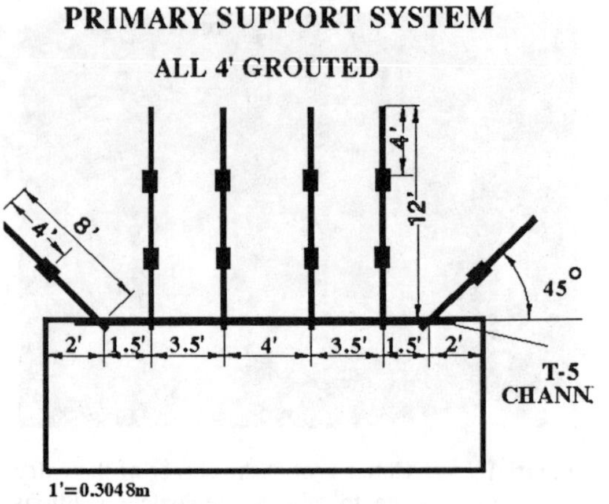

PRIMARY SUPPORT SYSTEM

ALL 4' GROUTED

4' 8' 4' 12' 45°

2' 1.5' 3.5' 4' 3.5' 1.5' 2'

T-5 CHANN.

1'=0.3048m

Figure 12. Designed primary roof support system tested.

Figure 13. Implemented the designed primary roof support system in the mine.

the ribs via angle bolts. A supplemental roof support system consisting of two superbolts, a truss roof bolt system, and two yieldable legs was designed in accordance to the schematics of Figure 14 and implemented in the mine (Figure 15). The designed roof reinforcement systems performed very well, kept the fractured strata together like a cantilever beam standing up in the gob zone (Figure 16). Analysis of the roof reinforcement systems were made and it was found to be over designed even though it was much cheaper than the ineffective system commonly used in this mine. The analysis also indicated that strata movements at the vicinity of the opening were very severe, far more than the extension limit of the bolts (Figure 17), therefore the lengths of the bolts were reduced

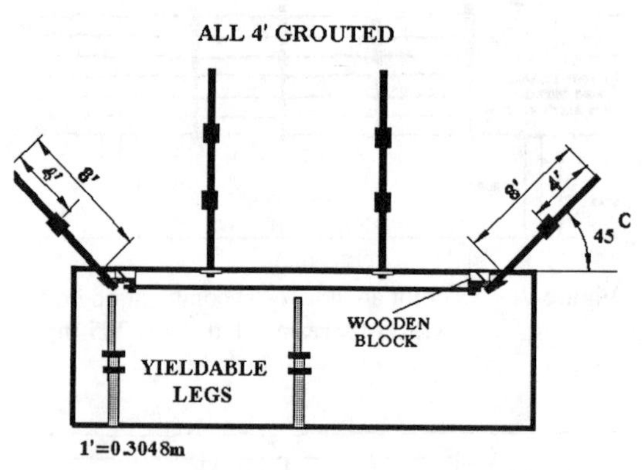

Figure 14. Designed supplemental roof support system (cutter between the rib and left side yieldable leg).

Figure 15. Implemented designed supplemental roof support system in the mine.

Figure 16. Performance of roof reinforcement system. Fractured strata is intact behind the face.

to avoid rupture of the long bolts below the anchoring horizon. Therefore, after a series of experiments, the primary and supplementary roof bolting systems were modified and combined into one system. This system was installed in one cycle, reducing the time duration for the roof to be left unsupported, hence limiting strata separation. Figure 18 shows schematics of the final combined roof reinforcement system which was utilized in the mine. The final roof reinforcement maintained roof strata stability during panel development as well as during extraction of the longwall panel. The cost of the new system was less than one-third of the old systems of roof reinforcements.

CONCLUSIONS

Under normal geologic and in-situ stress conditions each of the rock reinforcement principles will offer sufficient support to maintain roof strata stability, provided the reinforcement system is compatible with the geology of the roof strata. However the design rationale for the rock reinforcement system should be based on the understanding of site specific geological conditions and in-situ stresses which produce instability. Adverse geologic and in-situ conditions may require application of more than one principle of rock reinforcement. The reinforcement system should be designed to maximize the effectiveness of each principle and should be implemented as soon as excavation is made. A shorter period for unsupported roof limits strata movements and separation, hence increasing the effectiveness of rock bolting. The immediate application of rock reinforcement not only effectively increases interface friction resistance of the rock strata but this will also keep rock strata under confinement and constraint, limiting strata deformation during further development of additional pressure. Hence the integrity and self support of laminated/fractured rock strata is preserved.

ACKNOWLEDGMENT

This project in part was co-sponsored by Beth Energy Mines, Inc., West Virginia University's Energy and Water Research Center (EWRC), and the College of Mineral and Energy Resources. The author acknowledges the assistance of Mudassar Ahmad, research assistant in the Mining Engineering Department, in the preparation of this paper.

Figure 17. Severe strata movement at the vicinity of the opening, far more than the extension limit of the bolts, resulting bolt failure.

COMBINED SUPPORT SYSTEM

ALL 6' RESIN BOLTS

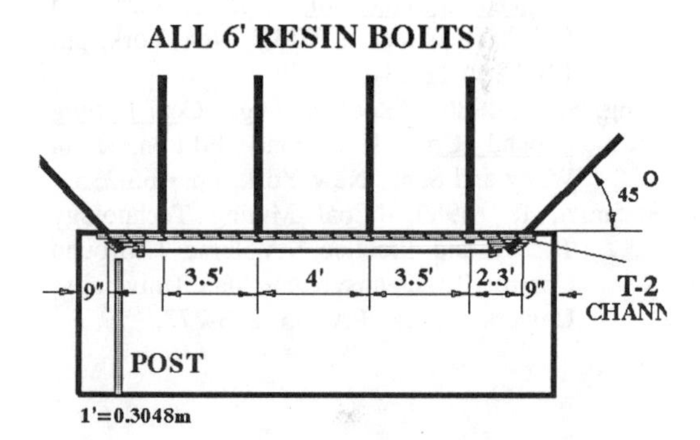

Figure 18. Combined designed primary and supplementary roof support systems installed in one cycle of operation and timber post was set under the cutter.

REFERENCES

Fairhurst and Singh, B., 1974. "Roof Bolting in Horizontally Laminated Rock," Engineering and Mining Journal, February, pp. 80-90.

Karabin, G. J., Jr. and Hoch, T. M., 1979. "An Operational Analysis of Point Resin-Anchored Bolting System," MESA, IR 1100, p. 14.

Khair, A. W., 1983. "Physical and Analytical Modeling of the Behavior of Truss Bolted Mine Roofs," Proceedings International

Symposium on Rock Bolting, A. A. Balkema, Rotterdam, pp. 125-142.

Neal, G. M., Haycocks, C., Townsend, J. M. and Johnson, L. P., III, 1976. "Influence of Some Critical Design Parameters on Roof Truss Support Capacity--A Preliminary Report," Proceedings, 17th U. S. Symposium on Rock Mechanics, Snowbird, Utah, August, Utah Eng. Exp. Sta.

Obert L. and Duvall, W. I., 1967. "Rock Mechanics and the Design of Structures in Rocks," Wiley, New York, New York, pp. 236-274.

Panek, L. A., 1957. "Anchorage Characteristics of Roof Bolts," Mining Congress Journal, November, pp. 62-64.

Panek, L. A., 1964. "Design for Bolting Stratified Roof," Transactions, SME/AIME, Vol. 229, pp. 113-119.

Panek, L. A. and McCormick, J. A., 1973. "Roof/Rock Bolting," SME Mining Engineering Handbook, A. B. Cummins and I. A. Givens (eds.), AIME, New York, pp. 13-125 to 13-134.

Peng, S. S., 1986. "Roof Bolting," Coal Mining Ground Control, Second Edition, John Wiley and Sons, New York, pp. 166-236.

Stefanko, R., 1980, "Coal Mining Technology Theory and Practice," Volume I Ground Control, The Pennsylvania State University, University Park, PA, pp. 223-277.

Geomechanics 93, Rakowski (ed.) © 1994 Balkema, Rotterdam, ISBN 90 5410 354 X

Transient and permanent rock-bolting: A computer-aided design scheme

A. Kidybiński
Central Mining Institute, Katowice, Poland

ABSTRACT: Permanent rock-bolting as a sole support system of coal-mine roadways offers non less than 70% materials costs savings when compared with arched openings and reduces approx. 80% transportation and repair costs. It requires however careful simulation and designing system to include effects of rheology, moisture and seismicity, as well as detailed analysis of roof-bed separation in time-lapse. Transient (temporary) bolting doesn't require such a comprehensive approach but offers substantial work easing especially on face-ends and during longwall salvages. Some important elements of computer simulation and designing programs developed by the author are discussed in brief.

1 GENERAL CONDITIONS

Contrary to the arched steel support of coal-mine roadways which may be considered as metal-made structure where steel properties are decisive, rock-bolting support appears to be reinforced rock-made structure and therefore rock-mass properties decide about its stability. It is widely understood that rock-bolting as sole support of coal-mine openings is applicable mainly in more competent rock formations such as in Australia and the U.S.A. while in other coal-producing countries where weaker rocks appear such as Germany, France, Britain, Poland, Russia and China roof-bolting was less often used and partially successful. In Poland it was anly recently that fundamental guidelines were formulated on the basis of international experience, for official approval of applying rock-bolting in collieries. The main of them are following:

- average compressive strength of rocks - non less than 15 MPa
- horizontal fractures - non closer than 0.1 m,
- soaking ratio 0.8 or higher.

Particularly important are requirements concerning designing methods for bolting patterns which have to be based on computer simulation of rock-strata behaviour under loads expected. This includes analyses of real roof-span, effect of time-lapse, moisture and seismicity as well as computer-aided generating of maximum extent of break-line within roof-strata for a period adequate to opening's maintenance time required. Tectonic stresses should be also analysed, if existing.

2 ROOF-STRATA TESTING

Coal-bearing formations are bedded (stratified) and their simulation requires analysis of non-homogeneous medium with precise vertical log of strength data. The most relevant measure of the rocks resistance to gravity forces seems to be bed separation resistance (BSR) defined as the strength against vertical tension. This parameter is measured by Hydraulic Borehole Penetrometer (HBP) in vertical borehole made through roof strata and punched at each 0.05 m altitude above an opening. This way 5 m long and almost continuous log of the rock resistances along vertical position is obtained which then may be used for sequential bed separation analysis.

BSR array and compressive strength of a rib material (coal mostly) are crucial local rock characteristics which should be known at each 100-400 m hoirizontal distance respectively to changing geology. But other parameters of regional type should be known as well for computer simulation. Among them following are most important:

- cleats or/and fracturing characteristics (RQD at least),
- rheologic parameters (creep/relaxation velocity versus stress/strain, or a full rheologic model with coefficients of elasticity, viscosity, viscoelasticity and limit conditions)
- strength deterioration with moisture (for all types of rocks involved),
- dynamic modulus of elasticity and a forecast of maximum probable seismic energy of events expected.

3. TRANSIENT BOLTING

There are many cases where short-time bolting may assist mining operations. Up-hanging of steel arches at longwall-face-end may be given as first example, Fig. 1. It's easing equipment displacing at the intersection by eliminating additional props and roof-bars, while one of the side-legs of an arch is timely removed to enable machine cutting. Widening of traditional arched roadway in coal for new longwall installation - may be another example, Fig. 2.

During salvage of longwall faces temporary roof-bolting may ease transporting of heavy equipment by replacing of standing supports by bolts and improving roof-safety. As one of options a special pre-driven opening to dissolve longwall operation may be considered, Fig. 3.

In all these examples time period required for bolts operation is not longer than days or weeks. During such a short time rock deterioration process due to rheology, moisture and weathering may be neglected and simple criteria chosen for bolting design and control, especially in these cases where bolts serve as a supplement to existing support. The length of bolts is usually established as a part of openings cross-section diameter (0.5 for example) and bearing capacity of partially-grouted bolts is controlled by simple pull-test. Break-line position simulation (see Fig.1) helps to settle the bolts properly inclined to vertical.

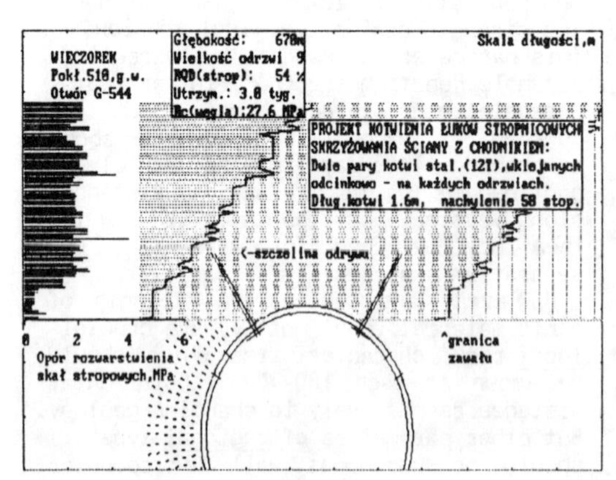

Fig.1. Up-hanging of steel arches at face-end

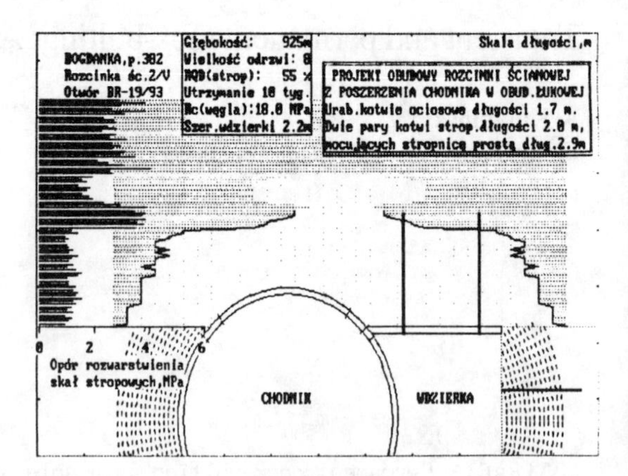

Fig.2. Widening of a roadway for longwall installation

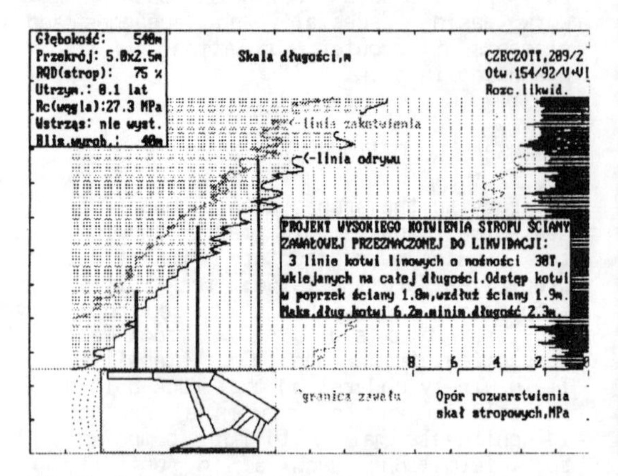

Fig.3. Longwall face salvage with cable-bolts

4 PERMANENT ROCK-BOLTING

Long-time stability requirement for an opening (months,years) poses the necessity of more sophisticated forecasting analysis which includes possibility of rocks deterioration. Main reasons for worsening of rocks quality are time-lapse (rheology, fracturing), moisture and seismicity. General stress redistribution should also be considered especially in terms of possible growth of effective opening width due to fractures in the ribs. From all these reasons resin fully-grouted bolts (full column) are most recommended for permanent bolting.

4.1 Effect of time-lapse

Long-time loading always reveals ability of rocks to creep. When unit load is high long beam of rock tasted in bending fails after some time of steady loading. It is important therefore to investigate rocks in creeping and to find viscoelastic and viscous parameters of their rheologic pattern. This, of course, due to long time of a study should be considered not as local but rather regional recognition. Nevertheless, once parameters mentioned above are known long-time projection of creeping process is possible by calculation. The author uses simple mathematical formula that relates final creep-deformation for the time-period of openings life required to the apparent strength of rocks in bending. This way, strength of rock is reduced proportionally to expected long-time deformation, separately for each type of rock appearing in the roof, and Burgers pattern is used as simplified rheologic model of rocks. The final growth of fracture net is also projected in time and another formula relating apparent RQD to the strength of a rock-mass as compared with rock-material is being used. Finally, the thickness and apparent strength of each separated rock layer is identified from the BSR log and calculation of roof-stability is performed goin step-by-step upwards. It is believed that this method (called Sequential Bed Separation method) simulates as close as possible behaviour of bedded formations in the field of gravity stress. In Fig. 4,5 and 6 position of beak-line in roof may be seen, representing the end of time-period required for openings life. On this basis the length of roof-bolts is calculated using well-known "a foot-over-bolting-horizon" rule.

4.2 Effect of moisture

Longer rock exposure within destressed zone in roof usally brings about additional water saturation and further strength reduction of rock. In order to evaluate this effect specific curves of strength versus maisture content should be known for each rock type involved. Whereas moisture percentage of rocks in roof after long time exposure is hard to predict, most of these curves have an asymptote for high moisture contents. It seems to be rational to assume that this value represents most critical conditions which may occurr within roof strata and use it in calculations as additional strength reducing factor.

4.3 Acting roof span

Penetration of fractures in ribs weakens the support (pillar) of roof-strata beams

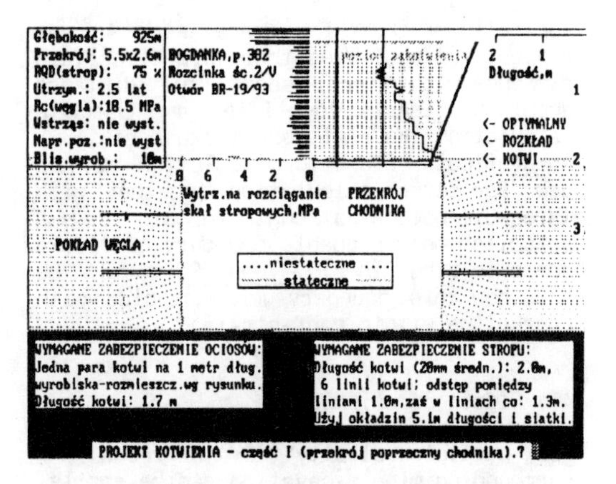

Fig. 4. Roof-and-rib-bolting in maingate ahead of mining

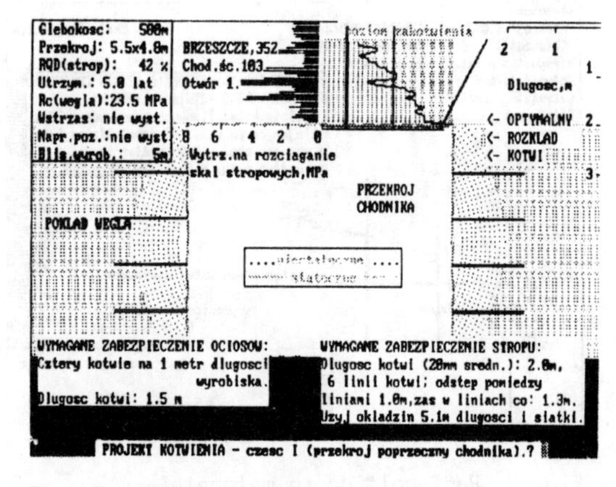

Fig. 5. Bolting in large-size opening

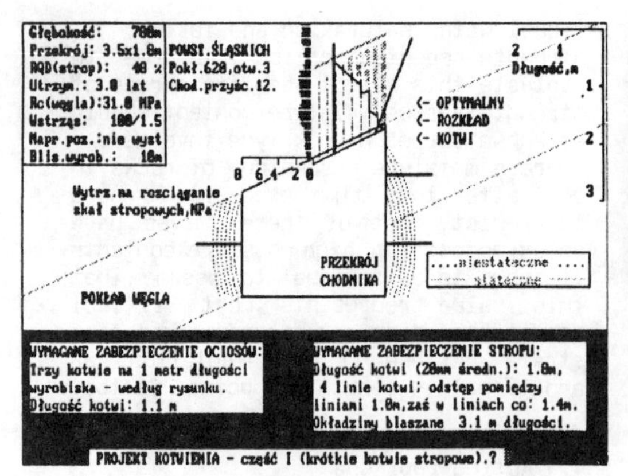

Fig. 6. Bolting of a gateroad within inclined coal seam

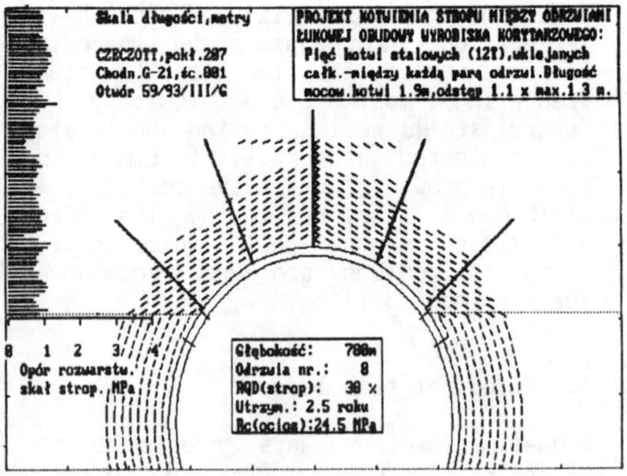

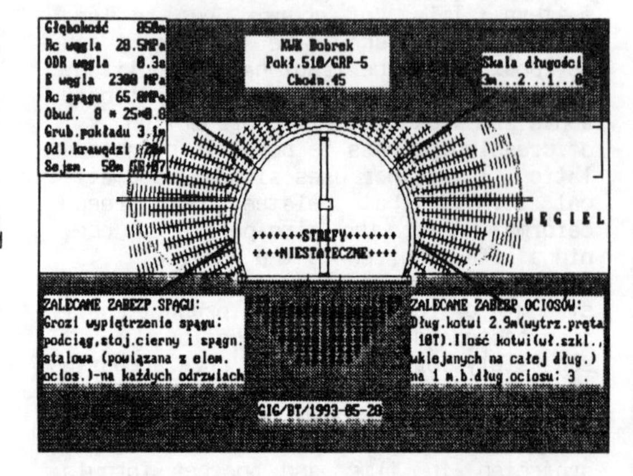

Fig. 8. Reinforcement of roof-strata between arches

and allows them for more sag. A real support for roof lies therefore at certain distance within coal seam. This distance should be clearly identified and used in calculations as real span instead of theoretical. Close-form solution may be used for stress calculation together with three-dimensional strength formula to establish most probable boundary of cracked zones in ribs. In Fig. 7 and 8 it may be seen how this boundary may coincide with break-line within roof-strata.

4.4 Effect of seismicity

Peak particle velocity (PPV) of rocks surrounding mine excavation can be easily found from seismic energy at source and

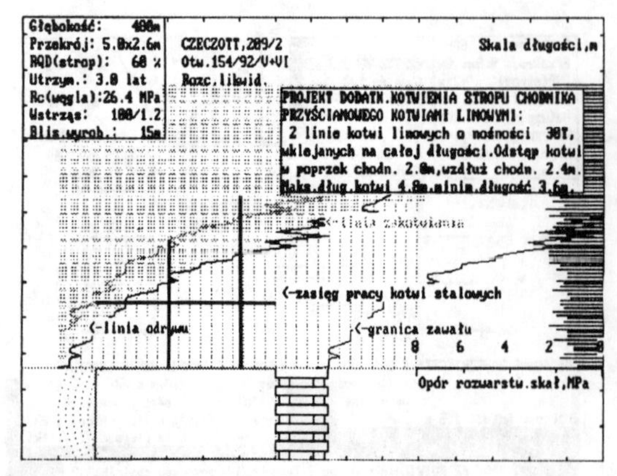

Fig. 7. Roof-bolting in maingate adjacent to goabs

Fig. 9. Reinforcement of tailgate to prevent coal-bumping

focal distance anticipated. Dynamic stress calculated thorefrom may be superimposed with effective static stress and additional range of fracturing found, see Fig. 9. Rock-bolting design should cover this additional range of fractures both in rib and roof-strata by applying of longer fully grouted bolts. Moreover, during calculations of bolts-spacing additional safety factor for dynamic stresses should be taken into account.
In seismic conditions special type of steel rebars are recommended, namely those with increased elongation before failure (25% at least).

5 FINAL REMARKS

Rock-bolting especially with fully-grouted

resin bolts proved to be reliable and safe support system both for underground mines and tunnelling. In Polish collieries it is now being intersively implemented with economic gains of approx. USD 200,000 per 1 km of a single roadway as compared with traditional steel-arched openings.

Geomechanics 93, Rakowski (ed.) © 1994 Balkema, Rotterdam, ISBN 90 5410 354 X

Quantification of the effective Coulomb and the Hoek-Brown parameters of the pre-reinforced rock mass

A. Mahtab, S. Xu & P. Grasso
GEODATA SpA, Torino, Italy

ABSTRACT: Our premise is that rock mass pre-reinforcement is equivalent to improving the strength of the rock mass by providing an additional confining pressure. The formulations of Coulomb and Hoek-Brown admit the effective cohesion, c, and effective s. The correspondence between the two formulations is shown. The calculation of effective c and s is illustrated through an example.

1. INTRODUCTION

The need for improving the rock mass resistance in advance of excavation is recognized in both civil and mining practice. Improving the cohesion of rock (Muir Wood 1979) and the angle of friction (Bischoff and Smart, 1975) have often been discussed in the literature. In particular, the influence of bolting in improving the cohesion of rock mass has been well documented (Egger, 1978; Wullschlager and Natau, 1987; Grasso et al., 1989; Labiouse, 1991). All these investigators have used the Coulomb criterion to demonstrate the effect of pre-reinforcement.

The Hoek and Brown criterion (Hoek and Brown, 1980, 1988), further modified by Hoek et al. (1992), is now also widely used. There is a need to relate the influence of bolting on the parameters of Coulomb and Hoek-Brown criteria; this is the principal objective of this paper. Another objective is to provide, for ready reference, the correspondence between the two criteria through a set of transformations (Hoek 1990, Mahtab and Grasso 1992).

2. FORMULATION OF EFFECTIVE STRENGTH PARAMETERS

2.1 Assumptions and Rationale

1) Rock reinforcement (grouted, untensioned bars) is installed around a circular tunnel in an isotropically stressed rock

2) Failure of rock is governed by either of two criteria:

- Coulomb (Fig. 1):
  - $\sigma_1 = 2c \tan\alpha + \sigma_3 \tan^2\alpha$
  - $\alpha = \pi/4 + \phi/2$,
  - c =cohesion
  - ϕ=angle of friction
- Hoek-Brown (Fig. 2):
  - $\sigma_1 = \sigma_3 + (m\,\sigma_3 C_o + sC_o^{\,2})^{1/2}$
  - m, s = rock characteristics other than the unconfined compressive strength, C_o, of intact rock

3) As a result of deformation (in the rock-reinforcement system), the reinforcing bars (bolts, cables) pick up tension, which acts as an additional confining pressure, $\Delta\sigma_3$ on the rock. The net result is

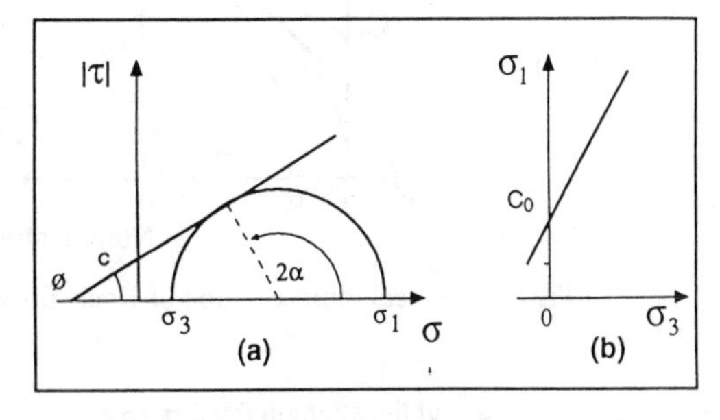

Fig.1: The Coulomb criterion.
(a) σ-τ plane.
(b) σ_1-σ_3 plane.

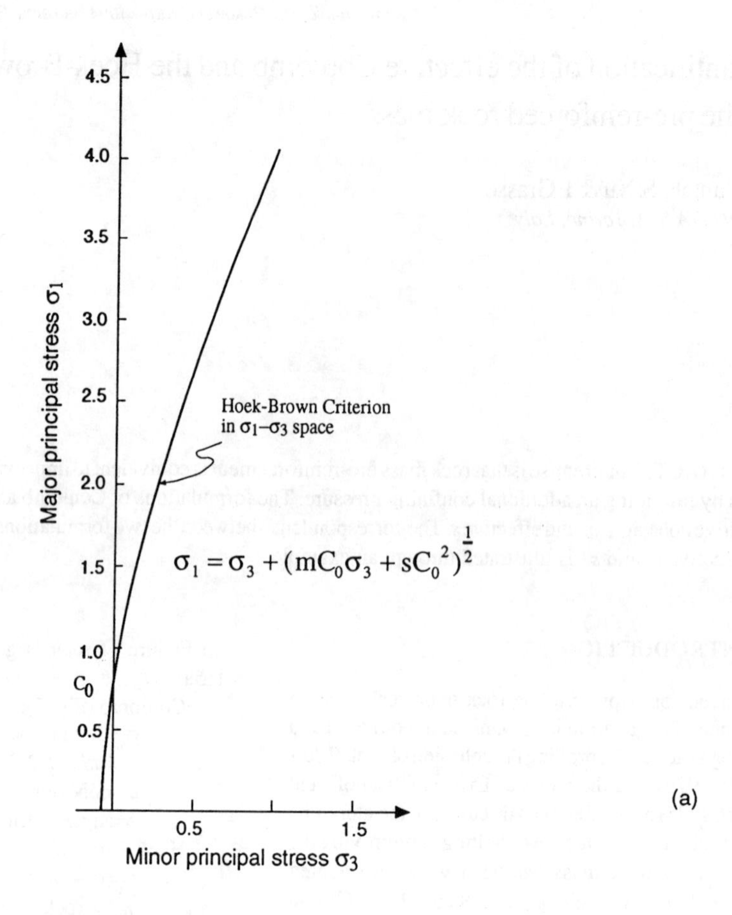

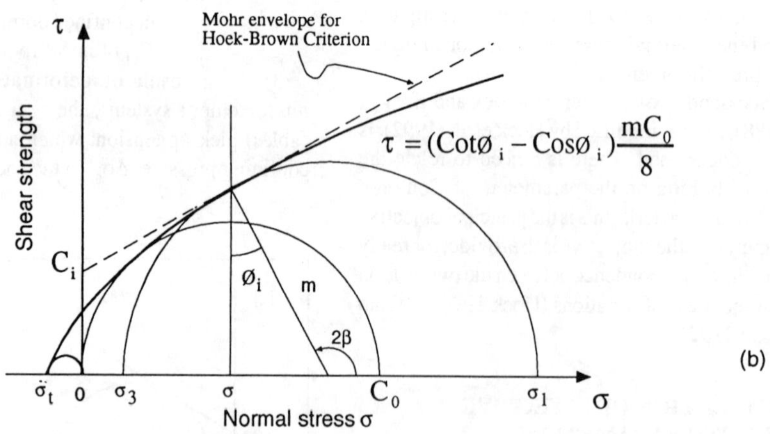

Fig.2: Graphical representation of Hoek-Brown criterion. (a) σ_1-σ_3 plane. (b) σ-τ plane.

an effective increase in the strength from σ_1 to σ_1^*.

4) The Coulomb parameters subject to modification are: cohesion, *c*, and angle of friction, φ; the unconfined compressive strength C_o being a function of *c* and φ.

5) The Hoek-Brown parameters subject to modification are *m* and *s*: empirical constants for a class of rock; the intact rock strength, C_o, is a constant.

2.2 *Effective Coulomb Parameters*

Beginning with the Coulomb failure criterion

$$\sigma_1 = 2c\tan\alpha + \sigma_3\tan\alpha \quad (1)$$

the effective cohesion, c^*, can be derived as a function of the confining pressure $\Delta\sigma_3$, supplied by the reinforcement.

The effective cohesion concept was first introduced by Daemen & Fairhurst, 1970, and Muir Wood, 1978, and was further developed and applied by Grasso et al., 1989, using the expression

$$c^* = c + \frac{\Delta\sigma_3}{2}\tan\alpha \quad (2)$$

The strength improvement due to c^* is illustrated in Fig. 3.

The concept of an effective friction angle, ϕ^*, first used by Bischoff and Smart (1975), is further developed in the following fashion.

Starting with eq.(1), and introducing a confining pressure $\Delta\sigma_3$ we have

$$\sigma_1 = 2c\tan\alpha + (\sigma_3 + \Delta\sigma_3)\tan^2\alpha \quad (3)$$

using ϕ^* as the effective angle of friction, eq.(3) transforms into

$$\sigma_1 = 2c\tan\alpha^* + \sigma_3\tan^2\alpha^* \quad (4)$$

where $\alpha^* = \pi/4 + \phi^*/2$, with the effective friction angle as:

$$\phi^* = 2\tan^{-1}\left[\frac{\left\{(c+\sigma_3\tan\alpha)^2 + \Delta\sigma_3 + \sigma_3\tan^2\alpha\right\}^{1/2} - c}{\sigma_3}\right] - \pi/2 \quad (5)$$

Note that ϕ^* is a function of the minor principal stress, σ_3.

2.3 *Effective Hoek-Brown Parameters*

The Hoek and Brown criterion is written as:

$$\sigma_1 = \sigma_3 + (m\,C_o\,\sigma_3 + sC_o^2)^{1/2} \quad (6)$$

where

- σ_1 = major principal stress
- σ_3 = minor principal stress
- C_o = uniaxial compressive strength of intact rock
- m = empirical constant ranging from 0.001, for highly disturbed rock, to about 25 for hard rock
- s = empirical constant ranging from 0, for jointed rock masses, to 1 for intact rock.

The additional confining pressure, $\Delta\sigma_3$, is introduced into the Hoek-Brown criterion, giving

$$\sigma_1 = \sigma_3 + \Delta\sigma_3 + [m(\sigma_3 + \Delta\sigma_3)C_o + sC_o^2]^{1/2} \quad (7)$$

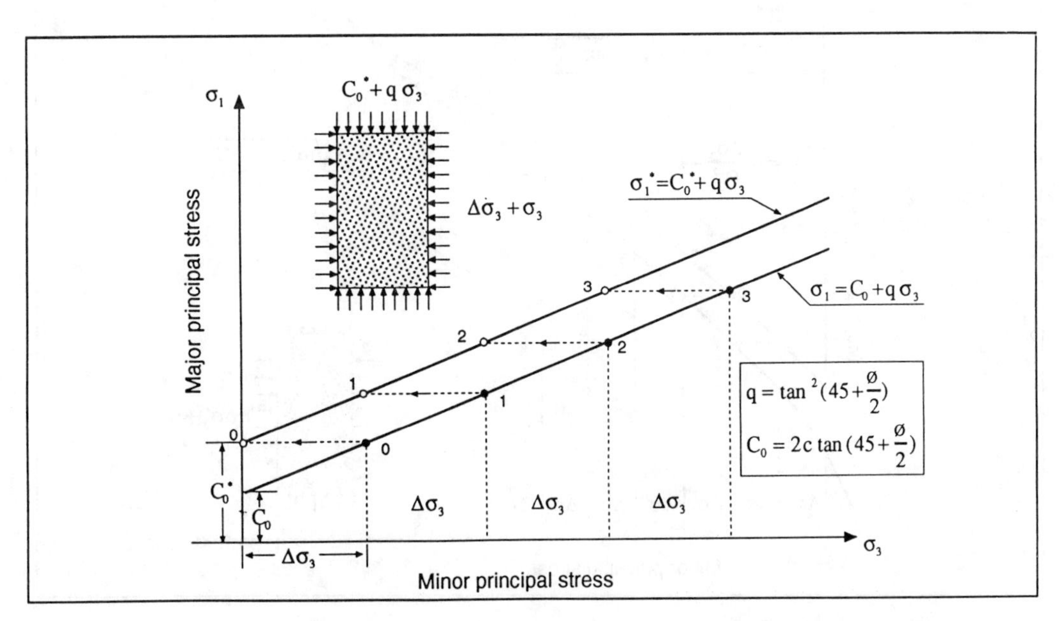

Fig.3: Graphical representation of effective cohesion, c^*, and Coulomb criterion.

If we consider the improvement of the rock in two alternative ways: effective *s* (or *s* *) and effective *m* (or *m* *), we have the following forms:
For effective *s*, rewrite eq.(7) as:

$$\sigma_1^* = \sigma_3 + \Delta\sigma_3 + \left[m\sigma_3 C_o + \left(m\frac{\Delta\sigma_3}{C_o} + s \right) C_o^2 \right]^{1/2} \quad (8)$$

or $\sigma_1^* = \sigma_3 + \Delta\sigma_3 + [m\sigma_3 C_o + s^* C_o^2]^{1/2}$ (9)

$$\text{where} \quad s^* = m\frac{\Delta\sigma_3}{C_o} + s \quad (10)$$

The strength improvement due to s^* is shown in Fig.4.
For effective *m* (or m^*), eq.(6) may be rewritten as:

$$\sigma_1^* = \sigma_3 + \Delta\sigma_3 + \left[m\left(1 + \frac{\Delta\sigma_3}{\sigma_3}\right)\sigma_3 C_o + sC_o^2 \right]^{1/2} \quad (11)$$

or $\sigma_1^* = \sigma_3 + \Delta\sigma_3 + [m^* \sigma_3 C_o + sC_o^2]^{1/2}$ (12)

$$m^* = m\left(1 + \frac{\Delta\sigma_3}{\sigma_3}\right) \quad (13)$$

Note that m^* is a function of the minor principal stress, σ_3.

3. CORRESPONDENCE BETWEEN CRITERIA OF COULOMB AND HOEK-BROWN

3.1 Estimates of Coulomb parameters from Hoek - Brown parameters

In a recent publication, Hoek (1990) has provided relationships (corresponding to 3 conditions) for estimating the Coulomb parameters (ϕ and c) from the Hoek-Brown criterion.

<u>Condition 1</u>

For a specified normal stress, σ, the shear stress, τ, is given by:

$$\tau = (\cot\phi_i - \cos\phi_i)\frac{mC_o}{8} \quad (14)$$

with the "istantaneous" angle of friction at (s,τ) defined by:

$$\phi_i = \arctan\,[1/(4h\cos^2\theta - 1)^{1/2}] \quad (15)$$

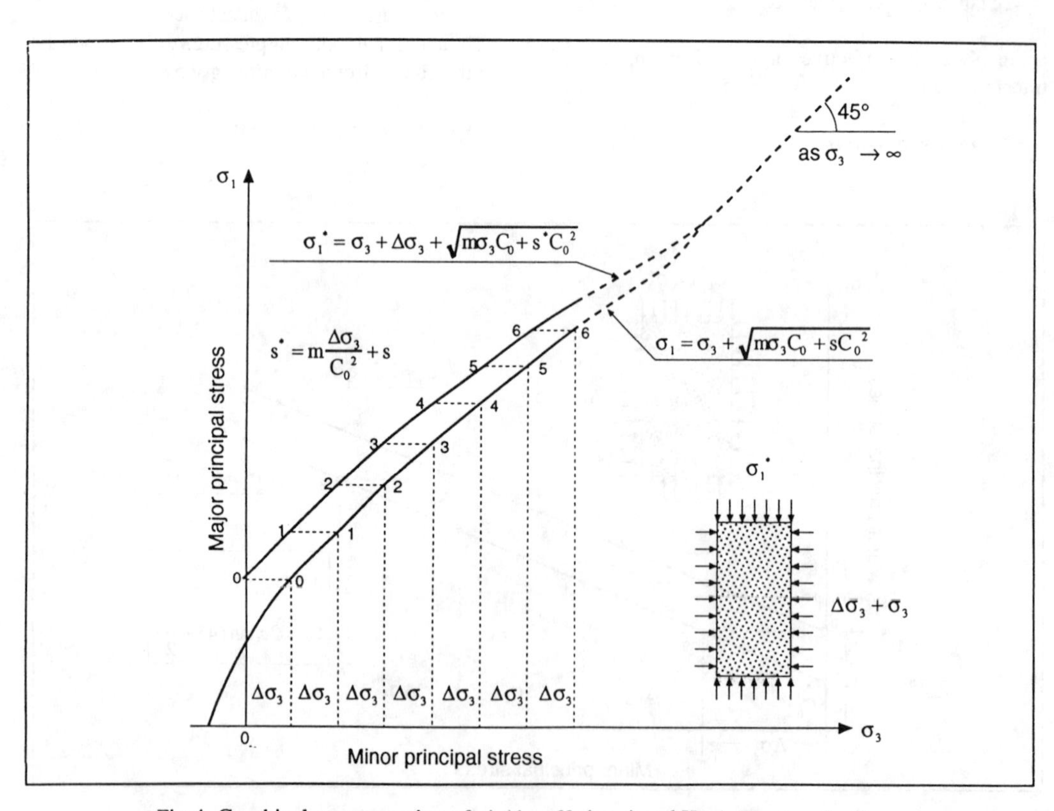

Fig.4: Graphical representation of s^* (the effetive *s*) and Hoek-Brown criterion.

and the "istantaneous" cohesion, c_i, corresponding to the point (s, τ) is given by

$$c_i = \tau - s \tan \phi_i \tag{16}$$

and where

$$h = 1 + \frac{16(m\sigma + sC_o)}{3m^2 C_o} \tag{17}$$

$$\text{and } \theta = 30 + \frac{1}{3}\arctan\left[1/\left(h^3 - 1\right)^{1/2}\right]$$

Referring back to the Coulomb criterion, the uniaxial compressive strength $C_{o(mass)}$ of rock mass with these ϕ_i and c_i values is given by:

$$C_{o(\text{mass})} = \frac{2c_i \cos\phi_i}{1 - \sin\phi_i} \tag{18}$$

Condition 2

For the situation where σ_3 is specified, the following relations are provided by Hoek (1990) to obtain the Coulomb parameters.

First, σ_1 is obtained from eq.(6) and input to the following calculations:

$$\sigma = \sigma_3 + \frac{(\sigma_1 - \sigma_3)^2}{2(\sigma_1 - \sigma_3) + \frac{1}{2}mC_o} \tag{19}$$

$$\tau = (\sigma - \sigma_3)\sqrt{1 + \frac{mC_o}{2(\sigma_1 - \sigma_3)}} \tag{20}$$

$$\phi_i = 90 - \arcsin\left(\frac{2\tau}{(\sigma_1 - \sigma_3)}\right) \tag{21}$$

The cohesion, c_i, is then obtained from eq.(16) whereas $C_{o(mass)}$ is obtained from eq.(18).

Condition 3

For another situation, where the uniaxial strength of the rock mass is the same for both criteria and may be assumed to be given by,

$$C_{o(\text{mass})} = \sqrt{s}C_o \tag{22}$$

the following procedure (Hoek, 1990) is used to obtain the Coulomb parameters:

$$\sigma = \frac{2sC_o}{4\sqrt{s} + m} \tag{23}$$

$$\tau = \sqrt{1 + \frac{m}{2\sqrt{s}}} \tag{24}$$

$$\phi_i = 90 - \arcsin\left(\frac{2\tau}{\sqrt{s}C_o}\right) \tag{25}$$

The cohesion, c_i, is again obtained from eq.(16).

3.2 Estimates of Hoek-Brown parameters from Coulomb parameters

The "converse" relationships have been derived (credited to Xu in Mahtab and Grasso, 1992) between the Hoek-Brown (HB) parameters, m and s, and the Coulomb (CU) parameters, c and ϕ, for the following three conditions. [The quantities assumed to be known are: C_o, associated with the intact rock, and c, ϕ, associated with the rock mass.]

Condition 1:

For a given s in the Coulomb criterion

$$\tau = c + s \tan\phi \tag{26}$$

$$m = \frac{8\tau \sin\phi(1 + \sin\phi)}{C_o \cos^3\phi} \tag{27}$$

$$s = \frac{1}{C_o^2}\left[\frac{4\tau^2 - mC_o(\sigma - c\cos\phi)(1 - \sin\phi)}{\cos^2\phi}\right] \tag{28}$$

Condition 2:

For a given σ_3

$$m = \frac{4}{C_o}\left[\frac{C_{o(\text{mass})}CU(c\cot\phi + \sigma_3)}{c\cot\phi}\right]\tan\left(\frac{1+\sin\phi}{\cos\phi}\right) \tag{29}$$

$$s = \frac{1}{C_o^2}\left[\left\{\frac{C_{o(\text{mass})}CU(c\cot\phi + \sigma_3)}{c\cot\phi}\right\}^2 - mC_o\sigma_3\right] \tag{30}$$

$$\text{where } C_{o(\text{mass})} = \frac{2c_i \cos\phi_i}{1 - \sin\phi_i}$$

Condition 3:

For $C_{o(mass)}$ HB = $C_{o(mass)}$ CU

$$s = \left[\frac{2c\cos\phi}{(1 - \sin\phi)C_o}\right]^2 \tag{31}$$

$$m = 4\sqrt{s}\tan\phi\left(\frac{1+\sin\phi}{\cos\phi}\right) \tag{32}$$

3.3 Selected Parameters for Correspondence

A comparison of the effective and original parameters/variables of the Coulomb and Hoek-Brown criteria is given in Table 1.

Table 1 Effective vs. original parameters /variable of Coulomb and Hoek-Brown criteria

Criterion	Effective parameters	Original parameters /variables
Coulomb	c * ϕ*	c, ϕ, $\Delta\sigma_3$ c, ϕ, $\Delta\sigma_3$, σ_3
Hoek-Brown	s * m *	s, m , $\Delta\sigma_3$ m , $\Delta\sigma_3$, σ_3

As seen from Table 1, and as noted earlier, the effective parameters ϕ* and m * depend on σ_3 and are, therefore, difficult to interpret and use for the comparison. In contrast, the parameters c* and s* depend only material properties and the incremental confinement, $\Delta\sigma_3$, introduced by the reinforcement and are, therefore, easy to interpret (as in Figs. 3 and 4).

3.4 Illustrative Example

The following example illustrates the concepts of effective cohesion and effective s through 2 steps:

•Select a set of Coulomb parameters and calculate c* for a given $\Delta\sigma_3$

•Estimate Hoek-Brown parameters and calculate s* for the same $\Delta\sigma_3$

The condition used for the correspondence is that the unconfined compressive strength of the rock mass is the same for both Coulomb and Hoek-Brown criteria.

1) The following Coulomb parameters are selected corresponding to a "disturbed" carbonate rock mass:

c = 0.2 MPa

ϕ=25°

$C_{o\,(rock\,mass)}$ = 2 c tan (45+ϕ/2)=0.6279 MPa

$C_{o\,(intact\,rock)}$= 63 MPa

Suppose we select a pattern of rebars to be installed in a circular tunnel for providing a confining pressure, $\Delta\sigma_3$=0.5 MPa, subsequent to the deformation of the rock mass around the tunnel. Then the effective cohesion is given by:

$$c^* = c + \frac{\Delta\sigma_3}{2}\tan\alpha = 0.5924\text{MPa}$$

2) The Hoek-Brown parameters corresponding to the original Coulomb parameters assumed above can be calculated using the transformations given in eqs.(31) and (32).

$$s = \left[\frac{2c\cos\phi}{(1-\sin\phi)C_o}\right]^2 = 0.0001 \quad (33)$$

$$m = 4\sqrt{s}\tan\phi\left(\frac{1+\sin\phi}{\cos\phi}\right) = 0.0293 \quad (34)$$

The effective s, corresponding to $\Delta\sigma_3$=0.5 MPa can be calculated using eq.(10).

$$s^* = m\frac{\Delta\sigma_3}{C_o} + s = 0.00033$$

The value of s* can now be used to calculate the σ_1*, of eq.(9).

4. CONCLUSION

The pre-reinforcement or requalification of weak rock mass through a pattern of grouted, untensioned bolts, has in practice reduced the stabilization pressures required for tunnels. The quantification of the reinforced effect can be achieved both for Coulomb criterion (effective cohesion) and for Hoek-Brown criterion (effective s). The effective parameters should, of course, be assumed to apply to the rock zone delimited by the length and pattern of the bolts. More experience is needed to suggest a choice between effective c and effective s. for design.

5. REFERENCES

Bischoff, J. A. and Smart, J.D. 1975. A method of computing a rock reinforcement system which is structurally equivalent to an internal support system. Proceedings of the 16th U.S. Symposium on Rock Mechanics, Univ. of Minn. ASCE, New York: 179-184.

Daemen, J.J.K. and Fairhurst, C. 1970. Influence of failed rock properties on tunnel stability. Proceedings of the 12th U.S. Symposium on Rock Mechanics, Univeristy of Missouri: 885-875.

Egger, P. 1978. Neuere Gesichtspunkte bei Tunnelankerung. Felsmechanik Kolloquium Karlsruhe, Clausthal: 263-276.

Grasso, P., Mahtab, A. and Pelizza, S. 1989. Reinforcing a rock zone for stabilizing a tunnel in complex formations. Proceedings of the International Congress on Progress and Innovation in Tunnelling, Vol. 2, Toronto: 663-670.

Hoek, E.and Brown, E.T. 1980. Empirical strength criterion for rock masses. J. Geotech. Engng. Div. ASCE, Vol. 106: 1013-1035.

Hoek, E.and Brown, E.T. 1988. The Hoek-Brown Criterion - a 1988 update. Proc. 15th Canadian Rock Mech. Symp., Univ. of Toronto: 31-38.

Hoek, E. 1990. Estimating Mohr-Coulomb friction and cohesion values from Hoek-Brown failure criterion. International Journal of Rock Mechanics Mining Sciences & Geomechanics Abstracts 27: 227-229.

Hoek, E., Wood, D. and Shah, S. 1992. A modified Hoek-Brown failure criterion for jointed rock masses. Proceedings of ISRM Symposium - EUROCK '92, Chester: 209-214.

Labiouse, V. 1991. Rock bolting in the rock-support interaction analysis. Proceeding of the 7th International Congress of ISRM, Vol. 2, Aachen: 1321-1324.

Mahtab, M.A. and Grasso P. 1992. Geomechanics Principles in the Design of Tunnels and Caverns in Rock. Elsevier.

Muir Wood, A. M. 1979. Ground behaviour and support for mining and tunnelling. Proceedings of Tunnelling '79, IMM, London: xi-xxii.

Wullschlager, D. and Natau, O. 1987. The bolted rock mass as an anisotropic continuum: material behaviour and design suggestion for rock cavities. Proceedings of the 6th International Congress of ISRM, Vol. 2, Montreal: 1321-1324.

Geomechanics 93, Rakowski (ed.) © 1994 Balkema, Rotterdam, ISBN 90 5410 354 X

Rockburst mechanisms and tunnel support in rockburst conditions

T.R.Stacey & W.D.Ortlepp
Steffen, Robertson and Kirsten, Johannesburg, South Africa

ABSTRACT: The range of rockburst source and damage mechanisms that are believed to occur in deep level hard rock mines will be described. One of the damage mechanisms considered to be particularly severe is that of ejection of blocks or slabs. From observations and interpretations of rockburst occurrences, typical ejection velocities in significant ejection events are estimated to be of the order of 5 to 10 m/s. Thoughts on source mechanisms associated with the ejection damage mechanism are presented.

The design of rock support capable of containing damage of this nature will be described. In addition the results of tests on yielding and conventional rockbolt elements, under simulated rockburst loading, will be described. These tests are very graphic in demonstrating the inadequacy of conventional rockbolt support elements in rockburst situations, and also provide useful design information.

1 INTRODUCTION

Rockbursts continue to be a major hazard to safety and production in many mines throughout the world. Rockbursts resulting from major seismic events, and with associated major damage, have been reported in Canada, the United States of America, Chile, South Africa, Russia, Poland, and India. Rockburst events have also occurred in many other countries. Yet, a solution to the problem appears to be no closer than it was some 20 years ago.

Distinction is made between rockburst source and damage mechanisms, since source and damage may be separated significantly in both distance and time. Particular attention is given to mechanisms of this type, which are considered to be the most damaging events. An engineering approach rather than a seismological approach is adopted. The paper deals exclusively with the hard rock mining environment, and excludes consideration of outburst events observed in the coal mining industry.

The question of support under rockburst conditions is also dealt with. Recent experiments, involving simulated rockburst loading of rockbolts, are described, which demonstrate very graphically that, for the more violent events, conventional support is ineffective and inappropriate. For such conditions it is necessary to make use of yielding support, and to base support design on energy principles rather than on stress-strength relationships.

2 ROCKBURST MECHANISMS

In the literature "rockburst" is used to describe a wide range of occurrences often without dealing with the physical nature of the phenomena. In this paper "rockburst" is defined as a seismic event which causes significant damage (Ortlepp, 1984). There are no constraints on the magnitude or type of seismic event, only that it must generate sufficient energy to cause violent damage in the opening.

The study of source mechanisms is essentially a seismological activity, but the development of methods to counter the damage mechanisms (rockbursts) is an engineering activity. It is not necessary to understand, fully, the source of a seismic event in order to develop and implement solutions to overcome the rockburst problem. Having said this, it must be noted that such solutions are not yet available, although some measures have been developed to reduce the effect of rockbursts.

2.1 *Source mechanisms*

To assist in the definition of conditions for which solutions must be developed, it is helpful to list what are believed to be the main types of source mechanism involved in rockburst events in tunnels. In roughly ascending order of energy magnitude, the main source mechanisms are considered to be as follows:

- strain bursting
- buckling
- face crushing
- virgin shear rupture in the rock mass
- reactivated shear on existing faults and/or shear rupture on existing discontinuities

A simplified basis for differentiation of these categories in terms of essential nature and seismic magnitude is given in Table 1, based on Ortlepp (1992a).

The essential natures of the first three source mechanisms are different from the last two. The significance of this is that for the first three the source and damage locations are probably coincident -the rock involved in the source is also involved in the damage. For example, strain bursting occurs right at the surface of the opening and is strongly influenced by the local shape of, and stress concentration on, that surface. The unstable failures represented by buckling and face crushing are also importantly affected by the openings in their immediate vicinity - that is, they cannot occur in the absence of the opening. In contrast, the last two mechanisms represent shear failure along a surface and the extent of this shear failure zone could be hundreds of metres. This surface may, or may not, daylight into an excavation. Such conditions are only likely to occur in association with large scale mining operations.

In the last two mechanisms the large energy release that results from the instantaneous relaxation of elastic strain initiated by the onset of movement, essentially determines the magnitude of the event. In addition multiple sub-sources of energy may result from "impacts" at minor steps or "jogs" or on shearing of smaller asperities along

Table 1. Suggested classification of seismic event source with respect to tunnels

Seismic Event	Postulated Source Mechanism	First Motion from Seismic Records	Richter Magnitude M_L
Strain-bursting	Superficial spalling with violent ejection of fragments	Usually undetected, could be implosive	-0,2 to 0
Buckling	Outward expulsion of larger slabs pre-existing parallel to opening	Implosive	0 to 1,5
Face crush	Violent expulsion of rock from tunnel face	Implosive	1,0 to 2,5
Shear rupture	Violent propogation of shear fracture through intact rock mass	Double - couple shear	2,0 to 3,5
Fault-slip	Violent renewed movement on existing fault	Double - couple shear	2,5 to 5,0

the shear surface during the progression of the movement front. This is illustrated diagrammatically in the series of sketches in Figure 1. An alternative possibility is shear (and impacts) not only along single shear surfaces, but also along sympathetic features as a result of stress transfer to these neighbouring structures. The time of occurrence, magnitude and characteristics of these sub-sources may be "hidden" in the much greater energy release and coda from the main event. However, it is possible that the sub-events, rather than the main event, may actually be the sources of corresponding rockburst damage. This is particularly the case if the sub-events are located closer to excavations than the main event zone. If this is the case, the seismic waves interacting with the excavations may be expected to be of higher velocity and frequency, owing to the short distance of travel, than the main event waves. In the monitoring of seismicity in mines, it is therefore most important that the seismic records can identify and differentiate between these sub-events and also "extract" them from the main event. If this is not done, attempts to associate all damage with the single origin of the main event will be blurred and often erroneous, leading to spurious conclusions.

It is thought probable that major perturbations involving the mechanisms sketched out above may sometimes be stretched out in time so that sub-events could be spaced seconds or even minutes apart. The single major seismic event would thus extend into a seismic episode equivalent to the "after shock" succession following a large earthquake.

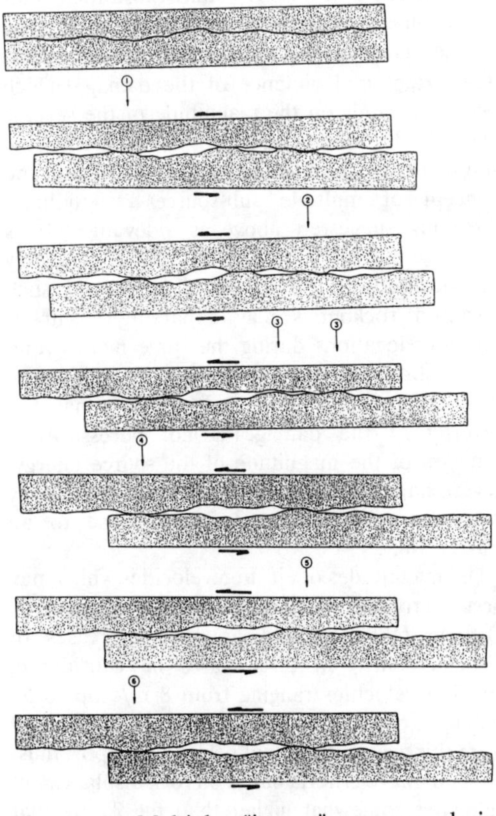

Figure 1. Multiple "impact" sources during shearing along a joint surface

2.2 *Damage mechanisms*

There are very few published definitive studies of the mechanisms of rockburst damage. Ortlepp (1992a) categorised observed types of damage, and this categorisation is used as the basis of the following treatment of rockburst damage.

Strainbursts

The typical geometry and characteristics of a strainburst event are summarised in Figure 2. Fragments of rock, usually in the form of thin plates with very sharp edges, are violently ejected locally from the rock surface.

The rockbursts which have been reported in Scandinavia (Broch and Sorheim, 1984, Grimstad, 1986, Myrvang and Grimstad, 1983) and New York (Binder, 1978) are believed to be of the strainbursting mechanism. Stacey (1989) reviewed the rock fracturing problems associated with boring in massive rocks, also believed to represent

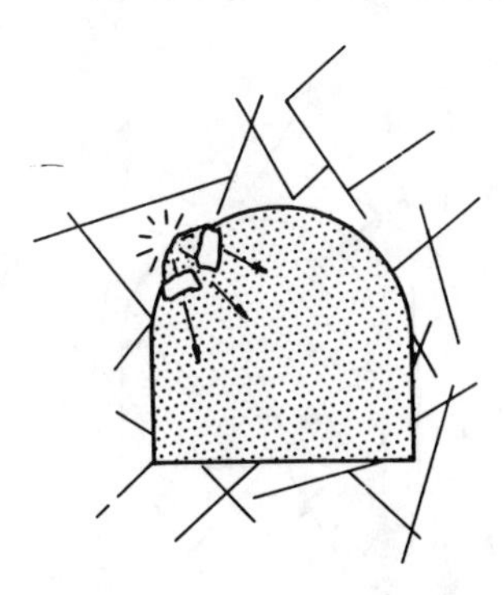

Figure 2. Strainburst damage mechanism

the strainbursting mechanism. The strainbursting mechanism in tunnels has been dealt with more fully by Ortlepp and Stacey (1992).

Buckling

The concept of a buckling rockburst damage mechanism is illustrated in Figure 3. Such a mechanism is most likely to occur in a laminated or transversely anisotropic rock mass. However, such a mechanism will not necessarily be manifested in the sidewalls - the damage could occur anywhere around the periphery of the opening, including the face, where the orientation of the geological structure is favourable for buckling instability.

The energy source for such a mechanism will be mainly the strain energy stored in the "plates" subject to potential buckling, but additional energy input may come from shear and compressive components of a seismic wave whose source is somewhat distant from the damage location. The latter is not an essential energy input, and the bifurcatory nature of buckling will result in the sudden release of the locally-stored strain energy. The locations of the major source mechanism and the damage mechanism will be coincident. An example of what is believed to be a buckling rockburst is described by Semadeni (1991).

Ejection

In concept this mechanism is manifested as the directional ejection of portion of the tunnel wall (or floor, roof or face) associated with a transient energy-wave (shock wave). Freedom of movement, and sometimes the shape of the ejected blocks of rock, are usually dictated by the presence of existing joints or induced fractures. The mechanism is illustrated in Figure 4.

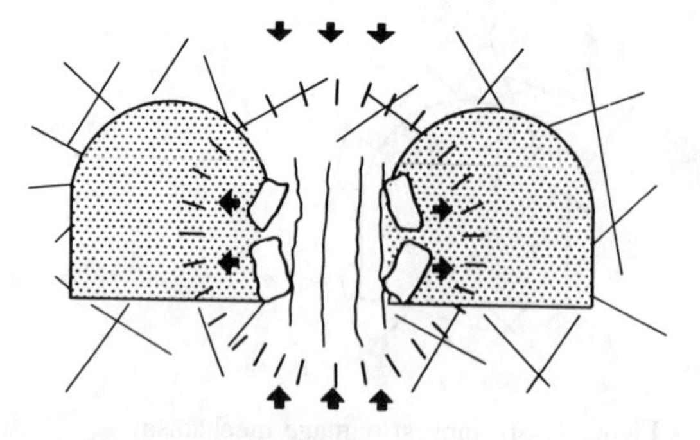

Figure 3. Pillar burst damage mechanism

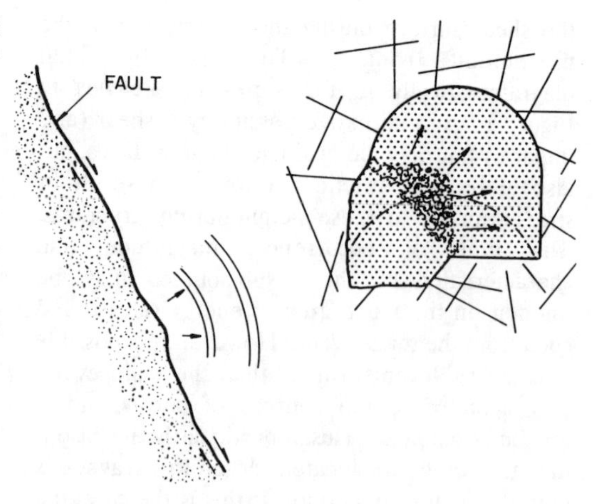

Figure 4. Ejection damage mechanism

The source of the energy which leads to this type of rockburst is a seismic event whose focus may be at some distance from the damage zone. The source and damage locations are not coincident. The extent and violence of the damage which occurs depends on the magnitude of the seismic event and the proximity of the tunnel to the source of the seismic energy. In this regard, the concept of multiple sub-sources at multiple locations suggested above is relevant. It is observed that ejections are often directional, may be localised in extent, and that several such localised rockbursts may apparently occur at different locations during the same main event. These observations are consistent with the concept of multiple sub-sources. It follows that the severity of the damage is not necessarily a function of the magnitude of the source energy; severe damage may result from a smaller energy event or sub-event located very close to an excavation.

The magnitudes of ejection velocities which may occur in rockburst events have been considered by Ortlepp (1993a) based on observed cases of displacement. In three cases he considered, ejection velocities ranging from 8 m/s up to 50 m/s were estimated. The conclusion that followed from this was that rational design of support must adopt damage criteria based on rock displacement velocities somewhat higher than the 3 m/s that has been recently assumed (Jager et al, 1990). It is suggested that ejection velocities in the range 5 to 10 m/s are more realistic for design of support to contain severe rockburst damage.

Other damage mechanisms

Three other damage mechanisms are described by Ortlepp (1992a). They include arch collapse, which is facilitated by the occurrence of well-defined geological structure or induced fracturing. Gravity is considered to be the main driving force, with seismic energy providing an enhancement of the gravity. Further mechanisms are that of implosive type damage, which has definitely been observed in mining tunnels, and an inertial displacement mechanism, which is a postulated mechanism only.

3 SUPPORT OF TUNNELS UNDER ROCKBURST CONDITIONS

The loading imposed on rock support during a rockburst can be significantly different in both magnitude and form compared with that imposed by quasi-static instability. Support requirements will be different for the different damage mechanisms identified above. However, the most severe loading in the case of tunnel or chamber-like openings will definitely be that imposed during ejection mechanism events. In this paper, consideration of tunnel support in rockburst conditions will therefore deal only with this severe loading condition. Support appropriate for this damage mechanism will almost certainly be able to contain damage due to the other mechanisms.

3.1 *Support against the ejection rockburst mechanism*

Wojno et al (1987) calculated that, to control severe rockburst deformations, a support tendon should be able to absorb 25 kJ of energy. This is equivalent to a 2m thick block being ejected with a velocity of 3 m/s. Estimates of velocities of ejection were given by Jager et al (1990) as up to 6 m/s, but it was suggested that, in most cases, 3 m/s was a reasonable value for support design purposes. This does not agree with the ejection velocity magnitudes suggested by Ortlepp (1993a) as mentioned above, and the velocities in the range 5 to 10 m/s suggested as design values. The appropriate design velocity will vary from one mining region to another, and it is therefore not realistic to prescribe a particular value. It is appropriate, however, to illustrate the effect of ejection velocity on support design for rockburst situations.

Consider a single block of rock of mass M = 2T ejected from the surface of a tunnel with an initial ejection velocity of 10 m/s.

Referring to Figure 5, in case (i) the block is supported by a yielding bolt of 22mm diameter whose load displacement characteristic is given in Figure 5(b)(i). In case (ii) the block is supported by conventional static rock support, namely a fully-grouted 22mm rebar of 190kN breaking strength.

The initial kinetic energy of the block is:

$$\begin{aligned} KE &= 0{,}5 \times M \times V^2 \\ &= 0{,}5 \times 2 \times 10 \times 10 \\ &= 100 \text{ kJ} \end{aligned}$$

Assume that the rebar grout bond fails for 100mm on either side of the block boundary, allowing 200mm of the bar to elongate to its maximum

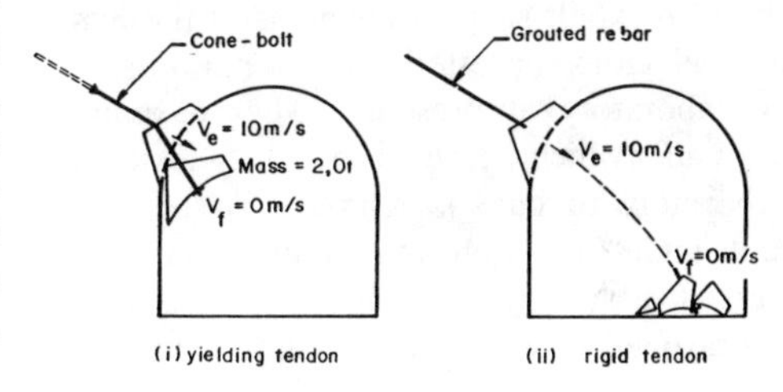

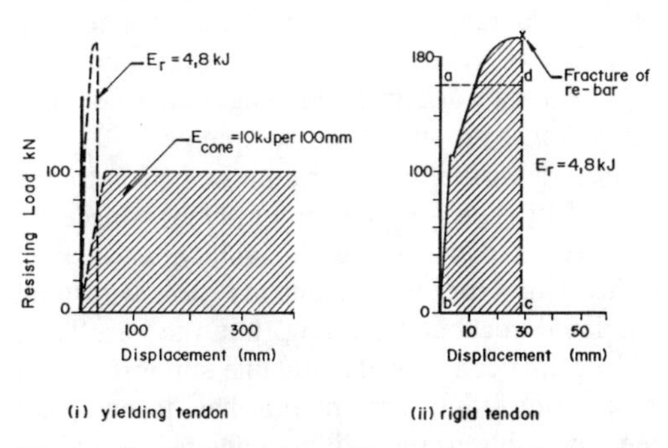

Figure 5. Simple dynamic analysis of ejection model of rockburst

possible extent (normally 16% under quasi-static loading). The energy absorbed in deformation of the steel is the area under the curve (ii) in Figure 5(b). This is approximately equal to the area abcd which is given by:

Energy in failing rebar = 160 kN x 0,03m
= 4,8 kJ

This is less than 5% of the kinetic energy imposed on the bar in the rockburst event. The velocity of the rock block after the bar is broken is only reduced to 9,7 m/s, which shows that the conventional support is completely inadequate to contain any such rockburst loading.

In contrast, the tensile behaviour of a yielding support element such as a cone bolt (Jager et al, 1990) is as shown in Figure 5(a)(i). The load displacement behaviour of such an element is as shown in Figure 5(b)(i), and 100 kJ of energy is consumed per metre of displacement. Therefore, in the rockburst event, the energy of ejection would be absorbed, and the rock block would be arrested completely after it had moved 1 metre.

3.2 *Impulse-load testing of rockbolt support elements*

Testing of rockbolt support elements at the high ejection velocities suggested above is not possible in the laboratory at present. The maximum velocity under which elements have been tested in the laboratory to date is 3 m/s. To test the elements under conditions more representative of rockburst loading, other methods must be used. By controlled post-split blasting of an underground chamber, Ortlepp (1969) carried out an impulse-load test of conventional and yielding tunnel support using de-coupled explosives as the impulse source. This test demonstrated the capability of yielding support to withstand very high energy dynamic loading whereas the conventional support failed completely.

Recently further consideration was given to this method of testing by Ortlepp (1992b), since, not only can it demonstrate the ability, or lack of ability, of rockbolt elements to withstand simulated rockburst loading, but the results of such testing can provide valuable support design information. Such information has, to date, not been available to the mining industry. A revised method of testing was suggested. This consisted of the explosive ejection of a reinforced concrete block secured to the sidewall of a tunnel by rockbolt elements. This type of test was suggested, based on the following:

- the observation that rockburst damage often involves the ejection of a metre thickness of rock
- the observation that rockburst damage is often directional
- the cost of the previous tunnel tests.

The concept of this revised test method has been applied recently in a series of tests on alternative rockbolt elements (Ortlepp, 1993b). In these tests the method was further revised, with the concrete blocks being constructed at the base of a quarry, and with the ejection force acting vertically. This approach had the following advantages:

- there were no restrictions on blasting in the quarry
- monitoring of the experiments was easier, and the block movement against gravity facilitated interpretation of results
- visual observation, television monitoring, and high speed filming of the tests was possible.

Comparative behaviour of two of the tests is shown in Figures 6 and 7. In Figure 6 a block, which was held down by five 25mm fully-grouted rebar elements (static resisting force 1350 kN), was ejected 4,7 m into the air. All of the rebar elements failed, three completely and two by pulling out of the anchorage. This height of ejection was 90% of that for an unsupported calibration block. In contrast, as shown in Figure

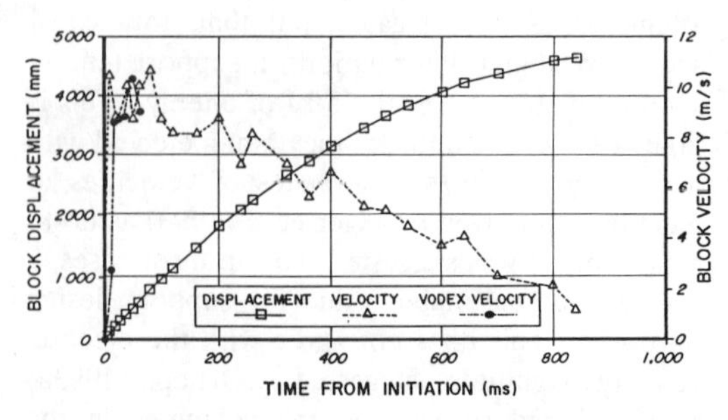

Figure 6. Ejection of block retained by 25 mm rigid rebar elements

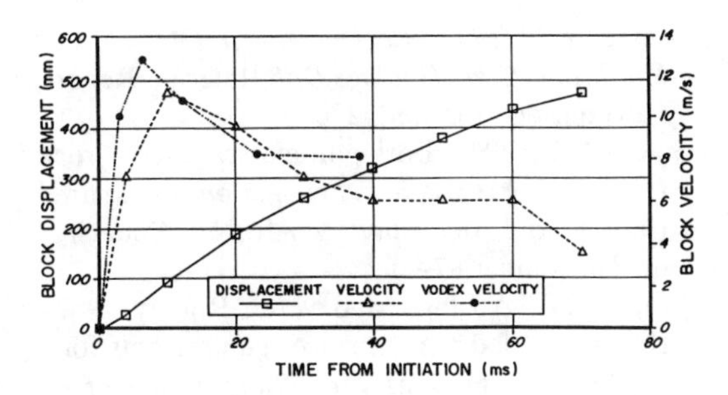

Figure 7. Ejection of block retained by 22 mm cone bolts

7, a block held down by five 22mm yielding cone bolts (resisting force 1105 kN) was only ejected a distance of 0,5m by the same explosive force. None of the cone bolts was damaged.

The results of these two tests are summarised in the following table.

It can be seen that the yielding cone bolts withstood an ejection velocity of 10 m/s without any distress, and, acting in tension, would have contained a rockburst ejection of this magnitude. It is estimated that the cone bolts absorbed 20 times as much energy as the "stronger" rebar elements, and, since they had yielded only 0,5m, they still had the capacity to absorb more energy.

4 CONCLUSIONS

Alternative mechanisms of rockbursts have been described. A clear distinction has been drawn between source mechanisms and damage mechanisms. Attention has been concentrated on one rockburst mechanism, that of rock ejection imposed by seismic waves. This damage mechanism commonly occurs in mining situations and manifests itself with extreme violence. In this case source and damage may be separated in both distance and time.

Rockburst ejections are also often observed to be directional in nature, and to involve typically a thickness of rock of about a metre. Significant ejection velocitites occur during rockburst events, and values in the range of 5 to 10 m/s are not unusual. Very much higher velocities are also believed to occur.

Under conditions of high ejection velocity, it is neither practicable nor economically possible to contain severe rockburst damage of the ejection type by increasing the strength of the tunnel support. Yielding support is essential, and using an energy approach, yielding support systems based on presently available components, can be designed to withstand rockbursts of major violence.

Based on the above observations, tests have recently been carried out using explosives to generate simulated rockburst loading. These tests have demonstrated the capability of available yielding rockbolt elements, in tension, to withstand ejection velocities of more than 10 m/s. Their capacity to contain large ejection energies up to 20 times greater than that which destroys stiff rockbolt elements of equivalent strength has also been demonstrated.

REFERENCES

Binder, L. 1978. Rockbursts in New York, *Tunnels and Tunnelling*, V 10, No 8: 15-17.

Broch, E. and S.Sorheim 1984. Experiences from the planning, construction and supporting of a

Table 2. Comparative performance of conventional and yielding rockbolts

Rockbolt Type	'Strength' of Rockbolts (kN)	Ejection Height (m)	Ejection Velocity (m/s)	Result
No Bolts	-	5,20	10,2	-
25 mm Re-bar	1350	4,65	10,2	3 Broken 2 Pulled out
22 mm Cone-bolt	1100	0,50	12,8	Completely Undamaged

road tunnel subjected to heavy rockbursting, *Rock Mech. and Rock Engng,* V 17: 15-35.

Grimstad, E. 1986. *Rock-burst problems in road tunnels,* in Norwegian Road Tunnelling, Pub. No 4, Norwegian Soil and Rock Engineering Association, Tapir Publishers, University of Trondheim: 57-72.

Jager, A.J., L.Z.Wojno and N.B.Henderson 1990. New developments in the design and support of tunnels under high stress, *Proc. Int. Deep Mining Conf.: Technical Challenges in Deep Level Mining, S. Afr. Inst. Min. Metall.,* Vol 1: 1155 -1172.

Myrvang, A. and E.Grimstad 1983. Rockburst problems in Norwegian Highway tunnels - recent case histories, in *Rockbursts - prediction and control, I.M.M.,* London: 133-139.

Ortlepp, W.D. 1969. An empirical determination of the effectiveness of rockbolt support under impulse loading, *Proc. Int. Symp. on Large Permanent Underground Openings,* Oslo, ed. T.L. Brekke and F.A. Jorstad: 197-205.

Ortlepp, W.D. 1984. Rockbursts in South African gold mines: a phenomenological view. *Int. Symp. Rockbursts and Seismicity in Mines, S.Afr. Inst. Min. Metall.,* Symp. Series No. 6: 165-178.

Ortlepp, W.D. 1992a. The design of support for the containment of rockburst damage in tunnels - an engineering approach, *Proc. Int. Symp. on Rock Support in Mining and Underground Construction,* Laurentian University, Sudbury, June 1992, Balkema: 593-609.

Ortlepp, W.D. 1992b. Implosive-load testing of tunnel support, *Proc. Int. Symp. on Rock Support in Mining and Underground Construction,* Laurentian University, June 1992, Balkema: 675-682.

Ortlepp, W.D. 1993a. High ground displacement velocities associated with rockburst damage, *Rockbursts and Seismicity in Mines,* ed R.P.Young, A.A.Balkema, to be published.

Ortlepp, W.D. 1993b. A dynamic test procedure for estimating performance of grouted rock studs under rockburst conditions. *J.S. Afr. Inst. Min. Metall.,* to be published.

Ortlepp, W.D. and T.R.Stacey 1992. Rockburst Mechanisms in Tunnels and Shafts, *Proc. Symp. Tuncon '92: Design and Construction of Tunnels,* Maseru, Lesotho, S. Afr. Nat. Council on Tunnelling: 83-89.

Semadeni, T. 1991. *Rockbursting in the Strathcona-Craig Haulage Drift,* Internal Report, Falconbridge Limited, 27p.

Stacey, T.R. 1989. Boring in massive rocks - rock fracture problems, *Proc. Seminar on Mechanized Underground Excavation,* S. Afr. Nat. Council on Tunnelling: 87-90.

Wojno, L.Z., A.J.Jager and M.K.C.Roberts 1987. Recommended performance requirements for yielding rock tendons. *Proc. Symp. Design of Rock Reinforcing,* S.Afr. National Group of ISRM: 71-74.

Strata mechanics
Lectures

Geomechanics 93, Rakowski (ed.) © *1994 Balkema, Rotterdam, ISBN 90 5410 354 X*

Research of static and dynamic load of long rock bolts

Uroš Bajželj
University of Ljubljana, Slovenia
Jakob Likar & Franc Žigman
Mining Institute Ljubljana, Slovenia

ABSTRACT: Rather extensive research of external axial force transfer along the loaded steel bolts elements to surrounding rock was performed in the mine in order to determine optimal anchoring length. In similar way we give the measurements results for dynamics loads on long cable bolts within front face blasting directly near measuring points. The presented results, determining first of all the time development and size of dynamic load, enable the project officers a more realistic planning. In second part of this report there are the results of "in situ" measurements giving the efficiency of rock bolting with measuring anchors with sensors on different depths respectively distances from the anchor head. The described measuring method enables to determine the so called neutral point . With the known neutral point it is possible to optimize the rock bolt system as single or combined supporting system.

1 RESEARCH OF CAPACITY ON STATICALLY LOADED CABLE BOLTS

Every use of anchoring system should be based on argumented decisions taking into account actual geotechnic conditions present on a determined area. To clear these problems we undertook comprehensive research of bearing capacity on different types of anchors. While we researched mostly the behaviour of passive anchoring they were built in according:

- Boring of holes with 38 and 52 mm in diameter
- Preparation and filling of boreholes with cement binders having water-cement ratio w/c = 0.3
- Pressing of bearing parts of anchors into boreholes filled with cement binders

We tested the one side spun cable bolts ø=15.2 mm (7 strains ø5 mm), double one side spun cable bolts 2x ø15.2 mm, one side spun cable bolts ø16 mm (9+1), and cross spun cable bolts ø28 mm.

The building-in of cable bolts was carried out with the Beril 325

Fig.1. Cable bolting machine BERIL 325

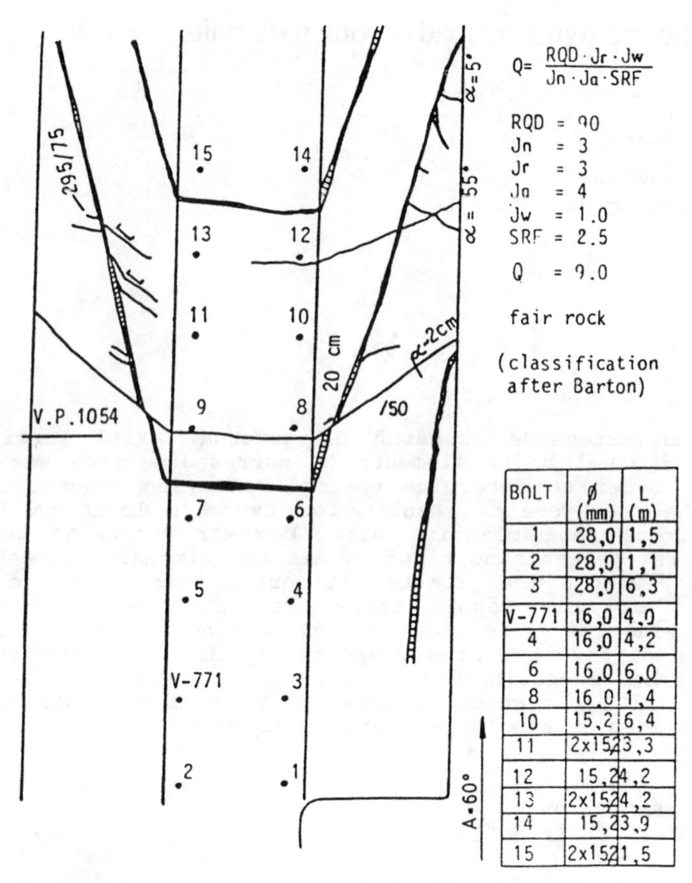

BOLT	Ø (mm)	L (m)
1	28,0	1,5
2	28,0	1,1
3	28,0	6,3
V-771	16,0	4,0
4	16,0	4,2
6	16,0	6,0
8	16,0	1,4
10	15,2	6,4
11	2x15,2	3,3
12	15,2	4,2
13	2x15,2	4,2
14	15,2	3,9
15	2x15,2	1,5

Fig.2. Ground-plan arrangement of test cable bolts of various types

cable bolting machine shown in Fig. 1. This machine was developed by the Mining Institute Ljubljana in cooperation with the Uranium Ore Mine Žirovski vrh and factory Belt.

With this machine it is possible to build-in single and double cable bolts up to 25 m long, without any bigger problem. The actual advantage of this machine is that the filling of binders is performed with low water-cement ratio under suitable pressure. Such a procedure enables filling of fissures and empty spaces in the rock with cement binders, what increases the strength of wider adjacent area.

1.1 Testing of bearing capacity of various types of statically loaded cable bolts

The research of bearing capacity on various types of statically loaded cable bolts was performed in the pit in the arrangement as shown in Fig. 2.

Each test was carried out with increasing of axial force on the cable bolt head and measuring of deformation. The biggest axial forces on the cable bolt head were nearly equal to the limit loads declared by the rope producers.

The results of statical tests are presented in Fig. 3, 4, 5 and 6. The difference between deformations at the same axial forces are evident, respectively different behaviour of various types of cable bolts can be expected.

1.2 Dispersion of axial forces with the cable length

Some testing cable bolts were

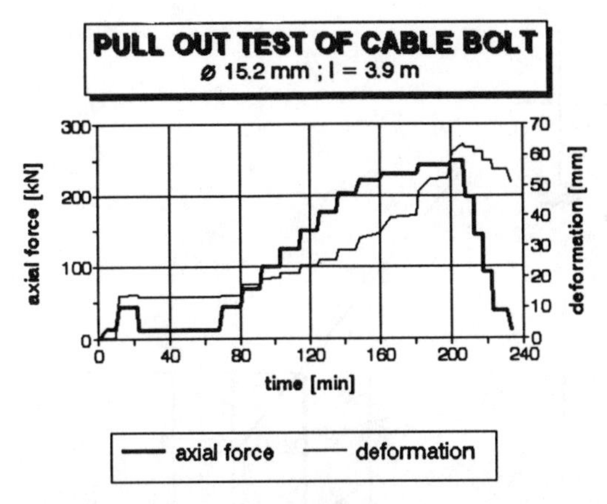

Fig.3. Results of pull out test for the cable bolt ϕ 15.2mm

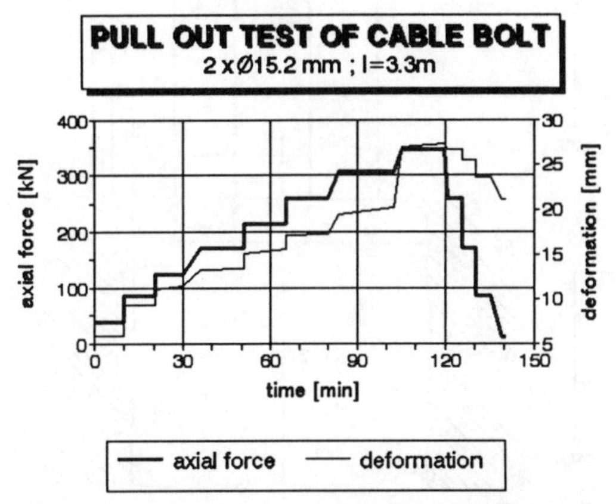

Fig.4. Results of pull out test for the double cable bolt 2 x ϕ 15.2mm

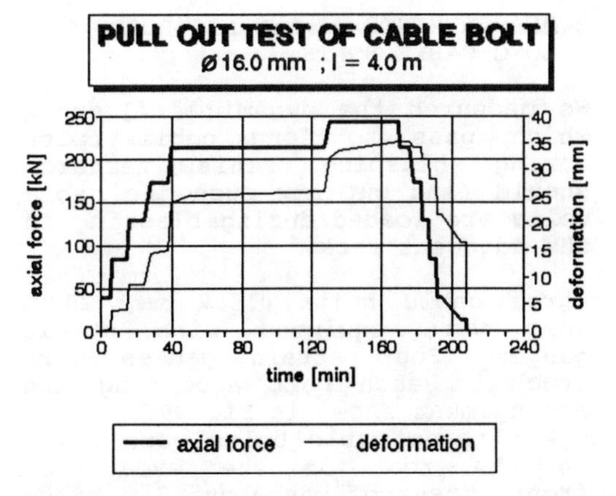

Fig.5. Results of pull out test for the cable bolt ϕ 16mm

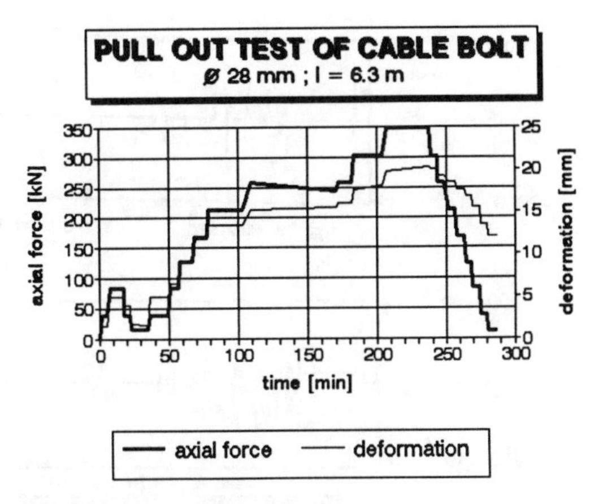

Fig.6. Results of pull out test for the cable bolt ϕ 28mm

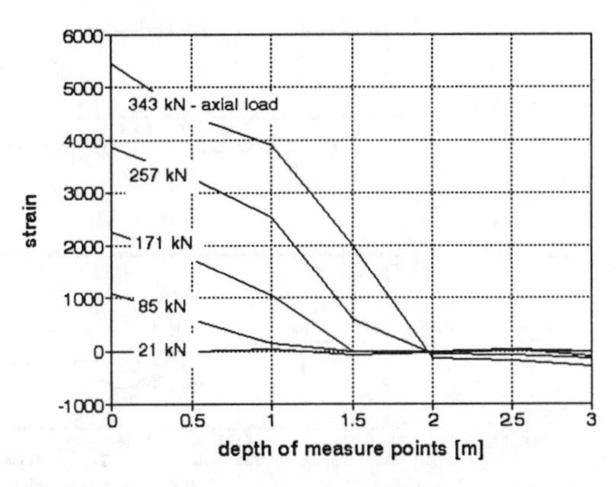

Fig.7. Dispersion of axial force along the bolt at different depths for the double cable bolt 2 x ϕ 15.2mm

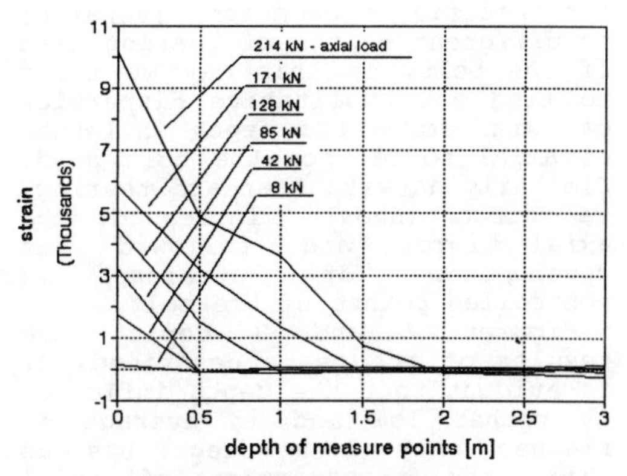

Fig.8. Dispersion of axial force along the bolt at different depths for the cable bolt ϕ 15.2mm

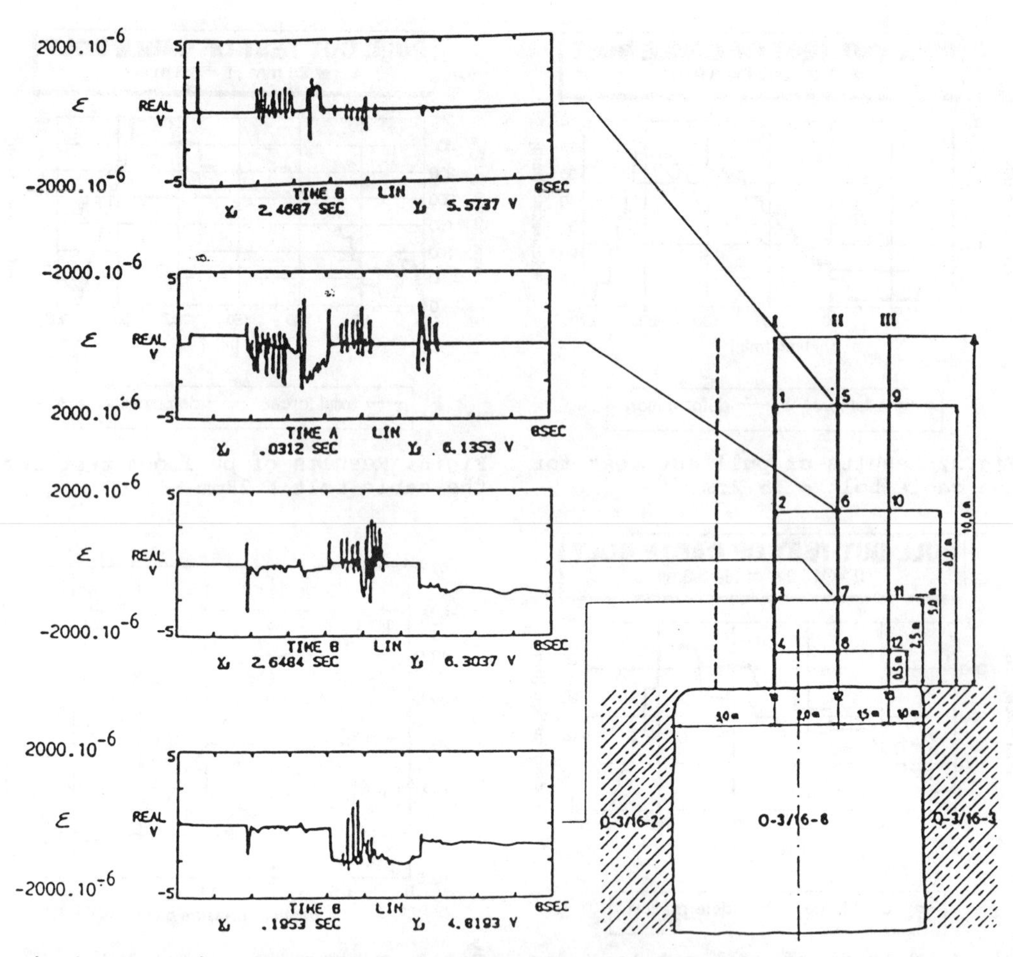

Fig.9. Measuring results of dynamic influences upon cable bolts during blasting

furnished with measuring devices for specific deformations (strains) on different depths of bearing part of the bolt. In this way we tried to find out qualitative dispersion of axial force and depth influence of axial force upon the bolt head. Similarly as with standard testing, we discontinuosly increased the axial force and followed the development of strains in controlled points of the bolt.

Figures 7 and 8 depict the results of the tests described. It is evident that the depth influence is rather low and in average it reaches only 2 m deep. Besides this, the transmission of axial force, along the bearing part of the bolt to the adjacent rock, is not linear at higher loads.

2 MEASUREMENTS OF DYNAMIC LOADS ON LONG CABLE BOLTS

We measured the dynamic influences which pass to long cable bolts during blasting. This research should find out how much the cable bolts are loaded during blasting in the adjacent rock.

Three cable bolts ø15.2 mm, 10 m long were equipped with strain gauges. Four strain gauges were stuck on each bolt according the arrangement shown in Fig. 9.

All three cable bolts were built in the roof of the excavation front. The roof was about 7 m high. The measurements of dynamically loaded bolts during blasting were performed with a 12-canals direct current bridge amplifier through which the electricity was supplied to 13-ways magnetophone. During blasting all dynamic changes on 12 strain gauges have been simultaneously registered. Figure 9 shows the results of measurements for determined measuring points along the cable bolts. The time of measuring is shown on the abscissa and the measured strains in μm/m on the ordinates. The times of loading are very short, they last some milliseconds.

The measured dynamic load upon cable bolts during blasting in no measuring point do not exceed 20 percent of the value determined as allowed bearing capacity of each bolt.

3 MEASUREMENT RESULTS OF MEASURE ANCHORS BUILT IN THE HRASTNIK 2 SHAFT

Anchors are the basic elements of supporting during the Hrastnik 2 shaft construction. The extention of shaft has been carried out in rocks, which could be damaged respectively their strength could be overloaded during construction.

During the shaft constructing we followed geotechnical parameters with various measuring methods. With measure anchors the efficiency and dispersion of axial force along the built in anchor should be determined, what can be the calculation basis for length of anchors used as supporting elements in shaft excavation.

On the 251 m depth we built in four measure anchors. They were positioned approximately in direction NW-SE respectively NE-SW. This orientation of measure anchors was conditioned by the tectonic structure of adjacent rock in order to measure the biggest influence upon the anchors.

For the construction of measure anchor we used ripped concrete steel ∅16 mm, 9 m long. Five pairs of strain gauges were stuck on each anchor in the way that between two diametrally opposite strain gauges the distance was 1.5 or 2 m. With such location of strain gauges the activations of anchors, respectively dispersion of axial force along anchor length, was

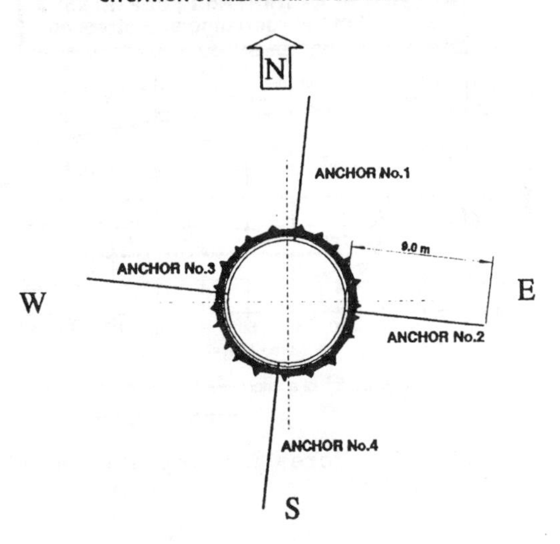

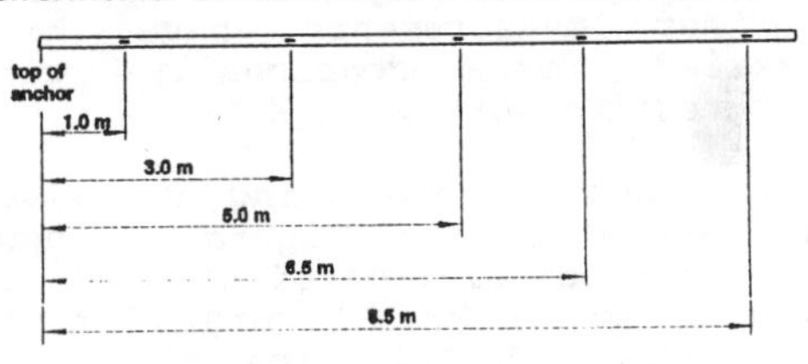

Fig.10. Position of measuring anchors and situation of strain gauges on measuring anchor

determined, including eventual bending of anchor in vertical direction. Figure 10 shows the arrangement of stuck strain gauges on each measure anchor. The strain gauges were stuck on distances (in m) 1, 3, 5, 6.5, and 8.5 from the shaft wall.

The strain gauges were connected to power cables in the way that the measurement performed across bridge amplifier and showing a positive value mean tensile strains respectively stresses, a negative values mean compressive strains (stresses).

Measure anchors were set into cleaned horizontal drillings filled with cement binder. The anchors were positioned in such a way that strain gauges lied on upper respectively lower side of anchor. Thus the vertical bending respectively movement of anchor can also be determined. The initial

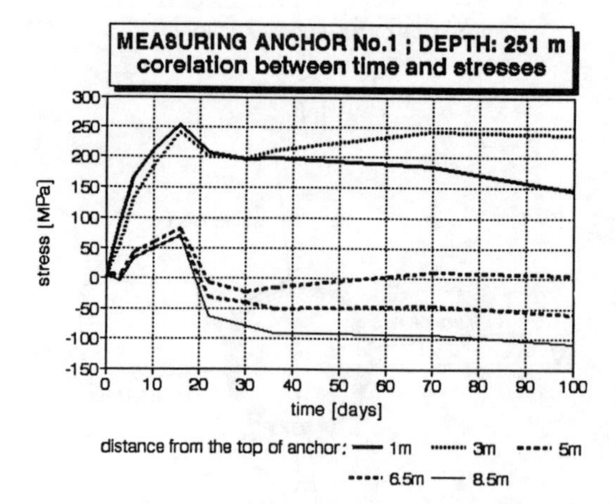

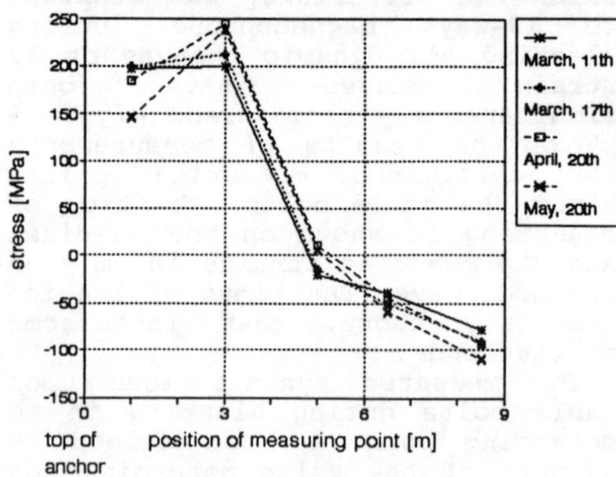

Fig.11. Time corealtion of stress in the anchor

Fig.12. Stress distribution along the anchor

readings were done immediately after positioning of anchors, further measurements have been carried out occasionally with progressing works.

The results of measuring are given for anchor 1 only. Figure 11 shows the time correlation of stresses in the anchor in MPa. Figure 12 gives the stress distribution along the anchor, for each measurement separately.

On the basis of measurements performed on measure anchors we can conclude that the shaft construction at the examined depths should be supported with at least 9 m long anchors. In order to reach the designed loading performance of anchors, a sufficient friction between the rock and the anchor should be secured by cement binder.

4 SUMMARY

The research results of bearing capacity on friction cable bolts respectively anchors loaded statically proved that the rope 15.2 mm (7x ϕ5 mm) has sufficient strength and rigidness necessary for builiding-in. An advantage is the possibility of building-in single or double friction cable bolts. The measurements of dispersion along the bolt length show that 65 percent of the load is taken over by the first meter of bolt length. The dispersion of axial force did not exceed 2 m of bolt length.

The results of measured dynamic influence upon the cable bolts during blasting proved that in no case the measured dynamic force in the bolt were as high as to cause plastic yielding of a bolt. The biggest measured strains reached 0.2 percent.

The measurements of activation of cable bolts respectively anchors in the Hrastnik 2 shaft proved that on the 251 m depth the load influence spreads six and more meters in the adjacent rock around the shaft. The measurement results confirmed that the built-in anchor systems help a lot to maintain relatively stable conditions along the shaft.

Geomechanics 93, Rakowski (ed.) © 1994 Balkema, Rotterdam, ISBN 90 5410 354 X

Geophysical assessment of the hydraulic injection process in coal seams under rockburst hazard

J.A. Dubiński
Central Mining Institute, Katowice, Poland

ABSTRACT : The hydraulic injection of coal seams under rockburst hazard is one of the basic methods for modifying their physical and mechanical properties in order to lower the energy storing ability of a seam. Rational application of the method requires an assessment of the process relative both to an extent of the changed structure zone and to the intensity magnitude of changes. Geomechanical complexity of the hydraulic pressurized injection process manifesting itself in the form of increased fractured porosity effect and increased coal seam moisture content effect requires for its assessment elaboration of a specific measuring procedure. Fundamental elements of the method based on the underground seismic measurement technique associated with new interpretation solutions and assessment criteria are characterized. An example of the method application in mining practice is shown.

1 INTRODUCTION

Protecting miners from one of the most severe natural hazards, which is mining tremors and rockbursts, requires continuous improvement in the existing methods of combating them. The used in practice anti-rockburst prevention is most often understood as an action against the excessive storage of energy in a rock mass.

According to Filcek (1980) the distribution of the storage zones is determined by the stress state in the rock mass and its physico-mechanical properties.

The purpose of a pressurized water injection into a coal body is to make such changes in its structure (increase of moisture) so that afore-mentioned energy storage ability should be reduced.

The water injection technique has widely been used in Polish hard coal mines for many years, but the methods for assessing its effectiveness have not frequently been utilized yet. It should be emphasized that the lack or limitation of such a control can lead to an unreasonable use of the water injection technique.

2 GEOMECHANICAL ASPECTS OF THE ASSESSMENT OF THE EFFECTIVENESS OF WATER INJECTION BY THE SEISMIC METHOD

The results of the extensive laboratory studies recently conducted (Kabiesz, 1988) indicate that in the case of coal seams having the water-absorbing structure an increase in its moisture results in a reduction of the potentiality for storing elestic energy and the natural proneness of coal to rockbursts.

Since the original porosity of coal is, as a rule, very low, an increase in effectiveness of the watering process of such coals can mainly by attained by increasing the microcrack and crack systems to produce large enough contact surfaces between the injected fluid and the coal substance. The low effectiveness of watering found in the zones of high seam effort, where the above joints'system is missing (stage of compaction), confirms the rightness of such an idea of the method.

The pressurized water injection process is a complex geomechanical phenomenon. It also results, besides the afore-mentioned moisture increase, from the fracturing increase effect caused by the pressurized fluid action (Dubiński, 1989).

By using the most practical seismic technique, for assessing the effectiveness of injection, the afore - mentioned geomechanical effects can lead to anisotropic variations of the measured velocity parameter.

3 CHARACTERISTICS OF THE SEISMIC METHOD FOR EVALUATION WATER INJECTION EFFECTIVENESS

It is assumed that an up to date method for controlling effectiveness of water injection must guarantee conclusive results. For this reason the methodological procedure should embody the following features :

-seismic measurements must be conducted in at least two measurement cycles, the first or base measurement being made prior to the process of water injection,

-measurements should be conducted using the technique of seismic transmission or profiling, the geometry of the system being desinged to cover the zone of potential extent of the water injection,

-digital processing of the seismic recordings ensures good identification of the arrival times of a seismic wave, both longitudinal or transverse, propagating in the coal seam.

The final result should establish the proportions of these two geomechanical effects in the total water injection process, i.e. increase in humidity and in fissure porosity of the given coal seam. Basing on the mutual relations between these two processes the seismic criteria for eveluating effectiveness of the water injection may be determined.

The initial functions describe the relation between velocity of propagation of the longitudinal and transverse seismic waves, and the physicomechanical parameters of the tested rock mass. In the relative form they may be written (Fajklewicz et al, 1972) :

$$v_p^*/v_p = \left[\frac{(1-\nu^*)}{(1+\nu^*)} \cdot \frac{(1+\nu)}{(1-\nu)} \cdot \frac{K^*}{K}\right]^{1/2} \qquad (1)$$

$$v_s^*/v_s = \left[\frac{(1-2\nu^*)}{(1+\nu^*)} \cdot \frac{(1+\nu)}{(1-2\nu)} \cdot \frac{K^*}{K}\right]^{1/2} \qquad (2)$$

where :

v_p, v_s - velocity of longitudinal and transverse waves,

ν - Poisson ratio,

K - modulus of volumetric compressibility,

* - index denoting values of velocities and parameters in the tested medium after the water injection process.

The simultaneous change in velocity of propagation of both types of seismic waves is expressed by the function $F(\nu)$:

$$F(\nu) = \left[\frac{(1-2\nu^*)}{(1-2\nu)} \cdot \frac{(1-\nu^*)}{(1-\nu)}\right]^{1/2} \qquad (3)$$

With low joints' density the function $F(\nu)$ is proportional to change in Poisson ratio, which increase in water saturation of the rock structure.

For the estimation of variations in joints' density use may be made of the observed effect of monotonic variation of velocity of transverse wave as a function of joints' density. Hence the function $G(\varepsilon)$ is a measure of degree of change

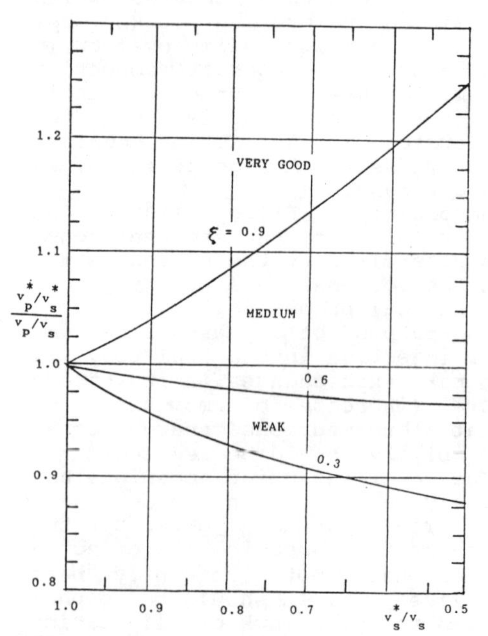

Fig. 1. Diagram for the evaluating of the degree watering effectiveness of coal seam.

in joints' density in the rock mass.

$$G(\varepsilon) = v_S^* / v_S \quad (4)$$

(ε - means joints'density).

Analysis of the mutual relations between functions $F(\nu)$ and $G(\varepsilon)$ reflects the action of the two mentioned geomechanical effects, i.e. changes of water saturation in the medium and joints' density, associated with the process of water injection to the coal seam.

In line with these studies (Wyllie et al, 1956) developed a kind of diagram, which when taking into account results of measurements in the mines provides a basic for deriving a modified seismic scale for evaluation of water injection effectiveness. The principal criterion parameter is here the magnitude ξ indicating degree of water saturation of the structure. Fig.1. shows this scale, developed when taking the following intervals of values of parameter.

Degree of water injection effectiveness	Value of parameter
0. None	less than 0.3
1. Weak	0.3 - 0.6
2. Medium	0.6 - 0.9
3.Very good	greater than 0.9

4 APPLICATION EXAMPLE

Seismic studies were conducted in seam 510 at the mine Katowice in a region where coal remnants were being water injected via 30 m long boreholes. The technique applied involved seismic transmission between the workings delineating the given remnants. Tests includes two measurement cycles timed for before and after the water injection to the coal seam. For interpretation of results a tomographic technique was employed, basing on the programme ATR-1/GIG. On the seismograms the phases corresponding to the longitudinal and transverse wave propagating in seam 510 were separated. On figs.2 and 3 may be seen distribution of anomalies v_P^*/v_P and v_S^*/v_S, respectively. Knowing the relations between these anomalies and making use of the scale described here for evaluation of effectiveness of water injection to the seam, then as a result may be obtained the distribution of zones of various degrees of water injection effectiveness. It is shown on fig.4.

5 CONCLUSIONS

The ensure rational utilisation of water injection and choice of its optimum parameters a wider use of

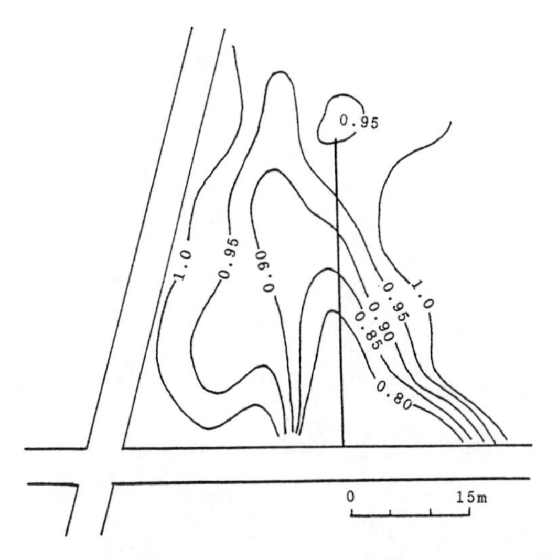

Fig.2. Distribution of the seismic velocity anomaly v_P^*/v_P arising due to water injection.

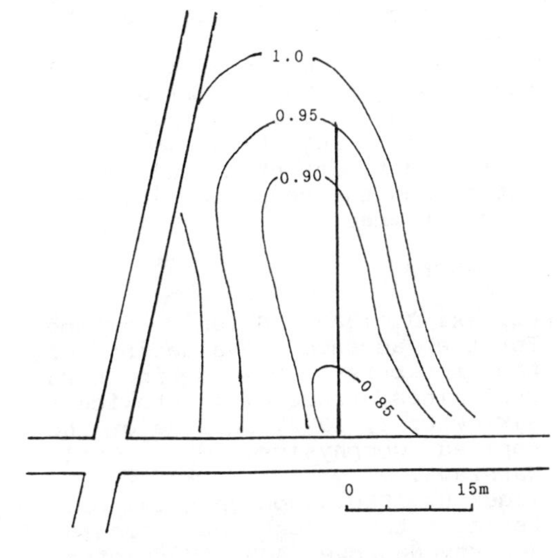

Fig.3. Distribution of the seismic velocity anomaly v_S^*/v_S arising due to water injection.

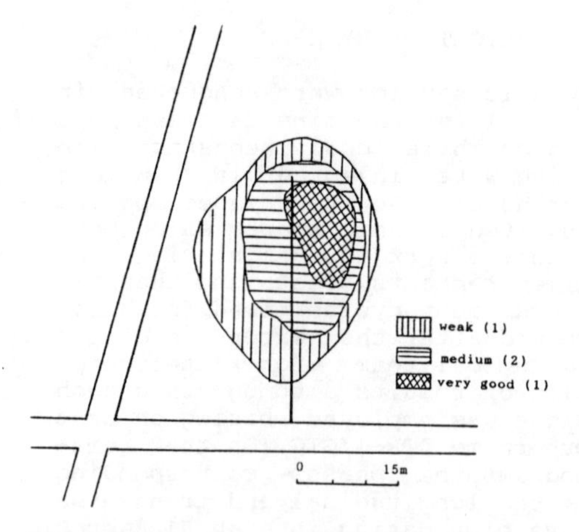

Fig.4. Distribution of zones of various degree of water injection effectiveness.

methods for controlling effectiveness of the process is required. The seismic method, based on recording anomalies of longitudinal and transverse wave arising in the medium due to the process of water injection, makes possible a conclusive evaluation of the geomechanical effects accompanying the water injection.

A seismic scale is proposed for evaluation of water injection effectiveness, the scale is divided into three degrees according to value of criterion parameter ξ i.e. water saturation of the coal seam.

The value of the seismic method for resolving the problem of evaluating effectiveness of coal seams' water injection has been confirmed by its practical application and the required information obtained.

REFERENCES

Dubiński J. 1989. Seismic method for the advanced evaluating of the rockbursts hazard in hard coal mines. Prace GIG: Katowice.

Fajklewicz Z. 1972. Outline of the applied geophysics. Wyd. Geol: Warszawa.

Filcek H. 1980. Geomechanical criteria of the rockbursts hazard. Zeszyty Naukowe AGH, Górnictwo. Vol.4. No 2: Krakow.

Kabiesz J. 1988. Change of the strength properties of carboniferous strata due to their watering Bezpieczeństwo Pracy w Górnictwie. No 4: Katowice.

Wyllie M.R.J. Gregory A.R. Gardner L.W. 1956. Elastic wave velocities in heterogeneous and porous media. Geophysics. Vol.21.

Geomechanics 93, Rakowski (ed.) © 1994 Balkema, Rotterdam, ISBN 90 5410 354 X

The effectiveness of destressing blasts in the Wujek coal mine

J.A. Dubiński
Central Mining Institute, Katowice, Poland
B. Syrek
Hard Coal Mine Wujek, Katowice, Poland

ABSTRACT : The Wujek hard coal mine ranks among the most afflicted with rockbursts mines in the Upper Silesian Coal Basin. Annually, several thousand seismic events of the energy E> 10^3 J occur there, of which the strongest ones reach the energy level 10^8 J. The destressing blasts carried out in both coal seams and roof rocks are designed to transform the rock mass structure and to liberate the accumulated energy by safe and controllable means. As a result the rock mass will be destressed. To assess the effectiveness of destressing the informative, geophysical parameters have been introduced, the distribution of which allows rational control of preventive measures. This possibility has been confirmed by the examples from the Wujek coal mine and the analysis of existing experiments allows to state some general rules to carry out destressing blasts effectively.

1 INTRODUCTION

The concussion and destressing blasting (CDB) is one of the basic ways of rockburst prevention. Such a blasting is carried out in a number of hard rock coal mines that exploit seams under rockburst hazard. The main objective of the CDB is the controlled release of as much of the elastic energy stored in a rock mass as possible without mining the coal body. As a result of the blasting a fractured destressed zone is being formed in the vicinity of an opening which leads to the shifting of a region of higher stresses deep into the coal body.

Because of the existence of divergent opinions on the purposefulness and effectiveness of the CDB research works designed to carefully analyse these problems from the rich observational and measurement material obtained and collected in the Wujek coal mine were undertaken. Certain more generalized statements were to be formulated from this analysis as well.

2 FORMULATION OF THE PROBLEM

As mentioned previously, the applicability of the CDB as a method for the active combating of rockburst hazard is a problem, at least controversial, if the stability of the coal seam -surrounding rocks geomechanical system is concerned. The opinions on the actions taken to improve the effectiveness of those blastings are ambigous as well.

Some case studies of performing very effective CDB and/or also a total lack of their effectiveness are known from the many years' experience. It is also well known that conducting the blasting with very large charges is expensive, time consuming and not always technically feasible.

The Wujek coal mine as being one of the most afflicted with rockbursts in the Polish coal mining industry has at its disposal an unusually rich material pertaining to this problem. In earlier papers (Mitrega & Syrek, 1987) the dependence of the seismic energy released during the blasting on the amount of explosives used for long walls with backfilling under the high rockburst hazard existing in that mine was presented. In another paper (Filipek et al, 1991) a similar analysis made for longwalls with caving was analysed.

The basic task of the present paper is an attempt to answer the following questions by making use of the previous and new data :

-when should the CDB be used,
-how should the blasting charge be selected in order that the blasting be most effective.

3 MINING AND GEOLOGICAL CONDITIONS AND THE FORMATION OF ROCKBURST HAZARD ON ANALYSED LONGWALLS

3.1 Longwalls with backfilling

In the analysis the CDB carried out in three different areas of the Wujek coal mine was utilized :

-longwall III in the Arkona section of the seam No.510,
-longwalls IV,V,VI in the south-eastern section of the seam No.510
-longwalls XIII,XIV,XXII,XXIII in the seam No.501.

All the above areas were under the state of very high rockburst hazard. In each case the first non-destressed slice of the seam was being mined. The seams No.501 and 510 are 5.5-8.0 m thick and lie at depths of 630-750 m. Between the seams No.501 and 510 (a distance of 40-60 m) and in the roof of the seam No.501 there occur mainly strong sandstones and arenaceous shales the layers of which are from several to about 30m thick. Over the seam No.416 that overlies the seam No.501 at a distance of about 50 m to 85 m the sandstone layers are thicker and occur more frequently.

On the analysed longwalls several hundred tremors of the energy of the order of 10^5 - 10^7 J have occurred as well as several thousand weaker events. The high rockburst hazard existing on the longwalls has been evidenced by the occurrence of 8 rockbursts.

3.2 Longwall with caving

The following 5 longwalls were analysed :

VIIIa, IXb, I, closing longwall in the seam No.501, closing longwall in the seam No.416.

These longwalls were driven in areas situated close to the afore-mentioned longwalls with backfilling. In each case the top, non-destressed slice of the seam was mined. The seam No.416 is about 6m thick and, thus, its thickness is close to those of the afore-mentioned seams No.501 and 510. Also, a deposition depth of the seam No.416 is similar, that is about 600 m. The high rockburst hazard existing on these longwalls has been evidenced by 4 rockbursts and very high seismic activity. On the longwalls under discussion the number of recorded seismic events of different energy is as follows :

Energy of tremors	Number
10^8	1
10^7	4
10^6	28
10^5	208

and anywhere from ten to twenty thousand weaker seismic events. The position of the analysed areas is schematically shown on Fig.1.

4 ANALYSIS OF RESULTS

On all the analysed longwalls, owing to the high rockburst hazard, the CDB has been carried out It was performed regularly by using the safe barbarite type explosives and the instantaneous detonators in boreholes of from 6 m to 10 m deep. Depending on mining and geological conditions and the hazard state single blasting charges varied from 20 kg to 280 kg. Taken as a whole, 1152 blasts on longwalls with backfilling and 1185 blasts on longwalls with caving were analysed. The dependence of the obtained seismic effect on the fired single blasting charge on longwalls both with backfilling and caving is shown in Fig.2. The visible major features of the curves are as follows :

-seismic effect induced by the same blasting charges is always smaller for longwalls with caving than for longwalls with backfilling,
-both curves for certain blasting charge reach the maximum level of efficiency and further enlargement of the charge is unnecessary.

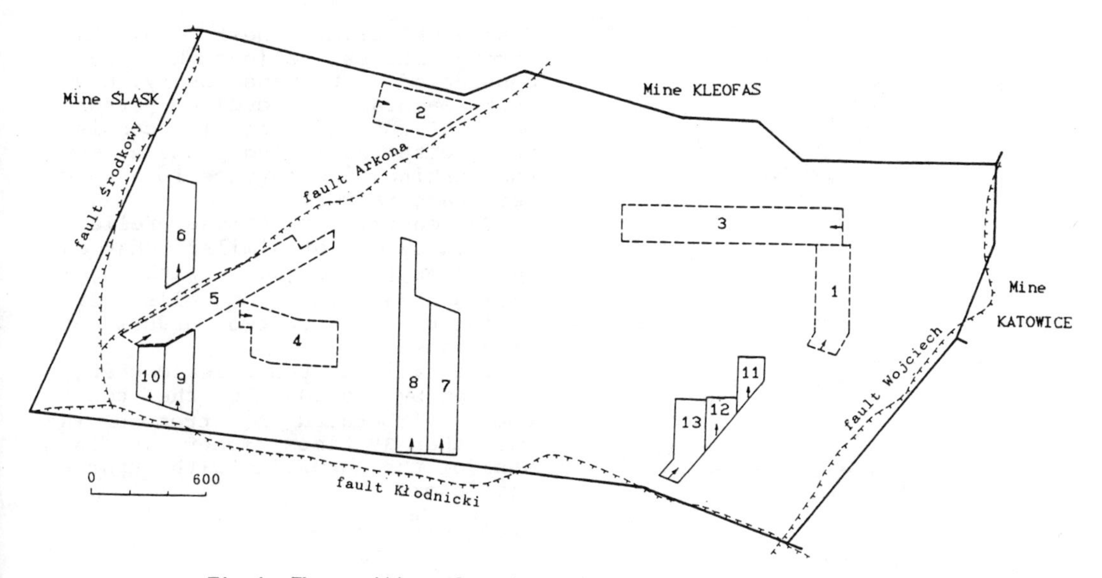

Fig. 1. The position of analysed longwalls in Wujek coal mine

Table 1. Effectiveness of the CDB on selected longwalls of the Wujek coal mine

Working	No of blasting	Obtained stress reliefs					
		No stress reliefs	10^3 J	10^4 J	10^5 J	10^6 J	10^7 J
Longwall VIIIa Seam No 501 - (1).	166 100%	143 86,0%	22 13,0%	1 0,6%	- -	- -	- -
Longwall IXb Seam No 501 - (2).	129 100%	21 16,0%	101 79,0%	7 5,0%	- -	- -	- -
Closing longwall Seam No 501 - (3).	92 100%	2 2,0%	82 89,0%	8 9,0%	- -	- -	- -
Closing longwall Seam No 416 - (4).	223 100%	34 15,0%	181 81,0%	7 3,0%	1 0,4%	- -	- -
Longwall I Seam No 501 - (5).	575 100%	91 16,0%	437 76,0%	47 8,0%	- -	- -	- -
Caving, total	1185 100%	291 25,0%	823 69,0%	70 6,0%	1 0,08%	- -	- -
Longwall III Seam No 510 - (6).	122 100%	19 14,0%	33 25,0%	50 45,0%	16 12,7%	3 2,5%	1 0,8%
Longwalls XIII, XIV Seam No 501 - (7,8)	384 100%	65 17,0%	221 58,0%	66 17,0%	24 6,5%	8 1,5%	- -
Longwalls XXII, XXIII Seam No 501 - (9,10).	326 100%	51 15,0%	146 45,0%	119 37,0%	8 2,4%	2 0,6%	- -
Longwalls IV,V,VI (SE) Seam No 510-(11,12,13)	320 100%	44 14,0%	231 72,0%	40 12,0%	4 1,5%	1 0,5%	- -
Backfilling, total	1152 100%	179 15,0%	631 55,0%	275 24,0%	52 4,5%	14 1,5%	1 0,08%

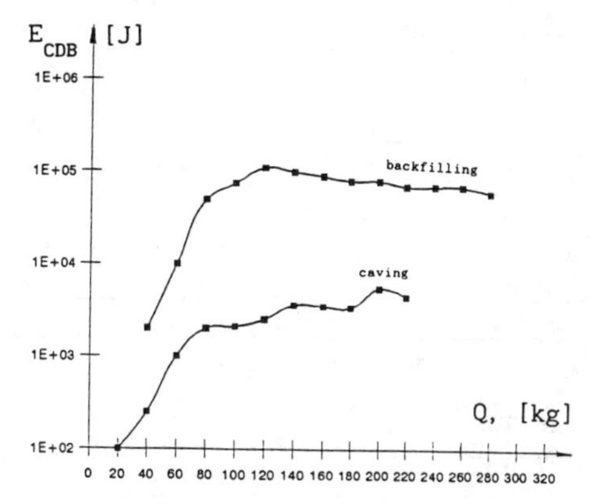

Fig. 2. The dependence of the seismic effect on the blasting charge

Data given in Table 1 provide quantitatively interesting information about the effectiveness of the CDB for each analysed longwall. Cumulative results illustrated graphically are shown in Fig.3 distinguishing the two extraction systems, with caving and with backfilling.

The measure of the effectiveness of CDB that is often mentioned in the literature (Dubiński et al, 1977; Skrzypek et al, 1987) is the ratio of the energy of tremors induced by the blasting to the total seismic energy emitted. Data given in Table 2 show esential differences between the extraction system with caving and the extraction system with hydraulic backfilling. In the case of the system with caving the above mentioned ratio is from a fraction of one percent to several percent, but in the case of the system with hydraulic backfilling this ratio amounts to even several dozen percent.

5 CONCLUSIONS

1. It has been found that the CDB for longwalls with caving driven

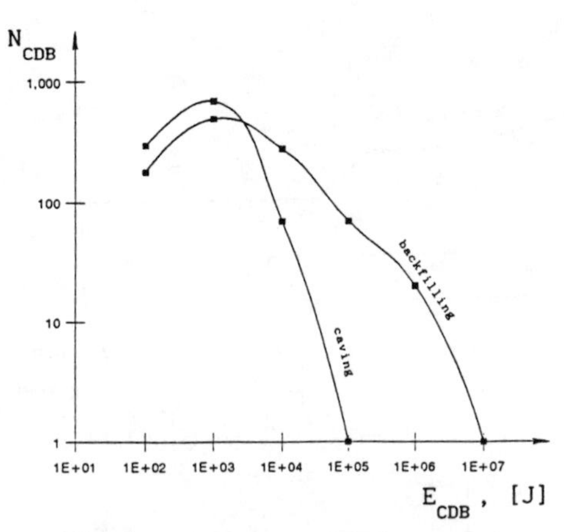

Fig. 3. The energetic-frequency distribution of tremors induced by CDB

Table 2. Comparison between the sum of the blasting induced energy ΣE_{CDB} and the total energy of tremors ΣE for the selected longwalls with caving and backfilling of the Wujek coal mine.

Working	ΣE	ΣE_{CDB}	% $\frac{\Sigma E_{CDB}}{\Sigma E}$
Longwall VIIIa Seam No 501 - (1).	$1,9 \times 10^7$ J	$9,4 \times 10^4$ J	0,5
Longwall IXb Seam No 501 - (2).	$3,3 \times 10^7$ J	$5,3 \times 10^5$ J	1,6
Closing longwall Seam No 501 - (3).	$2,1 \times 10^7$ J	$4,7 \times 10^5$ J	2,2
Closing longwall Seam No 416 - (4).	$1,9 \times 10^7$ J	$8,5 \times 10^5$ J	4,0
Longwall I Seam No 501 - (5).	$7,7 \times 10^8$ J	$2,8 \times 10^6$ J	0,4
Caving, total	$8,6 \times 10^8$ J	$4,87 \times 10^6$ J	0,6
Longwall III Seam No 510 - (6).	$5,4 \times 10^7$ J	$3,0 \times 10^7$ J	56
Longwalls XIII,XIV Seam No 501 - (7,8)	$7,1 \times 10^7$ J	$1,2 \times 10^7$ J	17
Longwalls XXII, XXIII Seam No 501 - (9,10).	$1,2 \times 10^8$ J	$7,6 \times 10^6$ J	6
Longwalls IV,V,VI (SE) Seam No 510 (11,12,13)	$2,3 \times 10^7$ J	$5,8 \times 10^6$ J	25
Backfilling, total	$2,68 \times 10^8$ J	$5,6 \times 10^7$ J	21

under high rockburst hazard is little effective compared to longwalls with backfilling where this effectivenees is considerably higher.

2. A minimum blasting charge needed for the CDB should not be less than 80 kg, but the changes in the charge from 80 kg to 220 kg have little effect on the changes in seismic energy of the resulting stress reliefs on longwalls with caving. For longwalls with backfilling the blasting charge of from 100 kg to 140 kg should be regarded as the best one.

3. For both the roof control systems, performing single fired big charge blastings next to the strong stress reliefs has been considered to be unnecessary. However, maintaining cyclic performance of the blasting in consistency with the mining process and the working face advance is a very important factor.

4. The energy of stress reliefs induced by CDB of similar charges is higher for longwalls with backfilling than for longwalls with caving by almost two orders of magnitude.

5. The ratio of the seismic energy induced by the CDB to the total energy liberated by tremors is more than 30 times higher in favour of the longwalls with hydraulic backfilling.

REFERENCES

Dubiński J., Gerlach Z., Kempny F. & Szot M. 1977. A Procedure for the selection of the best preven-ive measures and assessment of the effectiveness of using them under the conditions of mining the strongly bumping coal seam No 510 in the Wujek coal mine. Proc. IV Winter School of Rock Mechanics. Wisła.

Filipek M. Mitręga P. & Syrek B. 1991. An attempt of the assessment of the effectiveness of the concussion and destressing blasting performed on longwalls with caving under high rockburst hazard. Publ.Inst.Geophys.Pol.Acad. Sc. M-15 (235).

Mitręga P. & Syrek B. 1987. Dependence of the seismic energy released during the concussion and destressing blasting on the quantity of explosives used. Przegląd Górniczy. No 2.

Skrzypek Z. Kiedrowski J. & Gerlach Z. 1988. Geophysical assessment of the rockburst hazard state in workings and technological means of its reducing in the Katowice coal mine. Zeszyty Naukowe AGH. Górnictwo. No 141.

Geomechanics 93, Rakowski (ed.) © 1994 Balkema, Rotterdam, ISBN 90 5410 354 X

Mining induced seismicity in the Czech part of Upper Silesian Coal Basin depending on mining conditions

Petr Konečný
Institute of Geonics of the CAS, Ostrava, Czech Republic

ABSTRACT: Mining activities in the Czech part of the Upper Silesian Coal Basin are attended with dynamic phenomena. This phenomena in the Karviná area have been systematically observed since 1988 with the regional seismological network and the time, location, magnitude and energy of each registrated event have been reported. Undependent on the seismological observation the data concerning the intensity of mining have been collected. The influence of mining conditions have been analyzed.

1 INTRODUCTION

Mining activities in the Czech part of the Upper Silesian Coal Basin in particular in its Karviná area have been connected with rock burst occurrence for many years (Figure 1). The statistical data show, that the number of rock bursts (including the so called mikrorockbursts) does not immediately depend on the coal production either total or from the rockbursts prone seams. It is probably the result of intensive rock burst prevention used during mining of such seams. Seismological observation of dynamic phenomena was started in the year 1977, when the first seismic station was erected in a surface pillar of the First May Colliery in 1977. Within the period between 1983 and 1988 a local seismological network was constructed with 25 underground DSLA digital stations and two such stations on the mine surface. Additional data are supplied by stations with older type analogue equipment. Simultaneously a regional seismic polygon has been erected as an intermediate observation element designed especially for intense phenomena monitoring connected both with a national seismic system and with the network of underground seismic stations. This polygon was brought into operation in the year 1988. It was equiped with five three-component seismic stations with seismometers located in short boreholes of 30 m length. In the year 1990 it was extended up to 10 stations - 7 in short boreholes and 3 in underground workings.

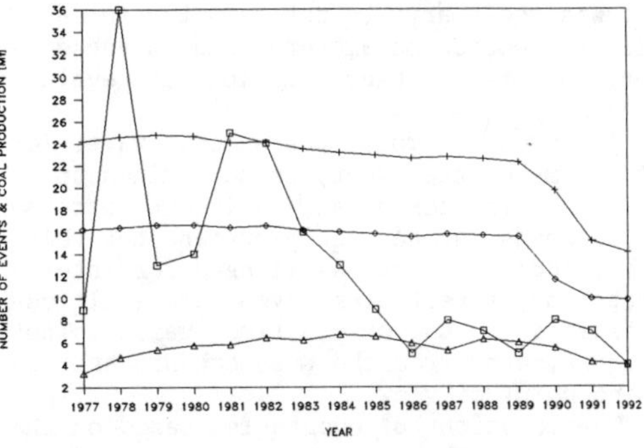

Fig. 1 The number of rock bursts (including microrocbursts) and the coal production in Karviná part of the Czech part of the Upper Silesian Coal Basin
□....number of rock bursts
+....coal production total
◇................Karviná seams
△................rockburst prone seems

Recently micronetworks of individual underground mines have been installed which are equipped either with digital HOÚ PCM3 apparatuses or with UGA-15/SL digital stations with a maximum of 15 seismic channels (Holub, Slavík, Kalenda, 1992).

All the above-mentioned networks are designed for observation of the Karviná part of the coalfield. The southern coalfield part of Frenštát-Trojanovice, where mining is planned for future, seismicity prior to mining activity is monitored by an independent network started in 1992 with five three-component surface stations with digital PCM3 apparatuses made in the Institute of Geonics of the Czech Academy of Sciences in Ostrava (Knejzlík, 1992).

The following pieces of knowledge are deduced from the mining induced seismicity observation realized mainly by the regional seismologic polygon and a special colnlection of data related to the intensity of mining and its conditions.

2 SOURCE PARAMETERS PROVIDED BY THE REGIONAL SEISMOLOGIC POLYGON

As it was mentioned above, the regional seismologic polygon was constructed in the period from 1988 to 1990. The polygon has been equipped with digital seismological stations delivered by Lennartz Electronics Company (125 Hz sampling). The quality of provided and interpreted data related to observed events - namely origin time, location, magnitude and intensity - depended of course on the phase of technical and methodical stage of development. Therefore it was necessary to estimate the reliability of source parameters used in subsequent analyzes, interpretations and evaluations.

The origin time of the seismic waves (or the time of the event) is determined up to 0.01 sec (Knotek et al, 1991). The accuracy depends on the equipment and the velocity model of rock massif used for interpretation, nevertheless even some small deviations in the origin time value is not significant from the geomechanic and mining point of view.

The location of events is based on the kinematic method. It is expected, that the accuracy of epicenter that coordinates determination (x,y) is in most cases better than 200 - 300 m. More unfavorable is the situation concerning the focal depth, where the accuracy is estimated to be about 700 m or worse (Knotek et al, 1991). This presumption has been tested with the help of the blasting for mining purposes.

The magnitude M of each event is determined in the following way:

$$M = \log A_{max} + \log d + \beta*(d-1)*\log e + K$$

where A_{max} = absolute value of the maximal amplitude vector of the seismic signal [ms-1]; d = epicentral distance [km]; β = damping factor of the maximal amplitudes (0.04 [km^{-1}]); e = the base of natural logarithm, K = fixation constant (5.202).

To test the reliability of magnitude values of events registered in the Karviná part of the Ostrava - Karviná Coal Basin

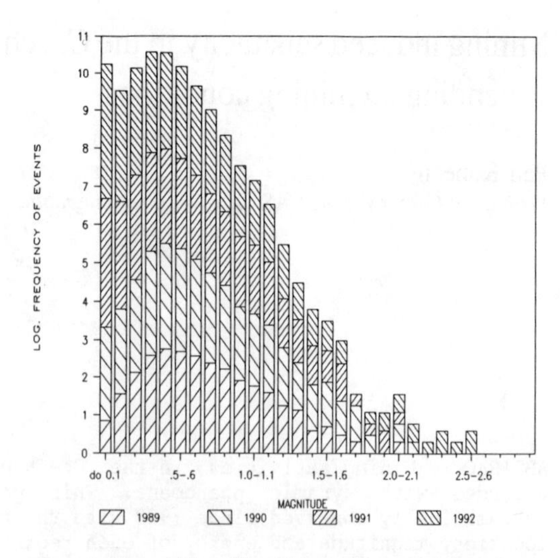

Fig. 2 The distribution of events according to their magnitude. Karviná part of OKB, 1989-1992

the analysis of a number of events N with the magnitude M was made. Referring N to one year (in the period 1989 to 1992) and a magnitude interval of 0.1 unit the results plotted on Figure 2 were obtained.

Comparing the results with the equation expressing the number of events N with the magnitude M

$$\log N = a - bM$$

where a,b are positive constants we obtain the following parameters (Tab.1):

Table 1.

year	a	b	range of M min.	max
1989	3.7	1.85	0.6	2.0
1990	3.45	1.50	0.5	2.2
1991	3.1	1.35	0.2	2.1
1992	3.4	1.67	0.2	1.9
ϕ	3.4	1.59		

The lower limit of magnitude mentioned in Tab. 1 is important when a subsequent evaluation, for instance the comparison of the mining induced seismicity of two different areas is made. For this purpose of course it is necessary to consider only events with magnitude higher than the lower limit for the analysed time period.

The events with magnitude greater then upper limit (Table 1) should be considered individually.

Similar processing was made with the value of energy of each event.

The energy is calculated according to the formula:

$$E = k * (A^p)^2 * d * e^{2*\alpha*(d-1)}$$

where E = energy of the event [J]; k = 1.633E7 [kgm.s]; A^p= absolute value of the P wave vector [$m.s^{-1}$]; d = epicentral distance [km]; α= damping factor [km^{-1}].

The distributions of registered events according to their energy in the studied period from 1989 to 1992 (the energy scale is expressed in log E, the interval of log E is 0.5) are demonstrated on Figure 3.

Similarly as in the case of magnitude it is possible to compare the results with the equation expressing the number of events N depending on their energy E

$$\log N = c - d \log E$$

where c,d are positive constants. The calculated parameters are in Table 2.

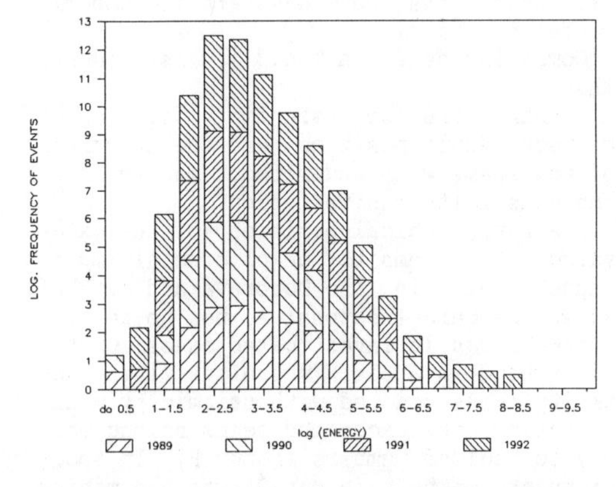

Fig.3 The distribution of events according to their energy. Karviná part of OKB, 1989.

Table 2.

year	c	d	range of logE min.	max
1989	4.7	0.67	3.0	6.0
1990	4.75	0.63	3.0	6.0
1991	5.1	0.73	2.5	6.0
1992	5.0	0.67	2.5	6.0
ϕ	4.9	0.68		

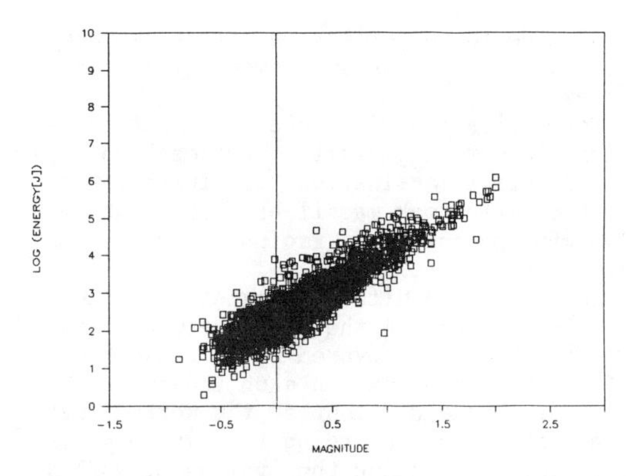

Fig. 4 Energy - magnitude relation for events in the year 1991

Similarly as in the case of the magnitude is the lower limit of the energy mentioned in Table 2 important for a subsequent evaluation. The events with magnitude larger than upper limit (Table 2.) should be considered individually. Finally the relation between magnitude and energy of events was analyzed (Figure 4). The equilibrium between magnitude M and energy E can be then written as follows:

$$\log E = 1.66 M + 2.29$$

where energy E is expressed in J.

The above described analysis of source parametres reported by regional seismological polygon from the point of view of their further use as a data set for consequent study of mining induced seismicity allows to make following conclusions:

- the accuracy of epicentre coordinates (x, y) is approximatly 200 - 300 m,
- the accuracy of depth determination is estimated to be about 700 m or worse, which is unsufficient for practical purpouses,
- the the number of events with magnitude in interval from approximately 0.5 to 2.0 or with energy in interval from 10^3 to 10^6 can be considered as homogene in the Karviná part of Ostrava - Karviná Coal Basin. The events with magnitude greater than 2 or energy greater than 10^6 should be considered individually.

3 DEFINITION OF INTENSITY OF MINING INTERVENTION INTO ROCK MASSIF AS SOURCE OF INDUCED SEISMICITY

In conditions of OKB where longwalling is the main mining method it can be presumed that a deciding effect on stress restructuring lies in the advancment of longwall faces.

For evaluation of intensity of intervention into rock massif a new methodology for numerical evaluation of mining intervention into rock massif has been elaborated for purposes of regional evaluation (Konečný 1988, 1990).

In connection with the existing seismicity evaluation in the OKB the whole area of interest is covered by a network of 250 x 250 m squares. In each square in individual calendar months the worked out area of seams according to mine map is subtracted including the values of worked thickness, kind of gob liquidation (backfilling or caving) and number of seam concerned. Apart from this additional information is registered which could be important for further processing as for instance depth of locality concerned, full face mining or slicing, eventually the number of slice worked and other comments are registered. For each seam and each square the so called mining intensity coefficient $I_{(K)}$ is calculated monthly:

$$I_{(K)} = P * m * k / 62500 \text{ [cm]}$$

where P is the magnitude of worked out area in a given month and square in m^2, m is the worked thickness of seam in cm, k is the coefficient of gob liquidation (in case of caving k = 1 and in case of backfilling k = 0,5).

Mining intensity coefficient $I_{(K)}$ is then in cm-dimension and expresses a fictive seam thickness in cm which should be mined over the whole area of square evaluated to recover an efficient volume of coal (considering effect of eventual backfill on induced stress zone and subsequent stress changes).

If more longwalls contact in the square evaluated the resultant $I_{(K)}$ coefficient is a sum of coefficients determined for individual longwalls. Naturally, if necessary the time interval can be reduced, in regional evaluations, however, the monthly evaluation of mining intensity usually appears as sufficiently detailed. The nature of $I_{(K)}$ coefficient indicates that when selecting longer time intervals or larger areas of several squares the $I_{(K)}$ coefficient could be added or averaged according to intentions of evaluation.

4. MANIFESTATION OF THE MINING INDUCED SEISMICITY IN DIFFERENT CONDITIONS

Using the above described data concerning mining intensity and induced seismicity it is possible to compare the situation in different geomechanical and geological conditions. The results of such analysis can be used for instance to estimate the risk of rockburst in different seams and parts of mine field.

To demonstrate an example of such evaluation, the mine field of the mine First May was chosen. This mine consists of three plants, whose mine fields are divided into blocks according to the tectonic structures. The mining in the different blocks has its specific manifestation in the rock mass failure and therefore also in induced seismicity. The intensity of mining and the number of events with energy larger than 10^3 J in different part of the First May mine field is presented on Figure 5. It is evident that there are blocks in which the mining is attended with seismic events in contrast to some other blocks, where the mining activity has no influence on seismic activity. The detailed examination shows us that these cases differ in the stratigraphic position of extracted seams (Tab. 3).

The depth and thickness of excavated seams in studied areas is in Tab. 3. too. In all these cases the seams are in subhorizontal position.

Comparing data in Tab.3 it is evident, that:

- there are no essential differences between depth position and the thickness of the seams with and without mining induced seismicity manifestation,
- stratigraphical position of the excavated seams seems to be of crucial importance. Mining in the Doubrava and Svrchní sušské members (Westphal A) has no seismic consequences in investigated cases. On the other hand in areas, where the mining activity is associated with seismicity manifestation, the excavated seams belong mainly to Sedlové members (Namur B). In Spodní sušské members (Namur C) was the mining in investigated period insubstantial.

The explanation of this phenomenon can be found in the geological and geomechanical properties of strata Sedlové members (Namur B). The thickness of these strata reaches up to 300 m. They consist mainly from sandstones and conglomerates. The thickness of different layers is very variable, some layers are very thick. They are characterized by sudden pinching and erosion too. From the mechanical point of view the rocks are usually hard (compression strength about 70-120 MPa) and elastic with tendency to brittle fracture. The coal has compression strength about 20 - 35 MPa and usually behaves elastic

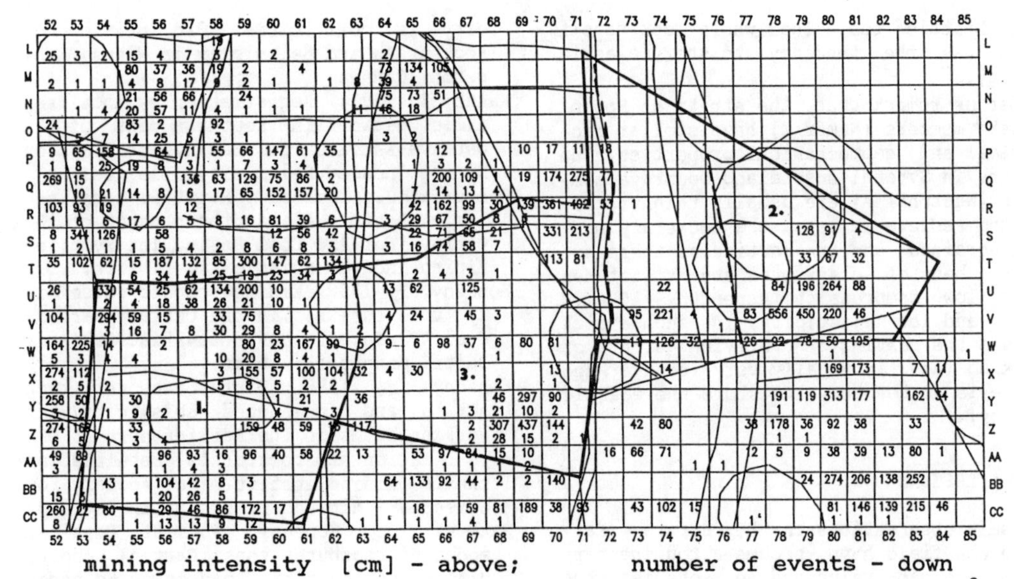

Fig. 5 The intensity of mining and the number of events with energy larger than 10^3 J in different part of the First May mine field (1989 - 1992)

Table 3.

plant	block	area	mining intensity total [cm]	number of events E>103J	stratigr. position of extract. seams (members)	portion [%]	seam thickness [cm]	depth [m]	annotat.
2	1	U73-V75	342	0	Doubrava	100	208	-307	
2	5+6	S77-V81	2342	0	Svrchní sušské	85	219	-358	
					Spodní sušské	15	292	-360	
							242	-355	M.V.
3	8	X67-AA70	1445	109	Svrchní sušské	1	271	-296	
					Spodní sušské	8	312	-295	
					Sedlové	91	220	-421	
							427	-420	M.V.
1	5	AA55-CC59	791	108	Spodní sušské	12	269	?	
					Sedlové	98	214	-336	
							266	?	M.V.
1	8	W58-Y82	810	80	Sedlové	100	349	-297	
1	1+9	U54-V61	1291	225	Sedlové	100	352	-447	

Footnote: Doubrava memb. - Westphal A
Svrchní sušské memb. - Westphal A
Spodní sušské memb. - Namur C
Sedlové members - Namur B, zone R

M.V. mean value

and brittle too. The whole rock massif then has the tendency to store elastic energy.

Let us remark that the strata of Spodní sušské members (Namur C) have similar geological and geomechanical properties. The layers in Svrchní sušské and Doubrava members (Westphal A) are usually thinner than in the sedlove members, the rocks are more fine-grained and in particular there are often beds of claystones and siltstones with low compression strength (40 - 70 MPa) and low elasticity. In such a rock massif the caving of overlying strata of working face is usually very regular and there is no tendency to store the elastic energy.

5 CONCLUSIONS

Seismic observations in the Ostrava - Karviná coalfield have become a routine part of mining geophysics and geomechanics when solving problems of rockbursts because they give objective data on development of rock massif failure process. During utilization of bulletin information - data on the time of occurrence, location of focal zone and energy of seismic events, especially when compiling sets expressing the number of events occurring within selected time intervals in longer time series, or when comparing various localities it is necessary to assess the information value of data on events of energy smaller than 1000 joules, or magnitude smaller than 0.5 as the efficiency of their monitoring may not be homogeneous. Their elimination from the sets processed by filtering them off will usually contribute to improvment of presentation of development of induced seismicity as it depends on mining activity advancment.

The comparison of mining parameters (intensity, excavated seam thickness, depth) with mining induced seismicity manifestation in different conditions shows us, that the dynamic fracturing of rock massif (with appearance of seismic events) depends crucially on the geomechanical and geological parameters determinating in particular the ability of rock mass to store elastic energy in consequence of advancment of underground workings. This support also the idea of regional prognosis of rockburst prone seams as it is practiced in Ostrava - Karviná Coal Basin.

REFERENCES

Holub, K., Slavík, J., Kalenda, P. 1992. Monitoring and analysis of microseismicity in coal mines. 3rd Int.Symp. on Rockbursts and Seismicity in Mines, Kingston, Canada

Kaláb, Z. 1992. Registration by local seismic network in southern part of the Ostrava-Karviná Coal Basin. Int.Symp. on Mining Induced Seismicity, Liblice.

Knejzlík, J., Zamazal, R. 1992. Local seismic network in southern part of the Ostrava-Karviná Coal Basin. Int.Symp. on Mining Induced Seismicity, Liblice.

Konečný, P. 1988. Mining-induced seismicity (rock bursts) in the Ostrava-Karviná Coal Basin, Czechos- lovakia. Gerlands Beitr. Geophysik, Leipzig.

Konečný, P. 1990. Disturbance of rock strata and in- duced seismicity when working seams in the Karviná part of OKR in relation to rockbursts, Habilitation thesis Mining Institute of Academy of Sciences - Mining University in Ostrava

Rakowski, Z., Konečný ,P. 1986. Physical model of rockburst zone. Partial Report, Mining Institute of Academy of Sciences in Ostrava

Rudajev, V. 1986. Rockburst seismics. Doctor thesis UGG Institute of Academy of Sciences in Prague.

Geomechanics 93, Rakowski (ed.) © 1994 Balkema, Rotterdam, ISBN 90 5410 354 X

Geomechanical problems in horizontal pipeline anchoring

P.P.Manca
Department of Mining and Minerals Engineering, University of Cagliari, Italy

A.Pala
Sardinian Acqueduct Board, ESAF, Cagliari, Italy

ABSTRACT: The paper is concerned with an anchoring system of a pressure pipeline, using a series of plates and friction piles in foundation soil of poor geotechnical quality. The solution illustrated here was employed for restoring the outlet of a tunnel section of the water mains following a burst caused by the faulty interaction between the existent anchoring block and the foundation soil. The rock formation concerned is composed of tuffites whose mechanical properties were further deteriorated by water spilling from the burst mains. The investigation which led to the remedial action suggested and the advantages achieved are discussed in the present paper.

1 INTRODUCTION

The choice and design of anchoring systems capable of resisting the forces acting on critical points of pressure pipelines is one of the most delicate problems in aqueduct planning.

In hydraulic terms the problem is relatively simple to define, even though the establishment of the critical geomechanical properties of the foundation soil on which the anchoring system rests is not an easy task.

In fact, soil deformability may sometimes produce settlement or sinking which, in spite of being compatible with block stability, cause the pipeline connected to the block to flex and may result in cracks and bursts.

In the case in question a slight sinking of the underlying soil caused cracks and major burst in some reinforced concrete tunnel section of the pipeline connected to the anchoring block.

Statical analysis of the slope at the outlet ruled out the possibility of the burst being caused by instability but that can be occured as the consequence of the spill water.

Remedial action consisted in constructing more deformable section and reducing the surface loads on the foundation soil by means of a system of piles which are also able to withstand the pipeline/soil movement.

2 THE ACQUEDUCT

The acqueduct, whose layout is shown in fig. 1, was built between 1969 and 1975 in northern Sardinia for supplying water to an industrial centre being developed at the time, as well as drinking water to neighbouring towns. With a water flow-rate of 2,200 l/s, the acqueduct is 49 km long with seven tunnel sections totalling 3600 m. The tunnels, lined with reinforced concrete, have a diameter of 2 m against the normal value of 1.4 m. Those parts of the pipeline before and after the tunnel sections follow the valley profile and the forces resulting from both variations (dimension and direction) of the acqueduct are absorbed by anchoring blocks (see cross-section in figure 2).

3 GEOLOGY

Geologically and stratigraphically the area is characterized by a sequence of grey-yellowish silty and/or sandy tuffites, distinctly stratified and overlain by a bed of ignimbrites (Spano and Asunis 1984, Barberi and Cherchi 1980). To facilitate the excavation of the tuffite and improve the tightness of the ignimbite roof, the tunnels were dug, wherever possible, at the contact.

In figure 2 is reported a typical stratigraphic cross-section too.

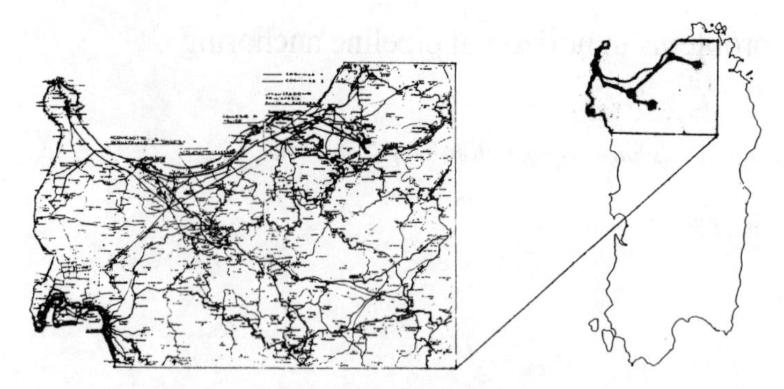

Figure 1. Sardinian map and the acqueduct layout.

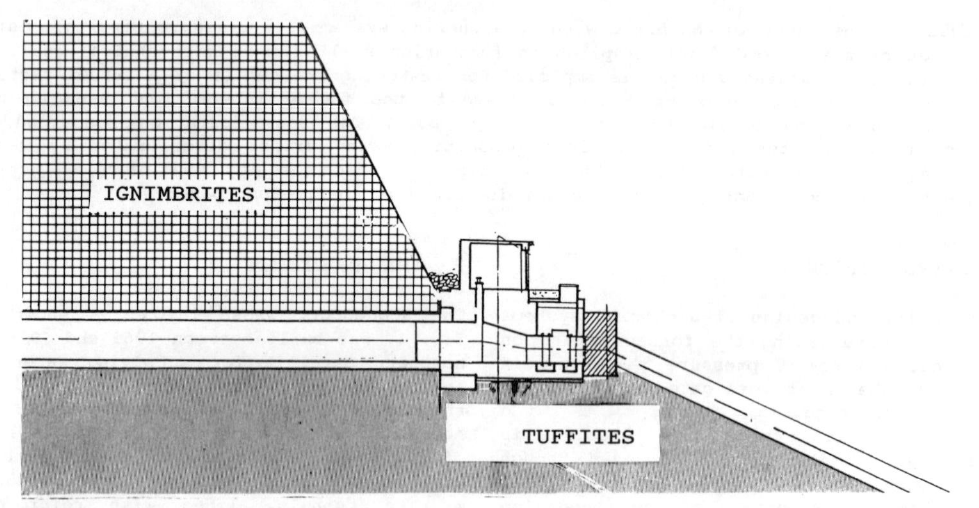

Figure 2. Cross-section of the main at the tunnel openings according to the original design showing the lithological series.

4 ANALYSIS OF FAILURE OCCURRENCES

Right from the construction stages, the tuffaceous formation on which a large part of the acqueduct lays proved particularly susceptible to the disturbances caused by the excavations and displayed a marked tendency to deform and expand in presence of water upon the contact.

During construction works this behaviour was responsible for a landslide and consequently one section of the aqueduct had to be rerouted.

An inspection carried out inside the aqueduct revealed that some ring shaped cracks in the tunnel lining near the outlets were the cause of leakage.

Remedial measures taken in the past had considered in repairing and simply sealing the cracks.

This situation continued until cracks appeared in the mains with leakage of pressure water and a first major burst took place in February 1977 (just two years after work had been completed).

At the time, the pipeline continuity was restored by placing a steel pipe inside the tunnel. However, no further action was taken as far as the other acqueduct sections were concerned, although the greater damage that had resulted from the complete break in the pipeline and the risks to people and things in the vicinity.

On the other hand, it was believed that this major incident could be attributed to geological-technical causes which would not happen again even though, once the pipeline had been repaired, the flowrate was reduced, for precautionary measures, from the predicted 2000 to 500 l/second.

Further incidents occurred over the

years though none of such great consequence. It was only in 1989 that a geotechnical investigation (Ciccu and Manca 1990) was conducted on the rock formations along the acqueduct route and a plan of action drawn up. It concerned all tunnel sections which the finding of the study had indicated as being equally at risk of breakage, and also those where no damage had previously occurred.

The above investigation indicated that though only sligth, the sinking that occured in the ground on tuffites upon which the anchoring blocks were founded was responsible for the pipeline bendig and for the resulting damage. In fact, the proneness of the foundation soil to yield, though it does not compromise the blocks, gives rise to flexural stresses in the concrete pipe rigidity connected to it.

Moreover the tipical ring-like cracks observed in several points along the pipe indicate that the first crack and subsequent water leakage was accompained by a decline in the mechanical properties of the anchoring blocks caused by a diminished block-soil friction. As a probable consequence of this the parts detached along the axis producing breaches of as much as a few centimetres.

The interpretation provided was borne out firstly by the fact that at all outlets where similar damages had been observed, the cracks, which affected the whole cross-section, only appeared after the first 15-18 m of the concrete section. This showed that these stresses increasing with distance were thus bending stresses, and that it was not only the simple case of a deterioration in the mechanical properties of the block.

Figure 2 shows the cross-section of the main at the tunnel openings according to the original design.

Another major burst occurred in late 1991 at different tunnel opening. This time the part affected lay only 50 m above the road used.

Water burst through a circular 3-4 cm crack in an orthogonal section of the tunnelled pipe suggesting that the head structure built at the tunnel opening might have undergone lengthwise settling. It was clear then that the structure had not been able to withstand the hydrostatic thrust of about 1000 kN acting at the time on the pipe section.

Apart from the damage to the conduit, the huge quantity of water that escaped before supplies could be shut off caused serious damage to the surrounding area and major stabilization works had to be undertaken both up and downstream from the tunnel opening.

In fact the mechanical properties of the tuffites through which the tunnel had been excavated, though not remarkable, were deterioreted completly by the 500 l/s of water whith spilled out during the 5-6 hours before it was eventually disconnected.

In addition to reduction of the material cohesion, voids and cavities of irregular and unpredictable size formed around the burst area.

5 SLOPE STABILITY ANALYSIS

Two analyses were carried out to assess stability in critical equilibrium conditions both before and after the burst, which were presumed to differ as the cohesion of the tuffites had altered to a considerable extent by the spill of water in the vicinity of the pipe.

The minimum values of the Safety Factor and the geometrical conditions of the critical surfaces obtained by the calculations are shown in Tables 1 and 2 and in figures 3 and 4.

Thus, the possibility that the burst had been caused by movements of the slope could be ruled out bv the results of table

Table 1. Stability analysis for the condition prior to the burst.

	Cohesion kPa	Internal friction angle
Tuffites	100	25°
Igninbrites	0	45°

Centre Coordinates X_c, Y_c [m] = 90.0, 92.5
Radius of slide surface = 91 m
Safety Factor = 1.526

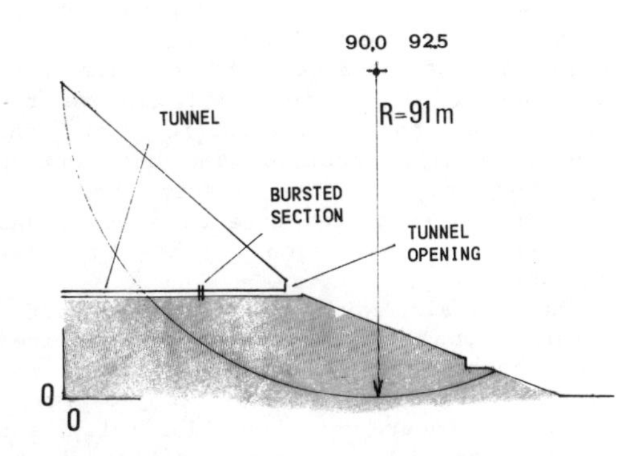

Figure 3. Stability analysis solution for the condition prior to the burst.

Table 2. Stability analysis for the condition after to the burst.

	Cohesion kPa	Internal friction angle
Tuffites	0	25°
Igninbrites	0	45°

Centre Coordinates X_C, Y_C [m] = 90.0, 92.5
Radius of slide surface = 72 m
Safety Factor = 1.154

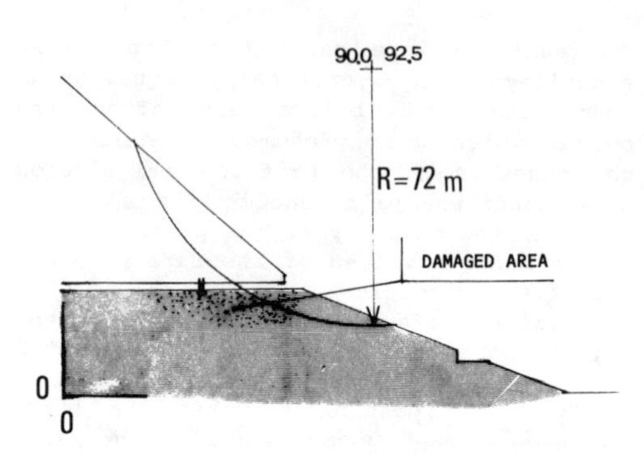

Figure 4. Stability analysis solution for the condition after to the burst.

1 indicating a SF greater than 1.5.

On the contrary, the results of table 2 indicated that the damage produced by the water (lack of cohesion), although reduce the radius of the slide surface to 72 m, is the cause of the reduction of the corresponding Safety Factor 1.154 and might have given rise to the onset of instability.

6 REMEDIAL ACTION

Because of the decline in the mechanical properties of the rock enclosing the water main it was necessary to diminish the stresses acting thereon and improve the resistance and deformability of the tunnelled pipe sections. The best option appeared to be the resort to deep foundations suggested here as a novel and interesting solution in similar situations.

The repair work was designed after careful consideration of the combined geotechnical and hydraulic aspects of the problem and had the following objectives:

1. To ensure greater elasticity and tensile strength in the end section of the pipe.

2. To ensure a greater tensile strength to the tuffites zone altered by the burst effects.

3. To alleviate the vertical loads conveyed by the anchorage system to the underlying ground.

4. To ensure that the thrust is transmitted to the ground according to a structurally functional solution and not based on the relative block-foundation soil friction only.

The repair work considered in reconstructing the metallic structure which ran along the hill slope and connected the second part of the tunnel to the conduit. Then the problem was tackled of designing a suitable anchorage system of this structure taking into account both the vertical and horizontal components of thrust. The original anchoring system was thought to be no longer effective especially in view of the major alterations undergone by the ground following the water spill due to the burst. For this reason the new anchoring method had not to depend on the axial tensile strength in the tunnel-pipe, on the variability of soil properties, on the low coefficient of friction on the platform of the end section.

Based on the above considerations, and accounting separately for the horizontal and vertical components of thrust the following measures were taken:

1. A reinforced concrete platform was built to counteract, by means of belts and ropes the vertical thrust component.

2. A system of reinforced vertical 25 m long piles was drilled and cast in situ to counteract the horizontal thrust component.

The results obtained are shown in figure 5.

Obviously, the system as a whole takes into account the poor and poerly known soil consistency and the low friction coefficient of the surface.

Although a careful statical analysis did not indicate it to be necessary (in that the thrusts are counteracted by the structures described above), the last 50 m of the concrete tunnel section of the water mains were lined with a pre-calendered inner metal casing. This was placed along the pipe axis so as to adhere as far as possible to the tunnel walls, thus reducing the gap to a minimum.

The new coaxial pipe laid in the innermost part of the tunnel was fixed by means of radial anchorage driven into holes drilled though the concrete and penetrating 50 cm into the surrounding rock. The anchors were then fixed with washers and welded to the metal casing thus sealing the hole.

Figure 5. View of the new solution main tunnel.

Similarly, during the finishing stages intakes were created for introducing under pressure a special mortar to fillin the gap between the tunnel and the metal casing. This stage of the work proved to be the most difficult one for the metal structure that had been carefully laid in place could not compensate for the unavoidable irregularities in the surface of the original tunnel-pipe.

7 CONCLUSIONS

A better consistency between stresses generated by an aqueduct system and the strength of the underlying soil formation is clearly obtainable by the use of pile fondation. Moreover, the geometrical characteristics (position and length) of the pile system can be designed to solve some possible slope instability problems.

The environmental effects of this new solution can also be considered as an important aspect especially in urban areas where the use of anchorage block system can negatively interfere with other town-plans and landscape features.

ACKNOWLEDGEMENT

The cooperation of the Sardinian Acqueduct Board, ESAF, is recognized. The research has been carried out within the programs of the National Research Council, CNR Cagliari Center.

REFERENCES

Barberi F., Cherchi A., 1980. Excursion sur le Mésozoique et le Tertiaire de la Sardaigne occidentale. *XXVII Congr. de la CIESME*: 127, 25 fig., Roma.

R. Ciccu, P.P. Manca, 1990. *Acquedotto industriale del Coghinas: relazione geotecnica relativa alle opere urgenti di ripristino*. Relazione inedita, 1990.

C. Spano e M. I. Asunis 1984. Ricerche biostratigrafiche nel settore di Castelsardo (Sardegna settentrionale). *In Boll. Soc. Sarda Sci. Nat.*, 23: 45-74, 1984.

Geomechanics 93, Rakowski (ed.) © 1994 Balkema, Rotterdam, ISBN 90 5410 354 X

Gravity monitoring of rock density changes caused by coal mining with regard to rock bursts forecasting

J. Mrlina
Geophysical Institute of the CAS, Prague, Czech Republic

ABSTRACT: Gravimetric monitoring was applied to determine density changes connected with strain fields variations within the rock masses, caused by coal mining. Observation stations were located on the surface and in three mine roads. During 13 periods of monitoring measurements the relation between anomalous changes of gravity and occurence of rock bursts was analysed. Certain positive correlation was found out after complex processing.

1 INTRODUCTION

During exploitation of coal seams in deep mines enormous stresses are generated in some parts of rock masses. Accumulation of elastic strain energy may result in rock bursts causing high risk for both miners and the mine itself with all the technology installed. Deformation of rocks induce density changes producing time-dependent gravity anomalies. The distribution of critical energy and dilatancy process always depend on local conditions, e.g. seam thickness, volume of exploited masses, remnant pillars position and state, general geological setting, etc.

2 GRAVIMETRIC MEASUREMENTS

As the first attempt in Czechoslovakia to observe gravity changes caused by deep coal mining, we suggested to apply gravity monitoring measurements in the ČSA mine in Karviná in the Ostrava-Karviná coal district. With regard to the local situation, 39 stations were placed on the surface above the R13933 projected longwall (seam 39) and also some stations in the mine roads of the floors No. IX (24 stations), X (52) and XI (48), see Fig. 1. According to the first results the measurements on the floor XI were abolished and on the floor X reduced.

Reference observations were carried out before the opening of the longwall and then 13 monitoring stages took place with approximately 3 weeks intervals. As there were no mass displacements near the stations and the gravity effect of exploited coal was negligible, the data processing was simplified to measured gravity evaluation, while surface subsidence taking into account (maximum 22 cm from February to November 1990). All observed relative gravity data were tied to the base stations situated at presupposed stable blocks.

3 DATA ANALYSIS

Complex analysis of the data had started with the examination of time-changes of gravity at every station, according to the character of which all the stations were devided into groups a - e (Fig. 1). Average time-changes curves for total surface group and each of a, b, c, d, e groups were assessed then (Fig. 4 a,b).

Theoretically, most of rock bursts should occur in the period after surface gravity maximum (mine roads minima), as this would be in a good accordance with the assumption of geomechanical parameters changes during a deformation process (Fig. 2). Bulk density is decreasing and porosity increasing in the stages III and IV. It was proved by Fajklewicz (1983), that the density may decrease even by -0.10 g.cm^{-3}.

The main question in our data interpretation was, whether such relation really exists. A few rock bursts prognostic cycles were determined ($\alpha, \beta, \gamma, \delta$) from the total surface curve and particular groups curves

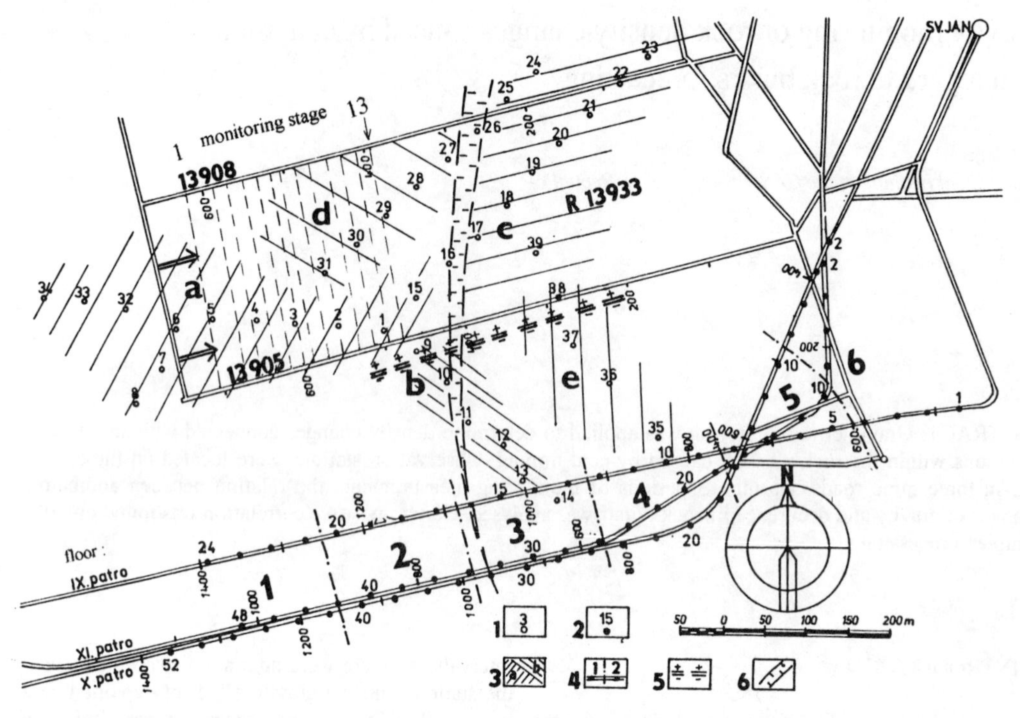

Fig. 1 Longwall R13933 with gravity stations on surface and in mine roads :
1 = surface gravity station, 2 = mine road gravity station, 3 = groups of surface stations, 4 = groups of mine roads stations, 5 = main boundary between blocks with different changes of gravity, 6 = fracture zone

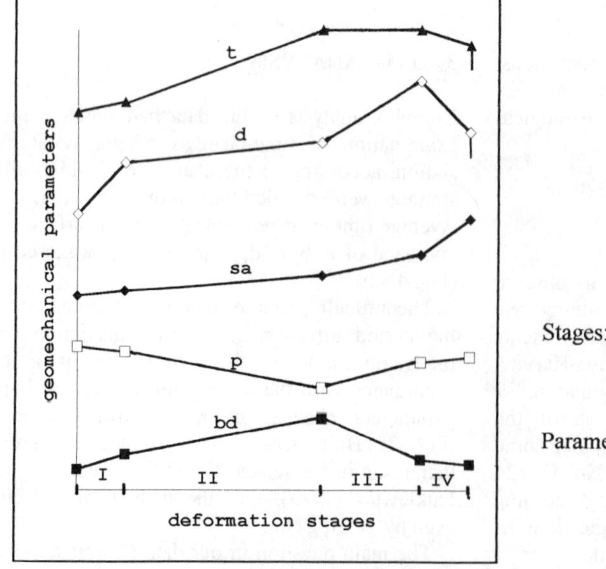

Stages: I - compression
II - elastic deformation
III - stabil dilatation
IV - non-stabil dilatation, rock burst
Parameters: t = tension
d = deformation
sa = seismo-acoustic
p = porosity
bd = bulk density

Fig. 2 Relative character of important geomechanical parameters changes during deformation process (adopted from Slavík 1992)

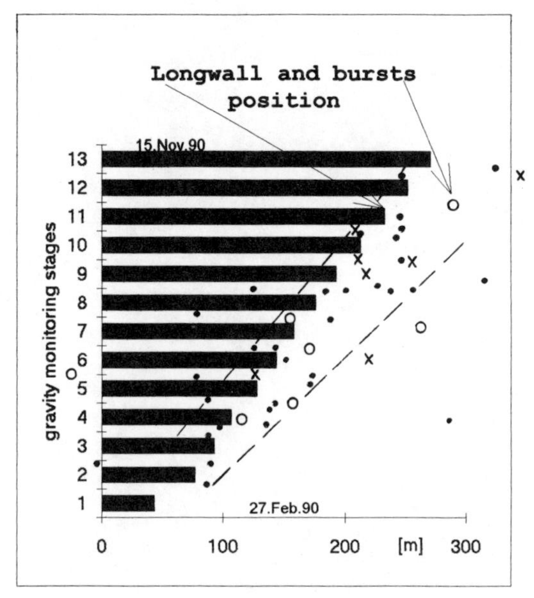

Fig. 3 Mining front during gravity monitoring campaigns 1 - 13 with rock bursts time-space relative positions. Zone of maximum bursts occurence is determined (dashed lines). Bursts devided according to their energy :

- • E> 10^5 J
- x E> 5.10^5
- o E> 10^6

(Fig. 4 c,d), with amplitudes 20 - 60 microGal (measurement accuracy of the Scintrex CG-2 gravity meter was about 10 microGal).

The essential stage was the analysis of the rock bursts time and space relation to gravity changes. It was found out that most bursts were really connected to time-dependent gravity anomalies !

Hypocentres were located prevailingly within an approx. 50-m-wide zone just forward of the longwall forehead (Fig. 3). All bursts epicentres (E>5.10^5 J) were related to corresponding surface gravity stations and marked on gravity curves (Fig. 4 c,d). In accordance with our presumption, most bursts occur on anomalous parts of the curves, especially along gravity decrease after the maxima (see the typical example at Fig. 4 d - cycle β). The correlation coefficient between interpreted gravity "prognostic" cycles and rock bursts occurence was quite high (+0.7).

4 CONCLUSIONS

With regard to our gravity monitoring data analysis it is possible to state, that a certain relation exists between time-dependent changes of gravity and rock bursts occurence in the coal mine. The changes are caused by rock density variations connected to geomechanical processes during deformation.

In contrast to the results of Fajklewicz (1983), when rock bursts occured not sooner than during

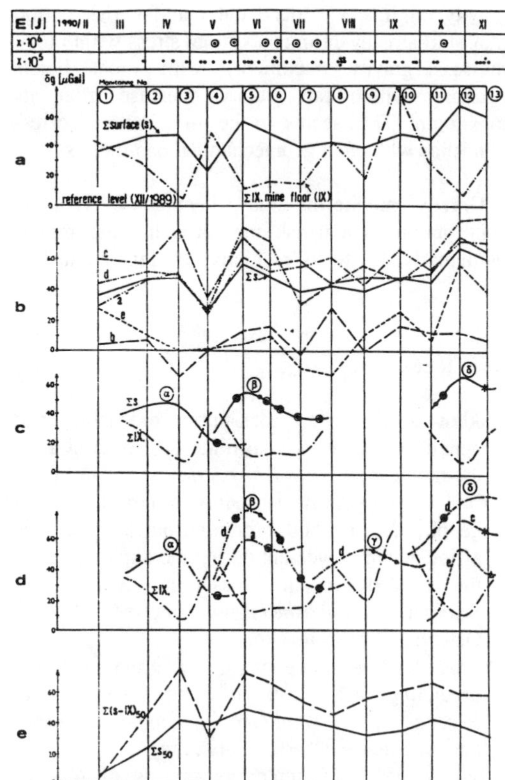

Fig. 4 Time overview of rock bursts, gravity changes curves and analysed prognostic cycles:

a - total time gravity changes on the surface and the floor No. IX

b - time gravity changes of particular groups of stations with similar character of variations

c - interpreted prognostic cycles α,β,δ according to the total gravity changes with rock bursts marks (only for E >5.10^5 and E >10^6 J)

d - interpreted prognostic cycles α-δ according to the particular groups gravity changes (only for E >5.10^5 and E >10^6 J)

e - time gravity changes of the stations within a moving zone ranging from the longwall forehead to +50 m forward of it and differential variations with regard to the changes on the floor No. IX gravity minima, we found out that the risk period may start already during maximum stress within rock masses (gravity maximum) with most bursts occuring within the gravity decrease after its maximum. This seems to be in a better correspondence with other geomechanical parameters.

Gravity monitoring may become a valuable contribution to anti-risk measures in coal mines, particularly due to its predictive character and low cost.

REFERENCES

Fajklewicz, Z. 1983. Rock-burst forecasting and genetic research in coal-mines by microgravity method. *Geophysical Prospecting,*31,p.748-765.

Mrlina, J. 1990. Gravimetry in complex of applied geophysical methods for anti-bursts measures efficiency improvement during the exploitation of the seam 39 in the block 3 in CSA mine in Karviná. *Non-published report* (in Czech), VVUÚ Ostrava, Czechoslovakia .

Mrlina, J. 1993. Application of gravimetry for predicting rock bursts in a black coal mine. *Proceed. of VII. Coal Geology Confer.*, 143-144. Fac. of Science, Charles Univ., Prague.

Slavík, J. 1992. Complex processing of seismological, seismoacoustic, geological and technological database of selected areas of the Ostrava-Karviná district focused on the burst phenomena prognosis. *Disertation report* (in Czech), Geophysical Inst., Prague.

Geomechanics 93, Rakowski (ed.) © 1994 Balkema, Rotterdam, ISBN 90 5410 354 X

Application of frictional pipe anchors for increasing the stability of different rock structures

N. Nikolaev, V. Parushev & R. Parashkevov
University of Mining and Geology, Sofia, Bulgaria

ABSTRACT: This paper is a summary of the application of frictional pipe anchors in Bulgarian mines experienced during the last ten years. A classification of rock structures is proposed viewing their interaction with the anchor support. A brief analysis is made of the schemes of the anchor support action in a different rock structures. An estimation of effect of the stabilization of rock mass using of the mechanical expansion anchors and frictional pipe anchors in made on the basis of the analytical relationships.

INTRODUCTION

The results of the observations and investigations done during the over tenth years application of the frictional pipe anchors (TFA) in different Bulgarian mines are summarized in this paper. TFA are applicated for supporting underground openings in the ore and coal mines and also in tunnel constructions. Experience and data are accumulated as a result of using TFA in wide range of rocks - from hard fractured up to soft layered. In addition laboratory and analytical investigation have been made. The achieved results and knowledge about the mechanism of the interaction of TFA with different rocks give a good basis for the creation of appropriate methodes for design of this type of support. The most important results and conclusions of these developments have been already reported of different scientific meetings. Nikolaev (1989), Nikolaev and Parushev (1984, 1986), Parushev (1986).

PRINCIPALES

By the introduction in mine practice of the new class of frictional anchors where the created in Bulgaria TFA belongs, the application field of mechanical anchors has been extended considerably. Practice shows that these anchors offer greater possibilities for insurance the stability of underground workings due to the complex action on the rock mass which includes effects undependable from the created initial forces. This is due to the property of TFA to counteract to the strain rock processes participating actively in these processes up to their full stabilisation without distructing the bond anchor - rock. This effect comes out from the force contact along the whole length of the anchor hole of the elastic pipe body possessing small stiffness. The second effect is due to the created stress fields which at appropriated orientation of the anchor alter favourably the rock mass state of stress.

According to the influence of these effects on the work of the anchor support, respectively on the way of design is required to be kept in view the structural properties of the rock mass exept its strength and strain indexes.

As an outcome of these considerations an appropriate classification of the most typical structural formations of the rock mass appearing in mine practice must be developed. This classification can be developed on the basis of the factors given in table 1.

Table 1.

Index	Rock structure	Structure disturbance 1	2	3
M	MONOLITIC	Fractured rock	Large blocks	Broken
L	LAYERED	Thick layers	Thinly layered rocks	Brocken

This table allows to be differed four typical structural groups according to effect of the anchor stabilization. Some of the already known schemes of action of anchor support can be applicated to any of these groups.

The rocks with indexes M1 and L1 belong to the first structural group, characterized with homogeneous rock mass softened by the presence of fractures. It must be outlined that in this case the fractures form not more than two systems (laying in two plains) which can not form blocks. The anchor's role is to decrease or to eliminate completely tension zones changing the state of stress around the opening which prevents the danger of creation of a third fracture system with eventual destabilization, of rock structure. Another effect aimed by anchors is the arch building in opening roof.

In the second structure group is considered large blocks M2 characterized by the formation of separate blocks in opening roof - at least 3 and most 5. In this case the most appropriate schemes for strengthening of the roof is arch building effect by the means of anchors or keying the loosening blocks.

In the third group is considered thin layered rocks L2. In this case the stabilization of the roof by anchors is based on the effect - compact bounded beam. The role of the anchors is to prevent the distruction of the bound cohesion between the single layers. This is realized by:

- increasing the frictional forces between layers using the active mechanical anchors;
- increasing the shear resistance along the laminating or
- combined action from two previous cases.

The last structure group includes broken rocks M3 and L3. Under broken rocks mass is understood broken rock mass which has lost its strength indexes of tension and minimum cohesion. The rock mass stability before reinforcing is based mainly of the friction between single pieces. The increase of the stability by means of anchors and other accessory support elements is realised by increasing the friction force between the pieces and keying effect in the arch, preventing the loosening of material.

One of the aims of this paper is to point out the differences of the action of TFA anchors on the rock mass in comparison with the mechanical expansion anchors in order to be used more rationally in mine practice.

From this point of view in the first two rock structure groups because of the fact that a prevailing role of their stabilization has the suspension effect it can not be pointed out a considerable difference between the two types of anchors. It can be accepted that the two kinds of anchors work almost equal with the difference that TFA stored at least the initial bearing capacity while it always decreases in the case of the expansion anchors.

ANALYTICAL ANALYSIS

A considerable difference in anchor action on the rock mass between the two kinds of anchors is established in the last two rock structure groups. Using some analytical relationships a comparative estimation has been done viewing the reinforcing effect of the two types of anchors.

As it was cleared the roof stability of thin layered rock structures is achieved by bounding with anchors in one common beam. In order to be fulfilled this condition it is necessary to equalize the acting shearing stesses which provoke separation of the single layers.

In the case of the comparative kinds of anchors this condition can be expressed by the following analytic dependences:

a) for expansion anchors

$$nP_eK_{te}\mu_r + C = \frac{3}{4}\frac{P.\ell_o}{\ell_a} \qquad (1)$$

b) for frictional anchors

$$nP_fK_{tf}\mu_r + n\ell_oP_f\mu_m + C = \frac{3}{4}\frac{P.\ell_o}{\ell_a} \qquad (2)$$

where n = density of anchor support, Number/ m^2; P_e, P_f = initial bearing capacity of the anchors respectively for expansion and frictional anchors; K_{te}, K_{tf} = coefficient taking into account the change of the bearing capacity of the anchor in the course of time, respectively for expansion and frictional anchors; μ_r, μ_m = coefficient of friction respectively rock - rock and metal - rock; C = cohesion between layers; ℓ_a = length of the anchor (hight of the beam); ℓ_o = width of the opening roof; P = rock pressure load.

The width of the beam is accepted to be 1 m.

Varying with density and length of anchors in equation (1) and (2) the required cohesion is obtained which the rocks must possess in order to be received stability of the system i.e. what part of the shearing stresses is taken by the rock. It is obvious that this factor is different for the two types of anchors when the density and the length is equal. The relation between these factors for the two types of anchors makes it possible to be done a comparative estimation of the reinforcing effect of the one type of anchors in comparison with the other i.e.

$$K_s = \frac{C_e}{C_f} \quad (3)$$

where K_s = comparative coefficient of reinforcing effect i.e. when $K_s > 0$ TFA are more effective and opposite; C_e = required rock cohesion when using expansion anchors obtained by equation (1) solved towards C; C_f = required rock cohesion when using frictional pipe anchors obtained by equation (2).
On fig. 1 are presented the graphical relationships of K_s in dependence of the density and the length of the anchors.

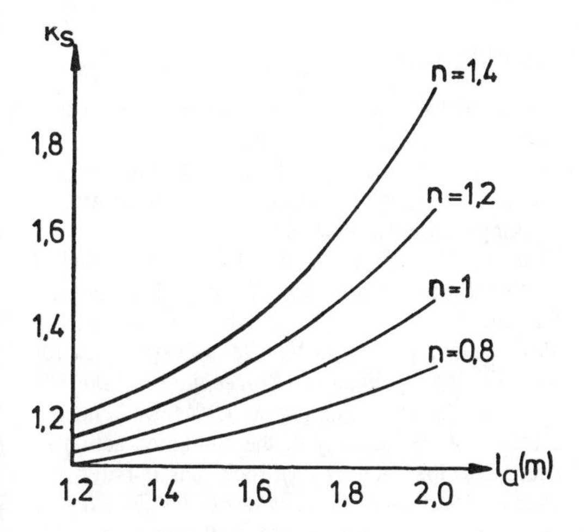

Fig. 1

It is obvious from the graphical relationships that for the investigated density and the length of anchors, TFA show greater reinforcing effect than the expansion anchors observed as when increasing the length and when increasing the density of anchors too.
The difference in anchoring effect of the two anchor types of support is increased particularly in the range of higher densities (steeper curves).
When rock mass is broken, in roof opening is build an arch of equilibrium whose abutments are set against the plains of the slipping inclined to the opening. In this case the load of rock pressure can be obtained according to the wellknown method of Terzagi (Kostner, 1962), transformed for the middle and deep mines:

$$P_v = \frac{H(0,5\ell_o - h\eta)}{\lambda\mu_r} \quad (4)$$

where H = density of rock; ℓ_o = width of opening; h = hight of opening; $\eta = \text{tg}\left(45 - \frac{\varphi}{2}\right)$; φ = angle of inner friction; $\mu_r \approx \text{tg}\varphi$; λ = coefficient of side resistance;

$$\lambda = \frac{1 - \sin\varphi}{1 + \sin\varphi}$$

For equalizing of the system we assume the most unfavourable case of distruction - slipping along the vertical plain. This condition is expressed by the equation (for length of opening - 1 m):

$$2P_H \frac{\ell_a}{\ell_o}\mu_r + C = P_v \quad (5)$$

where P_H = horizontal component of stress created by anchors.
Equations are worked out for obtaining the horizontal components of stress P_{he} and P_{hf} - respectively for expansion and for frictional anchors.

a) for expansion anchors

$$P_{He} = nP_e K_{te}\lambda \quad (6)$$

b) for frictional anchors

$$P_{Hf} = nP_f K_{tf}\lambda + \frac{nP_f K_{tf}\ell_o}{\mu_m} \quad (7)$$

For obtaining the second term for the right part of equation (7) is taken in account the known from the theory of foundation dependence for tightening of slightly bounded rocks by insurtion (hammering) of the piles which is expressed by accounting one half of stress acting in the hole wall with pile spaces under 2 m.
Applying the equation (6) and (7) in (5) and solving the mew equations towards C, we respectively obtain:

$$C_e = \frac{(0,5\ell_o + h\eta)}{\lambda\mu_r} - \frac{2P_e K_{te}\lambda\mu_r n\ell_a}{\ell_o} \quad (8)$$

$$C_f = \frac{(0,5\ell_o + h\eta)}{\lambda\mu_r} - nP_f K_{tf}\mu_r\left(\frac{2\lambda\ell_a}{\ell_o} + \frac{1}{\mu_m}\right) \quad (9)$$

As in the case with layered rocks we express the coefficient of anchor reinforcing effect by relation (3).

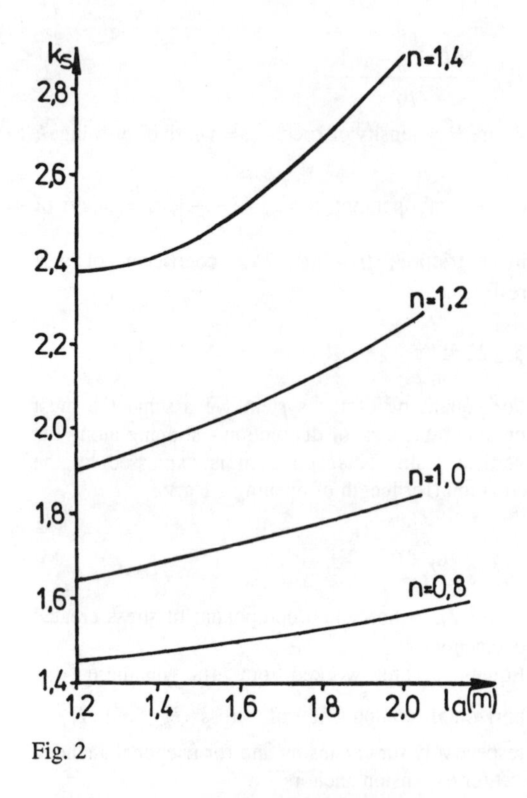

Fig. 2

At fig. 2 are presented graphical dependences of K_s from the length at different densities in broken rock structure. As it is clear from these graphical relationships in this type of rocks TFA are more effective than the expansion anchors.

In this case the influence of the density of anchors is expressed stronger - at density of 1,4 N/m^2 the reinforcing effect of TFA is abt 2,8 times higher than the conventional anchors.

CONCLUSION

On the basis of the proposed analysis about the work of TFA in comparison with expansion anchors can be drawn the following conclusions:

1. There is not considerable difference in reinforcing effect in the case of monolitic fractured and block rock structures except the fact that frictional pipe anchors never loose their initial bearing support.
2. Reinforcing effect of TFA is much greater in layered and broken rock structures. In some cases this effect is abt. 3 times greater than when applied expansion anchors.
3. Practical experience and the proposed analysis show that the initial bearing capacity of TFA has not a decisive influence upon the final effect of reinforcing.
4. TFA can be applied successfully in all up mentioned rock structures.

ACKNOWLEGMENT

The autors would like to express their gratitude to the National Fund "Scientific Investigation" - Ministery of Education and Science, and to the firm "Bulrock" which is a producer of anchors TFA for their support of finance and facilities for developing of the investigations in the field of anchor support during the last 5 years.

REFERENCES

Kastner, H. 1962. Statik der Tunnel und Stollenaues. *Springer Verlag.*

Nikolaev, N. 1989. Developing of Theory and Praxice of anchor support for underground workings. *Disertation.* Sofia.

Nikolaev, N., Parushev, V. 1984. Oval shaped frictional pipe anchor. *Rock Bolting.* Proceedings, Balkema.

Nikolaev, N., Parushev, V. 1986. Possibilities for rock bolting application. *Proceedings of the 9th Plenary scientific session of the IBSM.* Balkema.

Parushev, V. 1986. Sudy of the combined action of frictional pipe anchors in a rock massif befor and after its destruction. *Proceedings of the 9th Plenary Scientific session of the IBSM.* Balkema.

Strata mechanics
Technical notes

Geomechanics 93, Rakowski (ed.) © 1994 Balkema, Rotterdam, ISBN 90 5410 354 X

Mining activities and recent land subsidencies in Slovakia

J.Ďurove & M.Maras
Technical University of Košice, Slovakia

I.Lábaj
Lubeník Mine, Slovakia

P.Čičmanec
Nováky Mine, Slovakia

ABSTRACT: Our research was aimed at the analysis of land subsidencies in two mines in Slovakia. A lot of land subsidencies in these areas are in a close relation with underground mining activities. The mining influences on the surface presented in our paper are typical examples of different deposits in Slovakia. It is not simple to describe land subsidence processes and to acquire the corresponding models, but it is necessary to consider it mainly from the environmental and protection of the surface points of view respectively. Land subsidence intensity corresponds to the intensity of the underground mining. The research results make it possible to minimalize the negative influence of mining activity by the application of the principle: the mine advance in new fields of deposits.

1 INTRODUCTION

Besides the utilization of raw and energetic materials from domestic sources, in the extent minimally indispensable for the operation and development of the Slovak economy, the main strategic aim of the Slovak Republic (SR) is considered to be mainly in the safeguarding of the compatibility of this development with Acts on living environment. Influences of the underground extraction of mineral deposits on the surface, by the use of caving mining methods, consequencies of which are terrain subsidencies and corresponding rock movements, represent the danger, even the hazard for the occurence of collisions with ecological standards. The minimalization of dangers and removing of these hazards is therefore an important current task at the underground extraction of mineral deposits in the SR.

2 GEOMECHANICAL ANALYSIS OF LAND SUBSIDENCIES

Because of the limited capacity of this contribution we have concentrated ourselves only on two mining enterprises in the SR, which are at present relevant and there is a supposition of their future operation. In one case: we have concentrated ourselves on the magnesite - concretely on the Lubeník deposit, because Slovakia is the country with the global importance Mg deposits. The magnesite reserves in Slovakia are ca 600 Mt. In the second case: the important place, at the exploitation of primary fuel-energetic materials in the SR, belongs to the Nováky Brown Coal Mine.

On the surface at both of the deposits, with different mining methods, development of new geomorphological formations by caving of extracted underground spaces occurs. For the stability assessment of the mining area from the point of view of land subsidencies we have carried out the geomechanical analysis of the rock massif in both cases.

2.1 The Lubeník Mine

In the state raw materials' policy of the SR in the area of

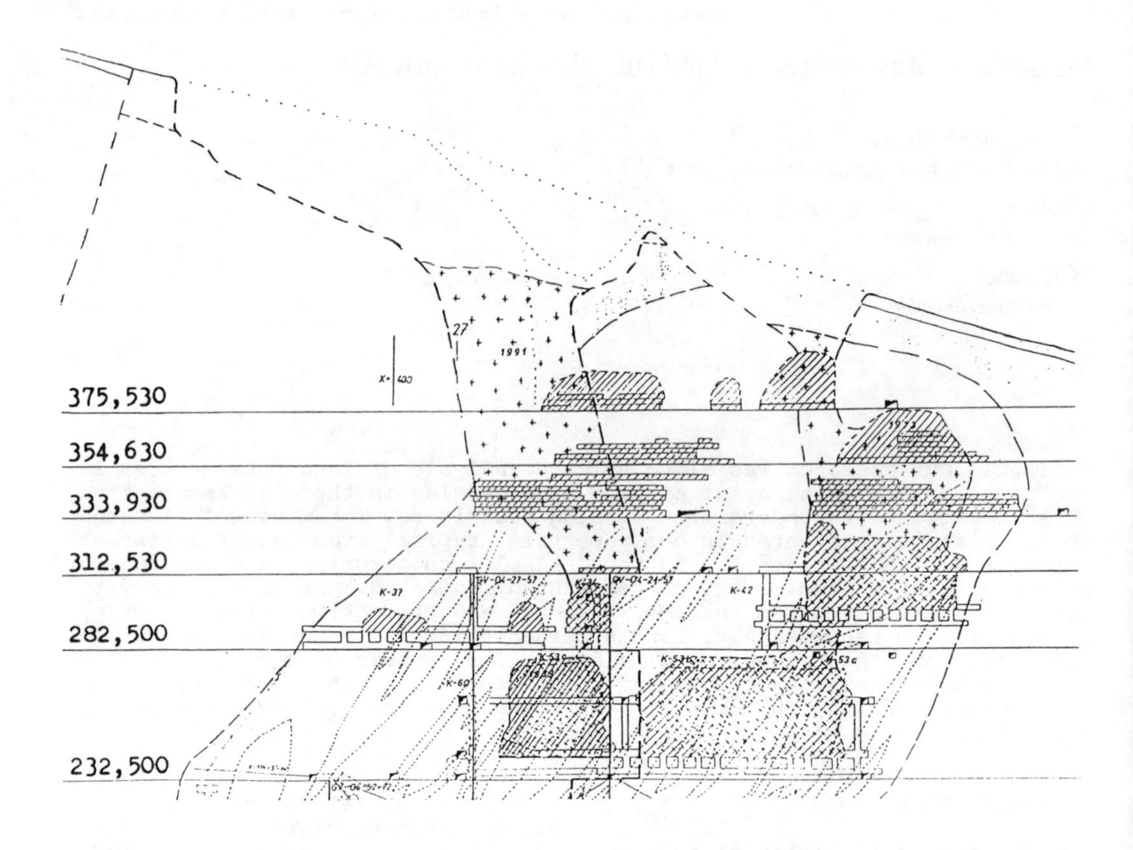

Fig. 1 Cross section of the Lubeník deposit

metal and magnesite industry, in the future the following mining areas are considered: Jelšava, Lubeník and Košice. At the current mine production of about 1.5 Mt/y the magnesite reserves will last for more than 100 years. By the modernization of the processing capacities at Lubeník and Lovinobaňa the assortment of products will be widened as well.

Furthermore, we will roughly describe the present technology of mining and liquidation with taking into account the overburden caving (Fiqure 1). We also will refer to reasons which lead to an unacpected oveburden caving into mine spaces with a significant effect on the surface in 1991 (Figure 2) as well.

As a good example of the rock strata with carbonate bodies in the Western Carpathians Mountains, which are characterized by a typical overlying sheet structure, is the morphology of the Lubeník magnesite deposit. Deposital filling consists of magnesite and dolomite. In the hanging wall and footwall of the deposit there are phyllites and graphitic phyllites respectively. The strength of the carbonate rocks is ca 5 times higher than at phyllites and 10 times higher than at graphitic phyllites.

At the Lubeník deposit the open chamber mining is generally introduced. After the extraction of the chamber (which means ca 50 % of the reserves), by the blasts of large extent, the technological pillars are disintegrated - liquidated and in that manner additional part of the reserves is extracted. At the same time it is expected that the caving of the extracted spaces occurs and a continual movement of the material being caved follows, which causes even surface subsidencies. From existing cavings at the deposit follows that these does not occur either at the same time or in short time interval after the

Fig. 2 Land subsidence after the sudden caving in 1991

liquidation of pillars, Grenda ('91). We suppose that the cause consists in not exactly defined positions of limestones in the hanging wall respectively unbroken remnants of own deposit body. In practice it means that the own liquidation of the block pillars is not in fact the liquidation of the extracted spaces. The continual caving up to the surface does not develop, and the danger of the sudden uncontrolled caving occurs. Figure 2 shows the surface subsidence after such a caving in 1991.

2.2 The Nováky Mine

In the Nováky Mine (NM) exploitation intentions we come out from the corresponding covering of the supposed brown coal needs in the SR till 2005 which were assessed by the state organs on the level of 4 Mt/y (3rd variant). In the 2nd variant the production decrease down to 2.7 Mt/y is being considered. Concretely for the NM it will mean the fulfilling of the costumers' requirements on the level of 1.1 - 1.3 Mt/y till 2005.

Furthermore, we will roughly describe the current mining technology in the Nováky Deposit (ND) with taking into account the overburden caving and also will refer to the corresponding problems. Right at the beginning it is necessary to be said that the so far procedure of the coal reserves (Figure 3) mining (incline) in the ND was chosen in order to maintain the maximal protection of the Bojnice Springs and also the prolongation of the service life of the real estate, built on the surface of the NM mine field, was taken into consideration.

At present the mining activity is carried out in mine fields where the overburden thickness achieves 300 - 400 m. The basic mining methods applied in the Nováky Mine are:

* Longwall method with the ripping of the roof block in the full thickness with the average technological recovery of ca 70 %. The seam thickness is 10 - 13 m.
* Panel faces (chambers) with the average technological recovery of ca 40 %. The seam thickness is 8.5 m.

The undermining this mining activity is characterized by the creation of continual subsidence hollows with the development of cracs on the surface.

It is known that all the mining effects on the surface are in close relation with the mining technology used. At the mining by the controlled block caving, the method used in the NM, the surface subsidencies occur, the size of which depends on various factors.The general theoretical principle of the secondary stress state in the surrounding of longwall faces under easy and regularly caving overburden is expressed by the Labasse's conception, Hatala (1985).

At present in the central part of the ND has been registered the 4.2m maximal terrain subsidence. In the given case, however, it is not the finite state because the total mine field "effective area" has not been extracted yet.

In the case of the decision according to the 2nd respectively the 3rd variant, a mining advance which would minimalize the mining influences on the surface (also the rescue of historically valuable object of the St. Andrew Apostle Church - a cultural sigth - is of a great importance) is being considered.

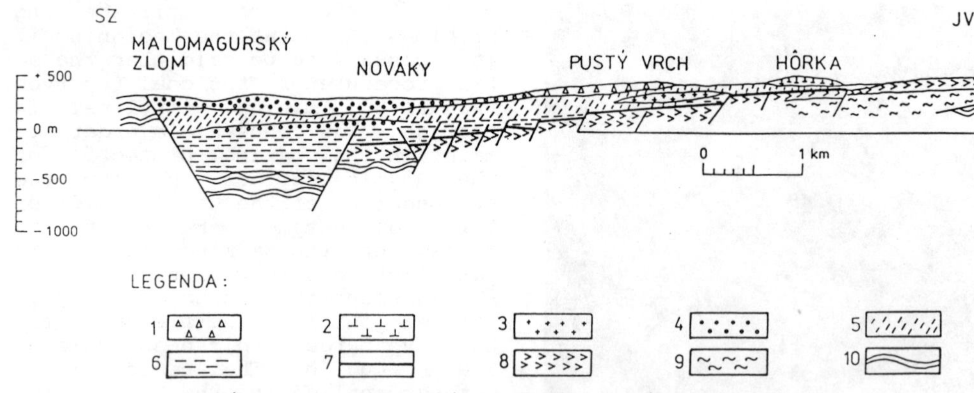

Fig. 3 Cross section of the Nováky deposit
1-andesite breccias and tufaceous breccias; 2-tuffs; 3-tuffits;4-gravels, gravel sands,sands; 5-clays; 6-overburden clays; 7-coal seam; 8-footwall tuffits; 9-Lower Miocene system; 10-Paloegene system

3 CONCLUSIONS

The Lubeník Mine:
From the mining-technical conditions and the geomechanical analysis of the stability situation at this deposit for the future follow the changes in the mining advance as well as in the method of the block pillars' liquidation. In addition to this we mean at the deposit to introduce the equipment for the automatic observation and evaluation of the critical stress states in the rock massif with the parallel warning ligts' signalization.
The Nováky Mine:
An endavour of the system approach to the solution of the mining activity consequences led us at present to the following: to chose such a mining conception which would enable its correction, i.e. on the basis of up-to-date information about the overburden movement, to use the corresponding mining startegy.

REFERENCES

Act of the ČSFR No 238/91 - Act on Wastes (Original in Slovak)
Ďurove,J., Maras,M. & Hroncová,E. 1992. Mathematical Model for Determination of Support's Load from Massif Based on Physical Model. *Proc. of the 2nd Czechoslovak Conference on Numerical Methods in Geomechanics, 2nd part*:167-170. Prague (Original in English).
Grenda,O. et al. 1991. Evaluation of the Real State of the Pillars' Liquidation in the Lubeník Deposit. OBV VÚ SMZ - Lubeník (Original in Slovak).
Machata,A. & Čičmancová,M. 1992. Organizational Structure of the Slovak Coal Mines in the Connection with the Transformation of the State Property. *Proc. of the Conference on Topical Problems of the Mining Industry.* Malenovice: VŠB Ostrava (Original in Slovak).
Somoláni,J. 1985. Solution of the Surface Effects of the Mining in the Nováky Deposit. *Proc. of the Conference on Geotechnical Problems of the Coal Mining and the Urbanistic Development of the Upper Nitra Area.* Nitrianske Rudno: ULB Prievidza (Original in Slovak).
Trančík,P.,Hatala,J. et al. 1985. Proposal of the Opening, Development and Extraction for the 2nd Level - South, ULB Baňa Nováky. *Final report of the RVT.* Košice: Technical University (Original in Slovak).

Geomechanics 93, Rakowski (ed.) © 1994 Balkema, Rotterdam, ISBN 90 5410 354 X

The system $H_2O - CO_2$ – NaCl in natural conditions of the Slaný basin

P. Kolář
Nihon University, Tokyo, Japan

P. Martinec
Institute of Geonics of the CAS, Ostrava, Czech Republic

ABSTRACT: Thermodynamic modelling of sudden outbursts of gases and rocks in the Slaný mine shaft is used to demonstrate the necessity to study phase equilibria of pore fluids in rocks at high pressures.

1 INTRODUCTION

Pore fluids of gases in equilibrium with mineralized waters form an integral part of rocks in natural conditions of sedimentary basins. While mechanical and pore characteristics of rocks are routinely studied (Martinec 1989a, Kolář 1991), the contribution of pore fluids on these properties has not been systematically investigated (Konečný 1993).

Preliminary experimental works have been carried out with rocks saturated with simple pore fluids, such as water and inert gases, e.g. helium, nitrogen (Gustkiewicz 1990, Lempp 1993). The results show that pore media can have a profound effect on deformation curves and stability of rocks and rock massif. In addition, documented are also numerous cases of a sudden release of unexpectably large amounts of gases from rocks, often accompanied with a destruction of mine works.

In the experimental simulation of saturated rocks under triaxial stress the role of pore media must not be underestimated. In particular, at temperatures and pressures typical for sedimentary basins the common pore fluid system H_2O - CO_2 - NaCl can undergo complex phase transitions.

The following thermodynamic analysis is used to demonstrate how the phase equilibria information can contribute to geomechnical analysis of anomalous rock behaviour.

2 GEOLOGICAL SITUATION

Two large outbusts of rock and gases occured during mining the Slaný mine shaft (Slaný hard coal basin, ca. 40 km WNW from Prague) in the depth of 870-992 m. During the outbursts ca. 10^4-10^5 m^3 of gas (99.5% CO_2) was released to the mine within tens of seconds.

The mining was conducted in Mirošov conglomerates and sandstones (Westfalian D) with low permeability and specific pore size distribution (Martinec 1989b). Geological documentation showed that the tectonic porosity of the rock massif was by order smaller that the primary porosity of the rock matter. A detailed petrographic analysis revealed that diagenetic cementation of the sediment caused carbon dioxide and variable amounts of water being closed in the primary pore system (Martinec 1989c). In-situ measurements of pore pressures indicated 5-15 MPa, the average temperature of the rock massif being 34°C.

The content of CO_2 in mine air was continuously registered during mining. Mine waters were sampled at several locations, the most dominant component being NaCl with the average concentration of 35 g/l. Petrographic analysis and mercury porosimetry were employed to determine density and primary pore volume of rocks. The average results for four typical rock sections of the mine shaft are summarized in Table 1.

3 THERMODYNAMIC DESCRIPTION OF THE SYSTEM H_2O - CO_2 - NaCl

Experimental phase equilibrium data for the pore fluid system H_2O - CO_2 - NaCl typically cover the areas of interest for chemical engineering (Markham 1941, Wiebe 1941) and the extreme conditions for geochemical

Table 1. Physical properties* of four rock sections of the Slaný mine shaft (875.7-888.3 m). Shaft diameter : 9.9 m.

No.	h [m]	d [t/m^3]	Vp [dm^3/t]	Vg [m^3]	t [°C]
1	3.2	2.44	37.2	6,500	33.5
2	3.1	2.29	61.3	3,400	34
3	3.2	2.28	63.6	6,200	34
4	3.1	2.32	61.5	5,500	34

*) h - section thickness; d - bulk density; Vp - pore volume; Vg - CO_2 production; t - temperature.

studies of metamorphic processes (Gehrig 1986).

A lack of direct experimental data in a large range of temperatures, pressures and concentrations can be substituted by reliable extrapolations. However, until recently an unified model capable to accurately describe the complex system containing both sub- and supercritical components in equilibrium with electrolyte solutions has not been available.

The following mixed model was employed for the description of the system up to 10 MPa and 20-40°C :

- The eight-parameter Benedict-Webb-Rubin equation of state (Novák 1972) was used for the calculation of state properties and the fugacity of pure carbon dioxide. The equation of state should be reliable up to twofold of the critical density of pure CO_2.
- The solubility of CO_2 in pure water was calculated by the approximative equation

$$\ln \frac{f^g_{CO_2}}{x^l_{CO_2}} = \ln H(T) + \frac{P\, v^l_{CO_2}}{RT}$$

where

x^l - mole fraction of CO_2 in the liquid phase,

f^g - fugacity of CO_2 in the gas phase,

R - gas constant (8.314 J/mol K),

P - pressure,

T - absolute temperature (Kelvin).

The partial molar volume of CO_2 in the solution v^l = 37.6 m^3/mol. This value is assumed to be independent on temperature and pressure (Malinin 1979).

The temperature dependence of the Henry coefficient H is given by

$$\ln H(T) = A + \frac{B}{T} + C \ln T + D T$$

The parameters of the function valid in the temperature range of 273 K > T > 353 K can be found in (Wilhelm 1977).

- The effect of NaCl on the solubility of CO_2 in water can be correlated by the empirical expression (Malinin 1979)

$$S_s = S_0 \, . \, 10^{-k_s m}$$

where

S_s, S_0 - solubility of CO_2 in mineralized and pure water, respectively;

m - molar concentration of NaCl,

k_s - salting-out coefficient.

The effect of pressure on k_s is neglected in the first approximation (Krajča 1977).

4 RESULTS AND DISCUSSION

The following two limiting cases were considered in the simulation of the CO_2 release in the Slaný mine :

a) Pore structure of the rock sections is fully saturated with the aqueous solution of NaCl (c_{NaCl} = 35 g/l). Carbon dioxide is in equilibrium with the solution at the temperature and pressure of the system.

b) Pore structure of the rock sections is filled with compressed carbon dioxide.

The results of the simulation expressed as the N.T.P. volume of CO_2 evolved from the individual rock sections are shown in Figures 1 and 2. It can be seen that for all pressure levels the amount of CO_2 dissolved in pore waters (Case a) is at least by order smaller than the corresponding amount of pure CO_2 (Case b). It should be noted that the amounts of CO_2 evolved for the Case b are in good quantitative agreement with the actual CO_2 productions registered during mining of the rock sections (cf. Table 1).

The reason for the sudden increase of CO_2 amounts at about 7-8 MPa is the occurence of the critical point of carbon dioxide (31.04°C, 7.384 MPa). In the vicinity of the critical point the compressibility of the fluid considerably increases.

Based on the simulation results, the mechanism of the sudden gas release in the Slaný mine can be formulated as follows:

The rock sections liable to sudden outbursts of rocks and gases are those with prevailing occurence of compressed free CO_2 in the pore system, i.e. those not completely saturated with mineralized water. At a sudden change of the equilibrium state by mining activity, ca. 10,000 - 30,000 m$_3$ of gas rapidly evolves from disintegrating rock matter in the immediate vicinity of the mine work. In a few cycles the degassing and disintegrating process propagates deeper into the rock massif until a new equilibrium is established.

A comparison of Figure 2 and Table 1 shows that the amounts of CO_2 registered in mining of individual rock sections correspond to pore pressures of 8-10 MPa. (Martinec 1991), studying the effect of water on mechanical properties of rocks from the Slaný basin, has found that pore pressures over 8 MPa can be critical for mechanical stability of rocks with water content over 2.5%.

5 CONCLUSION

An example of the thermodynamic analysis of a mining problem has been used to demonstrate the necessity to study phase equilibria of pore fluids at high pressures. Further applications can include exploiting fluid media from rocks and depositing waste industrial liquids and gases into deep wells.

Because of the complexity of most natural pore media, more advanced models should be further employed in the thermodynamic simulation of the systems. Simple and reliable methods applicable to extrapolations into regions beyond available experimental data are desirable.

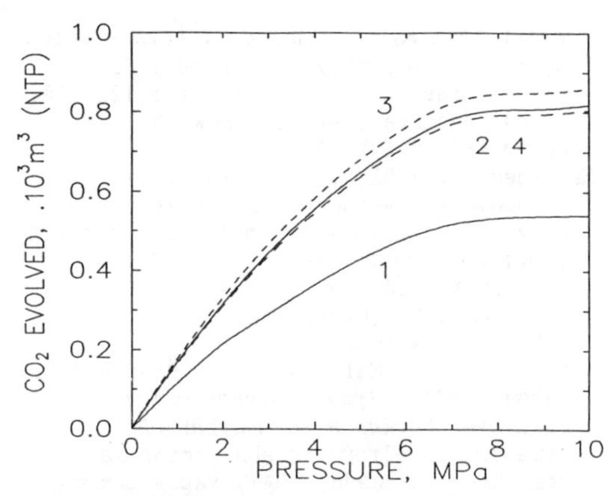

Fig.1 Gas amount evolved from individual rock sections. Case a - pore system saturated with mineralized water.

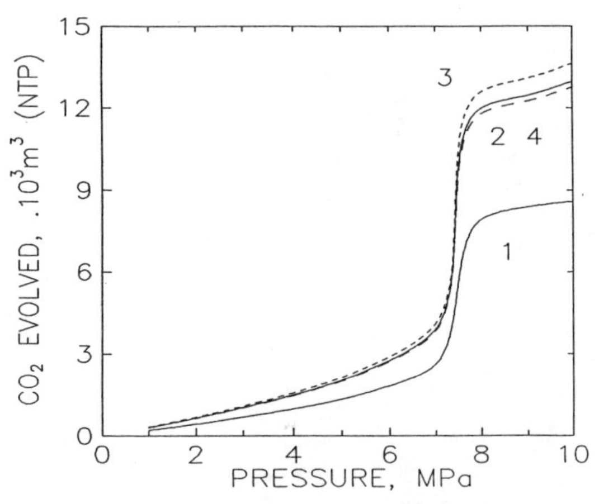

Fig.2 Gas amount evolved from individual rock sections. Case b - pore system filled with compressed CO_2.

REFERENCES

Gehrig, M., H. Lentz and E.U. Franck 1986. The system water - carbon dioxide - sodium chloride to 773 K and 300 MPa. *Ber.Bunsenges.Phys.Chem.* 90: 525-533.

Gustkiewicz, J. 1990. Deformacija i wytrzymaloczsz skal w trojosowym stanie naprzezenia z uwzglednieniem plynow porowych. *Strata as Multiphase Media* (Ed. by Litwiniczin J.). Institut Mechaniki Gorotworu PAN. Wydawnictwo AGH: Kraków.

Kolář, P., P. Martinec and P. Benš 1991. Methodological instructions for mercury porosimetry of clastic sedimentary rocks. *Ceramics* 35: 379-390.

Konečný P., P. Martinec 1993. Effect of geological features of rocks and rock massif on different stress types. Grant No. 105/93/2409. Ostrava: Institute of Geonics.

Krajča, J. 1977. *Plyny v podzemních vodách (jejich vlastnosti, průzkum a využití).* Praha: SNTL/Alfa.

Lempp, Ch. 1993. Influence of fluids on mechanical rock properties in low-porosity crystalline rocks. *Annales Geophysicae I*: Supplement I to Vol.11. Springer Int.Europ.Geophys.Soc.

Malinin, S.D. 1979. *Physical chemistry of hydrothermal systems with carbon dioxide.* Moskva: Nauka (in Russian)

Markham, A.E., K.A. Kobe 1941. The solubility of carbon dioxide and nitrous oxide in aqueous salt solutions. *J.Am.Chem.Soc.* 63: 449-454.

Martinec, P., J. Krajíček 1989a. *Properties of rocks from the Slaný hard coal basin.* Prof.Paper No.49. Ostrava-Radvanice: Coal Research Institute.

Martinec, P., P. Kolář 1989b. Petrologie a litologie obzoru mirošovských slepenců v jamách Dolu Slaný. Final Report of the Government Project P-01-125-808. Ostrava-Radvanice: Coal Research Institute. (in Czech)

Martinec, P., P. Kolář 1989c. Pórové charakteristiky hornin mirošovského obzoru a radnických vrstev ve slánské pánvi. *Fyzikálne vlastnosti hornín a ich využitie v geofyzike a geológii III.* (Ed. by Kapička, A., I. Túnyi): 125-128. Praha: Jednota československých matematiků a fyziků.

Martinec, P. 1991. Vliv vlhkosti na mechanické vlastnosti hornin slánské pánve. *Fyzikálne vlastnosti hornín a ich využitie v geofyzike a geológii IV.* (Ed. by Krišťáková, Z.): 46-51. Praha: Jednota československých matematiků a fyziků.

Novák, J.P., A. Malijevský, J. Šobr and J. Matouš 1972. *Plyny a plynné směsi. (Stavové chování a termodynamické vlastnosti plynů).* Praha: Academia.

Wiebe, R., V.L. Gaddy 1941. Vapor phase composition of carbon dioxide-water mixtures at various temperatures and at pressures to 700 atmospheres. *J.Am.Chem. Soc.* 63: 745-747.

Geomechanics 93, Rakowski (ed.) © 1994 Balkema, Rotterdam, ISBN 90 5410 354 X

Physical and mechanical properties of sediments of coal-rock series

A. Kožušníková
Institute of Geonics of the CAS, Ostrava, Czech Republic

ABSTRACT: The main objective of this work was the study of rocks containing different proportions of inorganic and organic matter in the series of coal – ashy coal – carbonaceous rock – rock with coal admixture – rock without organic admixture. These sediments have been characterized by their physical and mechanical properties and their petrology as well.

1 INTRODUCTION

In sedimentography and coal petrography, the composition and the properties of coal and rocks have been studied thoroughly, while little attention has been paid up to now to the so called transition rocks of the series of coal – ashy coal – carbonaceous rock – rock with coal admixtures – rock with impregnated organic matter – rock without organic admixture, especially to their physical and mechanical properties. Nevertheless, coal matter gives origin to important inhomogeneities in the rocks which affect the physical and mechanical properties of such rocks and their behaviour in the process of their disturbance and deformation as well.

In the work (Kožušníková 1992), a set of 110 samples including coal, transition rocks a rocks with coal admixture was studied. These samples were taken from all stratigraphic units of the Upper Carboniferous in different parts of the Czech part of the Upper Silesian Coal Basin. The study included macropetrographic description, determination of facial type of the given sediment and of its basic physical and mechanical properties (volume weight, velocity of propagation of ultrasonic longitudinal waves, uniaxial compressive strength, tensile strength determined by Brazilian test, Young modulus), micropetrographic analysis and chemical analysis to determinate the content of the organic matter in the rock. The results were elaborated using statistical methods.

2 RESULTS AND DISCUSSION

Basic statistical evaluation of the studied sets of rocks of varying proportions of organic matter and of different facial types yielded the following results:

1. The decisive factor for physical and mechanical properties of rocks is the ratio between the organic and the inorganic matter.

2. Volume weight increases starting from coal, going to the transition types of sediments and ending with sandstones with coal admixture. Maximum values of volume weight were obtained in sediments of the facies of fluvial channel, of the facies of alluvial plain and in rocks denoted as root horizon.

3. Uniaxial compressive strength increases with the increase of the content of inorganic matter in rock. The highest value were found in sandstones with coal admixture and the lowest ones in coal. The lowest strength was obtained in sediments belonging to the facies of swamp (i.e. in coal), the highest strength was found in sediments of the facies of fluvial channel, root horizon and alluvial plain.

4. Variations in the other studied physical and mechanical properties have a similar character. The velocity of propagation of ultrasonic longitudinal waves, the Young modulus and the tensile strength show a rise, the lowest being the values for coal and the highest the values for sandstones. Also variations in physical and mechanical prop-

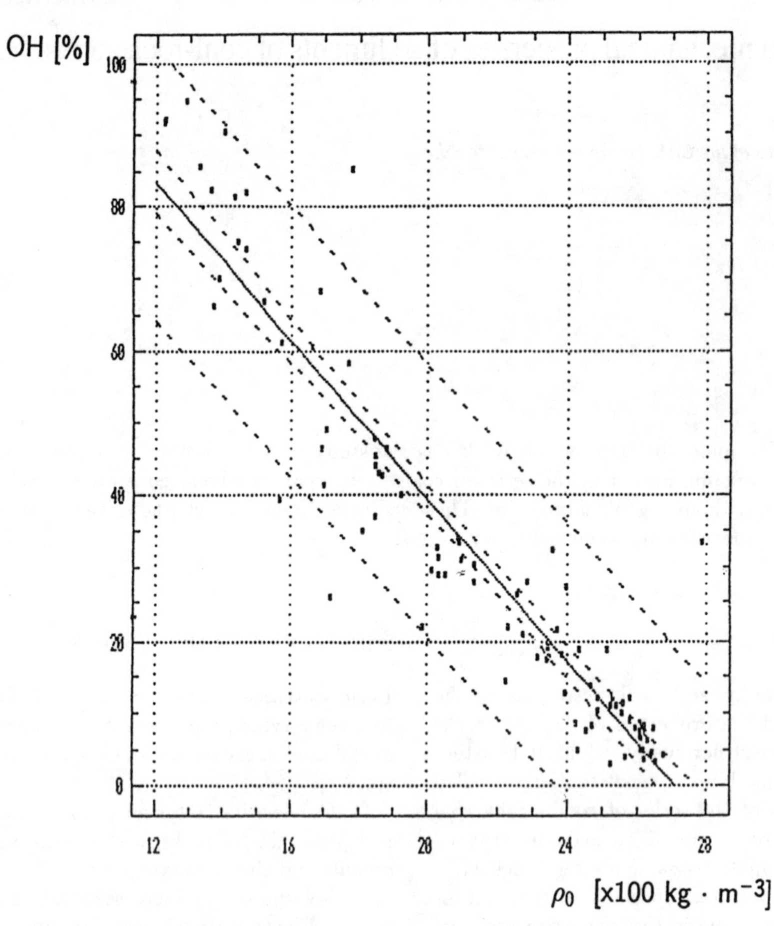

Fig. 1 – Dependence between the content of organic matter "OH" and volume weight ρ_0 (linear regression)

Table 1 – Classification of sediments on the basis of volume weight

CLASS	Content of organic matter [%]	Volume weight [kg · m^{-3}]
Coal and coal with inorganic matter	> 70	below 1450
Ashy coal	70 – 50	1450 – 1800
Carbonaceous rocks	50 – 20	1800 – 2350
Rocks with coal admixture to rocks without coal admixture	< 20	above 2350

erties of sediments belonging to different facies are similar to those found in the uniaxial compressive strength.

From the results of correlation and regression analyses follows that:

1. Linear dependence between the volume weight and the content of organic matter — Fig. 1 is very close (correlation coefficient = −0.93). With the increase of the content of organic matter, a linear drop of volume weight of rock occurs.

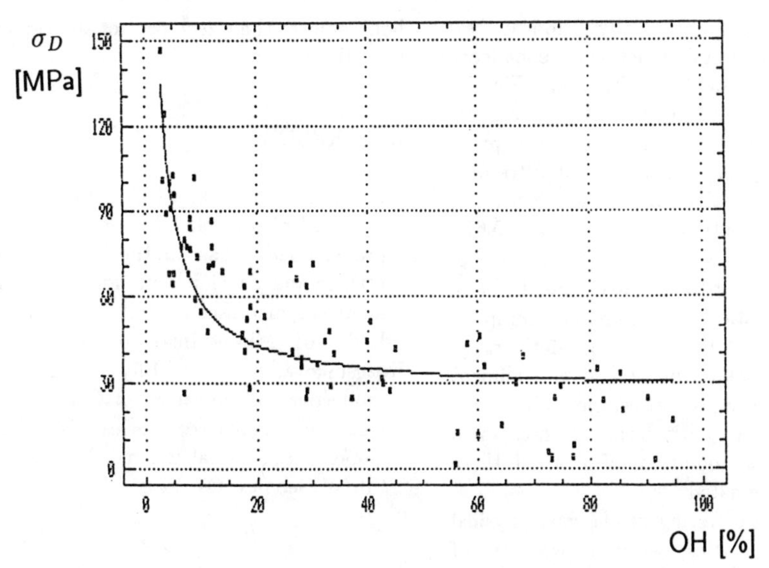

Fig. 2 – Dependence between the content of organic matter "OH" and the uniaxial compressive strength σ_D (hyperbolic function)

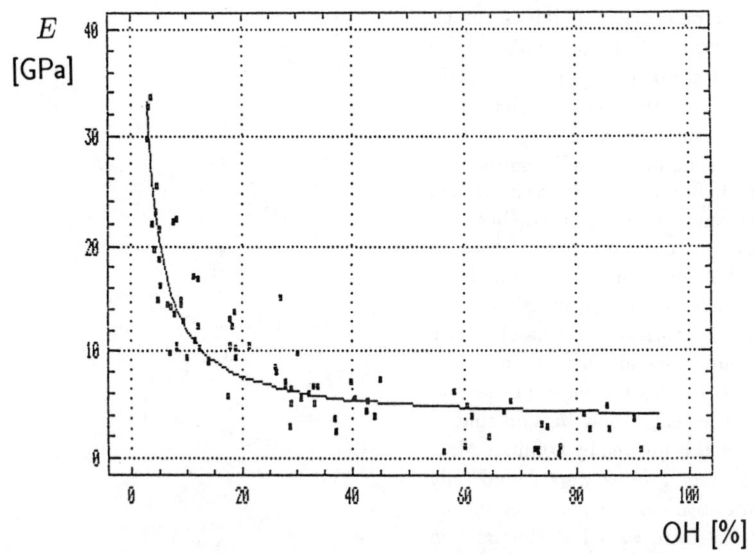

Fig. 3 – Dependence between the content of organic matter "OH" and Young modulus E (hyperbolic function)

Using Stepwise Variable Selection, no other type of regression function which would correspond better to the distribution of the data, was found. Thus, if it is necessary to make an estimate of the content of coal matter in rock, volume weight may be used as the decisive parameter. Tab. 1 shows the ranges of volume weights corresponding to the individual classes set out in dependence on the content of organic matter found by regression dependence between the content of organic matter and volume weight.

2. The relations between the content of organic matter and uniaxial compressive strength σ_D and between the content of organic matter and

Young modulus E are statistically significant, especially for the hyperbolic functional dependence ($r = -0.81$ and $r = -0.91$) — Fig. 2 and Fig. 3.

3. The relations between the content of organic matter and other physical and mechanical properties (the velocity of propagation of ultrasonic waves v_p, the tensile strength σ_{PT}) are of a lesser statistical significance ($r = -0.77$ and $r = -0.67$). Using the cluster analysis to find a relation between the content of organic matter and the basic strength characteristic (i.e. the uniaxial compressive strength σ_D and the tensile strength σ_{PT}) and between the facial type of a given sediment and the basic strength properties, the sediments of root horizon differed greatly from the rest, which is due to the special conditions of in which these sediments were originated.

In addition to the testing of classical physical and mechanical properties, also shear properties of some types of discontinuity planes of genetic origin found in the samples of the studied set, were measured. It follows from a comparison of shear properties of the studied genetic discontinuities and of discontinuities of tectonic origin (Fialová, et al. 1990) that genetic types of discontinuities have comparable values of shear strength at a certain normal stress, but a lower friction coefficient.

On the basis of micropetrographic analyses, typical microstructures found in the studies samples of rocks were described. In microscopic examination, not only microstructures were observed, but also contacts and transitions between coal matter and clastic sediments. On the contact between them, some peculiar phenomena were observed, such as accumulation of pyrite, corroded grains of quartz and quartz rims around coal laminas. As far as the micropetrography of coal matter found in the analyzed samples is concerned, it may be said that in the majority of cases, it is formed by kolinite. Macerals belonging to other groups were found only sporadically, an interesting occurrence was the association of inertodetrinite with liptodetrinite in the clayey matrix. In some cases, layers of coal, the maceral composition of which differed greatly, were found on the area of a single polished section.

3 CONCLUSIONS

Even through the validity of the results obtained is limited by the number of samples, the results give a summary of the characteristics and the properties of the above mentioned rocks found in the Carboniferous of the Czech part of the Upper Silesian Coal Basin.

REFERENCES

Fialová, V., Kožušníková, A., Poláček, J., Rusek, J. & A. Kolcun 1990. Strength and deformation properties of rocks from the point of view of their texture and development, identification of rock mass as a macroscopic unit. *Final Report* SPZV II-6-1/01. Mining Institute, Ostrava.

Kožušníková, A., 1992. Physical and mechanical properties and petrology of Carboniferous sediments of the coal-rock series (Czech part of the Upper Silesian Coal Basin). *PhD Thesis*. Institute of Geonics, Ostrava.

Geomechanics 93, Rakowski (ed.) © 1994 Balkema, Rotterdam, ISBN 90 5410 354 X

Arch support enhancement of close-to-wall gangways by roof bolting method

T. Majcherczyk & A. Tajduś
Academy of Mining and Metallurgy, Kraków, Poland

ABSTRACT: The main reason for arch section strengthening by rock bolting is the stability improvement of a face/gate road intersection and smaller gate road closure behind the face.
The types of gate road support strengthening depending on roof properties are described. Analytical and numerical methods of face/gate road are presented.
Site observations from mines are also reported.

1 INTRODUCTION

In order to provide stability of roof and walls of corridor openings there is commonly used in Polish coal-mining industry a yielding arch support. It applies also to close-to-wall workings used to remove the gotten, or for ventilation or transportation purposes. Working on a oncoming wall in a given exploitation field causes necessity to maintain one of the discussed close-to-wall workings functioning.

The biggest difficulties with the maintenance of a corridor close-to-wall working made in an arc casing occur close to a crossing with a wall. For the technological purposes /caused by working coal combine or plane, the moving of a wall conveyor/ it is essential to remove from two to five curving arches.
In order to improve the stability of a corridor working in an area of the wall being exploited as well as after finishing it, in Polish mines a new method of securing close-to-wall gangways has been introduced. The method is based on enhancing yielding arches by using rigid steel resin bolts glued in along all the length.

Utilizing the gained experience, in the article the choice conditions of the enhancing anchor casing have been quoted depending on mining-geological conditions occurring in the area of the discussed crossing of the gangway with the wall.

2 THE EVALUATION OF MINING-GEOLOGICAL CONDITIONS IN THE AREA OF THE DISCUSSED GANGWAY OPENING

The analysis of mining-geological conditions occurring in the environment of the the discussed gangway working is made basing on a site inspection in currently undertaken mining activities and documentation materials such as mining working maps, geological cross-sections as well as roof rocks profiles.

In order to define strength and deformation properties of the rocks existing in the roof a method of drilling 6-meter long core is used. The cores are drilled in some places of the corridor opening which is to be bolted. The number of drilled cores depends on the kind of roof, variation of its construction etc. The tensile and compressive strength are defined by penetration research in the drilled cores. From the acquired cores samples are made and in laboratory conditions strength and

deformation parameters are defined.
Moreover for the evaluation of the structural construction of the rock mass occurring in the roof up to the height of 6.0 m, a measurement of the Rock Quality Designation /RQD/ is done.

3 DETERMINATION OF DESIGN PARAMETERS FOR ROOF BOLTING IN CLOSE-TO-WALL OPENINGS

The main purpose of the undertaken calculations is the choice of parameters for roof bolting in such a way that the bolt with the steel arches /ŁP type/ would provide the stability of the rock mass in the surrounding of crossing of the gangway with the wall /and after passing the wall/ after removing traditional lifts and posts fitted often outside the wall.
It is essential that using resin bolts allowed the secure removal of a few oncoming steel arches. Additionally it is suggested to eliminate very troublesome and expensive cribs placed between the cavings and the gangway opening and the gangway opening and to replace them with two row of organs. For example, a typical so far building-up of a crossing of a wall with a gangway opening is shown on Fig.1., and a crossing enhanced by roof-bolting - Fig.2.
From the introduced assumptions arises the concept of utilizing enhancement of of an arc casing with an resin bolts.
In order to attach a roof arch

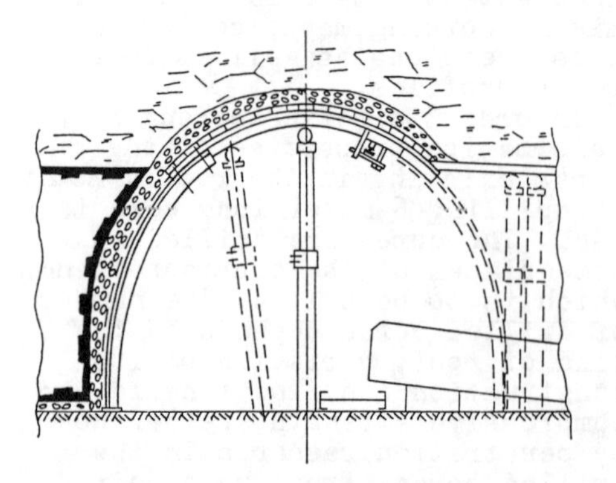

Fig.1 Wall and gangway crossing a traditional enhancement of support

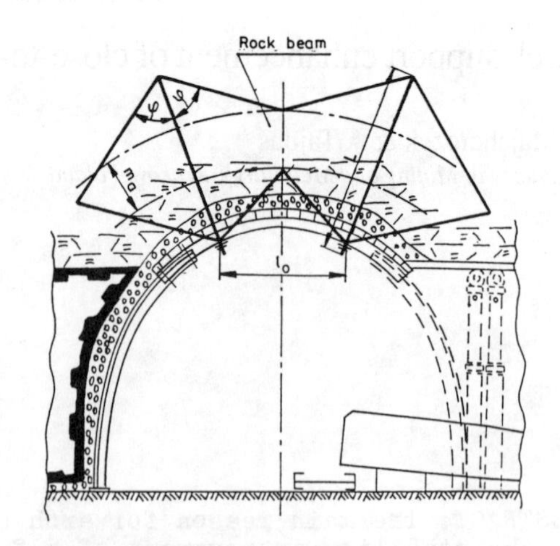

Fig.2 Wall and gangway enhanced by an roof-bolting

it is suggested to case the following:
a/ In the cross-section orthogonal to the main axis /running along the length of opening/in sets of two anchors distanced l_o from each other /Fig.2/, and four bolts to one roof arch departed from the vertical plain running along the main axis of an opening by an angle.
b/ In the cross-section along the main axis of the opening the spacing between the pairs of bolts from each other should be $S = S_b$

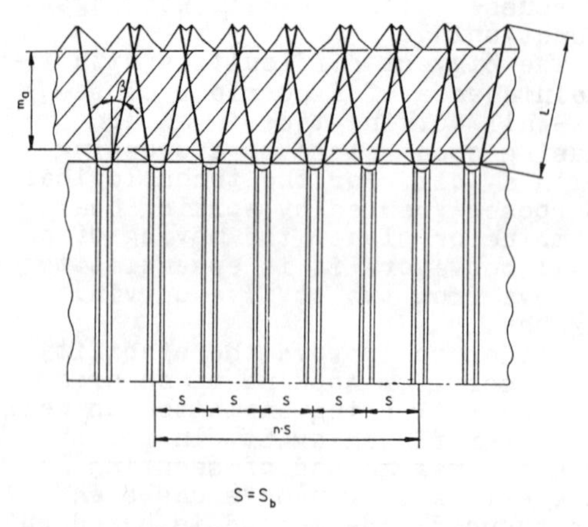

Fig.3 A cross-section along the main axis of a close-to-wall opening by roof-bolting

/where: S_b - bolt spacing/ accordingly with the spacing of the steel arches "S" /Fig.3./, but the departure of the bolts from the plain orthogonal to the main axis of an opening is accepted as .

Such a bolt mesh in the cross-section along the main axis of an opening should allow a creation of a flat rock beam.

An enhancement by roof - bolting should take place in front of coal face in a distance not exceeding 60 m /bolts are installed behind the zone of influence of the exploited wall/.

In order to design an enhancing roof bolts of a close-to-wall working, a following set of data is used:

a/ A shape and dimensions of a close-to-wall working, seam thickness, ways of eliminating exploited volume, etc..

b/ Parameters of geomechanical properties of rock layers remaining around a corridor opening and exploited seam.

c/ Detailed description of joint system around the working, discontinuity characteristics and their filling, orientation of joint system relating to the direction of running of opening and the dyrection of main components of the horizontal stress.

d/ Way of boring of a corridor opening.

e/ Ground water conditions of an opening.

f/ Predicted maintenance time.

Although those considerations apply only to roof bolts enhancing steel arches, the calculations for the roof-bolting are done as if there existed roof bolts. It is caused by the fact that at the crossing gangway-wall after detaching a few steel arches the stability of the roof is virtually assured by sufficiently assorted roof - bolting. One should also take under consideration that roof bolts are most of all installed when over the existing steel arches a broken zone has been created. Therefore the length of the bolts should reach outside the broken zone. In sake of that it is better to enhance the roof bolts just after the face of the bored gangway.

There exist a number of methods of determination of design parameters for roof-bolting. Its multitude expresses the difficulty of describing the phenomena of interaction between rock mass and bolts.

In the calculations done by the authors of the article for the Polish mines three different methods have been utilized:

a/ analysis methods:
- according to Z.T.Bieniawski
- supported rock beam,

b/ numerical method using finite elements or boundary elements.

3.1 Analysis methods

Bieniawski Method

In the Bieniawski method the asortment of the length of the bolt and its spacing depend on the value of RMR.

The value of RMR for the rock mass surrounding considered mining openings is defined according to widely known principles given by Z.T.Bieniawski [1], [2].

For each of the considered close-to-wall gangways a mean value of RMR coefficient is calculated using defined earlier data coming from specially drilled cores and geological profiles. If the value of the RMR coefficient is known on the considered region, in a mesh enough network, then a map is drawn showing the local layout of RMR coefficient in the seam roof.

Next, depending on the kind of the bolts to be used /mechanical or resin/ determination of the parameters of the bolts is done. The calculation procedures for both types of bolts have been described elsewhere [1], [2], [4].

Rock Support Beam Method

In the roof of an unsupported gangway working often occur tensile stresses causing the creation of the fracture zone in the roof rocks. As a result of bolting in the surrounding of each bolt a limited by a curved line area of compressive stresses is created /replacing the previously operating tensile stresses/. In a general case the curved-lined operating areas of bolts are ellipse-shaped which one axis is the length of the bolt. Simplifying, ellipse - action of the bolt can be replaced with rhombus. Different values of the acute angles for those rhombi are accepted. Generally in the western literature it is accepted that those angles are of 45° /Bieniawski, Hoek Brown/. Different value is given by Russian authors. According to Shirockov 8 the angle can be calculated from the formula:

$$\phi = \mathrm{arctg}\sqrt{2\cdot\xi} \qquad /1/$$

where:

ϕ - acute angle of the rhombus

ξ - side slit coefficient, $= \nu/(1-\nu)$

ν - Poisson coefficient

The strength of the support beam created as a result of bolting is analyzed in two orthogonal cross-sections of the opening: in the cross-section orthogonal to the main opening axis /Fig.2/, and in the vertical cross-section running along the opening /Fig.3/. The most dangerous place is the middle part of the roof of the opening in the vertical cross-section along its main axis, because with a certain value of tensile stress it is possible that the joints and discontinuities can grow and make progress. The maximum value of a vertical load "p_o", which could be carried by the rock, beam enhanced by bolts, can be calculated with formulas described with details in elaboration [8].
The load of rock beam "p_s" are the fractured rocks above.
The rock beam is going to carry the load of the fractured rocks if:

$$p_o = p_s \cdot n \qquad /2/$$

where:

n - safety coefficient

Basing on the calculations done according to Bieniawski method entry parameters of the roof bolting are defined and then it is checked if the beam created as a result of the bolting of the roof is going to provide the stability of the roof around crossing. The calculations are done for all the enhanced gangway openings.

3.2 Numerical calculations of the stability of the close-to-wall openings enhanced by bolts

Defining the state of stress in the surrounding of a crossing of wall opening with the gangway opening is a three-dimension problem and due to the easiness of building of a model, the boundary element method is utilized.
It is sometimes enough to treat the problem as a planar problem and then the state of stress, displacement and effort is determined in flat disc cut out from the most dangerous cross-sections as far as gangway stability is concerned.
From the up-to-date experience one can learn that a particularly endangered cross-section is the cross-section running orthogonally to the main gangway opening axis through coal seam and caving created after passing the wall, Fig.2. In this cross-section, for each of the analyzed gangway opening in which bolt enhancement is to be used, at least one numeric model is assorted which shows truly enough mining-geological conditions occurring in the closest surrounding of the opening.

In order to credibly calculate numerically it is essential to to know mechanical properties of the rock mass. The strength properties of rock layers are evaluated with a certain approximation by penetrating research.

From the laboratory research the deformation and strength properties of rock samples are defined in the surrounding of the considered mining openings.

It is very difficult to find interdependence between the properties of the rock mass and a rock sample, examined in laboratory conditions and it was subject to research for many scientists. Among others, Hoek and Brown [6] have proposed empirical criterion of failure allowing to determine strength and deformation parameters of rock mass knowing the strength parameters of rock samples and the RMR coefficient.

In order to illustrate the introducent above way of dealing with the problem, calculation undertaken for bolted gangway have been shown below in one of the Polish mines.

The considered close-to-wall gangway opening is made in steel arches support. It remains on a depth of 675 m. Coal seam is 1.3 m thick and directly in its roof there is a layer of shale about 4.4 m thick. Above there is fine-grained sandstone 28.8 m thick. The right hand seam has been exploited in caving way. So a caving and broken zone has been created reaching about 10 m /m- coal seam thickness/. It has also been accepted, that roof layers break over the caving at an angle of $\phi = 45^{o} + \rho/2$ /ρ - angle of internal friction of the roof rocks/. Floor rocks consist of alternating layers of shale, sand slate and sandstone. In order to define the geomechanical properties of the rocks remaining in the surroundings of the opening rock samples have been taken coming from core drillings in roof, floor and walls and also penetrating research has been done. There was also a detailed geological examination.
Basing on the analysis of the acquired data a mean value of RMR coefficient /RMR = 53.6/ along the considered gangway and the geomechanical properties of rock samples /Tabl. 1/ have been determined. Using the dependencies given by Hoek and Brown 6 geomechanical parameters of the rock mass have been defined. It has been additionally accepted that the bolts are 2.3 m long and they carry load of 0.12 MN.

Basing on the geological examination and the defined geomechanical rock mass properties a numeric model has been bulit /Fig.4/ and using the Finite Element Method the calculation has been done. A state of stress, displacement and effort according to Burzyński criterion of failure has been counted.

The outcome acquired have undergone a detailed analysis.
Due to their volume, only two main stress maps have been presented here and the cross-sections along the section in which it is suggested to implant bolts. The maps include the closest surroundings of the gangway opening and it is a segment including a square of a disc of the following dimensions: width 11 m, height 11 m including the gangway opening.
On figures 5 and 6 main stress maps have been shown sufficiently $\sigma 1$ and $\sigma 2$. From the figures it can be noticed that major compressive stress concentration has been created in the area of the left wall of the working. The maximum values of compressive stress $\sigma 1 = -28$ MPa and $\sigma 2 = -95$ MPa.
Occurring directly in the roof main stresses are of favorable values as far as opening stability is concerned. Above the roof of opening distanced about 5 m away there occur tensile stress which do not influence the bolted roof directly over the gangway opening.
Slight tensile stress occur also at the floor of the gangway what can lead to creating of a broken zone and a limited floor heavy.

Fig.7 shows the layout of rock mass effort along the designed bolts. According to the Burzyński criterion of failure they should be found mainly in the unbroken part of the rock mass.

The layout of effort along the bolts is as follows:
- Along the right bolt K2, for 0 m the effort reaches value 100%, then falls non-linearly to the value of 34% 1.2 m away and then rises up to the value of 92%,
- Along the left hand bolt K2, from 0 m to 2.0 m the rock mass is broken, after exceeding the distance of 2.0 m the effort diminishes to the value of 70% 2.5 m away.

From the considerations above arises that the bolts enhancing

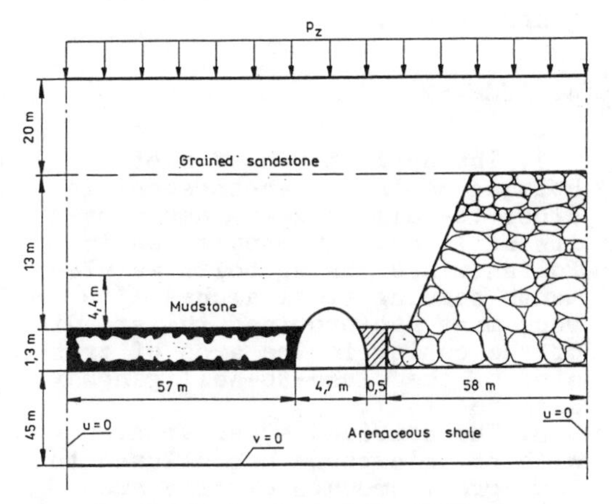

Fig.4 Numerical model of rock mass

Tabl.1

Rock	Young Modulus	Poisson Coefficient	Compressive Strength R_c MPa		Tensile Strength R_r MPa		Cohesion	Angle of int. fric.
	E MPa	ν -	Sample	Rock mass	Sample	Rock mass	k MPa	deg
Coal	3500	0.25	-	3.8	-	0.2	1.0	57
Sandy Shale	13000	0.22	83.7	11.97	9.46	0.5	1.66	59
Sandstone	15400	0.35	93.9	26.15	7.93	1.1	3.65	58.8
Sand Shale	14034	0.3	88.8	19.06	8.7	0.8	2.66	59
Organs	2250	0.25	-	3.0	-	-	-	-

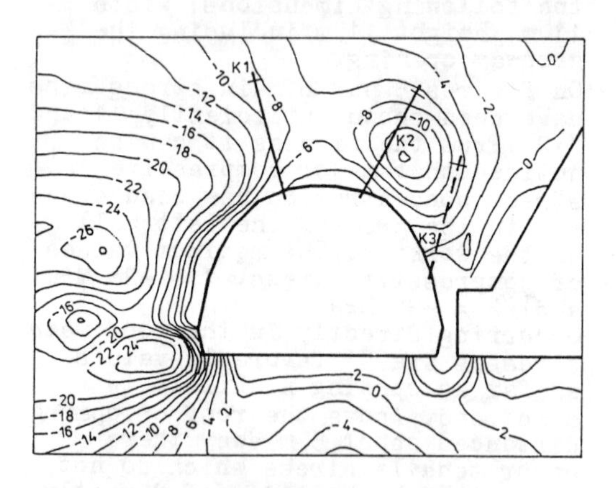

Fig.5 Main stresses distribution σ_1 - around a gangway opening

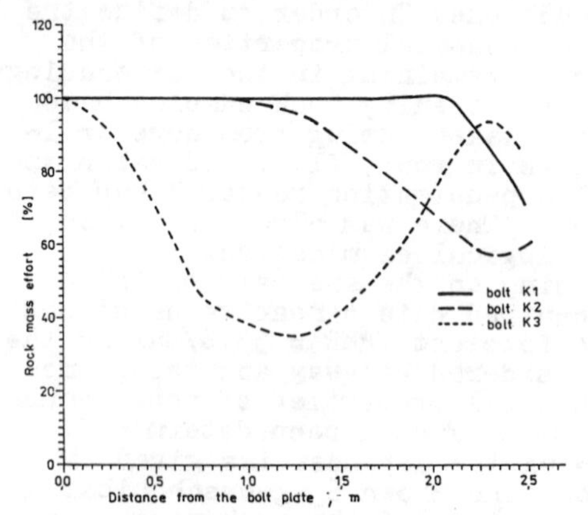

Fig.7 Rock mass effort along designed bolts

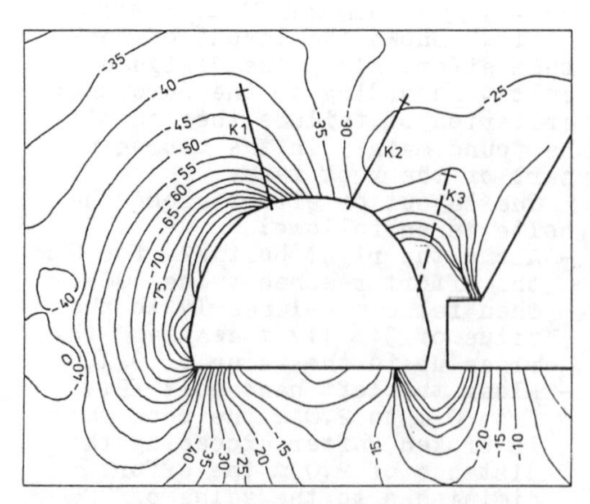

Fig.6 Main stresses distribution σ_2 - around a gangway opening

the rock mass secure its stability in the considered area of crossing gangway-wall.

4. SUMMARY

1. The application of roof - bolting casing as enhancement in close-to-wall gangways has proven its efficiency in many cases in Polish mines. Using bolts as elements bonding steel arches with rock mass has provided the stability of the casing in the area of crossing of the close-to-wall gangway with the wall.

2. The combined steel arches with bolts support has allowed to save great amounts of time and money. It has made possible to

eliminate posts used so far what has considerably improved the working conditions and the capacity of the crossing.

3. At present in a more than a dozen of Polish mines tested the steel arches combined with resin bolts support. Everywhere the tries have been successful. It gives new perspectives of applications of bolting.

5. REFERENCES

Bieniawski Z.T.1987. Strata Control in Mineral Engineering. A.A.Balkena, Rotterdam, Boston,

Bieniawski Z.T.1984. Rock Mechanics Design in Mining and Tunneling. A.A.Balkena, Rotterdam,

Daws G.,Hons B. Roof Bolting in Coal Mine Industry - Design and Performance. Wiadomości Górnicze 1/1992, pages 27 - 30, in Polish,

Daws G. Coal Mine Roof Bolting. The Mining Engineer 10/1987, pages 147 - 154,

Hoek E. Strength of Joined Rock Masses. Geotechnique 3/1983, pages 187 - 233,

Hoek E. Estimating Mohr - Coulumb Friction and Cohesion Values from the Hoek - Brown Criterion. Int.Rock Mech.Min.Sci. Geomech. Abstr., 3/1993 pages 227 - 229,

Kidybiński A. Stability of Coal - mine gate roads - A dynamic approach. Proc. of the Int.Conf. Geom.,Hradec, Checks. 24 - 26 Sept. 1991, page 9 - 16,

Shirockov A.P.,Pisliakow B.G.1988 Rasciot i wybor kriepi soprajażjennych gornych wyrabotok. Moscow, Nedra.

friable rocks used so far such has considerably improved the working conditions and the capacity of the crossing.

3. At present in more than a [illegible] of coal mines tests the steel arches combined with resin bolt support [illegible] have been successful. It gives new perspectives of applications of bolting.

REFERENCES

Franklin J.A. (1989) Strata Control in Mineral Engineering. A.A. Balkema, Rotterdam, Boston.
Bieniawski Z.T. (1984) Rock Mechanics Design in Mining and Tunnelling. A.A. Balkema, Rotterdam.
Daws G. [illegible] Roof Bolting in Coal Mine Industry - Design and Performance, [illegible] (1989), pages 67 - 70. [illegible]
Dees [illegible] Coal Mine Roof Bolting. The Mining Engineer [illegible] pages 147 - 152.
Hoek E. Strength of Jointed Rock Masses. Geotechnique 3/1983, pages [illegible]
Hoek E. Estimating Mohr - Coulomb Friction and Cohesion Values from the Hoek - Brown Failure Criterion. Int. J. Rock Mech. Min. Sci. Geomech. Abstr. (1990) pages 227 - 229.
Kidybinski A. Stability of Coal [illegible] approach. Proc. of the Int. Conf. [illegible] Sept. 1991, page [illegible]
[illegible]

Geomechanics 93, Rakowski (ed.) © 1994 Balkema, Rotterdam, ISBN 90 5410 354 X

Geotechnical classification of carboniferous rock mass with alteration of the 'variegated beds' type

P. Martinec
Institute of Geonics of the CAS, Ostrava, Czech Republic

ABSTRACT: The present paper proposes a geotechnical classification of the rock mass with alteration of the "variegated beds" type which is based on the verified geological, lithological, petrological and physical properties of the rocks and coal found in the above beds. The proposed classification can be used for the description of those parts of the rock mass where information from exploratory wells, boreholes drilled in the mine other workings is not available.

1 INTRODUCTION

In the Upper Carboniferous of the Upper Silesian Coal Basin, so called variegated or red beds denoted as such by (Dopita, Králík, 1973, Dopita, 1988) are found frequently. A deep oxidation of coal seams associated with the reduction of seam thickness up to their thinning out are characteristic features of these beds. They represent root zones in an originally thick oxidired weathering zone of Jurassic-Cretaceous age (Martinec et al., 1992, Krs et al., 1993) which was denuded to the level of the present paleorelief. The alterations found in the rocks and coal seams affect adversely the exploitation of coal, and especially the safety of the mining and the amount of the resources available for the mining.

From a geomechanical point of view, the properties of the altered rocks and coal are also affected which always manifests itself as a lack of stability of the rock mass. Those parts of the rock mass, which could not be mined out - as unstable act as pillars and represent a danger for the coal mining (concentration of stress, induced seismicity in zones prone to rock bursts and others).

2 VARIEGATED BEDS

As shown by Dopita (1988), the presence of variegated beds was found in all stratigraphic units of the Karviná formation (Namurian B, Westphalian A, B, C), the most affected by this type of alteration being the "saddle" seams and the Sušské seams.

In the southern part of the Upper Silesian Coal Basin, variegated beds are associated with the paleorelief of the Carboniferous. They have the greatest area at those sites where the Carboniferous strata are deposited subhorizontally near the surface of the overthrust mountains (Fig. 2A, C). With increasing depth, the extent of these beds diminishes, their boundaries are not sharp enough and are modelled in detail by the geological structure of the rock mass at the given locality.

Variegated beds reach, most frequently, the depth of 20 to 100 m, sometimes more than 150 m under the surface of the overthrust mountains (Střelec, Martinec 1992A, 1992B). The greatest thickness of these beds laid subhorizontally was found in the northern part of the exploratory field of Dětmarovice - Petrovice where it reached more than 200 m. At this site, the variegated beds are a continuation of the same beds in the adjacent Polish mines. The greatest depth of the variegated beds was encountered in Orlová structure where it attained as much as 600 m. A body of variegated beds follows here a synclinal turn, from overthrown to horizontal position and thins out several hundred metres eastwards.

To this weathering zone, the near-surface weathering zones of the Tertiary and the Quaternary are superposed, having a typical limonitization and kaolinization of the rocks (Martinec, 1989).

Table 1. Typical physical parameters for conglomerates, sandston siltstone and siltstone for Upper Carboniferous sediments from Upper S Basin

Type of alterations: grey - unaltered rocks, A - type of alteration 2.1., B - type of alteration see Chapter 2.1.

		σ_D [MPa]			σ_{TS} [MPa]			E [GPa]		μ [-]		n [%]		
		type of alteration								type of alteration				
		grey	A	B	grey	A	B	grey	A	grey	A	grey	A	B
Conglomerates		84 79-115	69* X	X	7 3,5-8	6,3* 3,9-9,9	X	23 17-26	17,3* X	0,2 0,18-0,21	0,2* 0,18-0,29	3,8 2,3-4,3	7,3* 4,2-12,3	X
Sand-stones	coarse gr.	84 65-110	73 52-85	X	8 6-9	7,5 5,7-8,1	X	19 17-21	16,7 15,2-21	0.19 0,17-0,22	0,23 0,18-0,25	4,2 1,8-5,2	5,3 3,5-10,3	X
	medium gr.	99 75-110	70 38-86	43 22-55	9 5-12	7,2 4,7-8,9	2,3 1,8-5,3	20 17-22	14,3 7,8-18,9	0,2 0,17-0,22	0,18 0,17-0,2	4,9 1,8-6,2	4,7 3,3-12,3	12 7-18
	fine gr.	112 89-121	70 39-102	52 15-63	12 6-14(?)	7,5 4,3-11,3	4,7 1,5-7,2	21 18-23	14,2 7,6-20,1	0,19 0,18-0,21	0,2 0,16-0,23	2,5 2,1-3,9	11,3** 3,5-15,2	10,2 6-16
Sandy siltstones		100 73-120	68* 22-73	110 70-140	12 6 - 14	6,3* 3,8-9,7	10,3 8,3-15	21 18-23	12,1* 5,3-14,5	0,19 X	0,2* X	1,1 0,8-2,1	3,8* 2,9-7,2	3 2,5-7
Siltstone		85 65-99	45* 19-55	105 X	10 5,5-12,3	5,1* 4,1-7,8	9,5 7,6-13	19 15-21	9,3* 4,8-18,2	0,2 X	0,19* X	2,8 1,8-3,2	3,5* 3,2-6.9	2,3 1,6-3,3

X - data are lacking
* - sporadic data
(?) - precarios data (perhabs anomalous data?)
σ_D - uniaxial compressive strength /MPa/
σ_{TS} - tensile stregth /MPa/

μ - Poisson ratio [-]
n - total porosity [%]

2.1 *Alteration of rocks*

In dependence on the changes in mineral associations, the following three basic types of alteration of rocks can be distinguished (Králík 1976, 1982, 1984).

A. Variegated (red) beds with a low degree of thermal alteration and a partially oxidized coal matter with an unaltered matrix of the sediment. Temperatures of (alteration) up to 400 °C.

This type of alteration is characteristic for a part of bodies containing variegated beds, i.e. for bodies from the rock mass between coal seams with a deep oxidation or from those sites where the coal seams were preserved with a lower degree of alteration. From a geomechanical point of view, it is necessary to stress some of the differences which are characteristic for the above beds in comparison with the rocks of the grey development:

- Opening of joints and bedding joints occurs. This leads to an easy loosening and caving of the strata from the roof of mine workings. The coefficient of permeability, both for water and gas, increases several orders.
- Along zones of joints or faults of reduced thickness, the alteration may penetrate deeply into the unaltered Carboniferous.
- At the level of rock texture, only small changes in the distribution of pores occur, i.e. an increase of the proportion of pores with a radius greater than 500 nm occurs. This secondary porosity is distributed very unevenly and reflects local changes in the texture of the materials found.
- Together with changes in the matrix, also the properties of the rocks change. The changes of the parameters are shown in Table 1.

B. Variegated beds with a mean degree of thermal alteration with completely oxidized impregnated coal matter. These rocks bear marks of thermal and locally also accompanying hydrothermal alteration. The temperatures of alteration are 400 to 1000 °C.

This type of alteration is characteristic for the immediate surroundings (roof) of the oxidized coal seams only. The thickness of the altered zone is of the order of metres to tens of metres.

From a geomechanical point of view, this is an extremely unstable rock mass, with layers of breccia type or with a strongly disturbed rocks, readily slacking and swelling. They constitute a marked mechanical and material discontinuity in the rock mass with a body of variegated beds. The transition to a body with a lower degree of alteration is continuous.

C. Hematitized basic rocks similar to volcanites, with a secondary hydrothermal alteration. These rocks are found rarely and are associated exclusively to places after the oxidation (burning out) of the coal matter.

2.2 *Alterations in coal seams*

The alterations of coal matter in the seams were studied thoroughly in a number of localities in the Upper Silesian Coal Basin from the point of view of coal petrology (Klika, Kraussová, 1992), geochemical (Klika et al. 1992a, Klika, 1992) and of liability to coal oxidation (Taraba in Klika et al., 1992).

In the body of thew variegated beds, coal matter in the seams is fully oxidized (burnt out). In contact zones, alterations of two types are found:
a) oxidation of coal matter,
b) thermal metamorphosis of coal without direct oxidation.

Four basic types of altered coal showing a greater or a smaller thermal and oxidation alteration are distinguished by Klika and Kraussová 1993, according to the extent of oxidation and thermal alterations. A summary of chemical and technological properties of coal, of the content of humic substances, of the reflectance of vitrinite and elementary analyses of coal for the individual coal types and subtypes is given by Klika (1991).

2.3 *Orientation of the development of coal seams in relation to the bodies of variegated beds*

The orientation of the coal seam relative to the variegated beds is an important circumstance. If the seam penetrates into the body of variegated beds perpendicular to its boundary, the contact zone is developed in the whole thickness of such seam and the influence of the alteration diminishes gradually. If the body of variegated beds is found in the roof of the seam, the contact zone is associated with the contact of the seam with the roof and is developed irregularly under such contact. The influence of the alteration diminishes in direction towards the floor. In detail, the development of the interface is broken and is influence both by the presence of certain rocks in the given places and by the degree of tectonic disturbance.

3 PROPOSAL OF GEOTECHNICAL CLASSIFICATION

It is difficult to present single characteristics for all bodies of variegated beds from a geotechnical point of view because locally, they exhibit a great variability as far as the presence of rocks, coal seams and relation to the actual relief of the Carboniferous are concerned. However, variegated beds have a number of common features (properties) which can be used for their classification. The inner structure of the bodies of variegated beds encountered in active mines and out of them is known very little. Only exceptionally, these bodies were opened by exploratory galleries of boreholes drilled in the mine. Their structure is known partially from exploratory wells from the surface. A basic geomechanical investigation of the properties of rocks and rock and rock mass encountered in the variegated beds was not implemented. More information is available on the properties of the contact zone, both from exploratory and mining works.

From a geomechanical point of view, the bodies of variegated beds and the adjacent contact zones represent unmined (left) parts of the rock mass such as:
- pillars with a fractured inner structured and altered properties,
- bodies which have a communication with the water bearing Lower Baden clastics or gobs and represent a secondary ground water body of mineralized and gaseous waters (Dvorský, 1989),
- near the surface of the Carboniferous, and increased accumulation of methane in the variegated beds was found.

The bodies of variegated beds together with the contact zone represent rocks and coal seams which have had an anomalous development. After uniaxial compressive strength intact rock material, drill core quality RQD of discontinuities, the condition of the discontinuities and the general conditions of ground water, by means of the RMS system of geomechanic classification, total ratings were determined.

Category I - Massif of Carboniferous unaltered rocks, grey development
rocks - without alteration
seams - without alteration
massif - tectonic faults with a slight hydrothermal alteration and mineralization, joints without rims
properties of rocks - depending on the petrographical type, grain size and degree of diagenesis
RMS system total - good rock (>75) to very good rock.

Category II - Altered Carboniferous massif (bodies of variegated beds)

Category IIa - in areas outside the seams affected by alteration and out of their surroundings, in a massif where "red" spots in the rocks are found but the existing coal seams are preserved morphologically and they exhibit oxidation or thermal alterations.
rocks - alteration of the A type,
seams - with oxidation or thermal alteration, their morphology and thickness preserved,
massif - disturbance (loosening) of the massif as a result al alterations in the vicinity of joints and bedding joints,
- increased permeability of the massif,
- presence of secondary collectors of water and gas - CH_4, CO_2, CO,
RNS system total rating - good rock (<58) to fair rock.

Category IIb - in the vicinity of strongly oxidized coal seams up to 3 to 4 times the thickness of the original seam rocks - alteration of B, C types, mostly around coal seams B,C, breccia associated with the caving (colapsing) together with secondary hydrothermal alterations (claying, zeolitization),
seams - strong oxidation of coal seam with loss of thickness up to its thinning out,
massif - discontinuities at the site of the thinned out coal seams with frequently clayed breccia associated with the caving,
- disturbance (loosening) of the massif as a result of alterations in the vicinity of joints and bedding joints,
- increased permeability of the massif,
- presence of secondary collectors of water and gas - CH_4, CO_2,CO,
RMS system total rating - fair rock (<37) to poor rock.

Category III - Contact zone between the body of variegated beds and the unaltered Carboniferous.

Category IIIa - thermal alteration of coal seams and rocks without oxidation,
rocks - without significant alteration as in Cat.I,
seams - strong, brittle coal with fracturing,
massif - no significant alterations,
RMS system total rating - good rock (>70).

Category IIIb - oxidation alteration of coal seams and rocks,
rocks - changes in colour, properties,
seams - oxidation of coal matter accompanied by humic acids,
- changes in physical and chemical properties of coal cause reduction of strength and loss of cohesiveness,
- liability of coal to spontaneous heating,
massif - disturbance (loosening) of the massif as a result of oxidation alteration of rocks (easy caving of the roof),

- increased permeability of the massif for water and gas - CH_4, CO_2, CO,
RMS system total rating - fair rock (>61) to good rock.

4 CONCLUSION

The information obtained up to now on the geological structure, the origin and the properties of rocks affected by alterations of the type of variegated beds shows that the bodies of variegated beds represent an anomaly in the structure of the Carboniferous rock mass. Such anomaly manifest itself in those cases when galleries and coal faces are opened in the contact zone (problems with unstable roof, selfheating) or when coal is mined in areas prone to rock bursts. Further, these bodies affect adversely mine ventilation and hydrogeology. In the process of the closing of inefficient mines, the bodies of variegated beds will have to be taken into account because they could act as ways of communication for mine water.

The proposed geotechnical classification of the rock mass with alterations of the type of variegated beds contains three categories and a number of subcategories which describe the alterations of rocks, seams and in discontinuities of tectonic or sedimentary origin.

REFERENCES

Dopita, M. - Králík, J. 1973. Red Beds im Obeschlesischen Steinkohlenbesken. - C.R. *7th Int. Congr. Geol. Stratigr. Carbonif. (Krefeld)*, 351 - 364, Krefeld.

Dopita, M. 1988. "Saddle" seams of Ostrava - Karviná Coal Basin, Thesis, Min. Univ. Ostrava (in Czech).

Dvorský, J. 1986. Protective pillars during the mining in the vicinity of dangerous collectors in OKR. *Uhlí*, 34, 2, 52 - 56 (in Czech).

Klika, Z. 1990. Geochemistry of red beds in Ostrava - Karviná Basin. 1. Rocks - main oxides of the elements. *Čas. Mineral. Geol.* Praha, 35 No. 4, 403 - 420 (in Czech).

Klika, Z.1992. Geochemistry of red-beds. *Thesis, Min. Univ. Ostrava,* MS.(In Czech)

Klika, Z., Taraba, B., Martinec, P., Dopita, M. 1991. Methodical procedure for the determination of protective pillars in coal seams affected by red beds. *Annex No. 4 to Directive of Ostrava-Karviná Mines No. 2/1991* of 24th Jan., 1991 (in Czech).

Klika, Z. et al. 1991. Directive for the determination of the contact of variegated beds with grey development of the Carboniferous in coal seams with a view to the determination of the protective pillar, *VVUÚ* (in Czech).

Klika, Z.- Kraussová, J. (1993): Properties of altered coals associated with Carboniferous red beds in the Upper Silesian Coal Basin and their tentative Classification. *Int. J. of Coal Geology*, 22, 1993, 217 - 235, Elsevier Sci. Publ. B.V., Amsterdam.

Králík, J. 1976. Clay minerals in the Carboniferous Red Beds of Ostrava-Karviná Coalbasin. 7th Conf. Clay Miner. and Petrology (Karlovy Vary), 263 - 272, Praha.

Králík, J. 1980. Red Beds in coal bearing sediments. *Sbor. věd. prací VŠB*, No. 1, Vol. XXVI, 1 - 18 (in Czech).

Krs, M. et al. 1992. Paleomagnetism and Paleogeography of Visean to Namurian of Upper Silesian Coal Basin, Czechoslovakia: a contribution to the EUROPROBE. *Geologica Carpathica*, 43,3,155-156, Slovak Academic Press Ltd., Bratislava.

Martinec, P., Krajíček, J. 1989. Properties of rocks of the Upper Carboniferous near the contact with the cover. *Proff. Paper No. 43. Coal res. Inst. Ostrava - Radvanice*, (in Czech).

Martinec, P. et al. 1993. Contribution to the genesis of "red beds" in the Czechosl. part of the Upper Silesian Coal Basin on the basis of petro-magnetic and paleo-magnetic study. In *Proc. VII Conf. Geology of Coal Deposits*, Charles University Praha, 131 - 134, (in Czech).

Střelec,V. - Martinec, P.1992. Investigations into hydrogeological and geomechanical conditions of mining under detritus. *TP 117/91, Coal mining under water bearing horizons in OKR*, MS, VVUÚ Ostrava - Radvanice, (in Czech).

Střelec,V.- Martinec P. 1993. Investigations into hydrogeological and geomechanical conditions of mining under detritus. ČSM mine, 4th block - *Project No. 117/91*, Coal mining under water bearing horizonts in OKR, MS, VVUÚ Ostrava-Radvanice, (in Czech).

Geomechanics 93, Rakowski (ed.) © 1994 Balkema, Rotterdam, ISBN 90 5410 354 X

Geotechnical zoning of crystalline rock mass

A. Matejček
Firm Geofos, Žilina, Slovak Republic

P. Wagner, P. Sabela & Š. Szabo
Comenius University, Bratislava, Slovak Republic

ABSTRACT: Heterogeneity of crystalline rock masses causes serious construction complications in this type of rock environment. The crystalline rock mass at the pumped storage power plant Ipeľ was subdivided into seven geotechnical types based on their geotechnical properties. This subdivision was the basis for the successful design of PSPP's Ipeľ underground structures.

1 INTRODUCTION

Engineering-geological investigation at several localities in the West Carpathians encountered widespread heterogeneity within the crystalline rock masses, caused by presence of tectonic fault zones and products of unequal weathering. Similar problems were also encountered in the area of designated for underground structures of pumped-storage power plant (PSPP) Ipeľ.

The PSPP Ipeľ is situated in the mountainous area of Central Slovakia (Figure 1). The rock mass consists of two rock complexes - the predominantly biotite granodiorites to granites in the western part, and the crystalline schists, migmatites and hybrid granodiorites in the eastern part. These two complexes are separated by the most significant tectonic element of the area the regional crush zone, which follows the River Ipeľ valley (Figure 1). The design assumes the construction of two reservoirs, connected by an underground hydraulic conduit located in the crystalline rock mass. The designed PSPP Ipeľ has installed capacity 710 MW. A large cavern for underground power station (30 m wide, 132.5 m long and 55.5 m high) needs to be constructed. Also, the question possible extension of the cavern construction had to be addressed. Solution to these problems required detailed informations about rock mass heterogeneity and properties.

2 EVALUATION OF ROCK MASS PROPERTIES

The study of engineering-geological and geotechnical rock mass properties for underground structures was based on the detailed evaluation of the 1070 m long exploration gallery, which was driven parallel to the proposed hydraulic conduit. Cross adits were driven in the expected locations of caverns (Figure 2). The sequence of underground exploration and evaluation was as follows:

- detailed documentation of galleries walls,
- continuous measurement of eletrical resistivity and micro-seismic refraction, and measurement of rebound hardness (by Schmidt hammer) along the all galleries length,
- laboratory tests of rock material properties.

The results of these studies enabled subdivision of the gallery rock mass area into detailed, quasi-homogeneous rock mass units. Their quantitative characterization, which lent support to recommendation for methods of excavation of underground structures, was carried out by:

- in situ load and shear strength tests,
- detailed study of rock mass

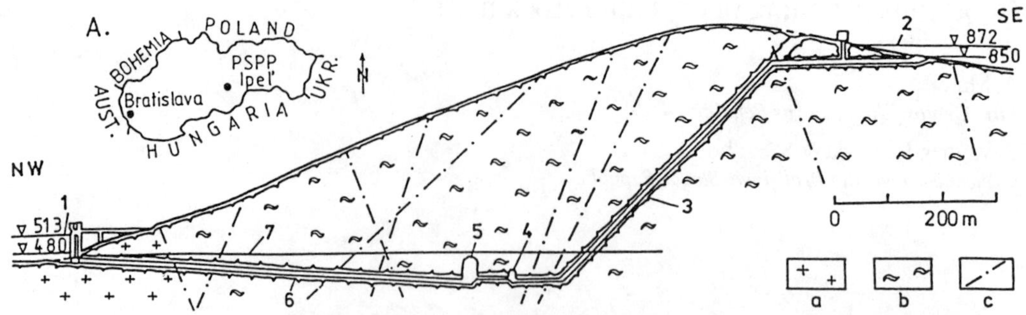

Fig.1 Schematic section of the Ipeľ pumped-storage power plant hydraulic conduit. A - location of PSPP Ipeľ, 1-lower reservoir (max.and min. working water level), 2-upper reservoir, 3-penstock, 4-cavern of spherical closures, 5-cavern of power station, 6-tailrace tunnel, 7-exploration gallery, a-blastomylonites, b-migmatites, c-significant faults

discontinuites.

Considering that rock mass properties are mostly controlled by the character of discontinuities, their parameters (orientation, spacing, roughness, etc.) were studied in detail at the chosen location of the galleries (Figure 2A). The huge database of discontinuities' parameters was analyzed by a specially developed computer program. The program output were the most important characteristics of rock mass properties, namely:

- structural engineering - geological characteristics, such as blockiness (expressed by the block size index, i.e.average size of the basic block of the rock mass) and loosening of rock mass (expressed by the index of loosening),
- special purpose geotechnical rock classifications, such as RQD (Deere & Miller 1966), determined from discontinuity spacing, NGI (Barton, Lien & Lunde 1974), RMR (Bieniawski 1979) and QTS (Tesař 1990). RMR and QTS classifications were used in direction of exploration galleries advance, and
- analytically determined values of rock mass deformation modulus according to Ruppenejt's empirical formula (in Rac et al. 1979).

3 GEOTECHNICAL ZONING

All the results were analysed and compared. NGI classification seemed as the most complex, sensitive and representative of the studied rock environment. Therefore its values were taken as the fundemental variables. From the practical point of view the QTS classification is very convenient due to its tight relationship to the technological classes of New Austrian Tunneling Method - NATM (Pacher et al. 1974 in Tesař 1989). This relationship was used to determine technological classes of NATM for cavern (30 m width) advance (Table 1). On the basis of all this information, seven geotechnical types of rock mass were defined.

Representative average values or range of values of the engineering - geological and geotechnical characteristics are given in Table 1. Assumed extent of defined geotechnical types in the horizontal plane of exploration gallery is shown in Figure 2B. The figure shows only this part of rock mass, where the construction of the cavern is planned. The geotechnical Types 1 and 2 consist predominantely of rigid granites and granodiorites of good physical condition. The geotechnical Type 3 includes all migmatities with bodies of paragneisses and Type 4 contains the same lithological types but with more closely spaced discontinuities. The geotechnical Type 5 represents significant tectonic faults and accompanying bodies of disrupted rocks. Type 6 includes water bearing disrupted zones. The geotechnical Type 7 represents disrupted and weathered rocks and soils within the regional fault zone. Type 7 was appeared in the first 200 m of

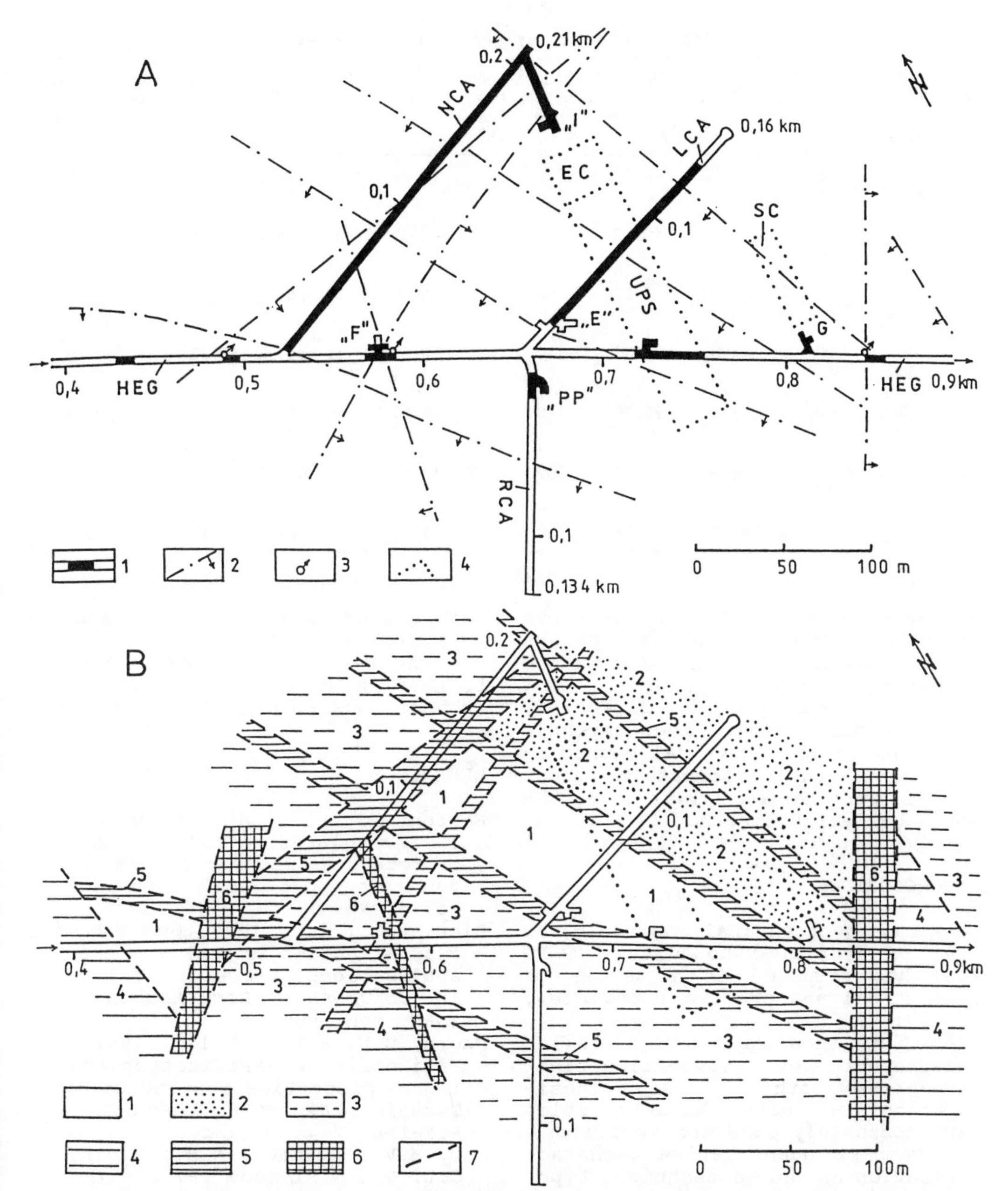

Fig.2 Geotechnical zoning of crystalline rock mass in the horizontal plane of exploration gallery. A - Location of gallery (top view) and input data. 1-parts of gallery, where the discontinuity parameters of the rock mass were studied in detail, 2-significant faults and their dip direction, 3-concentrated seepage in gallery, 4-designed underground structures, HEG-heading of exploration gallery, NCA,LCA,RCA-new, left and right cross adit, UPS-underground power station, EC-possible extension of the cavern, SC-cavern of spherical closures, "E","F","PP","I"-cross cuts for in situ tests. B - Geotechnical zoning. 1 to 6-extent of geotechnical (GT) types (their characteristics see Table 1) in the horizontal plane of exploration gallery, 7-boundries between geotechnical types.

Table 1. Quantitative characteristics of geotechnical types

GT type No.	Eletrical resistivity (Ohmm) AB=14m	AB=6m	Microseismic refraction (m/s)	Schmidt hammer values	Block size index (mm)	Loose-ning index (%)	RQD (%)	NGI	RMR	QTS	NATM (30m)	Deformat. modulus (MPa)
1	10000 15000	9000 12000	4200-5000	60-70	300	0,7	85-95	25-30	67-82	68-85	2.,3.	9500
2	7000 12000	4000 10000	3500-4500	50-60	400	1,1	70-85	19-25	64-73	67-75	3.	8000
3	5000 12000	2500 7000	2500-4500	40-55	300	1,8	65-75	11-19	56-65	60-67	3.	6000
4	4500 10000	2000 3500	3000-4000	35-50	250	1,1	50-65	5-11	55-60	55-62	3.,4.	3900
5	1000 2000	800 1200	2500-3500	28-35	180	3,4	40-50	4-5	44-50	45-55	4.	2800
6	2500 4500	1500 2500	2000-4500	25-40	200	2,3	40-50	3-4	48-53	50-55	4.	2600
7	<500	<500	1500-3000	22-32	200	4,1	35-45	1-3	35-45	38-45	4.,5.	1000

exploration gallery, and therefore it is not shown in Figure 2B. Values of deformation modulus, determined analytically and by in situ tests, are in some places different. Nevertheless, the incontestable advantage of the anlytical method is the continuous information about modulus deformation values along the whole galleries length.

4 CONCLUSIONS

Rock mass geotechnical zoning in the area of hydraulic conduit of PSPP Ipeľ has resulted in a series of theoretical and practical advances, namely:

- the technique of detailed measurement of discontinuities' parameters with derived classification rock mass characteristics was successfuly used and verified,
- complex quantitative characterization of the geotechnical types resulted in the recommendation for the most suitable methods for driving methods and supports for the various underground structures in the studied rock environment,
- assumed subdivision of geotechnical types in the horizontal gallery plane was the basis for the spatial subdivision of rock mass into geotechnicaly homogeneous units (after detailed evaluation of underground boreholes in the area of cavern),
- the results of above study can be used in design and investigation in analogous rock environment.

REFERENCES

Barton, N., Lien, R. & J.Lunde, 1974. Engineering classification of rock masses for the design of tunnel support. *Rock Mechanics, Vol.6*, No.4.

Bieniawski, Z.T. 1979. Geomechanics classification in rock engineering applications. *Proceed. of 4 th Int.Congress on Rock Mechanics, ISRM,* Montreux.

Deere, D.U. & R.P. Miller 1966. Engineering classification and index properties for intact rock. *Technical Report No. AFNL-TR -65-116,* New Mexico.

Rac, M.V., Ivanova, N.B., Rusin, G.L. & E.G.Slepcov 1979. Avtomatizirovannaja sistema obrabotki dannych (ASOD) po treščinovatosti gornych porod dlja inženerno-geologičeskich celej. *Inž. geologija, No.5* (in Russian).

Tesař, O. 1990. Classification of rock for underground construction and results of engineering-geological observations. *Proceed. of 6 th Congress IAEG in Amsterdam,* Balkema, Rotterdam.

Geomechanics 93, Rakowski (ed.) © 1994 Balkema, Rotterdam, ISBN 90 5410 354 X

Evaluation of rock mass behaviour based on levelling measurements conducted in the mine entries and on surface of ČSA colliery

V. Petroš
Technical University of Mining and Metallurgy, Ostrava, Czech Republic

ABSTRACT: Levelling measurements of fixed points located in the mine openings and on the surface can give good idea regarding the stress change and deformations in the rock mass. The paper presents an example of evaluation of levelling measurements during the exploitation of seam in highly stressed zone in the ČSA colliery of the Ostrava-Karviná coal field. The paper explains sudden changes in vertical displacement of observation points in gate road entries which are influenced by the left out pillars.

During the exploitation in highly stressed zone of the third sector of the ČSA colliery, a whole sequence of methods was used to observe the behaviour of rock mass. One of the methods was the monitoring of vertical displacement of the roof-bolt ends by levelling in both gate entries of the face and vertical displacement of fixed points on the surface above the face. The bolts were placed at an interval of 10 m in the gate entries and the levelling was conducted at an interval of three weeks. The height changes were calculated in comparison with the first levelling measurement, which was realized near two month before the beginning of the extraction. The levelling measurements was conducted by Mr. Miloš Ženč and Mr. Alois Borovec from Scientific Coal Research Institute, Ostrava-Radvanice. On the surface a line was monitored, as seen in figure 1.

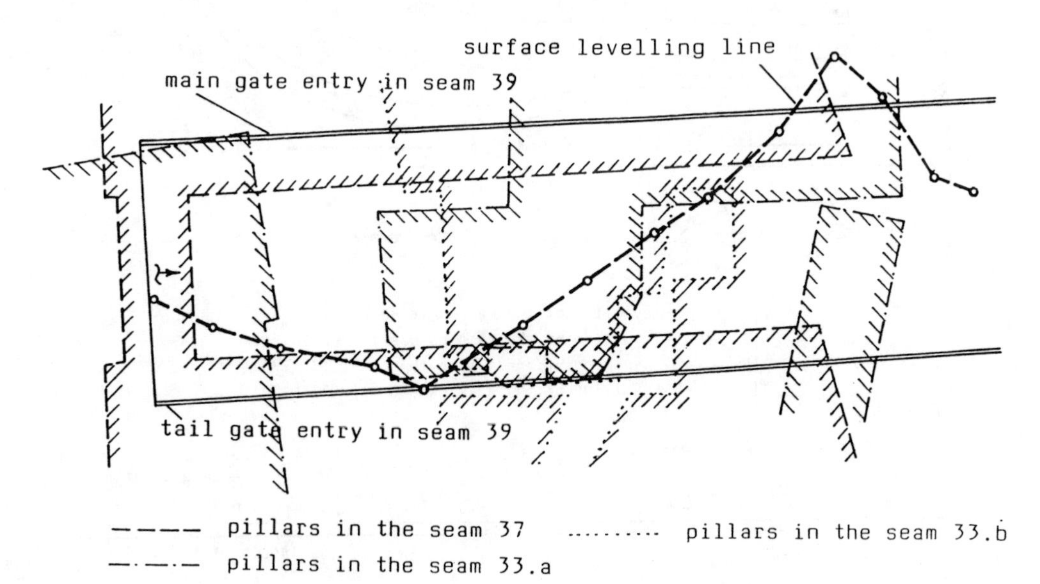

Fig. 1

Figure 1 also shows the pillars left in the upper seams. The seam number 39 situated around 20 m under the seam no.37, was extracted. In the seam no.37, a pillar was left near the middle of the exploited area nearly along the full length of the face and having a width of 150 m. The length of the exploited face is around 230 m, and that is why the gate entries are under the influence of the seam no.37. Other pillars are found at a height of more than 100 m.

The mining was continued for about two years with a constant face advance of 1,2 meters per day. Some results from the levelling measurement both in the mine and on the surface are shown in figure 2.

The curve in the upper part of the figure 2 show the development of surface subsidence along the observed line, which follows the direction of the advancing face. Under these curve are shown the position of the pillars in seams no.33a and no.33b. Next follows the position of the observed points in the gate entries during two different phases of extraction. Continuous and interruped lines give the level profile of the main gate entries, the others two lines express the level profile of the tail gate. At the lowest part of the figure 2 is shown the position of the faces during the observed time.

From the underground levelling measurements it is obvious that exploitation in this highly stressed zone influences the rock mass around the face practically for the full length of the mine entries (more than 700 m). The maximum vertical displacement of the observed points is found in the main gate entry around the distance of 560 m. With the approach of the face there occured great displacement of the points, which in turn means aggrevation of the stress situation at this place. This relates to an area where there are left out pillars in the upper seams, and substantially the short entries having measured great differences in vertical displacement we can explain the situation as follows. The rock mass under the pillar is compressed but the basic levelling was already conducted under this condition, therefore it was not detected - figure 3.

In figure 3 the basic levelling results are marked by an interrupted line. If the area is under the pillar then the real rock mass deformation can be expected as shown by continuous line. In the case the given area is influenced by an approaching face, the abutment stress zone is extended under the pillar (pointed line), which during the levelling manifests only by the gradual lowering of points at the stress

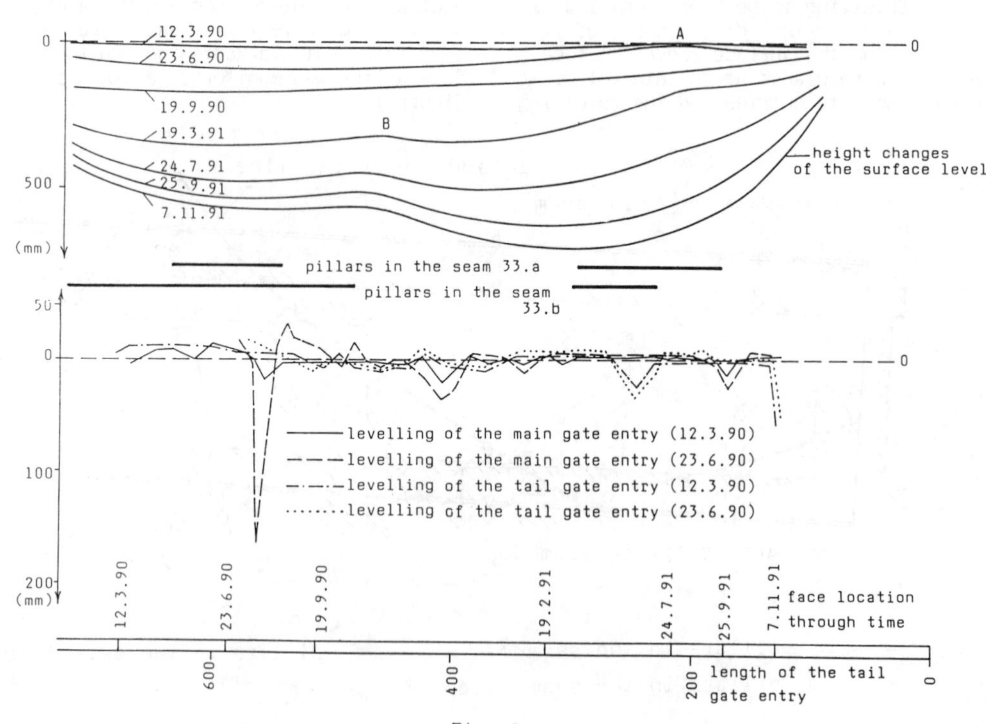

Fig. 2

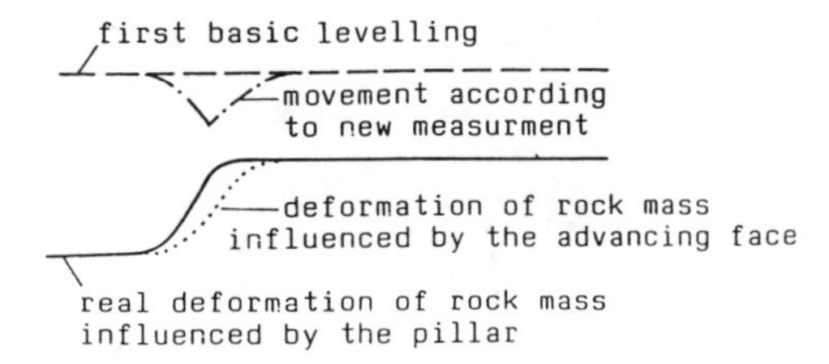

Fig. 3

boundary (point-line trace).

These sudden changes of the point positions on the mine entries is delimiting the area of original stress, which was within the rock mass before the realization of the basic levelling of the observed points. In places where vertical stress change has occured, there is also present a considerable shearing stress and often it provokes a shear disintegration. This effect was possible to observe utilizing seismic methods and the focuses of the seismic events was often located in these areas.

Soon after the complete advance of the mining face, movement was noticed at the observation points on the surface. At majority of the points a significant lowering of position was noticed, but in case of the point marked as "A" in the figure 2 an increase of 5 mm was recorded. Even in the next levellings when the measurements showed that all points went down, this point showed the smallest descent. This phenomenon continued till the face advanced to a distance of 200 m in the strike direction, when the left out pillars in the upper seams were undermined. In this moment a substantial crater was formed in the referred area. This event confirms that rock mass behaved here as a block and bivoted over the pillar left and only the undermining the pillar permitted the uniform subsidence of the surface (approximately beginning of september 1990).

This situation was again repeated in february 1991, as soon as the other pillars, marked in the figure 1, started to influence the massive. Again the point marked as "A" had a smaller descent than the rest of the points. The uniform subsidence of surface was noticed as soon as the pillars were undermined.

From figure 2, it is clear that in the area marked as "B" there is gradual decrease in lowering of points. Although this is not caused by an anomaly of the rock mass, it is rather an effect of the position of the face with respect to the line observed in the graph. A point with the smallest descend is above the border of the face (figure 1).

The rest of observations during the evaluation of results from the levelling measurements are alternating compression and relief of particular entries. The mentioned situation is clear from figure 4, when most of the length of the tail gate entry is more affected by the pressure (level profile shown by the interrupted line), the main gate entry is delineated by the situation alternately varies, both profiles come very close to each other. This means that when the tail gate entry is relieved then the main gate entry is

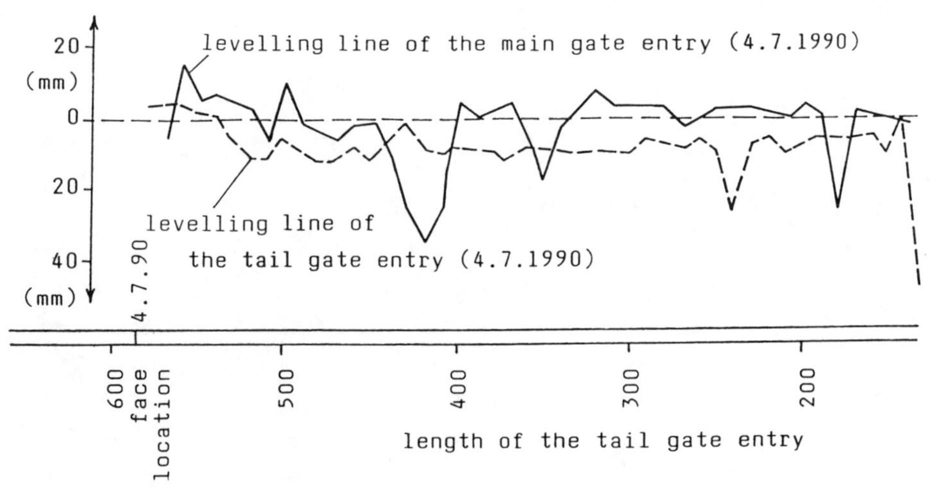

Fig. 4

compressed. At the same time we could observe a higher generation of sesmic energy always from the more compressed side of the face.

In conclusion it is possible to say that the measurement of fixed points in the mine openings gives a good idea of the stress changes occurring in the rock mass. Under the conditions of the measurement this was the only method which in advance showed the places of increasing stress, which at the same time manifestated a higher emission of seismic energy during the approach of face to the critical area.

Observation of surface points proved that rocks above the seam no.37, which are very competent sandstones of great thickness, behaved as a block.

Geomechanics 93, Rakowski (ed.) © 1994 Balkema, Rotterdam, ISBN 90 5410 354 X

Evaluation of stress measurements in the stowing area of a longwall face and the different influencing factors

V. Petroš
Technical University of Mining and Metallurgy, Ostrava, Czech Republic
J. Žák
Scientific Coal Research Institute, Ostrava-Radvanice, Czech Republic

ABSTRACT: The measurement of stress in the waste area offers valuable data about the stress distribution, mainly from point of view of future mining. This paper presents an example of stress measurements in stowed area of particular longwall face. There were different conditions around this face. In some parts of this area there were many pillars in the upper seams, on the other hand in some parts the upper seams were extracted. The paper show the distribution of waste area stress during two years of longwall mining and a results of observation for a period of 8 months after the extraction of the face.

The measurement of stress in the stowing material of a longwall face was made during the extraction of the seam 39 in the ČSA colliery. In one of the sectors mining in april 1983 led to a very strong rock burst at the superjacent seam 37. It is the first mining since that time. The face undermined a remnant longwall block in seam 37, being about 80 m longer. The extension of the face in the seam 39 was symetrical on both side of the pillar of the seam 37, therefore around the gate roads of the observed face a zone of 40m wide was under the influence of the extraction of the seam 37. During the exploitation, at a distance of 30-50 m a couple of pressure sensors was given to the stowing material. The pressure sensors were located close to the tail gate entry, in such a way that the first one was at a distance of 37-40 m from this entry and the second around 60 m from it. This then means that the first row of sensors was still under the influence of the superjacent extracted area and the second row was predominantly under the pillar of the seam 37. In this way were installed 9 pairs of pressure sensors but some of them were non functional from the begining. At the end of the coal extraction there were 15 functional pressure sensors. Their locations are shown in figure 1.

As shown in figure 1, the first pair of pressure sensors in the stowing material are still in the area which has been completely extracted in seam 37, the rest are located and distributed alternatively under the extracted area and under the pillar of the seam 37. The stress registered by each sensor was recorded at intervals of 5 minutes. Stress values from every sensor were feed to the computer, registering the time-stress relationship. From these values the daily characteristic was taken and was graphically presented by the computer in relation of the location of the sensor in the direction of the face advance. The values of the sensors located closer to the tail gate entry were joined with each other as well as the values of the sensors in the second row. Thus, there are two curves showing the development of the stress. Every day the computer made graphics of this situation during the full duration of the face activities, i.e., around 22 months. Some phases of the stress course are presented in figure 2.

Figure 2 shows by an interrupted line the stress course of the sensors close to the tail gate entry and with a continuous line the stress course of the distant sensors.

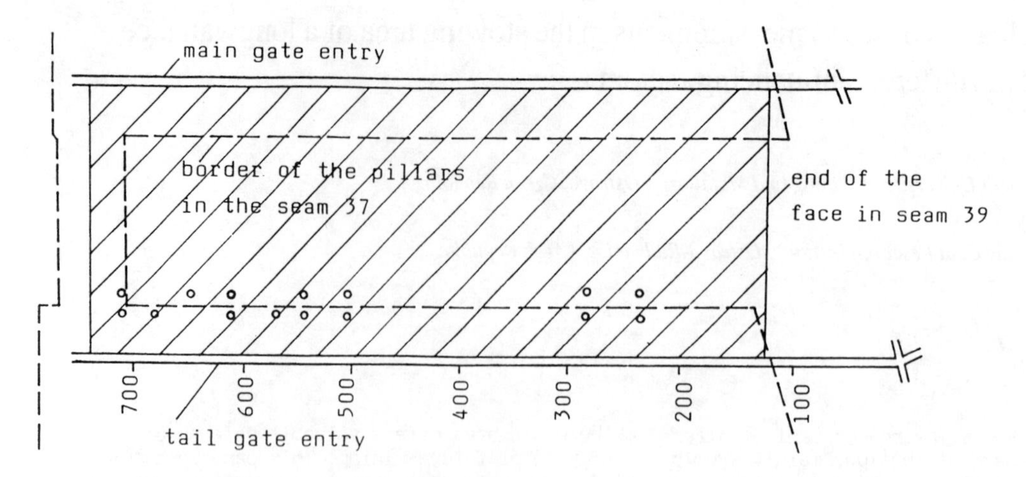
main gate entry
border of the pillars
in the seam 37
end of the
face in seam 39
700
600
500
400
300
200
100
tail gate entry

Fig. 1

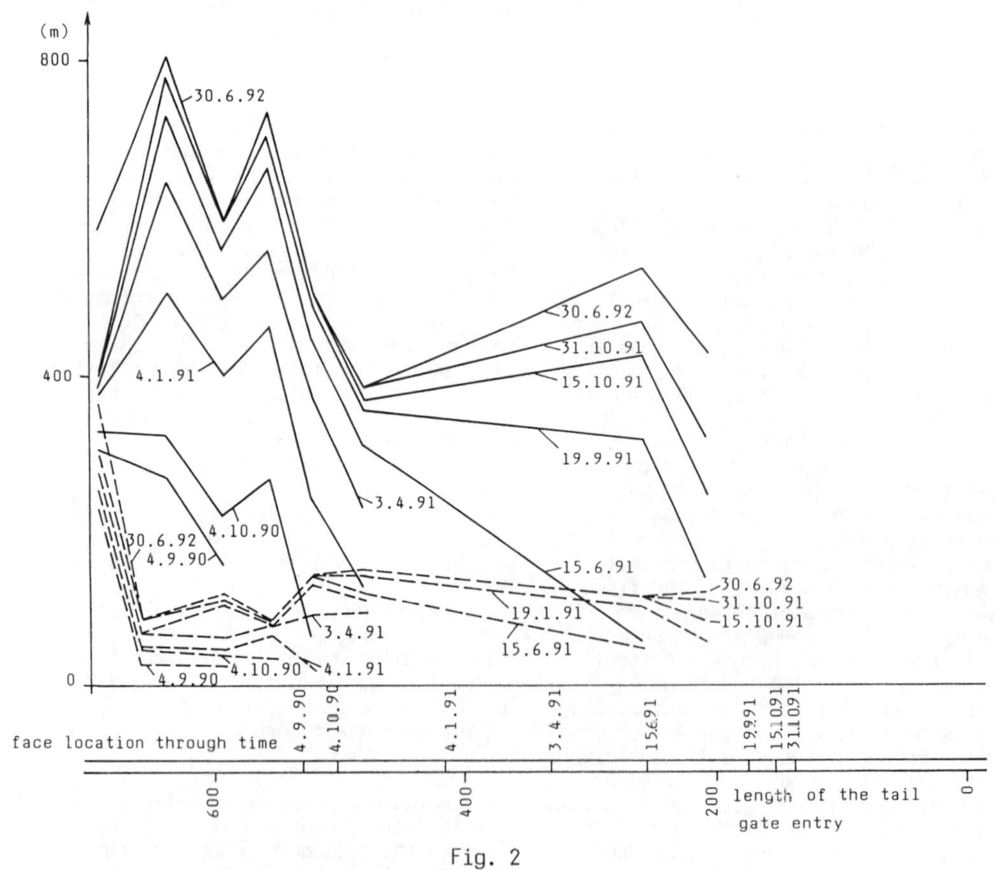
(m)
800
400
0
30.6.92
4.1.91
4.10.90
30.6.92
4.9.90
3.4.91
30.6.92
31.10.91
15.10.91
19.9.91
15.6.91
19.1.91
15.6.91
3.4.91
4.10.90
4.1.91
4.9.90
30.6.92
31.10.91
15.10.91
face location through time
4.9.90
4.10.90
4.1.91
3.4.91
15.6.91
19.9.91
15.10.91
31.10.91
600
400
200
0
length of the tail
gate entry

Fig. 2

The values are calculated as weight of the columns of roof rocks. On the "y" axis is given the height of rocks, which would initiate such stress by weight. The depth of the seam is around 800 m.

From figure 2 is clear the great difference between stress values from the two sensor rows, this is due to the influence of the superjacent extracted area and the compression caused by the pillar. Considerable influence to the difference between stress values is definitely due to the high strength of the roof rocks above the pillar of the seam 37. Smaller difference between values is shawn by the first pair of sensors. Both ot these sensors are located under the extracted area of seam 37. During extraction the values were very similar. Surely the small differences can be explained by the proximity to the border of the pillar in seam 37.

The other pressure sensors show a bigger difference between values in places which are further from the tail gate entry and closer to this entry. The faster and greater increment of stress was registered on the sensors situated far from the tail gate entry at a distance of 550--650 m. This effect is caused because area is stressed due to many pillars left in the upper seams. The maximal values of stress here corresponds to the weight of the whole roof column up to the surface. This means that under these pillars there is not protective effect of the exploitation of seam.

Stress value in extracted area generally is influenced by the distance of the points from the face and the time. The influence of the distance to the face is observed in the figure 2. It is clear that this influence is much more important in area situated under the pillar.

In area under the extracted part of the seam 37, during the advance of the face, there occured small changes of stress and this makes the curves to fuse together during some periods. The influence of time in the development of the stress in the exploited areas was possible to observe after the total extraction of the face- Observations were conducted for 8 month after the completion of the face. During this time the stress changes were registered mainly in the proximity of crosscuts and in the side of the extracted face. Higher increments of stress were registered in both sensors closer to the crosscuts under the extracted areas of the seam 37. This increment of stress close to the crosscut could be partially influenced by the neighbouring longwall face in seam 39, which along the time of the measurements had advanced around 100 m. In the side of the finished face the stress increased was recorded mainly in the last two sensors under the pillar of the seam 37.

In the extracted area of the seam 37 the stress was also increased only in the sensors nearest to the finished face. The rest of values in these stress sensors, except for the ones close to the crosscut, almost did not change at all. From the figure 2 it is also clear that around the tail gate at a distance of 280 m, the stress is increased under pillar. This place again responds to the presence of pillars in upper seams.

In conclusion it is possible to state that measurements in the waste areas give valuable data about the distribution of stress, mainly from the point of view of the next seam extraction in the same sector. If this measurements are enough it could be possible to establish the relationship between the increment of stress in a worked out area and the character of the roof rocks. Therefore, it could bring the investigation to a position, where without direct measurement in waste area, we could compute the stress from our knowledge of character of roof rocks. Since the magnitude of the stress in the waste is related also to the compression around the face, necessary knowledge could be obtained about the magnitude of the stress in the surroundings of long mine openings and specially of the faces.

Geomechanics 93, Rakowski (ed.) © 1994 Balkema, Rotterdam, ISBN 90 5410 354 X

Some views on the influence of tectonics on the occurrence of rock-bursts

J. Ptáček & L. Trávníček
Důlní průzkum a bezpečnost, a.s. Paskov, Czech Republic

The influence of disrupted zones or dislocations in general in a rock mass on the occurrence of rock-bursts is indisputable. However, the question often discussed is whether and which area next to dislocation affects the stress state of rock mass positively or negatively. Using examples of Ostrava-Karviná coal district, we would like to present in a brief contribution our geomechanical and geological views concerning the probable influence of tectonic dislocation on the occurrence of rock-bursts and on their probable mechanisms.

We have been analysing the influence of disrupted zones or dislocations in general on the distribution of stress in rock mass, and as a result of that, on the existence of the zones with potencital dangers of rock-bursts for almost 15 years. The use of the data of seismic observations in the Ostrava-Karviná coal district contributes to better location of foci of rock-bursts and to the understanding of their probable mechanism. Observations of seismic activity are provided by a network of surface and underground seismic stations.

Karviná region-the east part of Ostrava-Karviná coal district-represents an area of about 10x12 km. In Figure 1 there is its seismically most active part. The Figure shows the main faults and seismic phenomena (with emitted energy over 10^5 J). These seismic processes were recorded in the course of 5 years (1988-1993). The squares indicated in the Figure 1 (vertical with letters and horizontal with numbers) equal to 250x250 each. It correspondes with the precision of seismic location (in the x, y coordinates network).

In the Figure 1 there is obvious that seismic activity is different in each individual square (from 0 to 23 phenomena). That represents an energy-range from 0 to $1,7 \times 10^7$ J.

The distribution of seismic phenomena is irregular except for an area between two faults-Nepojmenovana and Jindřišská with a high concentration of seismic phenomena.

There are two types of areas without any seismic phenomena, as follws:

-the area without significant tectonic dislocations, e.g. area of squares K-59,62 and N-59,62 in the Figure 1

-the areas with dense network of dislocations, e.g. area of squares U-54,56, X-54,56, or U-61,64, X-61, 64 in the Figure 1

The irregularity of the distribution of seismic phenomena in the square-netvork is presented in the following table of exapmles of extremely high values of emited seismic energy (squares correspond to Figure 1):

square	number of phenomena	total energy 10^5	average energy 10^5
M-54	0	0	0
J-53	1	3,9	3,9
R-58	1	170	170
M-56	2	8,9	4,45
M-57	2	12,3	6,15
R-60	7	28,7	4,1
Q-60	23	93,1	4,05

From the analysis of the distri-

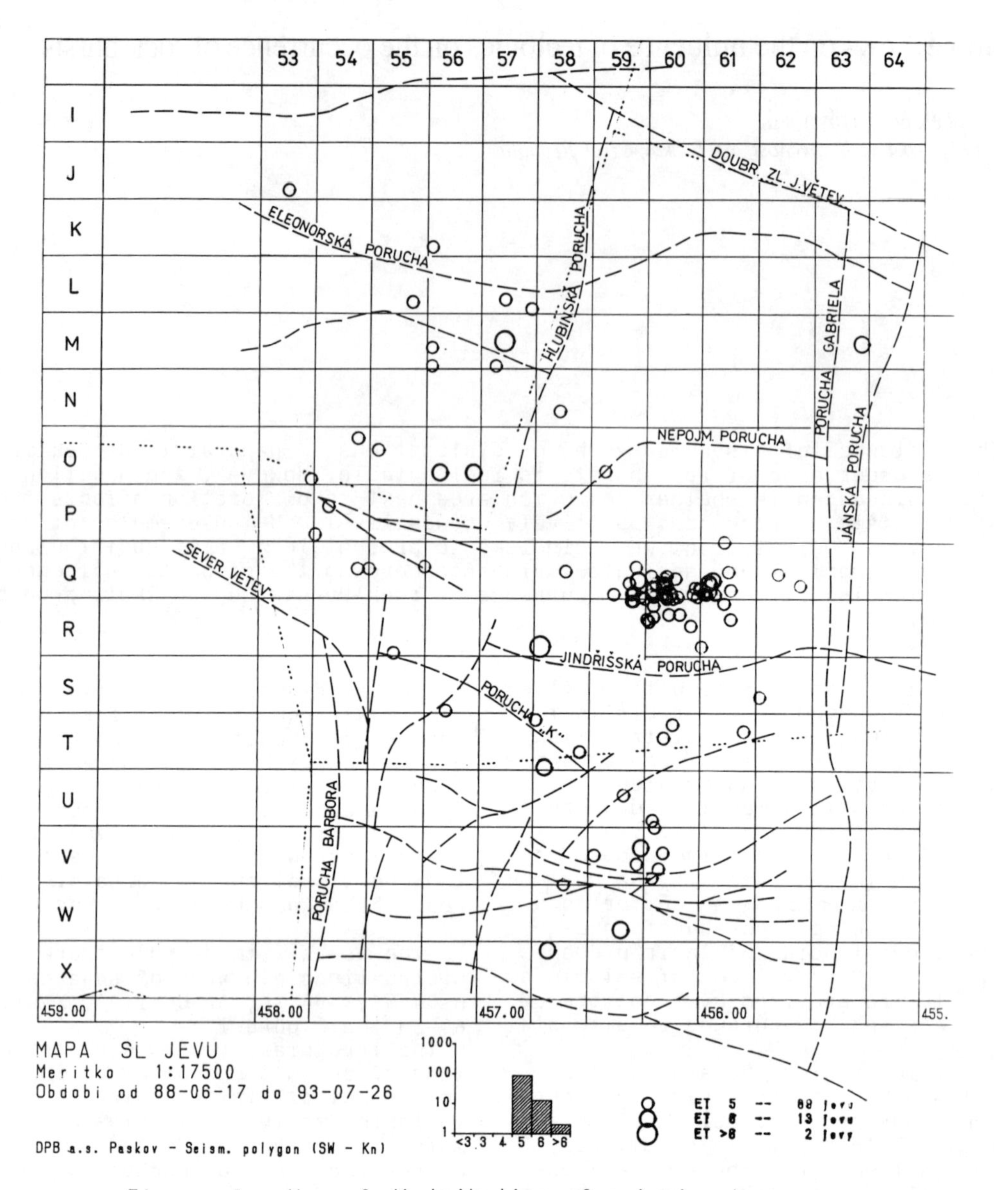

Figure 1 - Map of distribution of seismic phenomena

bution of seismic phenomena we deduced that isolated stochastic seismic phenomena of high energy (over 10^6 J) and average energy over 10^5 J) belong to stress-strain processes next to significant faults e.g.phenomena in the squares R-58 or M-57.

If we compare the distribution of seismic phenomena with tectonic structure of Ostrava-Karviná coal district,we can abstract the areas with high values of tectonic stress. In spite of the fact that we are aware that the mining process has a conclusive influence on the stress distribution in the rock mass, we also suppose that high value of stress in some areas can be refered to tectonic moves.

The areas with increased stress caused by tectonic moves were defined in some works of Department of geology of the Mining University in Ostrava (Grygar et al 1991).Thanks

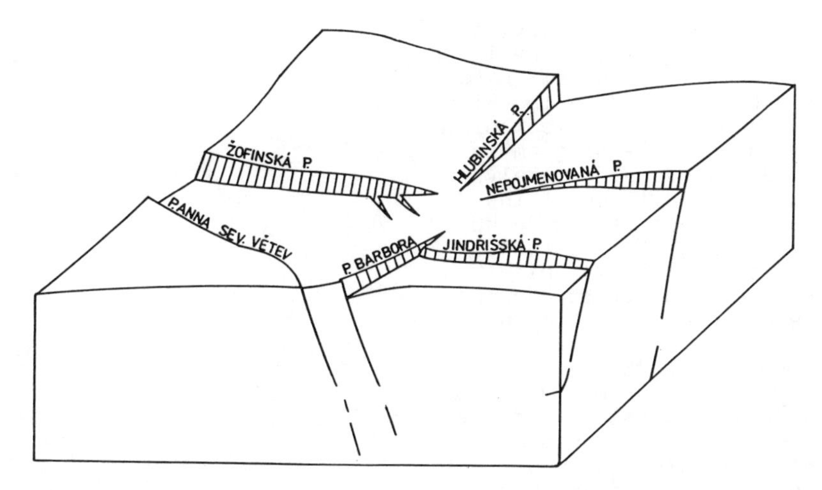

Figure 2 - Blockdiagram of Jindřich pull-apart zone

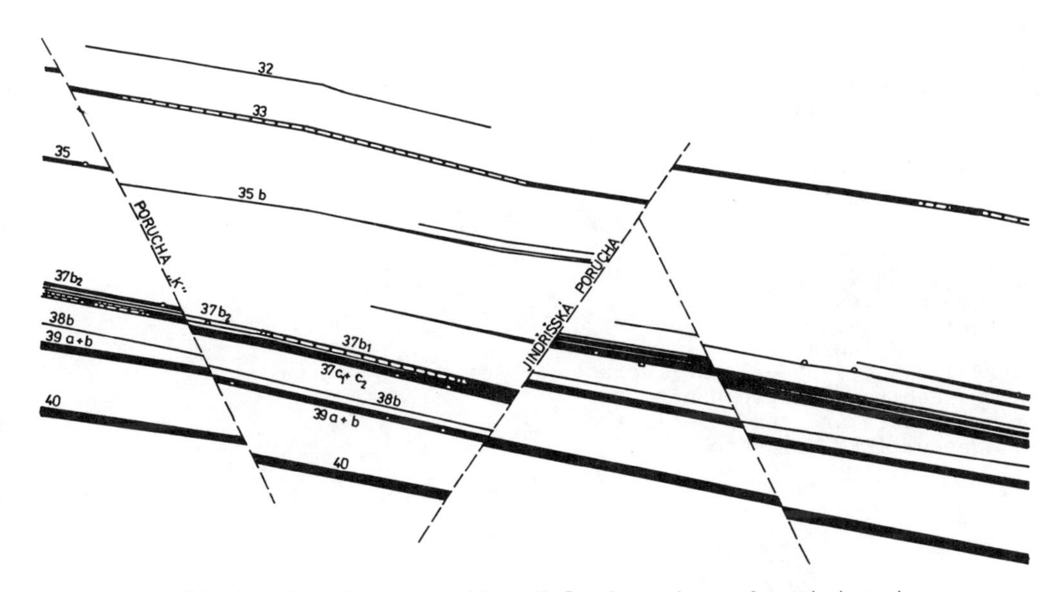

Figure 3 - Cross section N-S of region of rock-burst

to the cooperation with that Department we were able to upgrade our analyses. We define the fault areas with increased stress as follws:

-immediately next to the main fault in the lower (falling) bed

-parts without disruptions of small dimension

-regions of asymetric horst-structures

-the end of the main faults

-the areas in the pull-apart zones

We can present,on an example of rock-burst(see circle in the square R-58 in Fig.1)that occured this year in the coal mine CSA, the probable influence of some the above mentioned zones. The rock-burst occurred in a tectonic structure called Jindřich transtensile basin (Grygar et al 1991). That structure is a typical pull-apart zone including the end of the main faults Hlubinská and Barbora. On the basis of our analysis of rock-bursts we can consider it as the most complicated dynamic and tectonic structure of Karviná coal district. It has been confirmed through the seismic observation (see the squares R, Q-58, 62 in the Figure 1). The scheme of that tectonic sructure is shown in a diagram in the Figure 2. Based on the analysis of the natural conditions of the mentioned rock-burst,it was found out that asymetric horst-structure was next to the region in which the rokc-burst occurred. It is documented on the N-S cross-section in the Figure 3.

It was our aim to show some possibility of the use of combination of structure-tectonic analysis and seismic observations in the system of rock-burst protection. We know that the mining,the main influence of stress distribution in rock mass, covers up the influence of primary properties of rock-mass. We think, in spite of that,that the primary properties can be considered as a very significant influence in many cases of rock-bursts.

REFERENCES

Grygar,R.et al 1991.Metoda hodnocení strukturně tektonické stavby a paleonapěťových poměrů ve vztahu k prognóze otřesů. VŠB Ostrava.

Herget,G.1988.Stress in Rock, Rotterdam: Balkema.

Trávníček,L.1992. Wellenbilderanalyse desGebirgsschlagmechanismus mit der Apparatur Lennartz-Electronic, Nancy: ISRM Workshop.

Geomechanics 93, Rakowski (ed.) © 1994 Balkema, Rotterdam, ISBN 90 5410 354 X

Measurements of dynamic loads of roof support

S. Szweda
Silesian Technical University, Gliwice, Poland

ABSTRACT: Results of measurements covering the dynamic phenomena occurring in powered roof supports during rock mass tremors have been presented. The paper describes the applied measuring apparatus as well as the courses of underground tests. A resultant load transmitted by legs and vertical acceleration of a canopy have been determined. Courses of changes in loads of the support and accelerations of a canopy caused by tremors of the rock mass the energy of which ranged from $8*10^3$J to 10^6J as well as analogous courses recorded during stress relieving blasting have been presented.

1 INTRODUCTION

The investigation of physical phenomena occurring in a powered roof support during rock mass tremors and bumps is a reliable method for collecting the information which makes it possible to state the requirements to be met by a support designed for operation in underground workings exposed to bump hazards.

The determination of time curses of different physical quantities characterizing the operation of a support during a bump will also provide a possibility to determine parameters of mathematical model of phenomena occurring at that time in the support and in the surrounding rocks.

Underground measurements aimed at performing of these two tasks are carried out by the Institute of Mining Mechanization of the Silesian Technical University and by the KOMAG Mining Mechanization Center within the research project no 9 9031 92 03 financed by the State Committee for Scientific Research.

The course of measurements taken in the Zabrze-Bielszowice Hard Coal Mine as well as basic results of these measurements are presented in the continuation.

2 UNDERGROUND MEASUREMENTS

2.1 *Measuring apparatus*

In the main the measurements consisted in simultaneous determining of time courses of a resultant force transmitted by hydraulic legs and of vertical acceleration of a selected point of a canopy. Fig. 1 shows the way in which the measuring apparatus has been arranged on a support unit being subject to measurements.

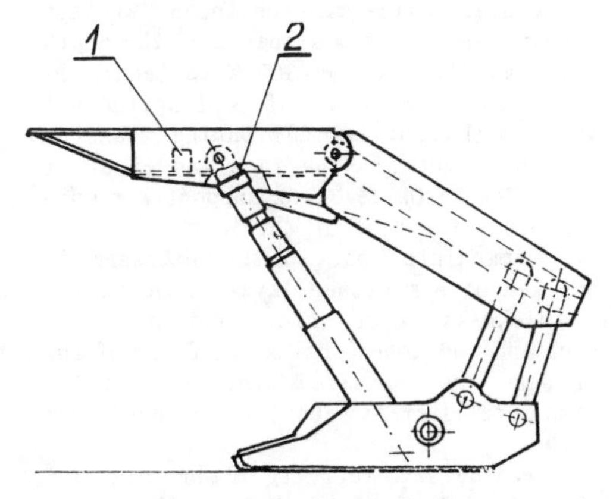

Fig.1 A diagram of the measuring apparatus arranged on a support unit. 1-acceleration sensing element, 2-force sensor.

The applied method for determining of dynamic loads of a support based on measuring of a force in a leg was described by Szweda (1991). The force value was measured by means of a strain gauge situated in a groove of the mechanical extension rod. After having been added up the measuring signal were sent through a telephone line to the surface and recorded on a magnetic tape.

The vertical acceleration of a canopy was measured with the aid of a sensor type KB-1a manufactured by BRUEL & KJAER. Measuring signals were transmitted to the surface through an independent slotted line constructed in an analogous way as the slotted line for measuring of forces in legs described above.

A frequency modulation of a signal with a carrier frequency of 10 kHz has been applied to transmitters and thus the influence of interferences existing in a communication network of the mine exerted on time courses of the measuring signals sent have been eliminated. Krasucki (1991) states that a carrier signal with a frequency of 10 kHz is being transmitted correctly by the communication network of a mine over a distance of up to 10 km.

Detailed information related to calibration of the applied measuring apparatus under laboratory conditions was given by Fober (1993).

2.2 *Course of measurements*

The measurements were carried out in the Zabrze-Bielszowice mine on the N 791 face mined in the 502 seam situated at the depth of 780 m. The face of 300 m in length is being mined by means of a longitudinal system with roof caving. During measurements a height of -the face was of about 2.4 m. The FAZOS 15/31 Oz support was used to protect the roof of the face.

The immediate roof in the 502 seam is composed of a mudstone layer of up to 1 m in thickness, over which sandstone 14 m thick and mudstone forming the floor of the 501 seam are deposited. The 501 and 502 seams are distant about 18 m one from another.

A great seismic activity of the rock mass was noticed in the face area. In the course of measurements carried out continuously for 12 days the mine seismic station re corded 174 tremors with the energy grater than 10^3J in the face area. There were 56 among those tremors the energy of which was higher than 10^4J.

These tremors did not cause damage to elements of the support and their results became evident in form of spillage of coal from the face and this phenomenon was less or more intensive. Some of the tremors recorded by the seismic station were not felt at all on the face. The strongest tremor recorded with the energy of 10^6J, the epicentre of which was located in the gob area at a distance of about 15 m from the place where the measurements were taken, belonged, for instance, to the tremors not felt.

Dynamic changes in loads of the support were observed during 12 tremors of the rock mass. Results of the measurements-their recorded courses are discussed hereafter.

3 RESULTS OF MEASUREMENTS

Considering the character of time course of forces in legs the recorded dynamic loads can be divided into three fundamental groups.

A group of loads characterized by an increased static load after a tremor is the most numerous. Fig. 2 shows e.g. time courses of a resultant force in legs as well as a course of vertical accelerations of a canopy caused by a tremor with the energy of $3*10^4$J.

A frequency interval, within which changes of accelerations were analyzed has been limited to 15 Hz because of the frequency of proper vibrations of a support unit (determined through measuring of mechanical impedance) amounting to 23 Hz.

The table 1 presents results of an analysis of the acceleration spectrum the course of which is shown in fig. 2.

Table 1. Result of an analysis of the acceleration spectrum shown in fig. 2 and coming within the interval 0.3 - 100 Hz.

Frequency	Hz	0.4	6.8	34.8	50.4	100.4
Acceleration	$\frac{m}{s^2}$	0.18	0.09	0.06	0.12	1.32

Time courses similar to those shown in fig. 2 have been recorded in 9 cases of

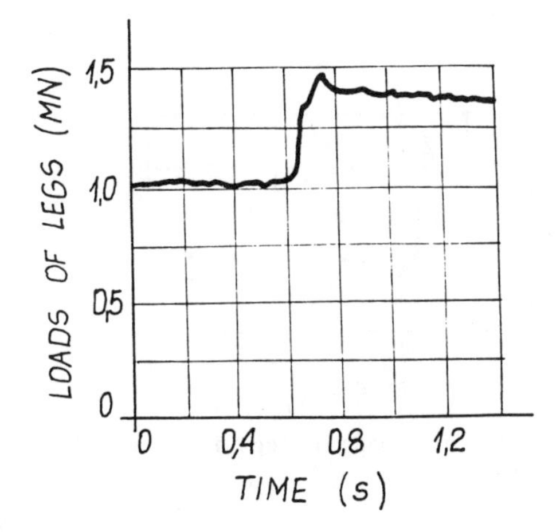

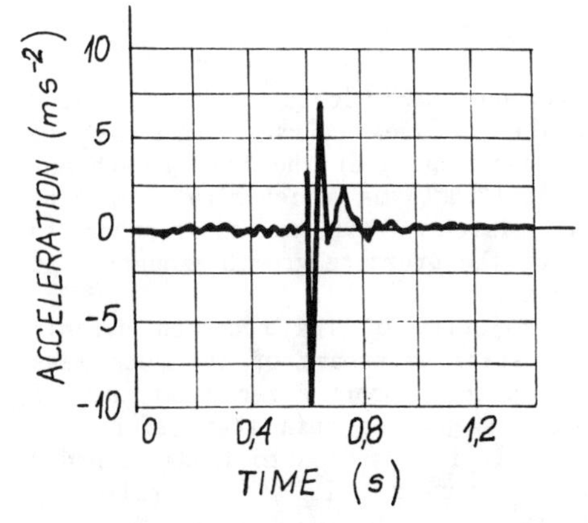

Fig.2. Changes in loads of legs and accelerations of a canopy caused by a tremor with the energy of $3*10^4$J.

tremors. Their epicentres were situated in the gob area relatively near the support unit, at a distance not exceeding 20 m from the face front.

The second group of the recorded changes in loads of the support unit is characterized by practically the same static load of the support before and after a tremor (see fig. 3).

During taking the measurements there were two tremors of this type. Their epicentres were situated on the face run at a distance of up to 70 m before the face front and about 200 m from the place of carrying out the measurements.

A characteristic feature of the third group of the recorded loads is a drop of static load of the support after a tremor.

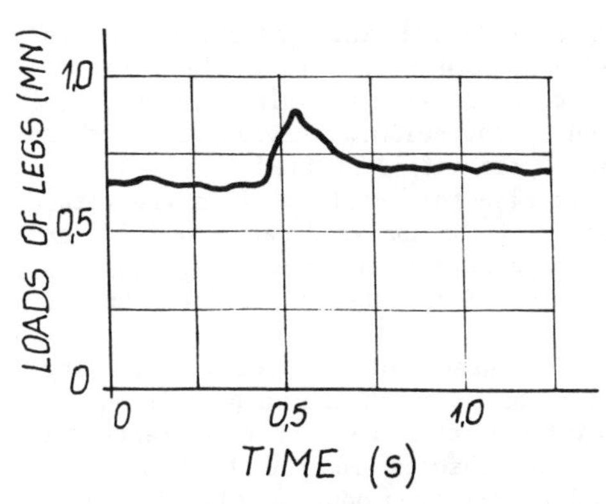

Fig.3. Changes in a force in legs caused by a tremor with the energy of 10^4J.

Such a course of changes in loads was caused by a tremor with the energy of $6*10^4$J situated in the gob area about 25 m after the face front and at a distance of about 60 m from the place of taking the measurements. A course of changes in the load caused by this tremor is shown in fig. 4 whereas results of an analysis of the acceleration spectrum are specified in the table 2.

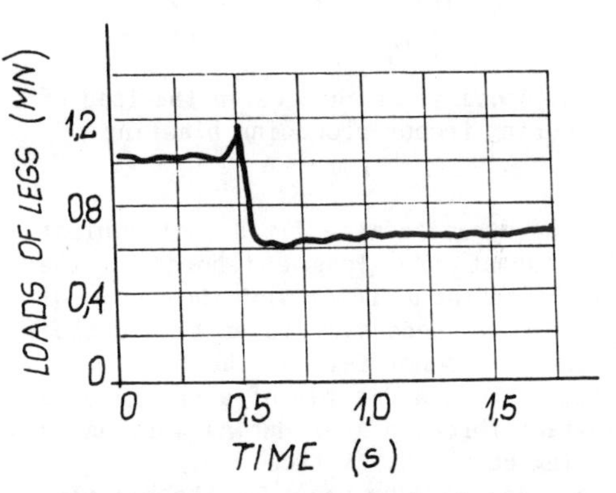

Fig. 4. Changes in the load of legs caused by a tremor with the energy of $6*10^4$J occurred in the gob area.

Table 2. Result of an analysis of the acceleration spectrum caused by a tremor with the energy of $6*10^4$J.

Frequency	Hz	0.4	6.8	50.0	62.4	100.0
Acceleration	$\frac{m}{s^2}$	42.6	0.11	0.15	0.12	1.37

Changes in the load which were analogous to those shown in fig. 4 took place in two other cases of tremors which were not recorded by the seismic station, so forcing of which was not identified.

It is also noticeable that in the vital majority of the analyzed cases of loads an occurrence of a tremor does not result in decreasing of the pressure exerted on the support.

The influence of a stress relieving blasting on a load of the support is not of an explicit character. For instance, a course of changes in the load of legs caused by tremor-producing blasting in a tailgate at a distance of about 40 m before the face front is shown in fig. 5.

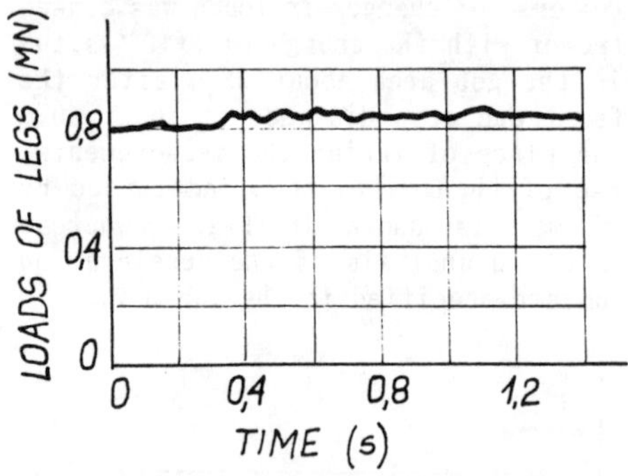

Fig. 5. A course of changes in the load of legs during tremor-producing blasting.

30 kg of dynamite were fired in two blast holes (about 30 m long and bored in the roof) producing a tremor with the energy of $2*10^4$J. It is necessary to emphasize that the tremor was not felt on the face.

Fig. 6 shows a diagram of a change of a resultant force in legs during working by blasting on the N 791 face.

Coal was mined by means of the blasting material in the period when the face front was within the range of action of an extraction edge of the 501 seam. The blasting material (of total weight of 70 kg) was disposed in 140 blast holes along the face and fired in three series. Force sensor recorded a change in the load of the legs at firing of the third series of blast holes situated nearest the place of taking the measurements.

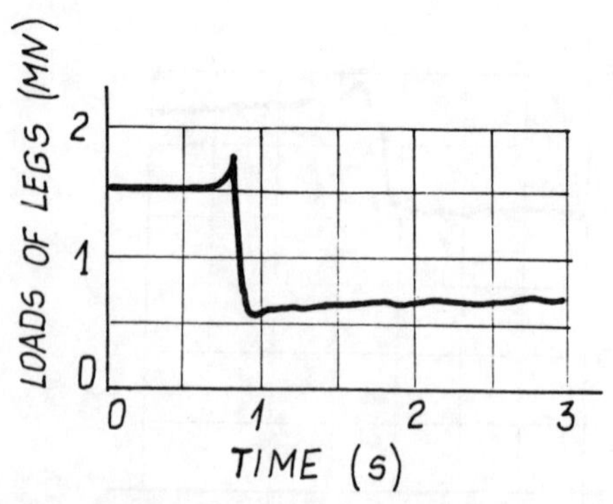

Fig. 6. Dynamic loads of legs during working by blasting.

4 RECAPITULATION

The recorded dynamic loads are characterized by relatively slow courses. This may be testified by speed of the load growth not exceeding 12.8 $kN*ms^{-1}$ which with a diameter of the piston of 200 mm corresponds to speeds of the pressure growth amounting to 407 $MPa*s^{-1}$.

In a majority of the recorded dynamic loads a static pressure of the rock mass exerted on the support after occurrence of a tremor is greater than that before the tremor. This refers also to loads recorded during a number of stress relieving blastings.

REFERENCES

Szweda S. 1991. The underground investigations of dynamic loads of a shield support. *Proc. Geomechanics'91*: 57-60. Rotterdam: Balkema.

Krasucki F.(editor) 1991. *Seminar of electrification and automation of mines. Mining telephone communication.* Gliwice: Skrypt Politechiki Śląskiej (in polish)

Fober S., Markowicz J., Szweda S. 1993. The method of underground measurements of parameters characterizing dynamic loads of mechanized roof support in mining conditions. *Z. N. Politechniki Sl. "Górnictwo" z. 210*: 39-48 (in polish, summary in english).

Geomechanics 93, Rakowski (ed.) © 1994 Balkema, Rotterdam, ISBN 90 5410 354 X

Influence of chosen factors on stability of mine shafts

A.Tajduś & M.Cała
University of Mining and Metallurgy, Department of Rock Mechanics, Cracow, Poland

ABSTRACT: Some factors which could influence the stability of mine shafts have been described. A procedure which lets estimate influence from excavation /using back analysis for Budryk-Knothe theory/ and ground water condition around shaft on shaft support stability has been introduced. As an example it has also presented the case of stability analysis for mine shaft in one of the Polish mines.

1. INTRODUCTION

Mine shafts are very intensively used excavations for all the time of its existence. That's why stability of mine shafts is necessary condition for well functioning of any underground mine.

2. CHARACTERISTICS OF CHOSEN FACTORS INFLUENCING MINE SHAFT STABILITY

2.1 Factors influencing shaft stability

Figure 1 shows group of factors which are necessary to take under consideration for proper estimation of mine shaft stability. For the proper estimation of shaft stability knowing of factors given below is required:

1. Lithology characteristics of rock layers around shaft /vertical intersection with named layers of rock and ground and its geomechanical parameters;
2. Hydrogeology characteristics of rock layers around shaft which requires consideration of follwing aspects:
- distinguishing of aqueous layers,
- detailed specifications of water horizons,
- susceptibility of aqueous layers to consolidation after water outflow,
- contraction of drying up rocks
- processes caused by rock drainage connected with loss of mineral substances /mechanical and chemical suffosion/,
  - underground erosion,
  - leaching of rocks,
3. Stress distribution in rock mass around shaft.
4. Shaft support characteristics /shaft diameter, way of driving, history of repairs and reinforcement of shaft support with reasons for applications/.
5. Detailed data about mining excavation around shaft /rate of disturbance of shaft pillar, way of excavation - fill, caving etc./.

There are two main factors influencing shaft stability:
- water outflow /drainage and drying/ from rock layers causing processes described above,
- mining excavation around shaft.

That's why deformations of shaft support and surrounding rock mass /displacement, strains, curvature, etc./ are sum of deformations caused by water outflow and excavations in a rock mass.

2.2 Influence of water outflow from rock mass on shaft stability

For estimating the influence of ground water conditions on the shaft stability knowledge about such a data as variability of water inflow

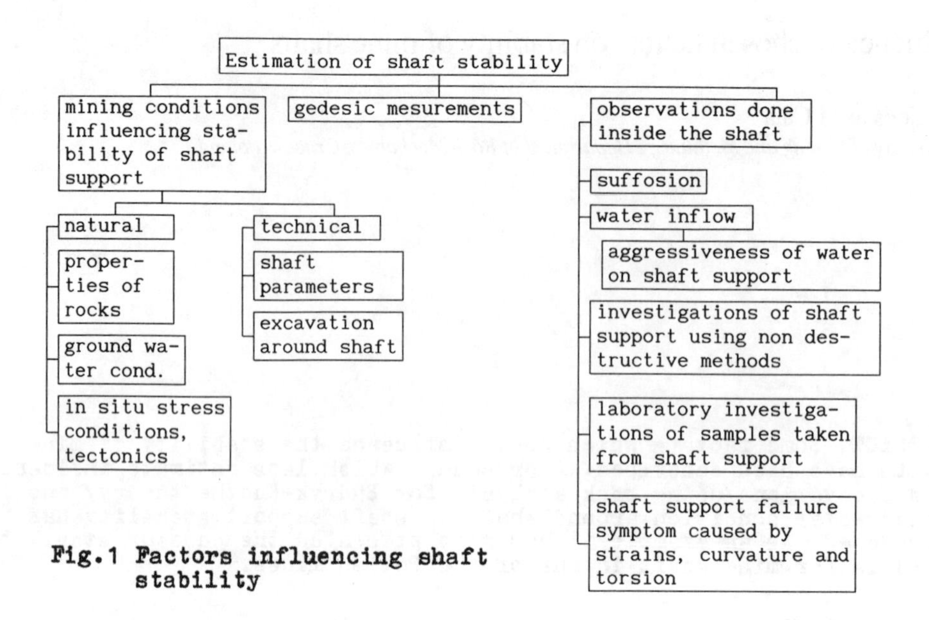

Fig.1 Factors influencing shaft stability

to the shaft and phenomena connected with it /suffosion, erosion, leaching, aggressiveness/ is necessary.

This data could be obtained from observations and measurements in the shaft. There are some unfavorable phenomena connected with water inflow to the shaft. One of this is consolidation of aqueous layers leading to condensation. The area of water outflow reaches outside shaft pillar region because of high degree of rock mass fracture which allows easier drainage. Shaft is immovable but rock layers which displace down cause /through pressure on support/ additional vertical compression forces acting on shaft support. These forces cause escalation of vertical displacement w and vertical strains ε_z in shaft support. When shaft crosses horizontal or weak - inclined layers, water outflow from rock mass do not effect horizontal displacement or horizontal strains and curvature. For estimating the value of compression forces created by water outflow, a scheme of shaft support loading showed on Fig.2 can be applied.

$$p_{z1} = \gamma_g \cdot h_1 \qquad /1/$$

$$p_{z2} = p_{z1} + p_w \qquad /2/$$

where:

p_w - vertical stress with hydrostatic lift taken into account,

$$p_w = \gamma_w /h_2 - h_1/ \qquad /3/$$

γ_g - unit weight of the rock mass,

γ_w - unit weight of the rock mass with hydrostatic lift taken into account,

p_{z1} - vertical stress on depth h_1 /roof of aqueous layer/,

p_{z2} - vertical stress on depth h_2 /floor of aqueous layer/,

Horizontal stresses acting on shaft support can be calculated as:

$$p_x = p_y = k_o \cdot p_z \qquad /4/$$

where:

k_o - coefficient of horizontal stress, for grounds it can be calculated as 1 :

$$k_o = 0.5 \cdot \sqrt[3]{D+1} \cdot [(1-\sin\varrho) / (1+\sin\varrho)]$$

ϱ - angle of internal friction,

D - shaft diameter,

Total pressure forces acting on shaft support can be calculated as:

$$P = \left[p_{x1}h_2 + p_{x2}(h_2 - h_1)\right] \frac{l}{2} \qquad /5/$$

where:
l - shaft support external perimeter: $l = \pi(D + 2g)$,

g - shaft support thickness,

p_{x1}, p_{x2} - horizontal stresses respectively on depth h_1, h_2.

Additional compressive force acting on shaft support on depth h_2 caused by water outflow is defined by:

$$Q = P \cdot \mu \qquad /6/$$

where:

μ - coefficient of friction between ground and support brick work.

Support brick work is rough so coefficient can be calculated as $\mu = 0.667 \cdot \varrho$ just like for retaining walls.

Additional compressive stresses in shaft support can be calculated as:

$$\sigma_z^d = \frac{4Q}{\pi \cdot (d_z^2 - D^2)} \qquad /7/$$

where:

$$d_z = D + 2g$$

Except additional compressive stresses σ_z^d originated from friction forces, there are also stresses from support weight which are equal:

$$\sigma_z^s = \gamma_s \cdot h_2 \qquad /8/$$

where:

γ_s - unit weight of shaft support,

Then total compressive stresses in shaft support on depth h_2 originated from vertical forces /without excavation influence/ can be calculated as:

$$\sigma_z = \sigma_z^s + \sigma_z^d \qquad /9/$$

Knowing the value of σ_z and making an assumption that shaft support is in elastic range, we can easy calculate the value of vertical strains from Hook's low.

Water which inflows to the shaft often transports small ground particles /mechanical suffosion/. That could be observed as seepages on shaft support or ground settling on constructions inside shaft. This is very unfavorable phenomenon and it causes creation of voids outside the support.

Chemical suffosion acts on similar principle but is connected with dissolving, washout and leaching rock and ground particles by water which inflows to the shaft. Both chemical and mechanical suffosion are difficult to estimate but shouldn't be neglected because created voids could lead to asymmetrical shaft support loading.

This can create tensile stress acting on shaft support.

Estimating ground water influence water aggressiveness level should be also taken under consideration. Some groups of chemical compound may cause corrosive damages of shaft support.

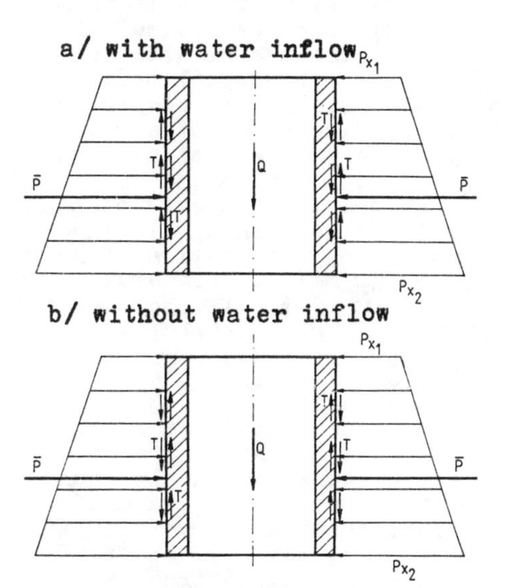

Fig.2 Shaft loading support scheme

2.3 Influence of excavation on shaft stability

For mine shafts stability protective pillars around them are created. Excavation shouldn't take place in pillar space but in Poland excavating coal resources from pillars is widespread. It's an effect of decreasing number of easy to excavate resources. Moreover it turn out that many protective pillars have been toosmall what is also a reason for shaft stability disturbance.

For estimating influence of excavation on shaft stability the Budryk-Knothe /B-K/ theory is commonly used in Poland. Main assumption of this theory are given elswhere [2], [3], [4]. One of the most important parameters of B-K theory is angle of main influence β, which estimates practical rich of excavation influence.
Protective pillar is designed before mining excavation so not well chosen β value can lead to too small dimensions of protective pillar and what follows that, to high shaft deformations caused by excavation. For each mining shaft regular geodesic measurements of ground subsidence in it's surrounding must be done. On that base using back analysis one can find main parameters of B-K theory, which are angle of main influence β and coefficient of horizontal stresses B. After collecting all data about coal seams excavated around shaft /depth, seam thickness, dimensions of excavations, angle β coefficient B etc./ using computer program [3] one can obtain deformation indexes of shaft: vertical and horizontal displacements w and u, declination of vertical axis T_z mm/m, curvature of vertical axis K_z [1/km] horizontal strains ε_x and ε_y [mm/m], vertical strain ε_z [mm/m].
Calculated values should be compared with geodesic measurements in chosen points on shaft support. Geodesic measurement should include: measurements of ground and shaft head frame subsidence, verticality of shaft axis, level of shaft of shaft hoist machine, girder strains etc. In case of significant differences between calculated and measured deformations calculations should be done again with changed angle β and coefficient B. Calculated deformations parameters couldn't be higher than allowable for shaft support. For example maximal value of vertical deformation for concrete support is ε_z = 1.6 mm/m, for concrete wall ε_z = 2.0 mm/m and for brick work ε_z = 2.5 mm/m, [5], [6], [7].

3. ESTIMATION OF PRESENT SHAFT STABILITY

For proper evaluation of shaft stability accurate analysis of shaft support conditions and possible damages of construction inside shaft are required. This analysis should include type and place of support damages, zones of water inflow to the shaft etc. Samples from shaft support for laboratory investigations should be taken.
Moreover, structural investigations of internal support damages could be done using acoustic - ultrasonic methods [5], [6], [7].
Detailed analysis of shaft support condition would let to find type of the deformations which caused damages, this are:

1. Excessive compressive vertical strains which cause: longitudinal support cracking, brick wall slide, distortion of shaft, crushing and bending of cage guide and girder breaking, crashing of vertical pipelines and tubes.
2. Excessive tensile vertical strains which cause: creation of vertical and horizontal cracks, breaking of pipelines and cage guides and falling of support fragments into the shaft.
3. Excessive shear strains which cause: creation of support shear, horizontal or askew displacement.
4. Excessive curvature and torsion which cause: cage wedging between cage guides, non uniform cage movement, faster using up of cage guides, shearing and removing of support fragments.

4. ESTIMATION OF SHAFT STABILITY IN MINE "X"

Shaft in mine X /diameter D=5.5m / has been driven in frozen rock mass due to poor ground water conditions till the depth of 50 m. Shaft support to the depth of 67 m is a brick work and from 67 m to 180 m is a concrete wall. Rock mass surrounding the shaft consist of layers: Quaternary formed by clay which reaches depth of 10.8 m, Tertiary formed by grey sandstone reaching depth of 148 m and Carboniferous formed by sandstones, shales and coal. There are three water bearing complexes: first water bearing complex is on the depth range from 9.1 to 10.8 m, water level is free and outflow could be estimated as few liters per minute; second water - bearing complex is on the depth range from 11.25m to 12.4m, water level is free, and similar to first complex water outflow could be estimated as few liters per minute /during shaft driving outflow was equal about 10 l/min/; third water - bearing complex is on the depth range from 16.0 m to 33.75 m, water level is free and water outflows measured to be about 120 l/min. Small suffosion has been observed in the shaft - shale deposits has been noticed at depth level of 42 m and 93 m. From investigations /done according polish standards/ it has been determined that water is non aggressive for shaft support. Started from 1970 local support peeling and after in 1990 dangerous horizontal cracking has been observed at the depth range of 78 m. Horizontal cracking zone has enclosed 3 m long shaft support interval. After investigations it turned out that concrete wall there has broken due to vertical compression /crushing of concrete wall support/. Since that time periodical observations on vertical line for support displacement measurement stated at the depth range from 66 m to 87 m has been carried out. Bench marks around shaft head frame has also been installed. Additionally vertical displacements among girders has been measured and level of shaft of shaft hoist machine has been checked. Estimations of support parameters has been carried out by laboratory investigations and acoustic - ultrasonic method. Brick work and concrete wall support except failure zone were found to be in good condition, without damages and its compressive strength ranges from $R_c = 8.8 - 9.2$ MPa. It has been calculated that support is able to carry acting static external loading /except failure zone/ with required safety coefficient. The question arises: what was the reason of shaft support failure at depth of 78 m because the way of repairing support is connected with it. Fig.3 shows diagram of shaft head frame subsi-

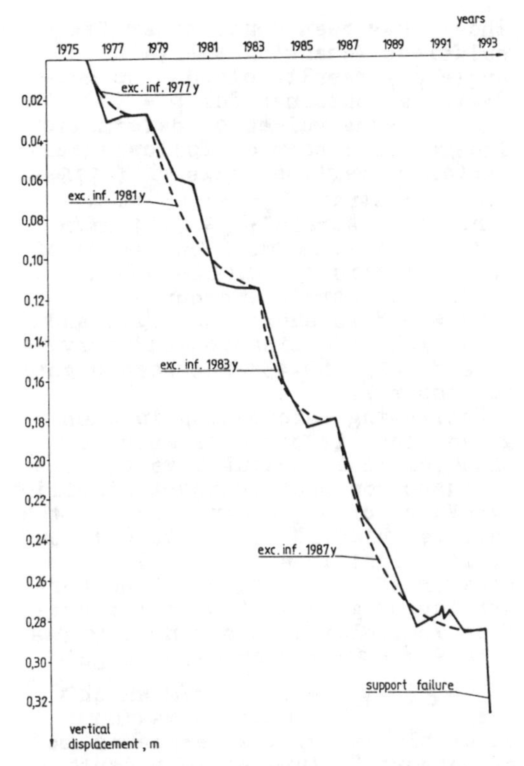

Fig.3 Shaft head frame subbsidence

dence during period from 1975 to 1992. Influences of particular excavations /1977, 1981, 1983,1987/ in shaft pillar region has been also signed.

From comparing increased values of shaft head frame subsidence with periods of excavation it could be assumed that excavation had significant influence on shaft deformations. Subsidences has been practically finished about three years from each done excavation. Influence of the last excavation /1987/ has stopped in the beginning of 1990. Then in july 1992 violent increase of head frame subsidence /reaching value of 30 mm/ has been noticed. As it has been stated before it couldn't be caused by excavation but most probably it has been the additional vertical stresses caused from water outflow. Measurements of displacements among girders confirms this thesis. Basing on values of geodesic measurements of surface subsidence back analysis using B-K theory has been done. After few calculations /changing the value of angle β / results closest to geodesic was obtained for $\beta = 51^o$. In failure zone values of deformation indexes have been as follows: declination of vertical axis T_z= 1.776mm/m vertical strain $\varepsilon_z = -1.65$ mm/m, horizontal strain $\varepsilon_x = 2.74$ mm/m and $\varepsilon_y = 1.11$ mm/m, curvature of vertical axis $K_z = 0.0028$ 1/km. Diagram of strain and curvature for $\beta = 51^o$ is showed on Fig.4 and fig.5 /also declination and curvature for $\beta = 63.43^o$ has been signed to compare/.

Estimating deformation indexes given above it could be assumed that vertical strains have highest influence on shaft support stability. Value of ε_z doesn t exceed the failure value for concrete wall but is its significant part - 78%. However after adding shaft support strains originated from water outflow /calculated according formulas 1 to 9 / - $\varepsilon_w = 0.48$ mm/m we get $\varepsilon_z^c = \varepsilon_z + \varepsilon_w = 2.12$ mm/m which exceeds maximum allowable value. That explains fact of support failure at 78 m depth. Due to repair support a damageable cartridge /could be wooden cartridge 0.5 m thick/ could be applied.

5. CONCLUSIONS

1. For proper shaft stability estimation of following phenomena should be taken under consideration: influence of excavation, influence of water outflow, present state of shaft support confirmed by mining and laboratory investigations. For estimating excavation influence the B-K theory can be applied [3].

2. Around each mine shaft and also inside shaft following measurements should be done: surface subsidence measurements, displacements and strains on chosen points of shaft and shaft head frame, verticality of the shaft, displacement among girders etc. Next calculated results of deformations should be compare with measured ones. When significant differences will arise parameters of B-K theory should be changed and water outflow should be also taken into consideration.

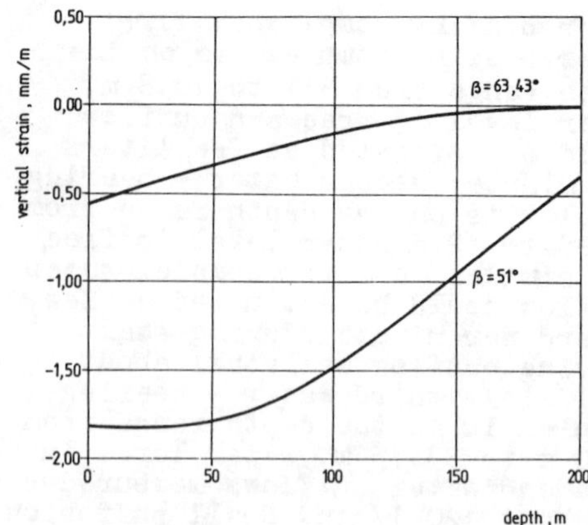

Fig.4 Shaft wertical strain

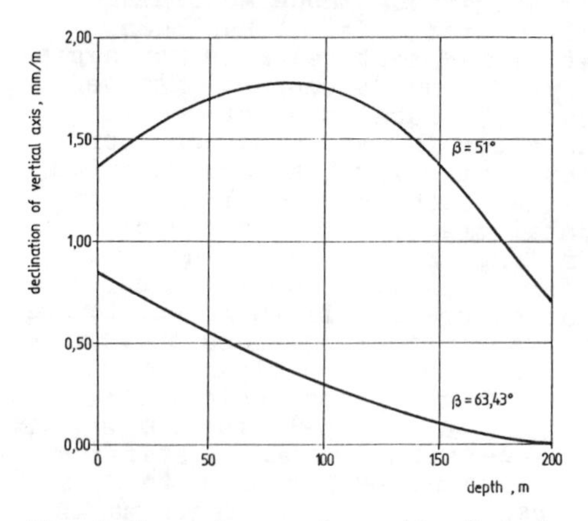

Fig.5 Curvature of vertical axis

3. Properly done shaft stability estimation let to find damages in shaft, cause of its arise and ways of repairs.

6. REFERENCES

Bieniawski Z.T. 1987. Strata Control in Mineral Engineering. A.A. Balkema, Rotterdam.

Drzęźla B. Information about computer programs for ground subsidence calculations. Przegląd Górniczy, 3/1974, in polish.

Flisiak J. 1989. Applications of computers for rock mass deformations forecasting. ZN AGH 1243, Górnictwo 142, Cracow 1989, in polish.

Knothe S.1984. Forecasting of mining

excavation influence. Śląsk, Katowice, in polish.
Strzelecki Z.,Witosiński J. Application of acoustic methods of control of supports and its interaction with rock mass. ZN AGH 552, Cracow 1975, in polish.
Tajduś A.,Witosiński J. 1992. Method of estimating capacity of concrete shaft support in dependence on its corrosive damages. Works of Commission of Mining and Geology, Górnictwo 29, in polish.
Tajduś A.,Witosiński J.1992. Application of non-destructive investigations results for estimating capacity of broken shaft supports. Works of IGiH PW 63, Wrocław, in polish.
Surface protection from mining damages; edited by M.Borecki, Śląsk, Katowice 1980, in polish.

Numerical methods
Main lectures

Geomechanics 93, Rakowski (ed.) © 1994 Balkema, Rotterdam, ISBN 90 5410 354 X

On non-monotonic stress path effect in numerical analysis of geotechnical structures

M. Doležalová & V. Zemanová
Institute of Geotechnics, Academy of Sciences of Czech Republic, Prague, Czech Republic

ABSTRACT: Analysis of monitoring and numerical calculation results shows development of non-monotonic stress and strain paths during monotonic loading and unloading of geotechnical structures. Stress paths calculated by FEM for different types of structures (embankment dam, shallow foundation, underground opening) are analysed in the paper with particular attention to the non-monotonic development of the shear stress level producing loading and unloading in shear. Correct simulation of soil stiffnes changes due to these paths is necessary for realistic displacement prediction. A check on capability of different constitutive models of geomaterials to account for this effect is made in the paper using unit response envelopes suggested by Gudehus (1979). Proper switch functions for variable moduli models and account for deviatoric hardening for elasto-plastic models are necessary for avoiding overestimation of displacements in numerical calculations of geotechnical structures.

1 INTRODUCTION

Due to structural changes of soils during loading and unloading the soil response is path dependent, i.e. depends not only on the stress state reached but also on the way leading to this state. To get a realistic displacement prediction, the stiffness changes associated with this path dependence should be taken into account by any constitutive model applied for geomaterials.

In this aspect two trends in the development of constitutive modelling can be noted. As to the theory a great number of sophisticated constitutive models aiming at covering all possible paths have been created during the last few decades. As to the computational practice, however, the simplest perfectly elastic and plastic models are mostly applied which neglect any path effect inside of the failure surface.

To narrow this gap between theory and practice, problem oriented constitutive models accounting for the most important strain and stress paths of the structure under consideration would be useful. In this way simpler constitutive models with better predictive capability can be obtained. For this purpose characteristic stress and strain paths of typical geotechnical structures should be investigated and the constitutive model response in these directions checked. The characteristic paths can be found by numerical calculations and by analysing field measurement results (Doležalová, 1976, 1993). A general check of constitutive relations in any direction can be done by numerical element tests using samples with homogeneous strain and stress distribution (Gudehus and Kolymbas, 1985).

Investigation of carried out in this direction by the authors showed development of non-monotonic strain and stress paths during monotonic loading and unloading processes. This is demonstrated in the paper using FEM not only for zoned dams as in the previous works (see Doležalová, 1993; Doležalová, Zemanová, 1994), but also for other types of geotechnical structures (shallow foundation and underground opening).

The monotonic increase/ decrease of the normal stress level is associated with non-monotonic development of strain anisotropy and shear stress level. Zones with higher shear stress level exert lateral expansion and make the adjoining zones work in condition of anisotropic compression, i.e. paths revealing unloading in shear occur. Development of non-monotonic path is also supported by irregularities of geometry and construction sequence of real structures. These features are important with respect to the marked stiffening effect of the paths inducing unloading in shear which was discovered by some non-standard laboratory tests not only for sand but also for clay. So overestimation of displacements can occur if these effects are neglected by the constitutive model used.

In the paper variable stiffness models and associated and non-associated elasto-plastic constitutive models are checked from this point of view using unit response envelopes suggested by Gudehus (1979). Imposing a unit strain increment of various direction on a soil sample the corresponding stress increment can be calculated according to the checked constitutive model. Assuming axi-symmetric sample and loading a lucid graphical representation can be obtained and the effect of path direction changes displayed.

So far the main conclusion is that proper switch function for variable moduli models and account for deviatoric hardening for elasto-plastic models are necessary for correct simulation of soil stiffness changes due to the development of non-monotonic stress paths producing loading and unloading in shear.

2 STRESS PATHS OF GEOTECHNICAL STRUCTURES

Strain and stress paths of geotechnical structures can be derived either from field measurement results or from results of numerical calculations. The first approach has been successfully used by the authors for embankment dams, but this possibility is rather exceptional. In this paper stress paths calculated by FEM for monotonically loaded (zoned dam, shallow foundation) and unloaded geotechnical structures (underground opening) are presented.

2.1 *Stress path groups*

Stress paths of real geotechnical structures are rather complex and so far even the most comprehensive experimental programs have been restricted either to stress paths with no or one sharp bend (Havlíček, Hroch, 1975 ; Havlíček, 1979) or to strain/stress paths with one or two sharp bends (Topolnicki, 1987). In this paper only stress paths with one sharp bend are analysed.

According to the above experiments considerable stiffness changes are associated with the path direction changes and certain stress paths form well defined groups. Such groups determined by combination of the switch functions sign$\Delta\sigma_{oct}$ and signΔi are shown in Fig.1. Here $\sigma_{oct} = (\sigma_1 + \sigma_2 + \sigma_3)/3$ denotes octahedral normal stress (compression +), σ_j (j = 1,2,3) principal stresses, $i = \tau_{oct}/\tau^{lim}_{oct}$ shear stress level, $\tau_{oct} = \frac{1}{3}[(\sigma_1 - \sigma_2)^2 + (\sigma_2 - \sigma_3)^2 + (\sigma_1 - \sigma_3)^2]^{\frac{1}{2}}$ octahedral shear stress and τ^{lim}_{oct} its limit value according to the failure surface applied. Similar stress paths groups defined by volumetric and deviatoric yield surfaces were distinguished by the double hardening elasto-plastic model of Molenkamp (1983) as it is shown in Fig.2. So analysis of occurrence and effect of such stress path groups seems to be useful for checking both, the variable moduli (zero order hypoelastic) and elasto-plastic constitutive models.

Thus in the following four stress path groups reflecting volumetric and deviatoric hardening are analysed and displayed with labels used in our FEM

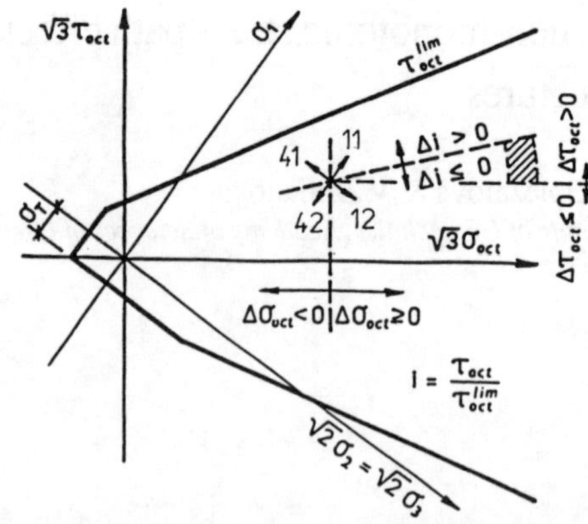

Fig.1 Switch functions and stress paths groups of path dependent variable stiffness model (Doležalová, 1985)

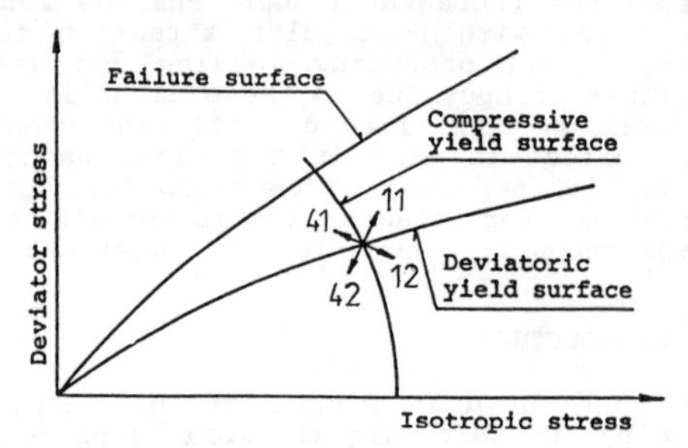

Fig.2 Failure and yield surfaces considered by double hardening model of Molenkamp, 1983

codes:

11 : $\Delta\sigma_{oct} \geq 0$, $\Delta i > 0$

12 : $\Delta\sigma_{oct} \geq 0$, $\Delta i \leq 0$

41 : $\Delta\sigma_{oct} < 0$, $\Delta i > 0$

42 : $\Delta\sigma_{oct} < 0$, $\Delta i \leq 0$.

2.2 *Laboratory test results*

To estimate the effect of these stress paths groups on soil behaviour, the above mentioned constants stress ratio (CR) triaxial tests of isotropically and anisotropically (K_0- paths) consolidated sand (Havlíček, Hroch, 1975) and silty clay (Havlíček, 1979) have been analysed. The characteristic features of these path groups are as follows:

- **11** monotonically decreasing soil stiffness which can be simulated by loading branch of standard triaxial tests with $\sigma_1 > \sigma_2 = \sigma_3$
- **12** marked stiffening effect depending on the relative shear stress level decrease ($\Delta i \leq 0$)
- **41** no results since these paths have not been included into the experimental program

42 abrupt stiffness increase which is well simulated by unloading branch of standard triaxial tests.

These conclusions were firstly published in Doležalová and Hoření (1982) and revised by Doležalová (1993). The findings allowed to explain some longterm discrepancy between field measurement and FEM calculation results so it was decided to analyse the biaxial test results of Topolnicki (1987) as well. This work has not been finished yet, but the first check for the path **12** fits the above conclusion.

2.3 *Zoned dams*

Strain and stress paths of these structures have been repeatedly analysed by the authors since 1976 and the results are summarized in Doležalová, Zemanová (1994).

A characteristic feature of the mechanical behaviour of zoned dams is the interaction of zones with different shear stress level. Zones with higher shear stress level (as a rule clay cores, but also transition zones) produce lateral expansion and make the adjoining zones work in condition of anisotropic compression. This interaction is associated by stress paths of loading in shear (11) for the laterally expanding zones and stress paths of unloading in shear (12) for the laterally compressed ones.

The development of these paths due to monotonic loading of a dam by self weight during construction and by water pressure during the first reservoir filling is depicted in Fig.3. Large zones with stress path **12** developing in the shells indicate stiffening of the fill and clarify why the use of oedometer test results recommended by some authors (Penman and Charles, 1971; Eisenstein and Law, 1979) for the shells resulted in realistic displacement prediction for this part of dam. Considering the field evidence, however, it is difficult to foresee what part of dam will be shared or compressed. This is the task of the constitutive model applied.

2.4 *Shallow foundation*

The behaviour of the subsoil (medium dense sand) under a shallow foundation loaded by monotonically increasing strip load was studied assuming lateral pressures coefficient at rest $K_0 = 0.54$ and 0.92. The development of stress path zones **11** and **12**, as well as the shear failure zones in vicinity of the footing are shown in Fig.4. The non-monotonic increase/decrease of the shear stress level and the associated paths can be noted again even for early loading stages presented here.

This indicates that unloading in shear occurs not only due to material inhomogeneity inherent to the structures like zoned dams, but also due to inhomogeneity induced by loading an initially homogeneous region. For shallow foundations the failure zones with no

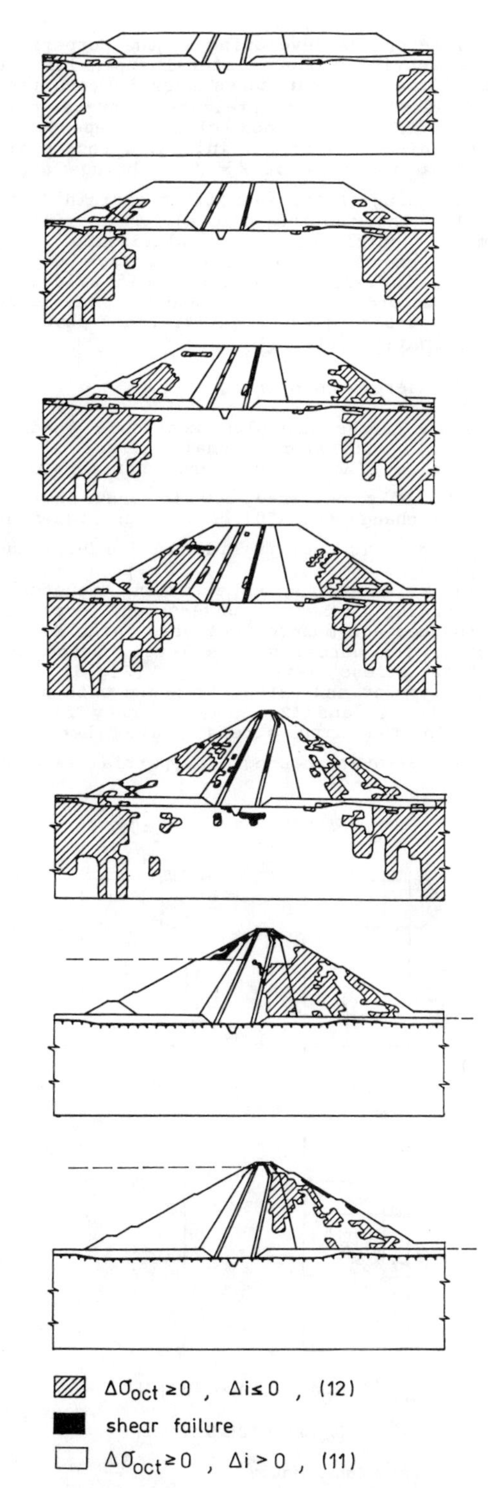

Fig.3 Zones indicating loading (**11**) and unloading in shear (**12**) during dam construction and reservoir filling

volume change developing at the corners of the footing produce lateral expansion and make the adjacent zones work in condition of anisotropic compression (stress path **12**). This phenomenon clearly depends on the anisotropy of the initial stress state and in the case of K_0= 0.54 brings about stiffening of the subsoil in certain depth below the footing (Fig.4). This finding may contribute to clarifying the development of so called "active depth" which is well known from field evidence, but according to the authors' knowledge has not been given a satisfactory physical explanation yet.

2.5 *Underground opening*

In the above examples monotonic loading caused monotonic normal stress level increase ($\Delta\sigma_{oct} >$ 0) and non-monotonic paths only occurred due to shear stress level change ($\Delta i \gtreqless 0$). A similar statement however, is not valid for underground openings.

Apart from the expectation, monotonic unloading of an underground opening produces non-monotonic stress paths due to both, the normal stress level change and shear stress level change (Fig.5). The development and extent of zones with paths **11**, **12**, **41** and **42** depend on many factors as in the situ stress state (here K_0= 1.5), excavation progress (partial face or

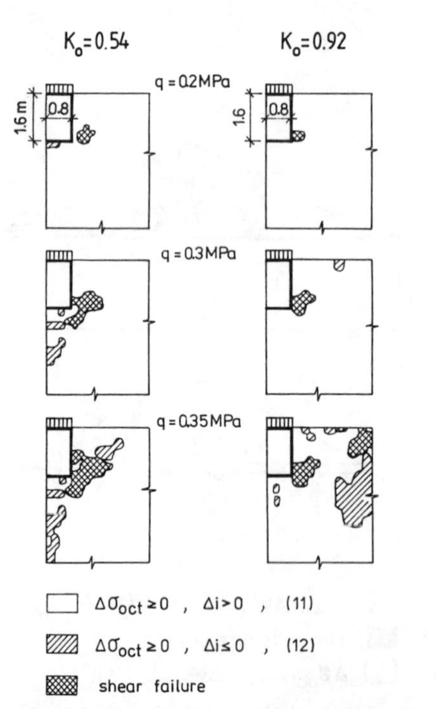

Fig.4 Influence of K_0 on the development of shear failure and zones of loading (**11**) and unloading in shear (**12**)

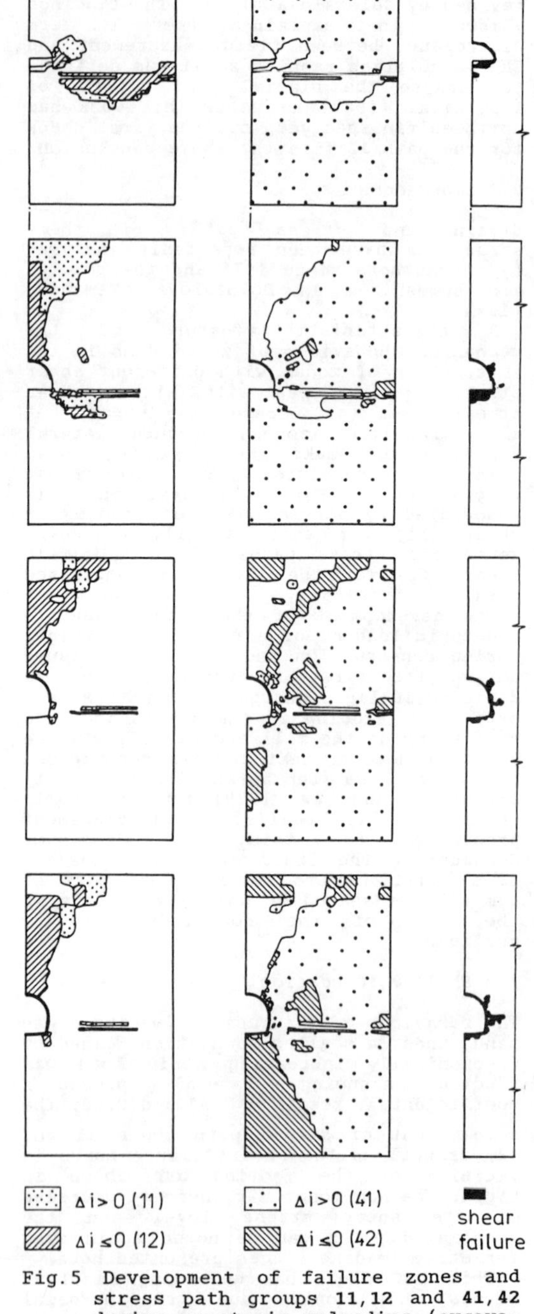

Fig.5 Development of failure zones and stress path groups **11,12** and **41,42** during monotonic unloading (excavation) of an underground opening

fullface excavation), construction technology applied for the temporary and permanent lining, etc. Similarly, the effect of various stress paths on the rock mass behaviour depends on the quality of the rock (here siltstone with uniaxial strength 50-100 MPa) with respect to the magnitude of the in situ stresses determined by the overburden height (here 900 m). Only stress paths bringing about structural changes of the material are meaningful for constitutive modelling. This should be checked by appropriate laboratory tests and the capability of constitutive models to account for them by numerical tests using unit response envelopes.

3 CHECK ON CONSTITUTIVE MODELS USING UNIT RESPONSE ENVELOPES

Numerical testing and comparison of incremental constitutive relations of geomaterials according to the methodology suggested by Gudehus, 1979 and Gudehus and Kolymbas, 1985 have been already published in detail in Doležalová, Boudík, Hladík, 1992 and in Doležalová, 1992. Here a short information on unit response envelopes and a check on some variable moduli and elasto-plastic models concerning their account for stiffening effect of path **12** are given.

3.1 *Unit response envelopes*

Numerical element test of a homogeneous soil sample allows to calculate the stress/strain response of a constitutive relation along any strain/stress path following either a laboratory test or a complex loading process which is characteristic of a real structure.

Since this path dependent behaviour can be only simulated by incremental laws integrated along the paths, only this type of constitutive relations is considered in the following.

The basic incremental law of hypoelasticity

$$d\sigma_{ij} = C_{ijkl} d\varepsilon_{kl} \qquad (1)$$

renders a general framework for treating different constitutive relations (linear, non-linear, elasto-plastic, etc.) of geomaterials. Eqn.(1) is homogeneous in time and the material matrix C_{ijkl} depends on the stress tensor σ_{ij} only.

Instead of following a particular path by a numerical test and plotting the corresponding stress-strain curve, it is more general and purposeful to compute the stress increments due to unit strain increments imposed on a soil sample in various directions. This generality concerning the path directions is of great importance for numerical analyses, where any path direction can be encountered in a material point during a loading/unloading process. The envelope determined by these stress increments in the principal stress space is called unit response envelope.

So assuming an initial state σ_{ij} and a constant strain increment $d\varepsilon_{ij}$ of different directions the corresponding incremental stress response $d\sigma_{ij}$ can be computed according to the checked constitutive law. A very simple and lucid graphical representation can be obtained for axi-symmetric stress state (σ_1, $\sigma_2 = \sigma_3$ and $\sigma_1 = \sigma_2$, σ_3) where the initial state is given by σ_1/σ_3, the strain increment by $\dot{\varepsilon} = \sqrt{\dot{\varepsilon}_1^2 + 2\dot{\varepsilon}_3^2}$ = const and $\alpha_{\dot{\varepsilon}}$ = arctg $(\dot{\varepsilon}/\sqrt{2}\dot{\varepsilon}_3)$ and the incremental stress response by $r_{\dot{\sigma}} = \sqrt{\dot{\sigma}_1^2 + 2\dot{\sigma}_3^2}$ and $\alpha_{\dot{\sigma}}$ = arctg $(\dot{\sigma}_1/\sqrt{2}\dot{\sigma}_3)$. In Figs. 6 a, b the imposed constant strain increment and a typical stress response envelope are shown in the triaxial plane of principal stress space. The strain and stress paths of some

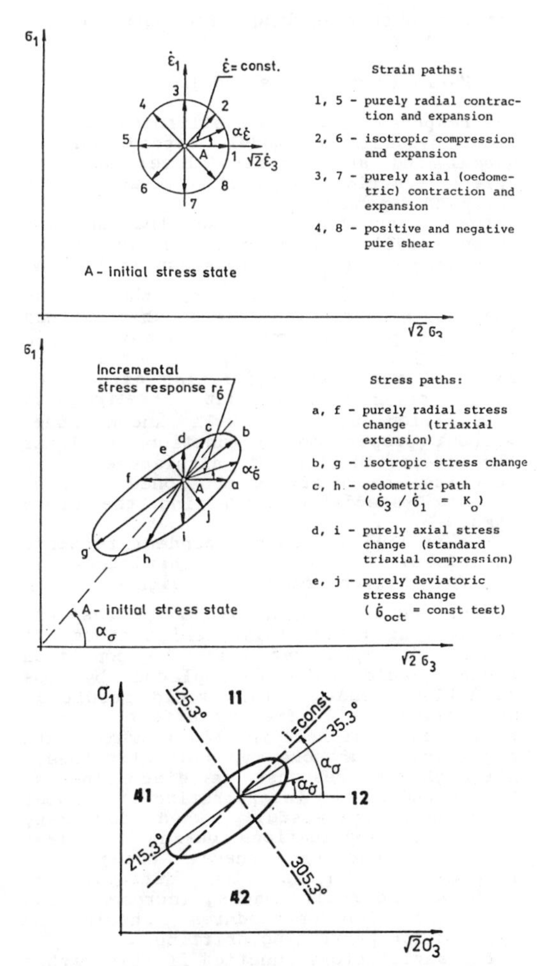

Fig. 6 Unit strain increments (a), typical unit response envelope (b) and stress path groups (c) in triaxial plane of principal stress space

laboratory tests are also displayed in this Figures. In Fig.6c the sectors corresponding to path groups **11**, **12**, **41** and **42** are depicted. As only rate-independent constitutive laws are considered, the overdot in the above relations and Figures simply denotes an incremental quantity.

Concerning the unit response envelopes the stiffening effect of path **12** appears as a stiff response $r_{\dot{\sigma}}$ in the directions $\Delta i \le 0$ or $\alpha_{\dot{\sigma}} \le \alpha_{\sigma}$ for $\sigma_1 > \sigma_2 = \sigma_3$ and $\alpha_{\dot{\sigma}} \ge \alpha_{\sigma}$ for $\sigma_1 = \sigma_2 > \sigma_3$. In these relations $\alpha_{\sigma} = \operatorname{arctg}(\sigma_1 / \sqrt{2}\ \sigma_3)$ corresponds to the stress state before the increment. Through Figs.7-9 the directions of α_{σ} and $\alpha_{\dot{\sigma}}$ are shown in the sector $\alpha^f_{\sigma} > \alpha_{\dot{\sigma}} > 35{,}3°$ where the stiffening effect of unloading in shear is as a rule neglected (α^f_{σ}- path direction corresonding with the failure surface).

3.2 *Variable stiffness models*

The response of variable stiffness (zero - order hypoelastic) models depends on the normal and shear stress level and the switch functions applied for accounting the path dependency.

The hyperbolic response (Duncan and Chang, 1970) with a switch function for unloading $\operatorname{sign}(\dot{\varepsilon}_1 - \dot{\varepsilon}_3)$ is shown in Fig.7a. A considerable decrease of the stress response with the increase of shear stress level is produced by this model. It is correct for the paths **11**, but not for the paths **12** where the stiffening effect had to be taken into account. The frequent application of this well known model without proper switch functions explains the overestimation of displacements in numerical calculations mentioned by many authors especially concerning the zoned dams.

The multilinear path dependent response (Doležalová, 1985) is depicted in Fig.7b. Switch functions $\operatorname{sign}\Delta\sigma_{oct}$ and $\operatorname{sign}\Delta i$ are used which divide the response envelope into four elliptical sectors. It should be emphasized that the condition $\operatorname{sign}\Delta i$ could not be replaced by the condition $\operatorname{sign}\Delta\tau_{oct}$. This would result in neglecting the stiffening effect of path directions hatched in Fig.1 which are significant for some types of structures. Although the response is discontinuous, the 2D and 3D FEM incorporating this model have been successfully used for many industrial applications during the last decade. However, these codes are recommended for solving deformational problems and small loading increments and self-correction procedures should be applied for preventing drifting.

The satisfactory function if this rather simple variable stiffness model especially regarding the displacement predictions is mostly due to the proper account for path dependence which is close to the concept of double hardening elasto-plastic models

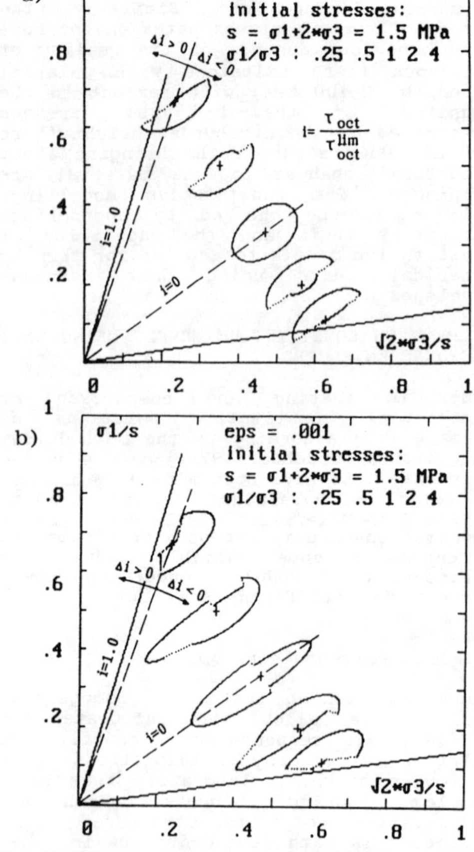

Fig.7 Unit response envelopes of variable moduli models
a) Hyperbolic response with unloading condition $\operatorname{sign}(\dot{\varepsilon}_1 - \dot{\varepsilon}_3)$
b) Multilinear path dependent response with unloading conditions $\Delta\sigma_{oct} < 0$ and $\Delta i \le 0$

(compare Figs.1 and 2). Stiffening effect of the path **12** is properly considered by this model (Fig.7b) and the shape of the unit response envelopes in proximity to the failure surface is similar to ones of elasto-plastic models accounting for deviatoric hardening (Figs.9a,b and Figs. 7 and 9 in Gudehus, 1979).

3.3 *Elasto-plastic models*

Elasto-plastic response with associated flow is shown in Fig.8 representing the behaviour of original (a) and modified (b) Cam Clay models (Roscoe, Burland, 1968). The response envelopes are continuous and convex and for the modified version also smooth. In Figs.8a and 8b the elasto-plastic response is compared with the elastic one yielding some information on strain energy dissipation. The Cam Clay models are incorporated in many commercial FEM codes (CRISP, ADINA and others) and can be effectively used for prediction of normally consolidated clay behaviour including pore pressure dissipation. The yield conditions, however, only controlled by volumetric hardening can produce overestimation of displacements if these models are used for frictional materials. This is evident from Figs.8a and 8b yielding weak stress response for conditions $\alpha_{\dot{\sigma}} \leq \alpha_{\sigma}$ especially in close vicinity to the failure surface (F).

Elasto-plastic response with non-associated flow rule recommended for frictional materials is represented in Figs.9a,b and 10 (Lade and Kim, 1988). In Fig.9a the elastic and the non-associated elasto-plastic stress responses are compared. In Fig.9b the check on proper account for the stiffening effect of the path 12 is made with positive result. In Fig.10 the yield (P) and plastic potential (G) surfaces corresponding to the initial

a)

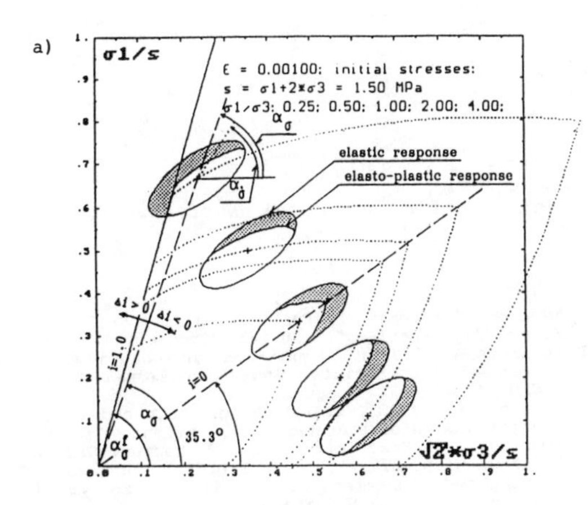

b)

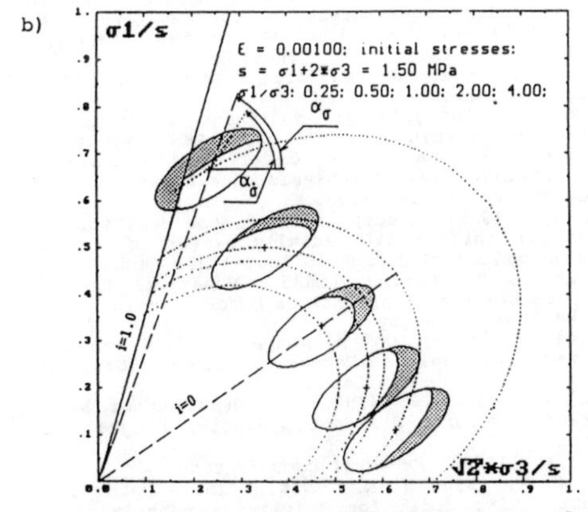

Fig.8 Unit responses of the original (a) and modified Cam Clay model (b) and their comparison with the purely elastic response

a)

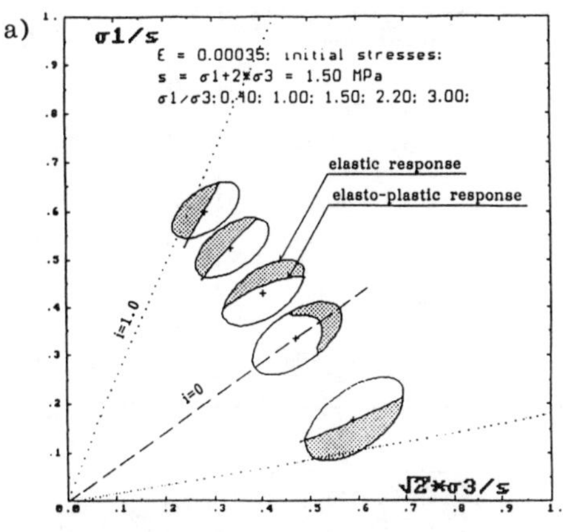

b)

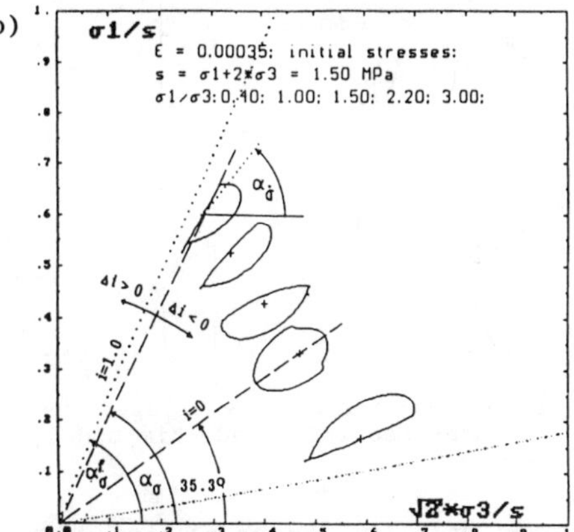

Fig.9 Unit response of the non-associated elasto-plastic model (Lade and Kim, 1988)
a) Comparison with the purely elastic response
b) Check of the model response in the path directions $\alpha_{\sigma}^{f} \leq \alpha_{\dot{\sigma}} \leq 35.3°$

states of the unit response envelopes are represented (Hladík, 1992). So using a yield surface, expanding in both, the direction of isotropic and isochoric compression, and a separate plastic potential surface for proper modelling of plastic (especially volumetric) strains, the response envelopes fulfil the requirements for a realistic displacement prediction.

4 CONCLUSIONS

Non-monotonic stress paths of some typical geotechnical structures and the capability of some constitutive models of geomaterials to account for their effect are analysed in the paper. The analysis allows to draw the following conclusions.

Monotonic loading of geotechnical structures produce non-monotonic stress paths concerning the loading in shear.

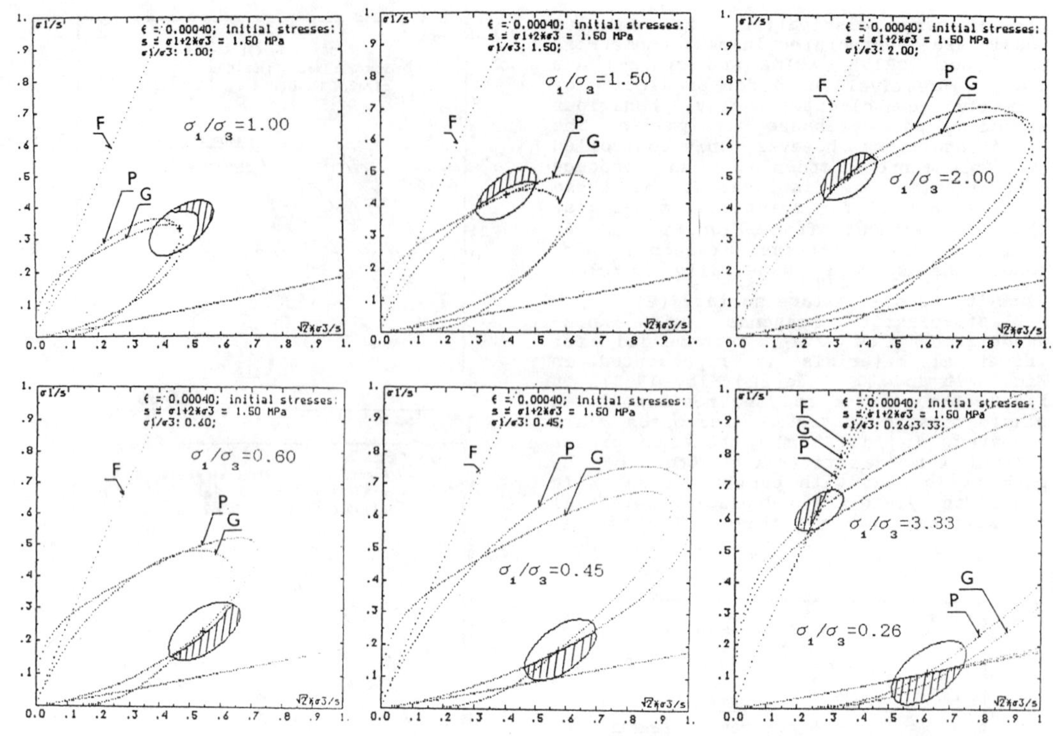

Fig. 10 Unit response envelopes vs yield and plastic potential surfaces for non-associated Lade-Kim model (1988)

Monotonic unloading is associated even a larger variety of paths: non-monotonic loading/unloading occurs not only due to shear stress level change, but also due to increase/decrease of the normal stress level.

The stiffening effect of deviatoric hardening seems to play a more significant role in realistic displacement prediction than it has been thought. Relevant switch functions for variable moduli models and account for deviatoric hardening for elasto-plastic models are necessary for proper consideration of this phenomenon.

Problem oriented constitutive modelling seems to be perspective concerning both, the correctness of the models and their simplicity as well. The related research should be extended in following directions: more systematic stress path analysis of a larger variety of geotechnical structures, laboratory test programs for characteristic path groups and path sequences considering different materials and initial states and numerical testing of further constitutive models using both, elements tests and solution of boundary value problems.

REFERENCES

Doležalová, M. (1976). Strain paths in embankments. *5th Budapest Conf. on SMFE*: 41-52. Budapest.

Doležalová, M. (1985). Description of a pseudo-elastic constitutive model. *Proc. of Conf. on "Use of Microcomputers for Solving Soil Mech. and Found. Eng. Problems"*: 130-141. Prague (in Czech).

Doležalová, M. (1992). Comparison of constitutive models of geomaterials. *2nd Czechoslovak Conf. NUMEG'92*: 176-182. Prague.

Doležalová, M. (1993). Unloading in shear and over-predictions in displacement calculations. *"Modern Approaches to Plasticity"*: 641-670. Horton. Greece.

Doležalová, M., Boudík, Z., Hladík, I. (1992). Numerical testing and comparison of constitutive models of geomaterials. *Acta Montana*: 33-56. Prague.

Doležalová, M., Hoření, A. (1982). Strain paths in rockfill dams - measurements, constitutive laws, FEM calculations. *4th ICONMIG*: 679-690. Edmonton.

Doležalová, M., Zemanová, V. (1994). On stress-strain behaviour of zoned dams. *XIII ICSMFE*. New Delhi. India (in print).

Duncan and Chang (1970): Nonlinear analysis of stress and strain in soils. *ASCE*. SM: 1629-1653.

Eisenstein, Z., Law, S. T. C. (1979). The role of constitutive laws in analysis of embankments. *3rd ICONMIG*. Vol. 4: 1413-1430.

Gudehus, G. (1979). A comparison of some constitutive laws for soils under radially symmetric loading and unloading. *3rd ICONMIG*: 1309-1323. Aachen.

Gudehus, G., Kolymbas, D. (1985). Numerical testing of constitutive relations for soils. *5th ICONMIG*: 63-81. Nagoya.

Havlíček, J. (1979). On the shape of State Boundary Surface of soils. *Research Report*. Stavební geologie. Prague (in Czech).

Havlíček, J. Hroch, Z. (1975). Deformation of sand. *Research Report*. Stavební geologie. Prague (in Czech).

Hladík, I. (1992). *Personal comunication*.

Lade, P. V., Kim, M. K. (1988). Single hardening constitutive model for frictional materials. I, II, III. *Computers and Geotechnics*. 5/307-324, 6/13-29, 6/ 31-47.

Molenkamp, I. F. (1983). Elasto-plastic double hardening model MONOT. *Delft-Geotechnics*. Delft. Holland.

Penman, A.D.M., Charles, J.A. (1971). Observed and predicted deformations in a large embankment dam during construction. *BRS Current Paper* 18/71.
Roscoe, K.H., Burland, J.B. (1968). On generalized stress behaviour of "Wet" clay, in Eng. Plasticity. *Cambridge University Press*: 535-609.
Topolnicki, M. (1987). Observed stress-strain behaviour of remoulded saturated clay and examination of two constitutive models. *Publications of the Institute of Soil and Rock Mechanics in Karlsruhe*. Vol.91.

Geomechanics 93, Rakowski (ed.) © 1994 Balkema, Rotterdam, ISBN 90 5410 354 X

Scale effects on rock mass deformability

P.H.S.W. Kulatilake
University of Arizona, Tucson, Ariz., USA

ABSTRACT: Due to the presence of joints, jointed rock masses show anisotropic and scale (size) dependent mechanical properties. The available literature on the topic up to 1992 shows that satisfactory procedures are not available in the rock mechanics literature to estimate the anisotropic, scale-dependent deformability properties of jointed rock. A brief description of a recently suggested technique which has the capability of estimating the anisotropic, scale-dependent deformability properties of jointed rock is given. Based on this technique, the following have been obtained: (a) 3-D plots to show the variation of deformability parameters of jointed rock with joint geometry parameters, (b) a relation between the deformability properties of jointed rock and fracture tensor parameters, (c) an incrementally linear elastic, orthotropic constitutive model to represent the pre-failure mechanical behaviour of jointed rock, and (d) some insight to estimate the representative elementary volumes (REVs) and REV property values with respect to the deformability properties of jointed rock. The new constitutive model has captured the anisotropic, scale-dependent behaviour of jointed rock. Limitations of the model are identified. Further work required to improve the model is mentioned.

1 INTRODUCTION

To arrive at safe and economical designs for structures build in and on rock masses, it is important to have a correct understanding of the deformability properties of jointed rock masses. Discontinuity networks in rock masses are complicated and inherently statistical. Due to factors such as the inability to simulate field conditions in the testing apparatus, material spatial variability, simplified assumptions used in test interpretations, scale effects, sample disturbances, and both instrument and human error, the geomechanical properties of discontinuities, intact rock and rock blocks estimated using laboratory or field tests are subject to some uncertainty. Therefore, it is difficult to accurately predict the deformability properties of jointed rock masses. Jointed rock masses usually show anisotropic and scale (size) dependent mechanical properties. The literature available up to 1992 shows that satisfactory procedures are not available in the rock mechanics literature to estimate the anisotropic, scale-dependent mechanical properties of jointed rock.

To obtain realistic results for jointed rock mass mechanical properties, many large volumes of rock of different sizes having various joint configurations should be tested under different stress paths at significant stress levels. In the laboratory, it is almost impossible to carry out an experimental program such as this. In the field, it would be very difficult, time-consuming and expensive. Therefore, it is no surprise that there is no such experimental program listed in the current literature.

In a recent paper, Kulatilake et al. (1993a) covered the available literature on the topic. In another paper, Kulatilake et al. (1992) have suggested a technique to perform stress analyses using the distinct element method (Cundall 1988; Hart et al. 1988) for rock masses with non-persistent joints to study the effect of joint geometry parameters (orientation, intensity, size, etc.) on the strength and deformability of rock masses. Using this technique, a fairly detailed investigation was performed to study the effect of joint geometry networks on the deformability of rock masses. A summary of the technique used and the results of the investigation are presented in this paper. For details the reader is referred to Kulatilake et al. (1992; 1993a) and Wang and Kulatilake (1993).

2 SUMMARY OF THE TECHNIQUE

Figure 1 provides the flow chart of the procedure used to study the effect of joint geometry parameters on the deformability properties of rock masses. Actual joints can be generated in two or three dimensions (2D or 3D), according to either a deterministic scheme or a stochastic scheme (Kulatilake et al. 1993b). A three-dimensional distinct element code (3DEC) is employed to perform these stress analyses because of its distinct features (for instance, the capacity of tackling large displacements and rotations of the rock block) which are especially useful in analyzing jointed rock masses. However, 3DEC requires that the problem domain be completely discretized into polyhedral blocks by the joint system. A typical non-persistent, actual joint network in 3D usually does not discretize a rock block into polyhedral blocks. Therefore, fictitious joints are introduced into each rock block to connect them with existing actual joints to discretize the rock block into small polyhedral blocks.

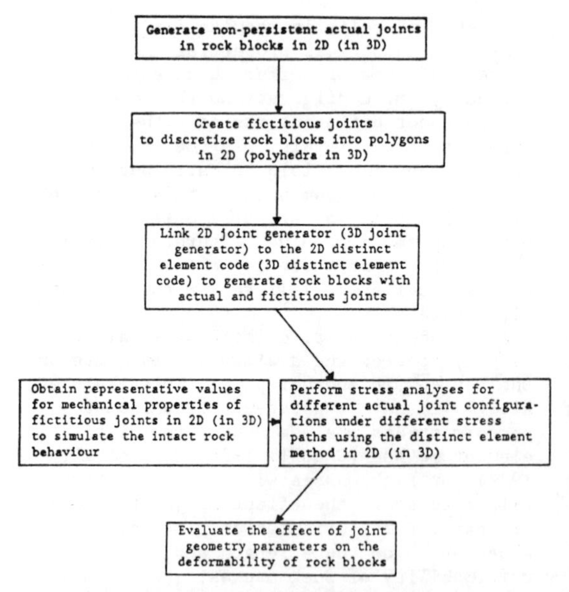

Fig. 1. Flow chart of the procedure used to study the effect of joint geometry parameters on the deformability properties of jointed rock.

For details concerning this connection procedure, the reader is referred to Wang and Kulatilake (1993). The question arises as to which deformation and strength parameter values should be assigned to the fictitious joints so that they behave like intact rock. A solution to this problem has been reported in a recent paper (Kulatilake et al. 1992).

3 EFFECTS OF JOINT GEOMETRY PARAMETERS OF NON-PERSISTENT ACTUAL JOINTS ON THE DEFORMABILITY OF ROCK BLOCKS

The final two steps of Figure 1 are related to this sub-topic. The procedures used and results obtained are given below.

3.1 Procedures

Cubical blocks of side dimension 2 m were chosen for this study. In these rock blocks, non-persistent actual joints were generated in a systematic fashion. In each rock block, a certain number of actual joints having a selected orientation and a selected joint size was placed to represent a joint set. Joints were considered as 2D circular discs. Joint center locations for these joints were generated according to a uniform distribution. In most blocks, only a single joint set was generated. In some blocks, two joint sets were generated. Blocks were generated with a number of different joint configurations. Fictitious joints were then introduced into each of these blocks to connect them with actual joints and to discretize the domain of the rock block into polyhedra.

The rock blocks were considered to consist of a granitic gneiss material (Hardin et al. 1982). The mechanical behaviour of both the intact rock and the fictitious joints was represented by linear, elastic, perfectly plastic constitutive models with the Mohr-Coulomb failure criterion, including a tension cut-off (Kulatilake et al. 1992). Parameter values given in Table 1 were assigned to the intact rock and the fictitious joints. For actual joints, the same constitutive model as for the fictitious joints was used along with the parameter values given in Table 1.

In order to estimate different mechanical properties of the jointed rock blocks, two different stress conditions were applied: (1) Each rock block was subjected to an isotropic normal stress first; then, in each of the selected three perpendicular directions, the normal stress was increased

Table 1 **Values Used for the Parameters of the Constitutive Models for Granitic Gneiss Intact Rock, and the Fictitious and Actual Joints**

Intact Rock		Joints	Fictitious Joints	Actual Joints
Parameter	Assigned Value	Parameter	Assigned Value	Assigned Value
Density (d)	2500 (Kg/m^3)	Joint normal stiffness (JKN)	7.5 x 10^6 MPa/m	6.72 x 10^4 MPa/m
Young's modulus (E)	60 (GPa)	Joint shear stiffness (JKS)	3 x 10^6 MPa/m	2.7 x 10^3 MPa/m
Poisson's ratio (ν)	0.25	Joint cohesion (jc)	50 MPa	0.4 MPa
Bulk modulus (K)	40 (GPa)	Joint dilation coefficient (jd)	0	0
Shear modulus (G)	24 (GPa)	Joint tensile strength (jt)	10 (MPa)	0
Cohesion (c)	50 (MPa)	Joint friction coefficient (tanφ)	0.839	0.654
Tensile strength (t)	10 (MPa)			
Friction coefficient (tanφ)	0.839			

until failure of the rock block (Fig. 2) to estimate the deformation moduli and the Poisson's ratios in the three perpendicular directions. (2) The rock block was subjected to an isotropic normal stress first; then, the shear stress was increased on each of the three perpendicular planes (Fig. 3) to estimate the three shear moduli. These stress analyses were performed using 3DEC.

From these analyses, the deformation properties of rock blocks with different deterministic combinations of joint size, number of joints, joint orientation and number of joint sets were obtained. The influences of the joint geometry parameters on the mechanical behaviour of jointed rock blocks were then investigated based on the results of the analyses.

3.2 Results

Figure 4a shows the effects of the joint size/block size and the joint density (number of joints per unit volume) on E_x/E_i when actual joints are oriented in a certain direction in the rock block. Figures 4b and 4c show the same effects on E_y/E_i and E_z/E_i, respectively. In the figure, E_i represents the intact rock Young's modulus and E_x, E_y and E_z are the deformability moduli of the rock block in the x, y and z directions, respectively. Figures 5a, 5b and 5c show the typical effects of joint size/block size and joint density on G_{xy}, G_{xz} and G_{yz}, respectively. In these figures, G_i represents the intact rock shear modulus and G_{xy}, G_{yz} and G_{zx} are the shear moduli of the rock block in the xy, yz and

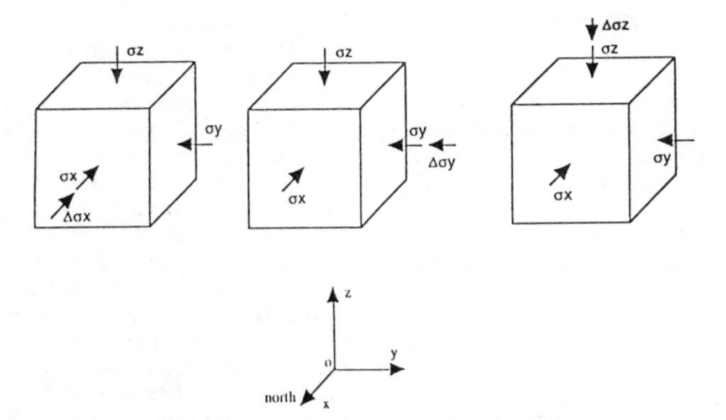

Fig. 2. Stress paths of first type used to perform distinct element stress analysis of rock blocks with actual joints.

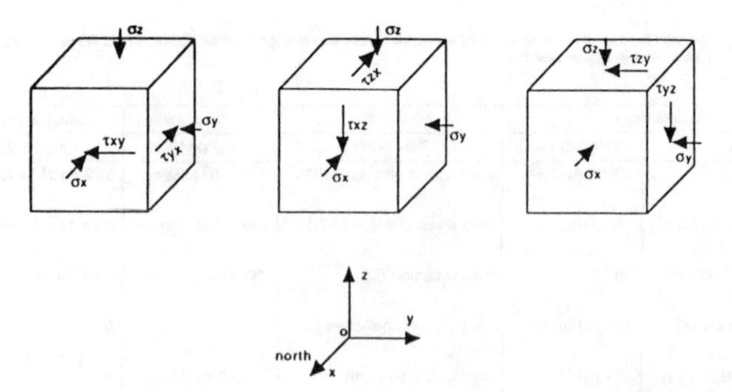

Fig. 3. Stress paths of second type used to perform distinct element stress analysis of rock blocks with actual joints.

zx planes, respectively. All the rock block moduli were determined at a stress level equal to 15% of the intact rock strength for illustrative purposes. However, these parameters can be estimated at any required stress level. These plots show that E_x/E_i, E_y/E_i, E_z/E_i, G_{xy}/G_i, G_{xz}/G_i and G_{yz}/G_i reach more or less asymptotic equivalent continuum behaviour (Representative elementary volume <REV> behaviour) with increasing joint density and increasing joint size/block size. Figures 4a through 4c together clearly show the anisotropy in the deformation modulus. Similar plots of E_x/E_i, E_y/E_i, E_z/E_i, G_{xy}/G_i, G_{xz}/G_i and G_{yz}/G_i were obtained for other joint networks considered in this study (Wang 1992). It was discovered that the asymptotic moduli values varied with the joint orientation. It was not possible to see any clear relationship between any of the Poisson's ratios and the joint geometry parameters (Kulatilake et al. 1993a).

4 FRACTURE TENSOR, ITS COMPONENTS AND INVARIANTS

To evaluate the combined effects of joint geometry parameters on the mechanical behaviour of jointed rock blocks, the concept of fracture tensor introduced by Oda (1982) was used.

For details about the fracture tensor, its components and invariants, the reader is referred to the paper by Kulatilake et al. (1993a).

5 RELATIONSHIPS BETWEEN ROCK MASS DEFORMABILITY PARAMETERS AND FRACTURE TENSOR PARAMETERS

For rock blocks having different joint configurations, the fracture tensor, its invariants and directional components were calculated using the equations given in the paper by Kulatilake et al. (1993a). The relationships between the deformability properties of the jointed rock blocks and (a) the first invariant of fracture tensor $I_1^{(F)}$ and (b) the directional component of the fracture tensor, were then investigated.

Figures 6 and 7 show the effects of $I_1^{(F)}$ on the deformation moduli and the shear moduli for a certain joint orientation. It can be seen that, generally, the $I_1^{(F)}$ has a great influence on the deformation and shear moduli properties of jointed rock masses. Each moduli shows a negligible change and an asymptotic equivalent continuum behaviour (REV behaviour) beyond a critical $I_1^{(F)}$ value. This relationship and the related critical value of $I_1^{(F)}$ were found to depend on the joint orientation and the direction of the deformability property considered.

In order to study the relationship between the deformability properties of rock blocks and the fracture tensor in detail, the components of fracture tensor were used. Figure 8 shows the relationship between the rock mass modulus in any given direction, E_m, and the component of fracture tensor in the corresponding direction, F_k. The best-fit power function obtained for the data, and the resulting multiple R^2 value are shown in the figure. When F_k is greater than about 2.5, the variation of E_m is very small and rock blocks show an equivalent continuum behaviour (REV behaviour). These results show clearly that, for all practical purposes, a possibility exists to estimate rock mass modulus in any given direction from a single power function by knowing the intact rock and the joint properties (from laboratory tests) and the joint geometry system (from field joint mapping).

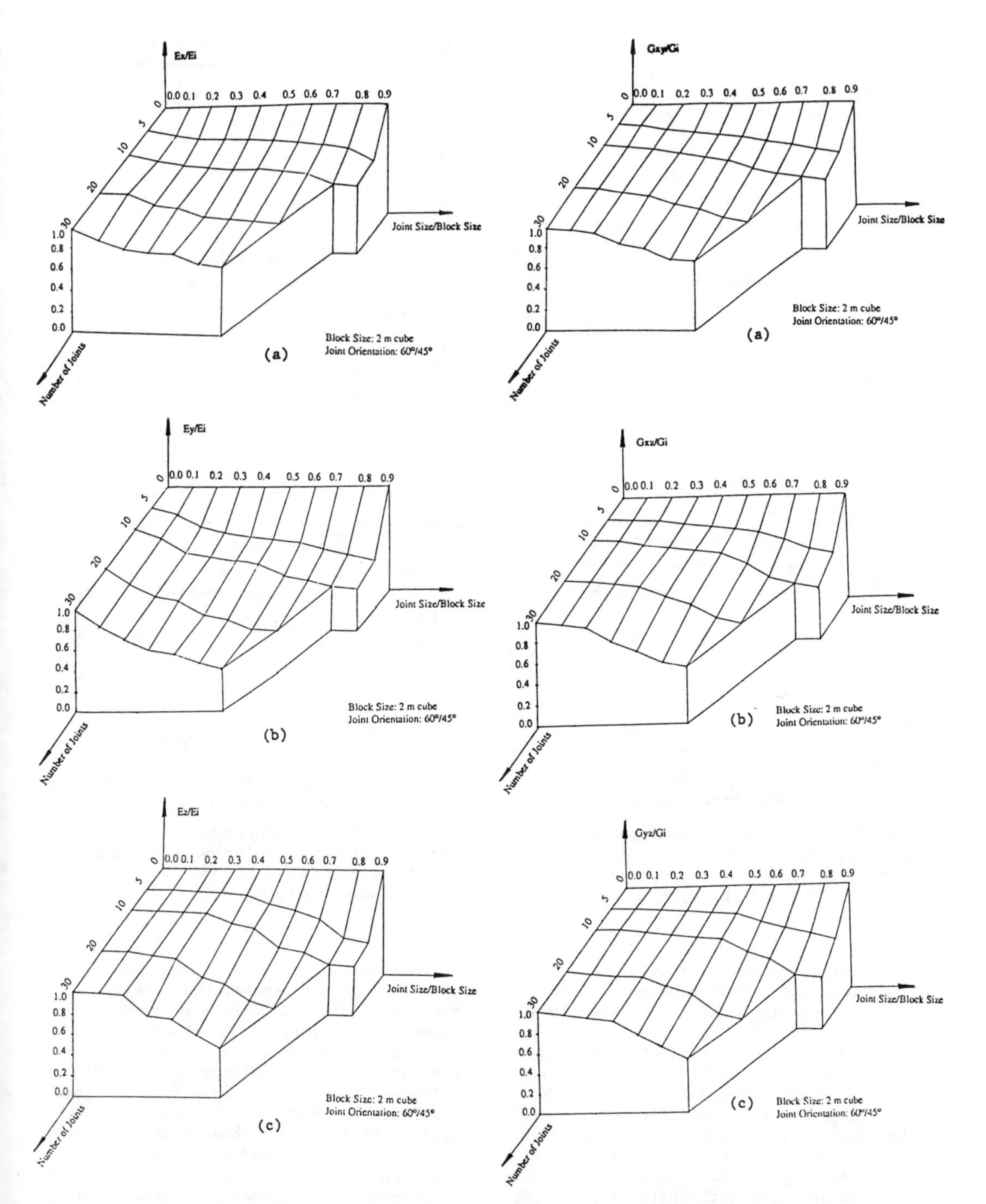

Fig. 4. Typical effects of joint size/block size and joint density on rock block deformability modulus (results from 60°/45° joint orientation): (a) on E_x/E_i, (b) on E_y/E_i and (c) on E_z/E_i.

Fig. 5. Typical effects of joint size/block size and joint density on rock block shear modulus on three perpendicular planes (results from 60°/45° joint orientation): (a) on G_{xy}/G_i, (b) on G_{xz}/G_i and (c) on G_{yz}/G_i.

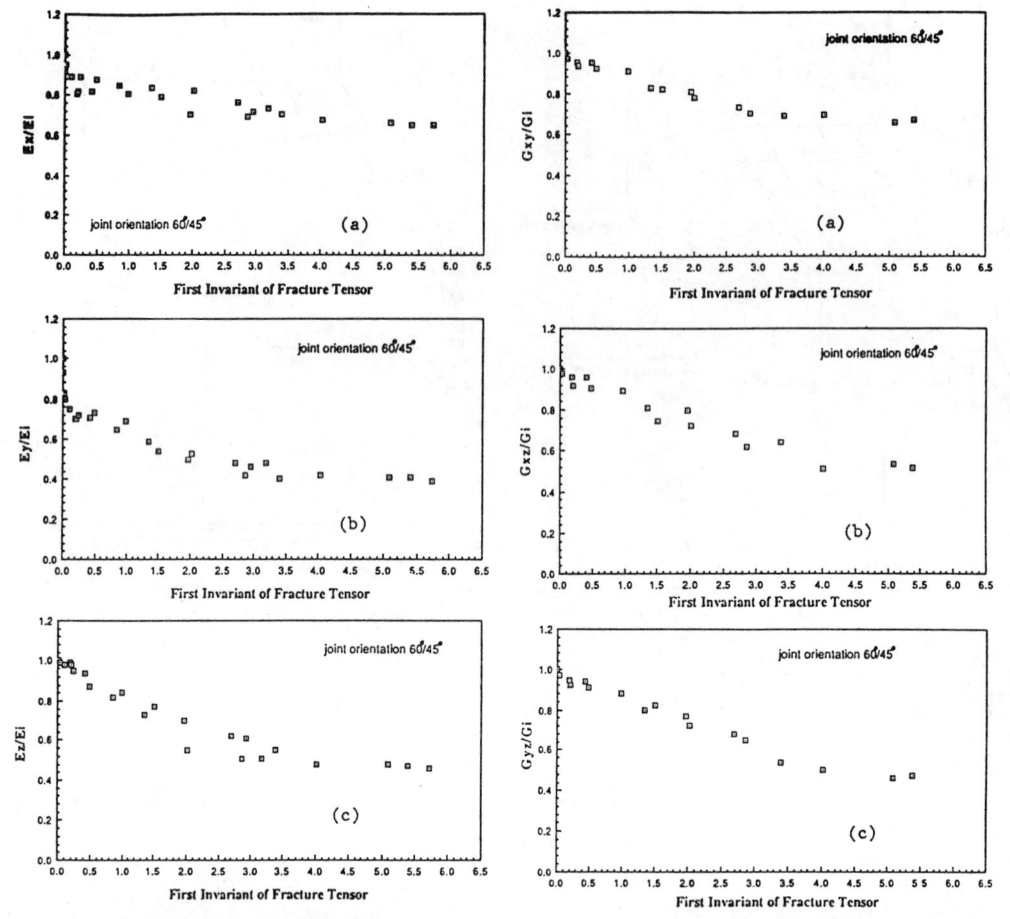

Fig. 6. Relationships between rock block deformational moduli and $I_1^{(F)}$ for 60°/45° joint orientation: (a) E_x/E_i vs. $I_1^{(F)}$, (b) E_y/E_i vs. $I_1^{(F)}$, and (c) E_z/E_i vs. $I_1^{(F)}$.

Fig. 7. Relationships between rock block shear moduli and $I_1^{(F)}$ for 60°/45° joint orientation: (a) G_{xy}/G_i vs. $I_1^{(F)}$, (b) G_{xz}/G_i vs. $I_1^{(F)}$, and (c) G_{yz}/G_i vs. $I_1^{(F)}$.

For the investigated rock blocks, it was found that the deformation modulus of the jointed rock block can be represented in different directions in three dimensions on the surface of an ellipsoid (Kulatilake et al. 1993a), where the direction of the radial vector of the ellipsoid gives the direction of the deformation modulus and the length of the radial vector provides a value proportional to the modulus value. The anisotropy of the rock mass modulus was found to increase with increasing $I_1^{(F)}$ (Kulatilake et al. 1993a).

Figure 9 shows the relationship between the rock mass shear modulus on any plane, G_m, and the summation of fracture tensor components on that plane. It can be seen that when $F_k + F_l$ is beyond about 4.0, a further increase in $F_k + F_l$ does not change G_m much. At this stage, the rock block shows an equivalent continuum or REV behaviour with respect to shear deformation. The best fit power function obtained for the data and the resulting multiple R^2 value are shown in the same figure. These results show that, for all practical purposes, the rock mass shear modulus on any plane can be estimated from a single power function, as long as the intact rock and joint mechanical properties (from laboratory tests) and the joint geometry system (from field joint mapping) are known.

The relationships in Figures 8 and 9 are used in the next section in setting up a constitutive model to describe the pre-failure behaviour of jointed rock masses.

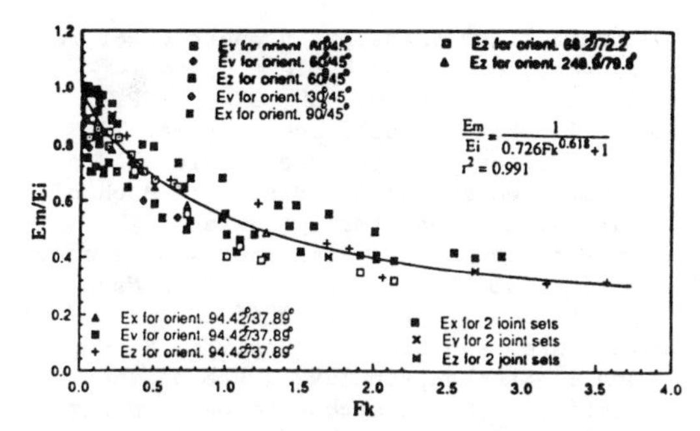

Fig. 8. Relationship between rock block deformational modulus in any direction, E_m, and the fracture tensor component in the same direction.

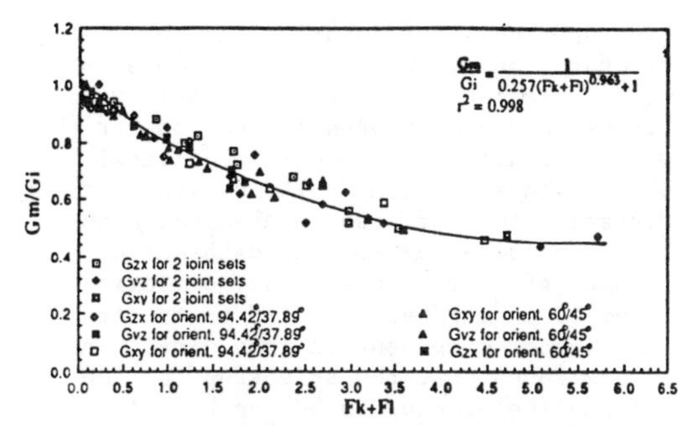

Fig. 9. **Relationship between rock block shear modulus on any plane, G_m, and the summation of fracture tensor components on that plane.**

6 A CONSTITUTIVE MODEL FOR ROCK MASSES TO REPRESENT PRE-FAILURE BEHAVIOUR

In the previous section, relationships were obtained between the deformability properties of jointed rock blocks and the fracture tensor components. Using these relationships, an incrementally linear elastic orthotropic constitutive model (Eq. 1) is suggested to represent the pre-failure behaviour of a jointed rock mass in 3D.

In Eq. 1, 1, 2 and 3 are the principal directions of the fracture tensor. The parameter values (a, b, m, n) in this constitutive model depend on (a) the stress level at which the mechanical behaviour of the rock mass is evaluated and (b) the constitutive models chosen for the intact rock and joints. For example, the equations given in Figs. 8 and 9 provide (a, n) and (b, m) values, respectively, applicable under the following conditions: (a) at the stress level of 15% of the intact rock strength, and (b) for granitic gneiss rock with the assumed constitutive models for the intact rock and joints. The rock mass Poisson's ratios can be assumed to vary between 0.5 ν_i and 2 ν_i, in which ν_i is the Poisson's ratio for the intact rock. In this model, the effect of joint geometry has been taken into account in terms of fracture tensor components; the scale-dependent and anisotropic behaviours of rock masses are incorporated. Note that the fracture tensor parameters are functions of the joint size distribution and the mean joint intensity of joints, which are in turn dependent on the rock block size. This explains how the scale effect is incorporated into the constitutive model. In addition, the constitutive model can be incorporated easily into various numerical techniques to perform stress analysis in jointed rock masses.

7 CONCLUSIONS

Joint geometry parameters of finite size joints showed a significant influence on deformability of jointed rock at the 3D level. It seems that a number of different combinations of joint density and joint size/block size values can provide estimations for REV behaviour. To estimate rock mass deformability properties, the fracture tensor seems to be very useful. Relationships found between deformability proper-

$$\begin{bmatrix} \Delta\epsilon_1 \\ \Delta\epsilon_2 \\ \Delta\epsilon_3 \\ \Delta\gamma_{12} \\ \Delta\gamma_{13} \\ \Delta\gamma_{23} \end{bmatrix} = \begin{bmatrix} \frac{aF_1^n+1}{E_i} & \frac{-\nu_{21}(aF_2^n+1)}{E_i} & \frac{-\nu_{31}(aF_3^n+1)}{E_i} & 0 & 0 & 0 \\ \frac{-\nu_{12}(aF_1^n+1)}{E_i} & \frac{aF_2^n+1}{E_i} & \frac{-\nu_{32}(aF_3^n+1)}{E_i} & 0 & 0 & 0 \\ \frac{-\nu_{13}(aF_1^n+1)}{E_i} & \frac{-\nu_{23}(aF_2^n+1)}{E_i} & \frac{aF_3^n+1}{E_i} & 0 & 0 & 0 \\ 0 & 0 & 0 & \frac{b(F_1+F_2)^m+1}{G_i} & 0 & 0 \\ 0 & 0 & 0 & 0 & \frac{b(F_1+F_3)^m+1}{G_i} & 0 \\ 0 & 0 & 0 & 0 & 0 & \frac{b(F_2+F_3)^m+1}{G_i} \end{bmatrix} \begin{bmatrix} \Delta\sigma_1 \\ \Delta\sigma_2 \\ \Delta\sigma_3 \\ \Delta\tau_{12} \\ \Delta\tau_{13} \\ \Delta\tau_{23} \end{bmatrix} \quad (1)$$

ties and the first invariant of the fracture tensor indicated dependence on the joint orientation. For deformability properties, REV behaviour was observed for $I_1^{(F)}$ values greater than a certain critical value. This critical value was found to be different for different mechanical properties considered here. The deformation modulus of rock blocks in any direction was found to be uniquely related to the fracture tensor component in that direction. It was found that the anisotropy of the deformation modulus in 3D can be represented through an ellipsoidal relationship. The shear modulus of the rock blocks on any plane was found to be uniquely related to the summation of the components of the fracture tensor on that plane. It seems possible to expect equivalent continuum behaviour or REV behaviour for rock mass deformability parameters when the corresponding fracture tensor component value is beyond a certain critical value. This critical value seems to be different for different mechanical properties of rock masses. An incrementally linear elastic, orthotropic constitutive model is suggested to describe the pre-failure mechanical behaviour of rock masses. This constitutive model has captured the anisotropic, scale-dependent behaviour of rock masses. In this constitutive model, the effect of joint geometry has been taken into account in terms of fracture tensor components.

In this study, actual joints were treated as smooth planar joints without any roughness. Actual joint constitutive behaviour was represented by a simplified model. For both single and two joint set cases, the same strength and deformability values were used for each joint. It is important to realize that these simplifications were used in arriving at the conclusions mentioned above. It may be worthwhile to study how the constitutive model obtained for jointed rock changes due to incorporation of more refined constitutive models for actual joints, and also due to specification of different strength and deformability values for the joints in different sets.

REFERENCES

Cundall, P.A. 1988. Formulation of a three-dimensional distinct element model - Part 1. A scheme to detect and represent contacts in a system composed of many polyhedral blocks. *Int. J. Rock. Mech. and Min. Sci.*, 25, 107-116.

Hardin, E.L. et al. 1982. A heated flatjack test to measure the thermomechanical and transport properties of rock masses. Office of Nuclear Waste Isolation, Columbus, OH, ONWI-260P.

Hart, R., P.A. Cundall and J. Lemos 1988. Formulation of a three-dimensional distinct element model - Part II: Mechanical calculations for motion and interaction of a system composed of many polyhedral blocks. *Int. J. Rock Mech. and Min. Sci.*, 25, 117-126.

Kulatilake, P.H.S.W., H. Ucpirti, S. Wang and O. Stephansson 1992. Use of the distinct element method to perform stress analysis in rock with non-persistent joints and to study the effect of joint geometry parameters on the strength and deformability of rock masses. *Rock Mech. and Rock Engrg.*, 25, 253-274.

Kulatilake, P.H.S.W., S. Wang and O. Stephansson 1993a. Effect of finite size joints on the deformability of jointed rock in three dimensions. *Int. J. Rock Mech. & Min. Sci.*, 30, 479-501.

Kulatilake, P.H.S.W., D.N. Wathugala, and O. Stephansson 1993b. Joint network modelling with a validation excercise in Stripa mine, Sweden. *Int. J. Rock Mech. & Min. Sci.*, 30, 503-526.

Oda, M. 1982. Fabric tensor for discontinuous geological materials. *Soil Mechanics and Foundations*, 22(4), 96-108.

Wang, S. 1992. Fundamental studies of the deformability and strength of jointed rock masses at three dimensional level. Ph.D. Dissertation, University of Arizona, Tucson.

Wang, S. and P.H.S.W. Kulatilake 1993. Linking between joint geometry models and a distinct element method in 3D to perform stress analyses in rock masses containing non-persistent joints. *Soils and Foundations* (in press).

Geomechanics 93, Rakowski (ed.) © 1994 Balkema, Rotterdam, ISBN 90 5410 354 X

Modelling of progressive damage evolution in rocks

Zenon Mróz & Maciej Kowalczyk
Institute of Fundamental Technological Research, Warsaw, Poland

ABSTRACT: The ultimate failure of rock is preceded by progression of consecutive damage modes producing very frequently periodic sequence of failure events. The cracking mode is associated with a set of oriented cracks inducing stress redistribution and orthotropic material structure. The crushing and post-buckling modes are associated with subsequent failure of cracked portions thus inducing total collapse.

The evolution of damage zones and the ultimate failure modes are discussed for several cases, namely
a) axisymmetric stress state around a borehole with internal pressure inducing progression of damage
b) interaction of a rigid punch with rock mass and the variation of penetration force is derived by considering progression of cracking, crushing and shearing modes.

1 INTRODUCTION

The present paper is concerned with the analysis of progressive failure modes in brittle materials, such as rock, ice, concrete, or ceramics. The constitutive models for such materials so far proposed are usually aimed to describe deformation response in the stable regime before reaching the maximal stress and post-critical strain localization, cf. Jaeger and Cook (1976), Derski et al (1989). However, in many technical problems such as rock crushing, cutting or drilling, ice plate interaction with off-shore platforms, etc., there is a need to account for all progressive failure modes with account for post-critical stress-strain response and localized slip or fracture modes. The present paper is addressed to such class of problems and simplified assumptions are introduced in order to generate analytical solutions. Our aim is to demonstrate that evolution of progressive failure modes induces periodicity of failure events with load-displacement curves exhibiting oscillatory character with development and termination of particular modes. The present paper is also aimed to illustrate general properties of solutions of boundary-value problems for brittle-plastic materials, namely periodicity of failure events, interaction of progressive failure and plastic flow, scale effect, etc. The major failure modes discussed are cracking, crushing, localized shear, buckling and post-buckling of cracked elements.

In Section 2, the analysis of progression of cracked and crushed zones in the axisymmetric case is discussed. In Section 3, the punch penetration problem is analysed in the case of plane strain and plane stress. The load deflection curve is derived and the consecutive failure events are discussed in detail.

2 CRACKING AND CRUSHING OF ROCK IN THE VICINITY OF BOREHOLE

Let us first consider an axisymmetric problem of a cylinder acted on by internal pressure and in the limit case of a borehole in an infinite continuum. Assuming the plane strain case, we assume the material to be in elastic, cracking or crushing state, depending on the actual stress state. Figure 1 presents the respective states within the cylinder. The detailed analysis was presented by Kowalczyk and Mróz (1988), following previous studies by Ladanyi (1967) and Helan (1984).

Assume the macrocracks to develop along radial directions and the cracked rock to be regarded

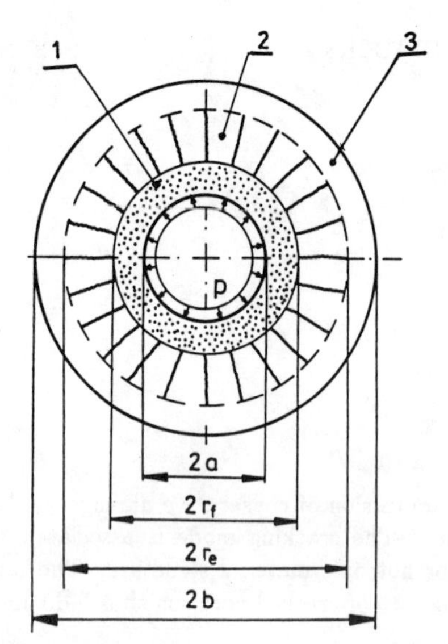

Figure 1: *Axisymmetric cylinder with different stress stage zones: 1) crushed 2) cracked 3) elastic*

as an orthotropic material of vanishing stiffness moduli in circumferential direction. The total strain components ε_r and ε_t in radial and circumferential directions are decomposed in the cracked zone as follows

$$\varepsilon_r = \varepsilon_r^e + \varepsilon_r^f \qquad \varepsilon_t = \varepsilon_t^e + \varepsilon_t^f \tag{1}$$

and in the crushed zone there is

$$\varepsilon_r = \varepsilon_r^e + \varepsilon_r^p \qquad \varepsilon_t = \varepsilon_t^e + \varepsilon_t^p \tag{2}$$

where $\varepsilon_r^e, \varepsilon_t^e$ are the elastic strains, $\varepsilon_r^f, \varepsilon_t^f$ are the fracture strains, $\varepsilon_r^p, \varepsilon_t^p$ are the plastic strains. It is assumed that in the crushed zone the material behaves like a granular isotropic medium, of specified residual cohesion C and the angle of internal friction φ. The equilibrium equations are

$$\frac{\partial \sigma_r}{\partial r} + \frac{\sigma_r - \sigma_t}{r} = 0 \tag{3}$$

and strain-displacement relations take the form

$$\varepsilon_r = \frac{\partial u}{\partial r} \qquad \varepsilon_t = \frac{u}{r} \tag{4}$$

The Hookes law for an isotropic material now provides

$$\begin{aligned} \varepsilon_r^e &= \frac{1+\nu}{E}((1-\nu)\sigma_r - \nu\sigma_t) \\ \varepsilon_t^e &= \frac{1+\nu}{E}((1-\nu)\sigma_t - \nu\sigma_r) \end{aligned} \tag{5}$$

In the cracked zone, there is

$$\varepsilon_r^f = 0 \qquad \sigma_t = 0 \tag{6}$$

Assume the cracked zone to contain many cracks propagating from the internal boundary $r = a$ to the interface boundary $r = r_e$. Using the Griffith condition of crack propagation, the potential energy release calculated at the progressing interface $r = r_e$ is assumed to be equal stress intensity (or fracture toughness) of n cracks, so that

$$nK_{Ic}^2 = \pi r_e \left\| \left[\sigma_t^2(r_e) \right] \right\| \tag{7}$$

where the bracket $\|[\ ]\|$ denotes the discontinuity at $r = r_e$, $[\sigma_t^2(r_e)] = \sigma_t^2(r_e^+) - \sigma_t^2(r_e^-)$.

The crushed zone follows the cracked zone for increasing internal pressure and the material satisfies the Coulomb yield condition

$$\begin{aligned} F_1 &= \sigma_t - \sigma_r + (\sigma_r + \sigma_t)\sin\varphi - 2C_0 \cos\varphi = 0 \\ F_2 &= \sigma_r - \sigma_t + (\sigma_r + \sigma_t)\sin\varphi - 2C_0 \cos\varphi = 0 \\ F_3 &= \sigma_r - S_t = 0 \\ F_4 &= \sigma_t - S_t = 0 \end{aligned} \tag{8}$$

where S_t denotes the tensile yield stress and C_0 is the initial cohesion. Here $F_1 = 0$, $F_2 = 0$ represent the shear mode, $F_3 = 0$, $F_4 = 0$ correspond to tensile mode of flow. In the crushed zone it is usually assumed that the residual cohesion vanishes, $C_r = 0$. The flow rule specifies the strain rates $\dot{\varepsilon}_r^p$, $\dot{\varepsilon}_t^p$, namely

$$\dot{\varepsilon}_r^p = \dot{\lambda}\frac{\partial g_i}{\partial \sigma_r} \qquad \dot{\varepsilon}_t^p = \dot{\lambda}\frac{\partial g_i}{\partial \sigma_t} \tag{9}$$

where $\dot{\lambda} > 0$ and g_i is the plastic potential,

$$\begin{aligned} g_1 &= \sigma_t - \sigma_r + (\sigma_r + \sigma_t)\sin\psi - 2C_g \cos\psi = 0 \\ g_2 &= \sigma_r - \sigma_t + (\sigma_r + \sigma_t)\sin\psi - 2C_g \cos\psi = 0 \end{aligned} \tag{10}$$

where ψ is the dilatancy angle and C_g is a constant. The potentials $g_3 = 0$ and $g_4 = 0$ are identical to $F_3 = 0$, $F_4 = 0$.

2.1 Solutions for specific zones

At the initial stage, the elastic zone occurs within the whole cylinder. The stress and strain distribution are governed by Lame equations and the critical pressure initiating cracking is

$$p_c = S_t \frac{(b^2 - a^2)}{(b^2 + a^2)} \tag{11}$$

where S_t is the characteristic tensile strength. The second phase corresponds to existence of two zones: elastic and cracked. In the elastic zone, the Lame solution applies with internal zone radius specifying the interface between cracked and elastic zones. The pressure acting on the interface is obtained from the solution for two zones and the crack propagation condition must be satisfied at $r = r_e$.

In the cracked zone we have

$$\sigma_r = \frac{A_1}{r} \qquad \sigma_t = 0$$
$$u = \frac{1-\nu^2}{E} A_1 \ln r + A_2 \tag{12}$$

where the integration constants follow from the radial stress and displacement continuity at $r = r_f$. Using the condition $\sigma_r(r_f) = p$ and the cracking condition (7), the relation between the radius of cracked zone r_e and the pressure p_e acting at $r = r_e$ is

$$p_e = K_I \sqrt{\frac{n}{\pi r_e}} \frac{(b^2 - r_e^2)}{(b^2 + r_e^2)} \tag{13}$$

Since at the onset of crack propagation there is $r_e = a$, $p_e = p_c$, hence in view of (11) and (13), there is

$$S_t = K_I \sqrt{\frac{n}{\pi a}} \tag{14}$$

The internal pressure within the tube is now explicitly related to the size of the cracked zone, namely

$$p = S_t \sqrt{\frac{r_e}{a}} \frac{(b^2 - r_e^2)}{(b^2 + r_e^2)} \tag{15}$$

The relation $p(r_e)$ exhibits limit point at the critical crack length r_{ec} and the loading process is unstable for $r_e > r_{ec}$.

As in the cracked zone there is a uniaxial stress state $\sigma_r \neq 0, \sigma_t = 0$, the crushing starts when the pressure p exceeds the compressive strength S_c. Thus for $p > S_c$, the crushed zone propagates from the internal cylinder boundary.

The next stage corresponds to existence of crushed, cracked and elastic zones. At the interface between crushed and cracked zones there is $p_f = S_c$ and for the radius $r_e = r_{ec}$, the radius r_f attains its maximum. The equilibrium evolution of states is achieved by assuming progression of cracked and crushed zones. The pressure acting at $r = r_f$ now is

$$p_f = S_t \sqrt{\frac{r_e}{r_{fk}}} \frac{(b^2 - r_e^2)}{(b^2 + r_e^2)} \tag{16}$$

The details of further progression are discussed by Kowalczyk and Mróz (1988). Here, we only present the evolution of zones for

$$\frac{S_t}{E} = 0.001 \qquad \frac{C_0}{E} = 0.0012$$
$$\nu = 0.3 \qquad \varphi = \psi = 20^\circ \tag{17}$$

In the case $b/a = \infty$, the progression of both cracked and crushed zones is stable and the respective diagrams are shown in Fig.2. Stability conditions for the case of progressing damage interfaces were considered by Dems and Mróz (1985).

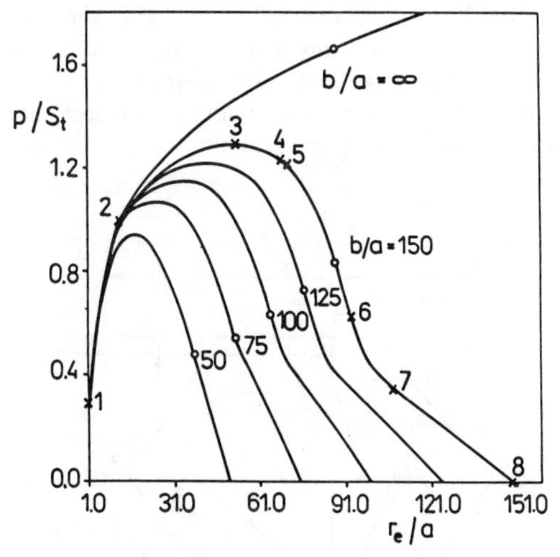

Figure 2: *Evolution of cracked zone*

3 PUNCH PENETRATION INTO PLASTIC-BRITTLE MATERIAL

The axisymmetric solution of the preceding section will now be used to construct a simplified solution of punch indentation in plane strain and plane stress cases. When a rigid punch is penetrated into an elastic semiplane, the stress concentration zones at punch edges induce localized cracking and crushing beneath the punch, cf. experimental data in papers by Pang et al (1989), (1990), Swain and Lawn (1976), Tokar (1990), Wagner and Schümann (1971), Wijk (1989), Lindqvist and Lai (1983). It is assumed that a crushed zone of radius r_f is formed with the hydrostatic stress state. This zone acts as a pressure loading on the remaining material thus inducing cracking and subsequent shearing or buckling of the cracked material blocks. It is assumed that the cracked zone is bounded by a circle of radius r_e and radial lines $\tau = \pm\alpha_f$, Fig.3. The cracks are assumed to propagate along radial directions, so the axisymmetric solution can be assumed to predict the radius r_e of the cracked zone. For simplicity, we neglect the progressive crushing and growth of the initial crushed zone of radius r_f. The subsequent failure mode within the plane develops in a form of localized shearing along velocity discontinuity lines, inducing motion of material toward the free surface. For an elastic-plastic material the stage of development of localized shear bands should be considered. However, we assume that the failure mechanism develops instantaneously similarly as in rigid-plastic materials. On the other hand, in-plane stress state, the cracked beams are assumed to buckle in the out-of-plane mode and subsequently they deform in a post-buckling stage until total failure of beams occurs due to bending fracture of end cross-sections. The post-critical stage is analysed by assuming the beams to be rigid with end cross-sections connected to the foundation by nonlinear springs of specified characteristics. These simplified assumptions are introduced in order to generate the analytical solution before more accurate incremental analysis will be provided.

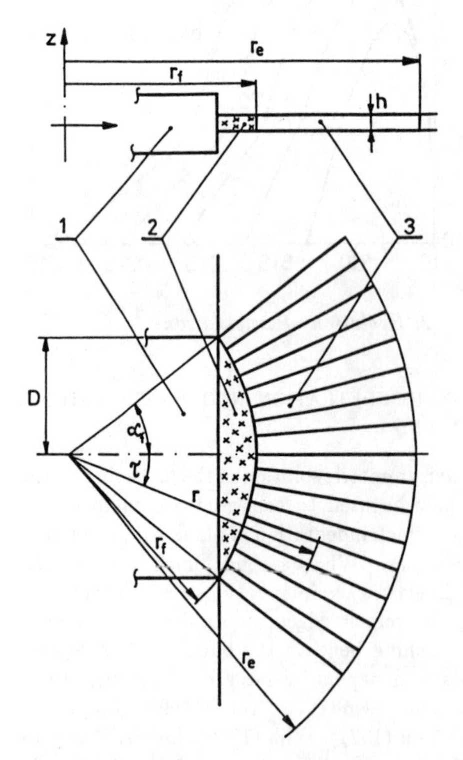

Figure 3: *Simplified punch indentation model:* 1) *punch* 2) *crushed zone* 3) *cracked zone*

3.1 Progression of cracking zone

Assuming the crushed zone to be fixed and specified by the radius $r_f = a$, the axisymmetric solution will be applied setting $b \to \infty$. The cracking condition now is assumed in the form

$$\sigma_t(r_e) = S_t\left(\frac{r_f}{r_e}\right)^{\kappa} = S_t\left(\frac{a}{r_e}\right)^{\kappa} \tag{18}$$

where κ is a positive exponent. For $\kappa = 0$ we obtain the strength condition and for $\kappa = \frac{1}{2}$ the energy condition. By selecting proper value of κ, one can account for the fact that now cracking develops within an annular segment of unspecified angle α_f and number of cracks. The constants A_1 and A_2 occurring in (12) are now obtained from displacement and radial stress continuity conditions, so that

$$\begin{aligned} A_1 &= -S_t r_e\left(\frac{r_f}{r_e}\right)^{\kappa} \\ A_2 &= \frac{1+\nu}{E} S_t r_e \left(\frac{r_f}{r_e}\right)^{\kappa}[1+(1-\nu)\ln(r_e)] \end{aligned} \tag{19}$$

The pressure p_e is calculated from (18). Since $p_f = -\sigma_r(r_f)$ the pressure in the crushed zone beneath the punch is

$$p_f = S_t\left(\frac{r_f}{r_e}\right)^{\kappa-1} \tag{20}$$

For $0 < \kappa < 1$, the growth of the fracture zone requires monotonic growth of punch pressure. Hence for some value $r = r_{ec}$, a new failure mode is to develop within the fractured zone.

3.2 Shear failure (plane strain case)

The second failure mode is assumed in a form of limit failure mechanism developed both in cracked zone and undamaged zones, typical for limit analysis, Fig.4. The shear planes A_0A_1, A_0A_2, A_1A_2 and A_2A_3 constitute kinematically admissible failure mechanism. However A_0A_1, A_0A_2, A_1A_2 are passed through the cracked zone of reduced cohesion. On the other hand, A_2A_3 passes through the undamaged material so the initial cohesion and angle of friction should be used in calculating the dissipation rate. The balance of rate work and internal dissipation now is, cf. Mróz and Drescher (1968)

$$2DpV_0 = 2Cl_2V_2\cos\varphi \tag{21}$$

where D is the punch half-width, l_2 denotes the length of the discontinuity line A_2A_3, V_0 is the punch velocity and V_2 is the velocity of block $A_1A_2A_3$. In writing (21) we neglect the residual cohesion on A_0A_2 A_2A_1 and A_0A_1 assuming it to be much smaller then C_0. Using hodograph and geometric relations, we obtain

$$p = C_0\left(\frac{u_f}{D} + \frac{\sin\alpha_1\cos\alpha_f}{\cos\alpha_0\cos(\alpha_0-\alpha_1-\alpha_f)}\right) \\ \frac{\cos(\alpha_0-\varphi)\cos(\alpha_0-\alpha_1-\alpha_f+2\varphi)\cos\varphi}{\sin(\alpha_1-2\varphi)\sin(\alpha_2-2\varphi)\cos(\alpha_2-\alpha_f)} \tag{22}$$

where the angles α_0, α_1, α_2 and the displacement $u_f = u(r_f)$ are shown in Fig.4. The critical pressure is obtained by determining the minimum of $p(\alpha_0, \alpha_1, \alpha_2)$.

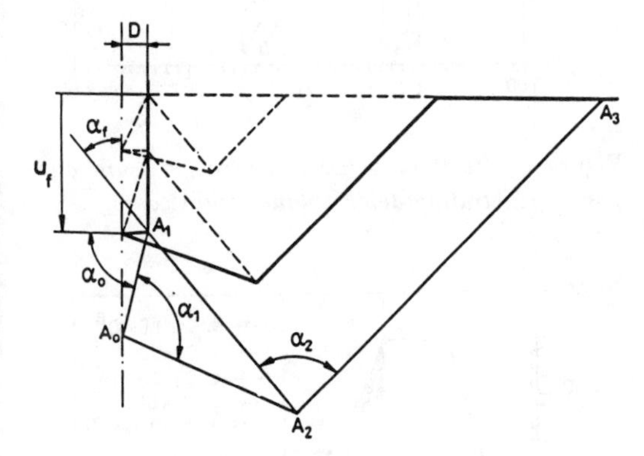

Figure 4: *Kinematically admissible failure mechanism (third cycle in plane strain case)*

3.3 Buckling failure (plane stress case)

In the case of plate thickness h small with respect to punch width $2D$, the plane stress condition prevails and the cracked material can be regarded as a set of tapered beams supported at $r = r_e$. The out-of-plane buckling mode of beams is therefore considered. The post-buckling response induces localized cracking and softening at $r = r_e$, which involve decreasing load acting on the beam.

Denoting $k = r_f/r_e$, and introducing the plane stress elastic constants $\nu_n = \nu/(1+\nu)$, $E_n = E(1+2\nu)/(1+\nu)^2$, we obtain the punch displacement and pressure in the form

$$u_f = \frac{S_t r_f}{E_n\sqrt{k}}[(1+\nu) - \ln(k)] \qquad p_f = \frac{S_t}{\sqrt{k}} \tag{23}$$

Denote by b_e and b_f the widths of cracked beams at $r = r_e$ and $r = r_f$, by $F = p_f h b_f$ the force acting on beam of length $l = r_e - r_f$. Depending on geometric parameters the buckling mode can occur within the plane or out-of-plane. The corresponding critical taper ratios are denoted by k_t and k_z. The following relations specify k_t and k_z at buckling

$$\frac{b_e(k_t)}{48l^2(k_t)}\theta_t(k_t)E_n h^3 - p_f(k_t)hb_f = 0 \\ \frac{b_e(k_z)}{48l^2(k_z)}\theta_z(k_z)E_n h^3 - p_f(k_z)hb_f = 0 \tag{24}$$

where θ_t and θ_z denote the buckling coefficients provided by Życzkowski (1956) in the tabular form. It follows from (23) that the critical taper ratio k_c and buckling mode are determined by the condition $k_c = \max(k_t, k_z)$. The values of b_e and b_f are

$$b_e = \frac{b_f}{k} \qquad b_f = \frac{2}{n-1}\alpha_f r_f \tag{25}$$

where n denotes the number of cracks and $\alpha_f = \arccos(D/r_f)$ is the opening angle of the cracked zone.

The post-buckling analysis is carried out by assuming the buckled bar as rigid with the inelastic hinge support at $r = r_e$. The bending moment M_g at the hinge depends on the rotation angle ϑ according to the relations

$$M_g(\vartheta) = \begin{cases} C_0\vartheta & \text{if } \vartheta \le \vartheta_0 \\ C_0\vartheta_0 - C_1(\vartheta-\vartheta_0) & \text{if } \vartheta_0 < \vartheta \le \vartheta_1 \\ 0 & \text{if } \vartheta_1 < \vartheta \end{cases} \tag{26}$$

where ϑ_0 and ϑ_1 denote critical values of ϑ indicating the onset and termination of the hinge failure process. Here C_0 denotes the hinge stiffness in the elastic stage and C_1 is the postcritical softening modulus. The value of C_0 is obtained from the value of the critical buckling load, $C_0 = F(k_c)l(k_c)$. The value of ϑ_0 is obtained by assessing the rupture stress of material fibers in tension, so that

$$S_t = -6\frac{F(k_c)l(k_c)\vartheta_0}{b_e h^2} \tag{27}$$

where compressive stress was neglected. The value of ϑ_0 evaluated for ice is of the order 10^{-5}. Higher values can be obtained by accounting for the compressive stress, thus

$$S_t = -\frac{F(k_c)}{b_e h^2}\vartheta_0[h\mathrm{ctg}\vartheta_0 + 6l(k_c)] \tag{28}$$

The value of ϑ_1 corresponds to $M_g(\vartheta_1) = 0$. The punch displacement and stress vary in the post-critical stage according to the relations

$$\begin{aligned} u_f &= \frac{S_t r_f}{E_n\sqrt{k}}[(1+\nu) - \ln(k_c)] + l(k_c)(1-\cos\vartheta) \\ p_f &= \frac{M_g(\vartheta)}{l(k_c)hb_f\sin\vartheta} \end{aligned} \tag{29}$$

When $\vartheta = \frac{1}{2}\pi$, the consecutive loading cycle commences and the periodic sequence of cracking, buckling, post-buckling modes develops.

3.4 Discussion of theoretical solutions

For numerical analysis, the following parameters were assumed

$$\begin{aligned} &S_t = 0.7MPa \quad S_c = 5MPa \\ &E = 10GPa \quad \nu = 0.3 \quad \varphi = 20^\circ \end{aligned} \tag{30}$$

typical for ice, cf. Sunder and Connor (1984). The problem parameters are D, h, n, α_f and C_1. The experimental data indicate that for brittle materials the critical stress for punch indentation is of the order of p_c/S_c = 4–10, cf. Wijk (1989), and the cracked zone radius reaches the values r_e/D = 2–8, cf. Pang et al (1990). For the plane strain case, the cracked zone and punch pressure evolution are shown in Fig.5–6 as functions of punch displacements u_f/D for different values of the angle α_f of cracked zone (assumed $\kappa = 0$). It is seen that the predicted values are within the range of experimental observation. For specified values of α_f,

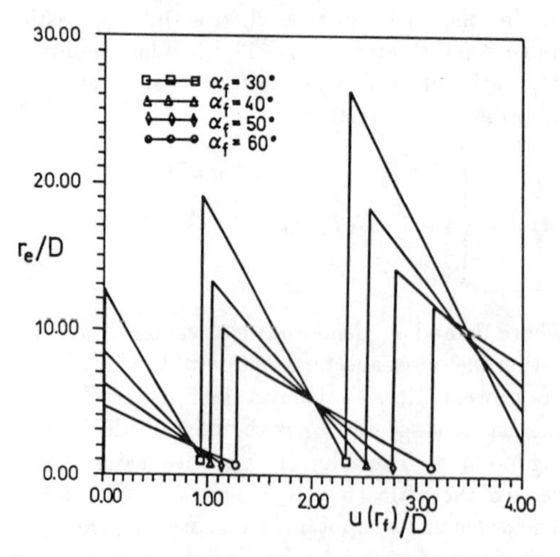

Figure 5: *Relation between cracked zone radius and punch penetration depth (plane strain case)*

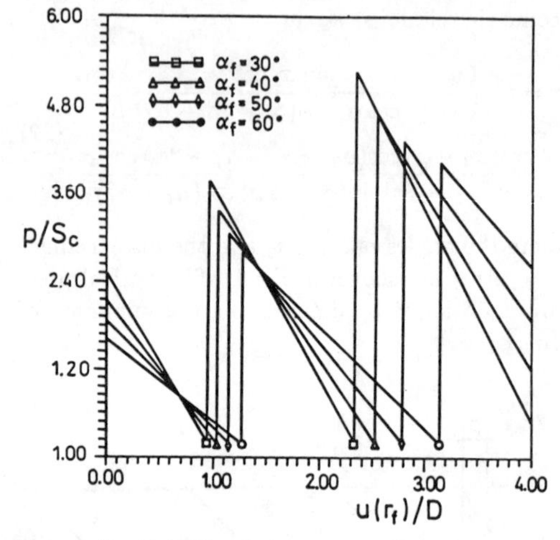

Figure 6: *Relation between punch pressure and punch penetration depth (plane strain case)*

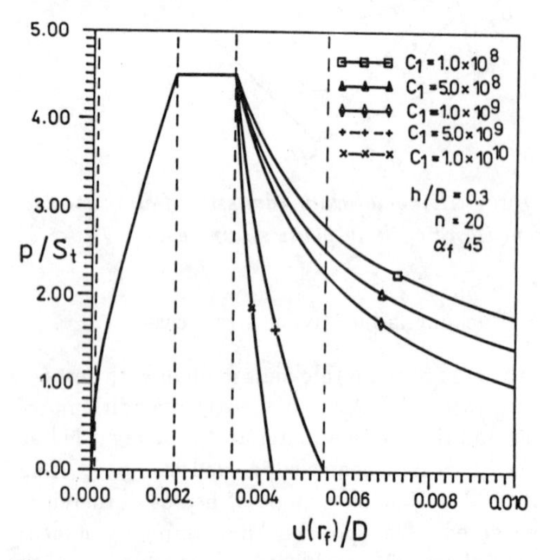

Figure 7: *Influence of hinge softening stiffness on relation between punch pressure and punch penetration depth (plane stress case)*

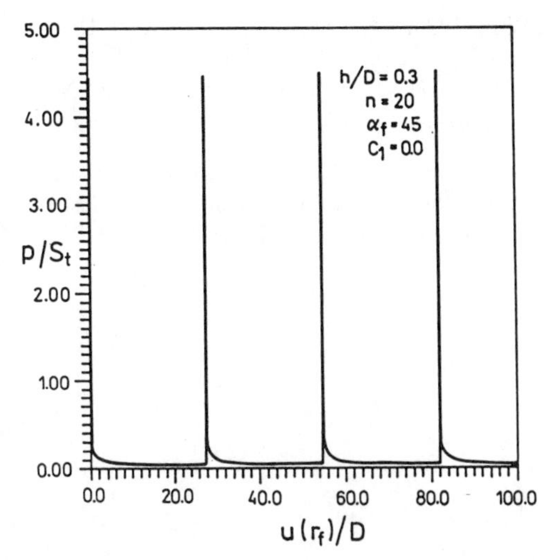

Figure 8: *Relation between punch pressure and punch penetration depth (plane stress case)*

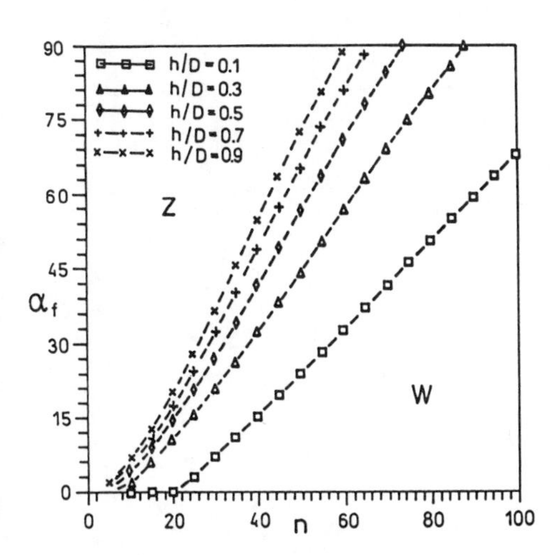

Figure 9: *Effect of crack number on variation of buckling mode*

the growth of cracked zone and maximal pressure is due to growth of the length of shear band in the final failure mode. Figures 7–8 show the respective curves for the plane stress case. The experiments carried for ice plates indicate that the maximal stress is reached at small punch displacements with subsequent drastic reduction of pressure for progressive punch penetration, cf. Michel and Toussaint (1977). The same character of load-displacement diagram was predicted by the present model. The effect of crack number on variation of buckling mode from in-plane to out-of-plane is illustrated in Fig.9. Here the plot of lines $k_t = k_z$ are shown with in-plane buckling (W) occurring below the curve and out-of-plane buckling occurring above the curve (Z). The effect of buckling mode leads to a new idea of damage mechanism in brittle materials. The buckling of cracked material was also considered by Bažant et al (1993).

4. CONCLUDING REMARKS

The present analysis provides the insight into the progression of failure modes in brittle materials. These are termed as cracking, crushing, buckling, post-buckling and shearing modes. The analysis of consecutive modes and of their interaction provides the resulting force-displacement diagram, usually of oscillating character, indicating periodicity of mode evolution. The qualitative assessment provides results confirmed by experimental data and observation. A more refined analysis is needed predicting number of cracks and their evolution. However such analysis would be much more complex.

REFERENCES

Z. P. Bažant, F. B. Lin, H.Lippmann, 1993, "Fracture Energy Release and Size Effect in Borehole Breakout.", *Int. J. Numerical & Analytical Methods in Geomechanics*, Vol. 17, No 1.

N. G. W. Cook, M. Hood, F. Tsai, 1984, "Observations of Crack Growth in Hard Rock Loaded by an Indenter.", *Int. J. Rock Mech. Min. Sci. & Geomech. Abstr.*, Vol. 21, No. 2, p.97-107.

K. Dems, Z. Mróz, 1985, "Stability Conditions for Brittle-Plastic Structures with Propagating Damage Surfaces.", *J. Struct. Mech.*, Vol. 13, No. 95.

W. Derski, R. Izbicki, I. Kisiel, Z. Mróz, 1989, *"Rock and Soil Mechanics"*, Elsevier.

R. T. Ewy, N. G. Cook, 1990, "Cylindrical Opening in Rock.", *Int. J. Rock Mech. Min. Sci. & Geomech. Abstr.*, Vol. 27, No. 5.

K. Hellan, 1984, "An Asymptotic Study of Slow Radial Cracking.", *Int. J. Fracture*, Vol. 26, p.17-30.

J. C. Jaeger, N. G. W. Cook, 1976, *"Fundamentals of Rock Mechanics."*, John Wiley & Sons, New York.

M. Kowalczyk, Z. Mróz, 1988, "Analysis of the cracking and crushing mechanism around the opening in brittle materials", (in Polish), *Archives of Mining Sciences*, Vol. 33, No. 4, p.403-439.

B. Ladanyi, 1967, "Expansion of Cavities in Brittle Media.", *Int. J. Rock Mech. Min. Sci.*, Vol. 4, p.301-328.

P. Lindqvist, Lai Hai-Hui, 1983, "Behaviour of the Crushed Zone in Rock Indentation.", *Rock Mech. Rock Engng.*, Vol. 16, p.199-207.

B. Michel, N. Toussaint, 1977, "Mechanisms and Theory of Indentation of Ice Plates.", *J. Glaciology*, Vol. 19, No. 81, p.285-300.

Z. Mróz, A. Drescher, 1968, "Limit Plasticity Approach to Some Cases of Flow of Bulk Solids.", *Transactions of the ASME*, p.1-8.

S. S. Pang, W. Goldsmith, M. Hood, 1989, "A Force-Indentation Model for Brittle Rocks.", *Rock Mech. and Rock Engng.*, Vol. 22, p.127-148.

S. S. Pang, W. Goldsmith, 1990, "Investigation of Crack Formation During Loading of Brittle Rock.", *Rock Mech. and Rock Engng.*, Vol. 23, p.53-63.

S. S. Sunder, J. J. Connor, 1984, "Numerical Modeling of Ice-Structure Interaction", *MIT Report.*

M. V. Swain, B. R. Lawn, 1976, "Indentation Fracture in Brittle Rocks and Glasses.", *Int. J. Rock. Mech. Min. Sci. & Geomech. Abstr.*, Vol. 13, p.311-319.

G. Tokar, 1990, "Experimental Analysis of the Elasto-Plastic Zone Surrounding a Borehole in a Specimen of Rock-Like Material Under Multiaxial Pressure.", *Engng. Fracture Mechanics*, Vol. 35, No. 4-5, p.879-887.

H. Wagner, E. H. R. Schümann, 1971, "The Stamp-Load Bearing Strength of Rock an Experimental and Theoretical Investigation", *Rock Mechanics*, Vol. 3, p.185-207.

G. Wijk, 1989, "The Stamp Test for Rock Drillability Classification.", *Int. J. Rock Mech. Min. Sci. & Geomech. Abstr.*, Vol. 26, No. 1, p.37-44.

M. Życzkowski, 1956, "Calculation of critical forces for elastic tapered beams", (in Polish), *Rozpr. Inż.*, Tom IV, Zeszyt 3, p.367-412.

Geomechanics 93, Rakowski (ed.) © 1994 Balkema, Rotterdam, ISBN 90 5410 354 X

Newton-like methods for the solution of geomechanical problems

G. Swoboda, M. B. Reed & I. Hladík
University of Innsbruck, Austria

ABSTRACT: We discuss the implementation and performance of several Newton-like solution algorithms, for large-scale elasto-plasticity finite element analyses. The iterative schemes used are: Truncated Newton, Quasi-Newton, limited storage QN and element-by-element QN. Among the aspects examined are: the effect of inexact computation in the Truncated Newton algorithm, the influence of the line search accuracy and the overwriting strategy employed and number of updates stored in the limited storage QN methods. It is shown that the element-by-element QN method is particularly efficient and has a number of important advantages.

1 INTRODUCTION

Many iterative algorithms have been developed by numerical analysts for the unconstrained minimization of a smooth function $f(x), x \epsilon \Re^n$; this is equivalent to solving the nonlinear system of equations

$$g(x) \equiv \nabla f(x) = 0.$$

However, to date the more recently-developed algorithms have not been widely applied in the solution of finite element problems involving material or geometric nonlinearities. As the power of computer hardware available to engineers continues to increase rapidly, more complex meshes (increasingly often three-dimensional) are being used, and the inclusion of an efficient method of equation solution becomes essential. Direct solution methods, such as Gaussian elimination, require too much data storage to be practical in large problems. For solution of large linear systems, the iterative method of conjugate gradients (CG) with preconditioning has proved very successful. This paper attempts to answer the question: are there modern methods for the solution of nonlinear systems which can be easily implemented and will perform well in an already-existing comprehensive finite element package?

Finite element analysis of many nonlinear problems, e.g. plasticity, leads to a system of nonlinear equations of the form

$$K(u)u - f = 0 \tag{1.1}$$

where $u, f \epsilon \Re^n$, and $K(u)$ is an $n \times n$ matrix depending on the vector u of unknowns (usually displacements). In this paper we shall deal with the solution of (1.1) by Newton-like schemes; these can be expressed as follows. Suppose that at the kth iteration, u_k and the residual

$$r_k = f - K(u_k)u_k$$

are known. Then compute Δu_k by solving

$$B_k \Delta u_k = -r_k. \tag{1.2}$$

The new iterate u_{k+1} is given by

$$u_{k+1} = u_k + \lambda_k \Delta u_k \tag{1.3}$$

where the steplength λ_k is found by a line search algorithm.

The choice of the matrix B_k depends on the solution scheme. The simplest scheme is the Initial Stiffness (IS) or Modified Newton-Raphson method. Here, at each iteration the linear system (1.2) is solved with $B_k = K_{el}$, where K_{el} is the linear elastic stiffness matrix for the mesh. Once the decomposition of K_{el} is performed, subsequent iterations require only $O(n^2)$ operations, but convergence of the iteration is only linear. Also, the data defining the decomposition of K_{el} must be held in core or read from file at each iteration, and for very large meshes the amount of data involved is too large to make this an attractive option.

Another method commonly used is the Tangent Stiffness (TS) method, where $B_k = K(u_k)$. Here, a new linear system must be solved at each step, and for large problems this is itself best done iteratively, by preconditioned conjugate gradients (SSOR preconditioning is used in this paper). The CG iterations are halted when

$$\|B_k \Delta u_k + r_k\| \leq q_k \|r_k\| \tag{1.4}$$

involving the user-defined tolerance q_k. As the CG iteration is only the inner loop of an outer Newton-like iteration, it is not necessary to set q_k small and solve (1.2) exactly at each step. When the CG iteration is halted before reaching convergence, the resulting scheme is called a Truncated Newton or Inexact Newton method. One way of choosing q_k is described in Blaheta & Kohut (1991).

2 QN METHODS

The classical Newton-Raphson method takes the Jacobian of (1.1) as the matrix B_k in (1.2). In a nonlinear finite element analysis the cost of evaluating the Jacobian is prohibitive. There is a class of minimization algorithms, called Quasi-Newton (QN) methods, in which a matrix is constructed which increasingly resembles the Jacobian (or its inverse) as the iteration proceeds; these methods have superlinear convergence properties. Full details of these methods can be found in texts such as Dennis & Schnabel (1983) and Fletcher (1989). We will summarize the form where the inverse of the Jacobian is approximated; then (1.2) is replaced by

$$\Delta u_k = -H_k r_k \tag{2.1}$$

so that only a matrix-vector multiplication is required at each step. Define

$$s_k = u_{k+1} - u_k \text{ and } y_k = r_{k+1} - r_k.$$

Then the matrices H_k are required to satisfy the Quasi-Newton equation

$$H_{k+1} y_k = s_k \tag{2.2}$$

at each step. Starting from an initial approximation H_1, H_{k+1} is constructed from H_k by a rank one or rank two update. There is a unique symmetric rank one update which satisfies (2.2), namely

$$H_{k+1} = H_k + \frac{1}{\alpha_k} u_k u_k^{\mathsf{T}} \tag{2.3}$$

where

$$u_k = s_k - H_k y_k \text{ and } \alpha_k = u_k^{\mathsf{T}} y_k;$$

we will call this Davidon's method. The Broyden family of rank two updates can be defined in additive form as

$$H_{k+1} = H_k + \frac{1}{\alpha_k} u_k u_k^{\mathsf{T}} - \frac{1}{\beta_k} v_k v_k^{\mathsf{T}} + \gamma_k w_k w_k^{\mathsf{T}} \tag{2.4}$$

where

$$u_k = s_k, \; v_k = H_k y_k,$$
$$\alpha_k = u_k^{\mathsf{T}} y_k, \quad \beta_k = v_k^{\mathsf{T}} y_k,$$
$$w_k = \frac{1}{\alpha_k} u_k - \frac{1}{\beta_k} v_k.$$

We will consider only the most widely-used member of this family, namely BFGS (where $\gamma_k \equiv \beta_k$).

So far, we have not defined the starting QN matrix H_1. This should be an approximation to the inverse of the Jacobian of (1.1). In the application of these methods to f.e. plasticity problems, we have available to us the elastic element stiffness matrices making up the global elastic stiffness matrix K_{el}, which would be the Jacobian for the linear elastic problem. One possibility is therefore to take the starting matrix H_1 to be K_{el}^{-1}. In this case, there will be the same amount of work per iteration as with the Initial Stiffness method — i.e. the solution of a global equation involving the decomposed K_{el} (Geradin *et al* 1981) — but with superlinear convergence properties. However, as has already been observed, in large meshes it is not practical to hold the global decomposition in core. An alternative, requiring just one global vector of storage, is to take

$$H_1 = [\operatorname{diag}(K_{el})]^{-1}.$$

3 LIMITED STORAGE QN

The main disadvantage of the classical QN methods is that the $n \times n$ matrix H_k must be stored. In limited storage variants of QN, only the starting matrix H_1 (which is often diagonal) and a limited number of updates, $u_1, u_2, \ldots, u_m$, are stored, and when a product such as $H_k y_k$ is required it is constructed by (in the Davidon case)

$$H_k y_k = H_1 y_k + \sum_{i=1}^{\min(k,m)} \frac{u_i^{\mathsf{T}} y_k}{\alpha_i} u_i. \tag{3.1}$$

Several strategies have been proposed for continuing the algorithm once all storage has been filled. The simplest is to delete all updates and re-start the iteration with H_1. Alternatively, the new update can be overwritten on one of the existing ones. Buckley & LeNir(1983) propose overwriting the latest (most recent) update, while Nocedal(1980) has produced an algorithm in which the oldest update is overwritten. If, in the Davidon case, the jth update u_j is overwritten by the new update u_k, then in order to still satisfy the QN equation (2.2) the new update must be defined as

$$u_k = s_k - H_k y_k + \overline{\alpha} u_j \tag{3.2}$$

where $\overline{\alpha} = u_j^\top y_k / \alpha_j$. This is equivalent to construction of the normal update in (2.3) but omitting the jth update in the definition of H_k. The same approach can be applied to the Broyden family (2.4), and to unsymmetric rank one updates such as Broyden's "good" update.

If only one update is stored, all three strategies above are equivalent. This class of "memoryless QN" methods has been called Secant-Newton (SN) methods by Crisfield(1980). The Davidon SN method, for example, is characterized by

$$u_k = s_k - H_1 y_k. \tag{3.3}$$

Secant-Newton methods have been successfully used in engineering finite element applications by Crisfield and others. As in the previous section, we will take $H_1 = K_{el}^{-1}$ or $H_1 = [\mathrm{diag}(K_{el})]^{-1}$.

4 ELT-BY-ELT QN

A drawback of the above methods is that they do not take any advantage of the particular structure of finite element problems, in which the global matrix $K(u)$ is assembled from a large number of element stiffness matrices, each of which involve only a few of the unknowns. That is, (1.1) can be written

$$\sum_{e=1}^{ne}(K^e(u)u - f^e) \equiv -\sum_{e=1}^{ne} r^e(u) = 0 \tag{4.1}$$

where ne is the number of elements, and the summation is the normal finite element assembly process. A QN method for this situation, the Partitioned QN method, has been proposed by Griewank & Toint(1982), and its application to finite element problems will be termed here the Element By Element (EBE) QN method.

In this method, we construct the matrices B_k to resemble the Jacobian of (4.1) (rather than H_k resembling its inverse), in the form of a sum of element matrices

$$B_k = \sum_{e=1}^{ne} B_k^e \tag{4.2}$$

The B_k^e are required to satisfy the QN equation on an element-by-element basis:

$$B_{k+1}^e s_k = y_k^e, \quad e = 1, 2, \ldots, ne. \tag{4.3}$$

The Davidon EBE QN update, for example, is

$$B_{k+1}^e = B_k^e + U_k^e \tag{4.4}$$

where

$$U_k^e = \frac{1}{\alpha_k} u_k u_k^\top,$$
$$u_k = y_k^e - B_k^e s_k,$$
$$\alpha_k = s_k^\top u_k.$$

At each step, the global system

$$\sum_{e=1}^{ne} B_k^e \Delta u_k = -r_k \tag{4.5}$$

must be solved, and this is performed by Conjugate Gradients with diagonal preconditioning. Note that this involves the matrices B_k^e only in matrix-vector products as on the left hand side of (4.5), and these are formed element-by-element, so that the global B matrix is never assembled.

There is an important advantage in taking the starting element matrices B_1^e to be the linear elastic element matrices K_{el}^e. In this case, there is no update of B_k^e to be made while the element e remains elastic. This follows from the fact that (by setting $K^e(u) = K_{el}^e$ in the definition of y_k) for an elastic element e:

$$y_k^e = K_{el}^e s_k^e. \tag{4.6}$$

Substituting (4.6) and (4.4) into (4.3), we see that the update U_k^e must satisfy

$$U_k^e s_k = (K_{el}^e - B_k^e) s_k \tag{4.7}$$

so that if $B_1^e = K_{el}^e$ we can take $U_k^e = 0$, until plasticity starts.

Similarly, in a mesh involving boundary elements to model the far field (which remains elastic), the boundary elements will not become involved in the updating procedure.

5 RESULTS

The algorithms described above were implemented in a comprehensive finite element code FINAL (Swoboda 1993) and their performance tested on a medium-sized 2D elasto-plasticity problem. The program was run in double precision on a MicroVax 3100.

In order to compare the convergence performance, a strict equilibrium condition was employed for the solution of (1.2), namely

$$\|r_k\| < 10^{-6}\,\|f\| \qquad (5.1)$$

in each load step.

Two line search strategies were employed — a simple one of successive halving of λ until $\|r_{k+1}\| < \|r_k\|$, and a more sophisticated scheme based on the algorithm in Fletcher(1989) and using a *regula falsi* technique to search for a zero of $r^T \Delta u$ (Papadrakakis & Ghionis 1986) until the Wolfe-Powell conditions are satisfied. The second Wolfe-Powell condition

$$r_{k+1}^T \Delta u_k \geq \sigma\, r_k^T \Delta u_k \qquad (5.2)$$

contains the parameter $\sigma, 0 < \sigma < 1$, which controls the accuracy of the line search. The values $\sigma = 0.99$ (very weak) and $\sigma = 0.5$ (normal) were tried. Each line search was only allowed to make a restricted number of tries; a re-start with the initial matrix B_1 or H_1 was enforced if the search was unsuccessful. A re-start also occurred if the new search direction Δu_k was uphill, i.e. if

$$r_k^T \Delta u_k \geq 0.$$

The initial trial steplength $\lambda_k^{(1)}$ in each line-search was taken as 1.0.

The test example is a strip footing problem. Six-noded triangular elements are used. The mesh has 36 elements and 174 nodes. A Drucker-Prager ideal plasticity model is employed. The elasto-plastic material constants are $E = 10^4$ KN/m^2, $\nu = 0.17, c = 10$ KN/m^2, $\alpha = 20°$. Four loading steps were employed to bring the model close to the limit stage so that a significant amount of nonlinearity would arise.

The algorithms tested were the following:
- Initial Stiffness, using decomposition of K_{el} by a direct, profile subroutine;
- Tangent Stiffness, using exact or inexact CG iteration with SSOR preconditioning at each step;
- full QN (Davidon and BFGS versions);
- limited storage versions of the QN algorithms, storing 1 (equivalent to Secant Newton), 3, 6 and 10 updates, and using the re-start, latest or oldest overwriting strategies;
- the element-by-element QN method (Davidon and BFGS versions), using $B_1^e = K_{el}^e$ and CG iteration with diagonal preconditioning at each step.

Performance was measured in terms of the total number of (outer loop) iterations needed over all load steps, and the total number of evaluations of r performed (the latter depending on the line search as well as the iterative algorithm) — and, where appropriate, the total number of (inner loop) conjugate gradient iterations needed. Total c.p.u. times are also quoted for comparison. Table 5.1 shows these data for the different algorithms, and for a weak line search ($\sigma = 0.99$ in (5.2)). The limited storage QN algorithms are represented by the Secant Newton (SN) forms. In the Tangent Stiffness and EBE QN methods, an 'exact' CG solution was performed at each step. In the SN and full QN algorithms, the starting matrix was

$$H_1 = [K_{el}]^{-1}.$$

The alternative choice for H_1, namely inverting the diagonal of K_{el}, was implemented, but found to be completely inadequate for efficient solution. Even in the full QN BFGS algorithm, only the first load increment (which needed 4 or 5 iterations for the other algorithms) required 32 iterations for solution, and it was impossible to obtain solutions for further increments. The Davidon algorithm was even worse, requiring over 100 iterations in the first load increment.

It is interesting to note that the Davidon QN and SN algorithms are superior to their BFGS counterparts; this is in contradiction to the generally-held opinion (derived from experience with general mathematical optimization), but in line with other results observed in continuum mechanics finite element applications — see e.g. Geradin *et al*(1981) and Reed(1992). The second observation is that the various algorithms based on satisfying the Quasi-Newton equation all perform very favourably in comparison to the Tangent Stiffness method. It must be remembered that the amount of work involved per iteration varies greatly from one algorithm to another, as does the amount of storage required. Because of the structure of the FINAL program, the residual evaluations were the most expensive part of the computations, and this is reflected in the c.p.u. times quoted. On this basis, it is the element-by-element BFGS algorithm which performs best.

While the Secant Newton algorithms performed well (considering that they store only one update) in terms of iterations, they required a large number of function evaluations during the line searches. This occurs because the initial steplength of 1.0 is not appropriate for SN iterations (or for limited storage QN during overwriting). Buckley & LeNir

(1983) have shown that in these circumstances the algorithm is equivalent to a preconditioned conjugate gradient iteration (in the BFGS case), and recommend an initial steplength appropriate to CG algorithms. A suitable choice exists for minimization problems (Fletcher 1987), but it is not straightforward to apply this to solution of nonlinear systems. We remark only that there exists the potential to substantially reduce the number of residual evaluations needed in the SN algorithms, by modifying the initial steplength.

In Table 5.2 the performance of the three overwriting strategies in limited storage QN is compared as the number of updates stored is increased until the full QN performance is reached. The Davidon algorithm is used. Surprisingly, neither 'latest' nor 'oldest' overwriting strategy is clearly superior to the simple re-start strategy in which all update information is deleted periodically. It is possible for a performance close to that of the full QN algorithm to be achieved with the storage of only a small number of updates. Similar results have been reported by Reed(1992), also for the BFGS algorithm.

In Table 5.3 the effect of using an inexact CG solution is examined, for the Tangent Stiffness and EBE QN algorithms with a very weak line search. While inexact solution does reduce substantially the number of CG iterations needed, it tends to require more evaluations of the residual. The adaptive strategy (taking $\eta = 0.5$) is that of Blaheta & Kohut(1991), who reported only a small increase, if any, in the number of tangent stiffness iterations required with inexact, as compared to exact, CG solution. A possible explanation for this can be seen if the results in Table 5.3, which are totals over four separate load increments, are broken down into performance over individual load increments; this is done (for iterations and residual evaluations) in Table 5.4. While the performance is almost the same in the early load increments, when little nonlinearity is present, the inexact algorithms perform much more poorly as the amount of nonlinearity increases and failure of the rock or soil mass is approached. (In some cases the sums over the load increments slightly exceed the totals in Table 5.3; this is because allowance for wasted computation before a re-start has been taken into account in the final totals.)

In Table 5.5 the effect of the line search algorithm is examined. While use of a very weak line search does increase the number of iterations needed to reach convergence, the total number of evaluations needed is significantly less for all the methods.

Table 5.1: Overall performance of algorithms.

algorithm	iterations (evaluations)	cpu time mins:secs
Initial Stiffness	301(301)	23:10
Tangent Stiffness	92(99)	8:35
QN DFP	98(122)	10:48
QN Davidon	48(73)	5:35
SN DFP	69(267)	27:23
SN Davidon	76(111)	9:00
EBE QN BFGS	51(63)	5:57
EBE QN Davidon	51(70)	6:36

Table 5.2: Effect of overwriting strategy.

Davidon QN (storing m updates)			
	iterations(evaluations)		
m	restart	latest	oldest
1 (SN)	76(111)	76(111)	76(111)
3	52(64)	74(88)	50(73)
6	53(81)	61(72)	47(76)
10	47(77)	47(78)	48(71)
full QN	48(71)	48(71)	48(71)

6 CONCLUSIONS

Of the algorithms tested, the element-by-element BFGS method with a very weak line search performs most efficiently. The classical Davidon QN method is equally good (Table 5.5), but this requires storage of a large number of update vectors, and solution of a global matrix equation with the decomposed K_{el} (which must also be stored) at each iteration. Attempts to reduce this storage and computation by using a diagonal starting matrix H_1 were wholly unsuccessful. Limited storage versions of the QN algorithms, based on overwriting, are also unable to compete with the element by element (EBE) QN approach.

The QN and SN algorithms performed more poorly as the loading proceeded, and the difference increased between the true Jacobian and the linear elastic K matrix which was used for establishing H_1. An interesting possibility would be to take

$$H_1 = [K(u)]^{-1},$$

that is, to use the Tangent Stiffness matrix at the start of the current load increment, and then to solve the resulting system at each iteration by preconditioned conjugate gradients. This will be tested in future research.

Table 5.3: Effect of inexact CG solution.

algorithm	iterations/evaluations/total CG iterations exact ($q_k = 10^{-4}$)	inexact ($q_k = 10^{-1}$)	adaptive
Tangent Stiffness	92/99/3270	107/136/1387	146/358/1409
EBE QN Davidon	51/70/2091	85/181/1572	131/320/1608
EBE QN BFGS	51/63/1835	68/100/1046	144/376/1573

Table 5.4: Effect of inexact CG solution, by load step.

algorithm	iterations (evaluations) load increments 1+2+3+4 exact ($q_k = 10^{-4}$)	inexact ($q_k = 10^{-1}$)	adaptive
Tangent Stiffness	4+11+31+46	5+13+37+55	4+20+60+90
	(4+11+32+52)	(5+15+47+72)	(4+31+124+226)
EBE QN Davidon	3+8+15+25	4+10+31+40	4+11+43+73
	(3+8+18+43)	(4+10+77+107)	(4+14+105+230)
EBE QN BFGS	3+9+16+23	5+9+21+33	4+11+67+62
	(3+9+16+37)	(5+9+34+58)	(4+11+215+197)

Table 5.5: Effect of line-search.

algorithm	iterations(evaluations) simple l-s	$\sigma = 0.99$	$\sigma = 0.5$
Initial Stiffness	301(301)	301(301)	65(336)
Tangent Stiffness	93(105)	92(99)	58(201)
QN DFP	89(166)	98(122)	64(162)
QN Davidon	47(61)	48(73)	44(88)
SN DFP	71(200)	69(267)	48(320)
SN Davidon	78(121)	76(111)	66(173)
EBE QN BFGS	52(69)	51(63)	56(103)
EBE QN Davidon	56(83)	51(70)	59(100)

For problems in which a significant degree of nonlinearity arises, it appears that a strategy of inexact CG computation in each iteration is not advantageous overall, and that significantly fewer iterations and evaluations will be required in total if the CG iterations proceed to convergence. This applies to the Tangent Stiffness and EBE QN algorithms.

A further advantage of the EBE QN method for large problems, is that it is very well suited for implementation on a parallel processing computer. The updating of each element QN matrix, equation (4.4), is completely independent, and these computations could be done in parallel. The solution of the global system (4.5) by preconditioned conjugate gradients can also be parallelized to a large extent, in formation of the matrix-vector product element-by-element, and in use of EBE preconditioners (Bartelt 1989).

The amount of storage required by the EBE QN method, while much less than the classical QN algorithms, is still significant, since the updated element QN matrices are stored, in addition to the elastic element stiffness matrices (which are required for stress calculations). Since the element matrices are of small dimension, there is little advantage to be gained in an EBE QN overwriting strategy. However, the EBE QN algorithm could be implemented in Secant Newton form, storing only one update vector in addition to the elastic stiffness matrix for each element. This will be the subject of a paper to the next IACMAG Conference.

REFERENCES

Bartelt, P. 1989. *Finite Element Procedures on Vec-*

tor/Tightly Coupled Parallel Computers. Zürich: Verlag der Fachvereine.

Blaheta, R. & Kohut, R. 1991. Fast iterative methods for the solution of problems with nonlinear behaviour of materials. In Z. Rakowski (ed.), *Geomechanics 91.* A.A.Balkema Rotterdam.

Buckley, A. & A. LeNir 1983. QN-like variable storage conjugate gradients. *Math. Prog. 27*: 155-175.

Crisfield, M. A. 1980. Incremental/iterative solution procedures for nonlinear structural analysis. In C. Taylor, E. Hinton & D. J. R. Owen (eds.), *Numerical Methods for Nonlinear Problems; Vol. 1*, p.261–290. Swansea, Pineridge Press.

Dennis, J. E. & R. B. Schnabel 1983. *Numerical Methods for Unconstrained Optimization and Nonlinear Equations.* Englewood Cliffs: Prentice-Hall.

Fletcher, R. 1987. *Practical Methods of Optimization, 2nd ed.* Chichester: J. Wiley.

Geradin, M., S. Idelsohn & M. Hogge 1981. Computational strategies for the solution of large nonlinear problems via Quasi-Newton methods. *Comp. Struct. 13*: 73-81.

Griewank, A. & Ph.L. Toint 1982. Partitioned variable metric updates for large structured optimization problems. *Numer. Math. 39*: 119-137.

Nocedal, J. 1980. Updating quasi-Newton matrices with limited storage. *Math. Comput. 35/151*: 773-782.

Papadrakakis, M. & G. Pantazopoulos 1993. A survey of quasi-Newton methods with reduced storage. *Int. j. numer. methods engng 36*: 1573-1596.

Reed, M. B. 1992. Newton-like methods with limited storage, for the solution of elasto-viscoplasticity problems. *Int. j. numer. methods engng 35*: 223-240.

Swoboda, G. 1993. Programmsystem 'FINAL' FE Analyse linearer und nichtlinearer Strukturen unter statischer und dynamischer Belastung. Version 6.6. Universität Innsbruck.

Numerical methods
Lectures

Geomechanics 93, Rakowski (ed.) © 1994 Balkema, Rotterdam, ISBN 90 5410 354 X

Contribution to the problem of differences, which arise by using various geotechnical models for underground excavations

Josef Aldorf & Hynek Lahuta
Mining University of Ostrava, Czech Republic

ABSTRACT: The contribution presents the results of comparative solutions of stability models of the circle and arch shape underground openings. The stability was surveyed by "weightless" model and "heavy half plane" model. The authors proposed condition load coefficient adopting the results gained from simplified "weightless" model.

The first step to evaluate the stability of an underground opening is to formulate the conditions determining the calculation model of geotechnical situation that reliably reflects all crucial features from both geotechnical and technological point of view.

At present the creation of calculation model is based above all on making use of developed computer technique that is capable to perform large and complicated numerical (FEM, BEM) or analytical models.

The analytical models of underground openings may appear more advantegous because the amount of input data required by these models is many times smaller than by numerical ones. This fact causes that the analytical models are quicker in the phase of their preparation. They are more flexible for solving of many parametrical situations which can reflect various influential factors. The applying of analytical methods that evaluate the stability of underground openings is based on the assumption to consider the stability issues as contact problem between rock massif and lining, that could be solved for example by Kolosov-Muschelishvili's formulas. Being acquainted with the conformal transformation relationship of shape of opening it is possible to utilize these formulas for solution of any opening shape and complicated technological situation.

The set of boundary conditions, assuming the equal rock massif and lining displacements on the contact, and the gravitational stress state make up the fundament of solution. The boundary condition for contact of rock massif and lining can be defined as follows:

$$\sigma_r^m = \sigma_r^v - \sigma_r^{(p)}; \quad \tau_{r\varphi}^m = \tau_{r\varphi}^v - \tau_{r\varphi}^{(p)}$$

$$u_r^m = u_r^v \quad ; \quad u_\varphi^m = u_\varphi^v$$

On internal surface of lining the following stresses are supposed:

$$\sigma_r^v = 0$$

$$\tau_{r\varphi}^v = 0$$

The gravitational stress state in the infinite distance from the axis of opening is assumed to be:

$$\sigma_v^{(p)} = -\gamma \cdot h \cdot \alpha = q$$

$$\sigma_h^{(p)} = -k_b \cdot \gamma \cdot h \cdot \alpha = \lambda \cdot q$$

$$\tau^{(p)} = 0 \qquad \text{(see fig. 1c)}$$

$\sigma_r^m ; \sigma_r^v$ - radial normal stress in rock (m) and lining (v)

$\sigma_\varphi^m ; \sigma_\varphi^v$ - tangential normal stress in rock (m) and lining (v)

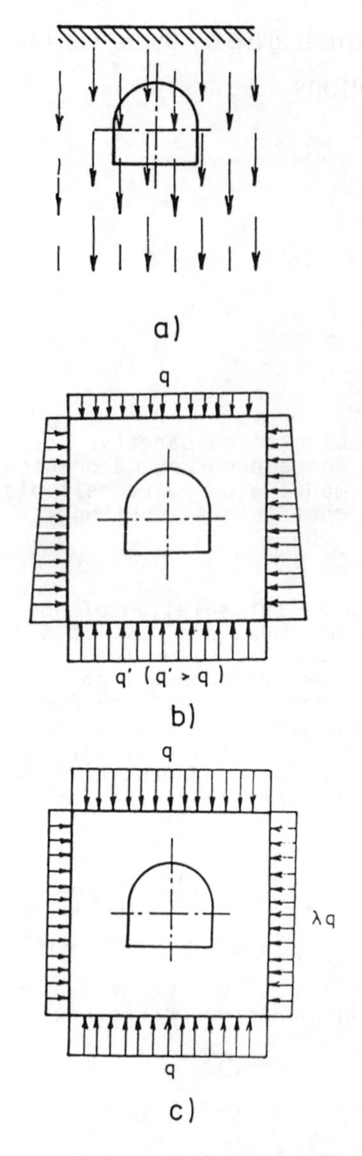

a) - the real calculation half plane
b) - the equivalent "endless" plane without weight
c) - the calculation "endless" plane without weight

fig. 1

u_r ; u_φ - radial and tangential displacement of rock (m) and lining (v)
τ - shear (tangential) stress in rock (m) and lining (v)

$\sigma_v^{(p)}$; $\sigma_h^{(p)}$; $\tau^{(p)}$ - vertical, horizontal and shear stress in intact rock massif
h - depth of opening under surface
γ - unit weight of rock
α - coefficient expressing some influences of material, technology etc.

So formulated problem corresponds with model shown on fig. 1b and 1c. It is possible to solve this model (without influence of weight) in described way very effectively, but it does not corresponds with real geotechnical conditions demonstrated on fig. 1a (the influence of rock weight, depth of opening). However this model is often applied for solving problems in this "shallow" conditions, because model on fig. 1a is very difficult to solve analytically for real opening shapes (with exception of circle shape). The question is: What mistake do we commit by this simplification, how is possible to eliminate this mistake by engineering methods and for which cases are both models practically equivalent?

On account of engineering purposes these relations were surveyed on comparative models for two types of opening shapes, circle and arch form, under the conditions given on fig. 1 and for $\gamma \neq 0$, h/D from range (0,20). There were especially examined the dependences of perpendicular displacements that were assumed as the most important and easy obtainable factor reflecting the lining load under the above mentioned conditions.

Accepting the model on fig. 1a as the basis (100= effect) the crucial points displacements on the opening shape (roof, wall) are related to the weightless model results (fig. 1c) and they show the value differences for weightless model under the condition h/D < 15 - 20 (fig. 2-4).

The evaluation of these results have provided the possibility to formulate the recommendation to set up the load coefficient (γ_z) reflecting real load state. This load coefficient adapts the load lining results - the stresses

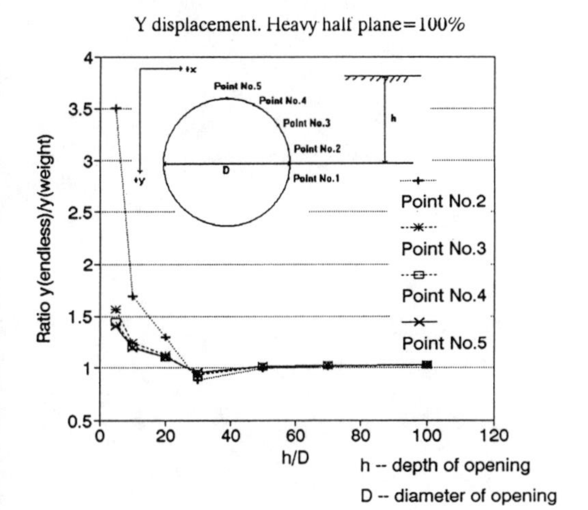

fig. 2

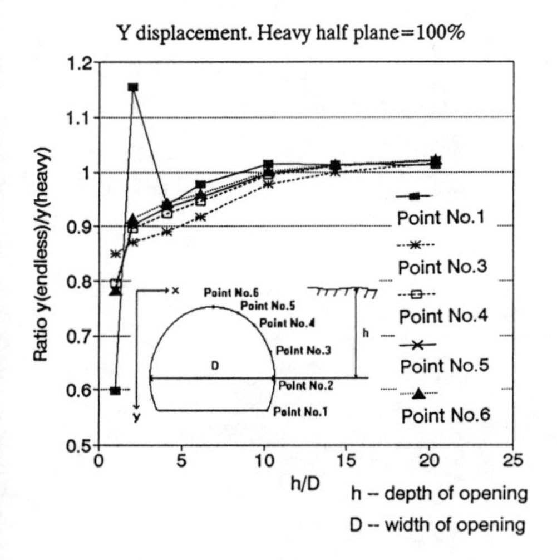

fig. 3

developing in lining - of weightless model in dependence on localization ratio h/D.

For the first approach the load coefficient μ_z is received as: the circle shape

$(h/D) < 6$	$\mu_z = 0{,}6 \div 0{,}8$
$6 < (h/D) < 12$	$\mu_z = 1{,}05 \div 1{,}1$
$(h/D) > 12$	$\mu_z = 1{,}0$

the arch shape

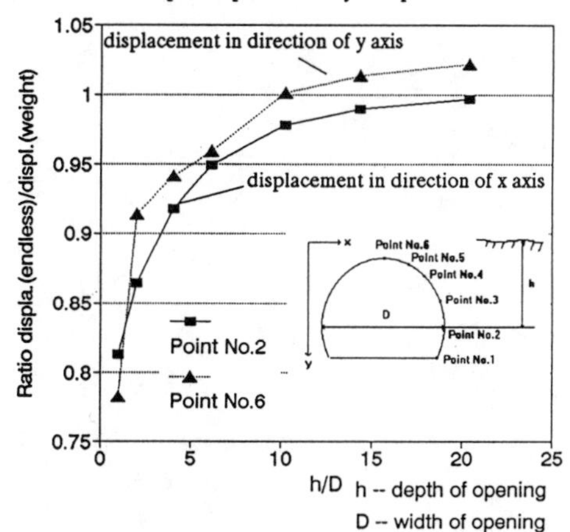

fig. 4

$(h/D) < 2$	$\mu_z = 1{,}3 \div 1{,}6$
$10 > (h/D) > 2$	$\mu_z = 1{,}05 \div 1{,}1$
$20 > (h/D) > 10$	$\mu_z = 0{,}95 \div 0{,}98$
$(h/D) > 20$	$\mu_z = 1{,}0$

It is apparent from the review that the arch shape is more sensitive and less suitable than the circle one. The arch shape effects for the extremely low thickness of overlying strata (h/D < 2) have a reverse character in comparision with the circle shape. This conclusion confirms the static suitability of circle shape of openings in case of low thickness of overlying strata.

Literature:

Bulyčev, N.S.: Mechanika podzemnych sooruženij. Nedra, Moskva 1982. (Mechanics of Underground Structures).

Aldorf, J.: The influence of time-dependent deformational properties of concrete on the load a loading capacity of Suport of underground working. Geomechanics 91 - A.A.Balkema, Roterdam.

Geomechanics 93, Rakowski (ed.) © 1994 Balkema, Rotterdam, ISBN 90 5410 354 X

Regular grids and local grid refinement

R. Blaheta, R. Kohut, A. Kolcun & O. Jakl
Institute of Geonics of the CAS, Ostrava, Czech Republic

ABSTRACT: The paper concerns several issues: regular structured grids, grid refinement, composite grids, iterative solution of problems on composite grids, adaptive refinement. A progress in these issues results in a possibility to create a reliable and efficient finite element software well suited for the solution of problems of geomechanics. An implementation of the above ideas in the finite element software GEM22 and an example of solution of a practical problem are shortly described.

1 INTRODUCTION

The main topic of our paper is how to perform a local grid refinement by a composite grid technique. The related question where local grid refinement should be performed will be discussed elsewhere.

In this paper, we describe the use of a composite grid technique which involve both discretization and special iterative procedure for the solution of composite grid problems. This technique enables very naturally and efficiently to perform local grid refinement within finite element software which use regular grids. Moreover, this technique enables successive adaptive refinement of an initial coarse grid.

2 REGULAR GRIDS

By regular grids we shall understand grids constructed by a mapping from uniform rectangular grid onto the given domain.

This construction supported by a suitable preprocessing procedures is more flexible than one might first expect and fits especially well for many geomechanical problems as for example problems of longwall mining, slope stability, etc. For these reasons regular grids are exploited in our finite element software GEMxx for the solution of 2D and 3D stress analysis problems, see (Blaheta, Dostál 1988), (Blaheta et al. 1989, 1993).

Regular grids have many advantages. They enable the use of simple structures for both input and output data, therefore the procedures for pre- and post-processing are simple and very illustrative. Additionally, the stiffness matrix can be stored economically according to some regular storage scheme. For the solution of corresponding linear system by a band solver we have very good ordering of unknowns, for the use of iterative solvers we can easily and efficiently vectorize the multiplication of vector by the stiffness matrix. All these facts influence favourably the efficiency of computations. The possibility of vectorization and parallelization is also one reason for increasing interest in regular grids. All these gains in efficiency are even more important when we want to solve nonlinear problems.

There may be also two types of difficulties with regular grids. They are not suitable for certain geometries, but this is not our case. The second difficulty occurs when we need a local refinement of the grid. This paper shows how to overcome this difficulty by the composite grid technique.

3 LOCAL GRID REFINEMENT AND COMPOSITE GRIDS

The grids with approximately equal density everywhere give usually nonuniform error distribution. That means that either the accuracy is insufficient in some places or that some computational work is wasted in the solution on too dense grid in other places. Therefore, we need a local grid refinement.

For regular grids the local grid refinement can be performed simply by the way of prolongated refinement depicted schematically in Fig. 1.

Comparing with (more) uniform grid we reduce
- some unknowns,
- the bandwidth.

Comparing the refinement of Fig. 1 with corre-

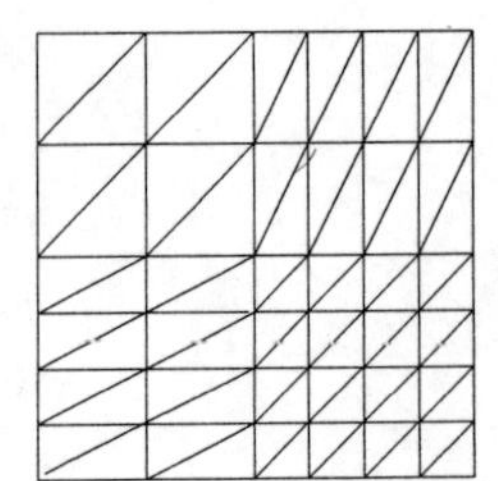

Figure 1: Refinement in one quadrant

sponding uniform grid, we can see that the number of unknowns is reduced to 1/2 and the bandwidth is reduced to 3/4. This results in a saving of nearly 75 % of the computational work when band solver is used and 50 % for iterative solver (the condition number is not reduced).

The long triangles which appear in this type of refinement need not be criticized in itself (Křížek 1991), but the degrees of freedom are generally not used economically from the point of view of approximation.

The alternative way how to perform local grid refinement is the use of composite grid, see Fig. 2. The finite element space now consists of all functions which have prescribed polynomial behaviour on individual elements and are continuous in the whole region. The last demand implies a constraint in some nodes on the inner boundary of the area of refinement.

The use of composite grid gives a higher reduction of unknowns. On the other hand a regular storage scheme of the stiffness matrix is affected. This drawback can be overcome by a special iterative procedure which will be described in the next section. Moreover, this procedure brings further advantages, e.g. the adaptive refinement possibility.

4 ITERATIVE SOLUTION OF COMPOSITE GRID PROBLEMS

We can start with well-known and very illustrative

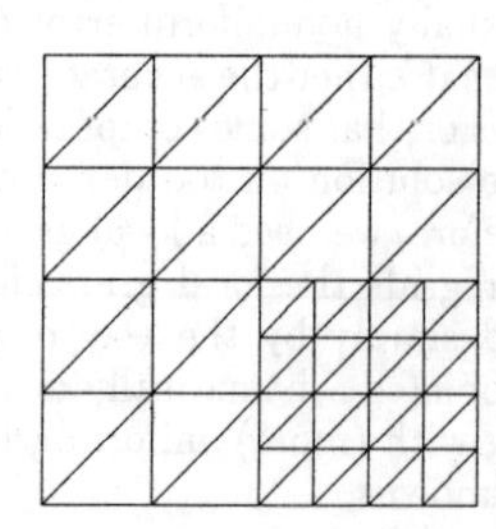

Figure 2: A composite grid

procedure for approximation of the composite grid solution.

Suppose that we solve boundary value problem in the domain Ω with the boundary $\partial\Omega$ which consists of the parts Γ_1, Γ_2, etc. on which various types of boundary conditions are prescribed. Let Ω_H be a coarse grid in Ω. Further, let Ω' be a part of Ω where a refined grid Ω'_h is required. The boundary $\partial\Omega'$ then consists of the parts $\Gamma'_e = \partial\Omega' \cap \partial\Omega$ and the internal boundary $\Gamma'_i = \partial\Omega' \setminus \Gamma'_e$. Obviously $\Gamma'_e = \bigcup \Gamma'_k$, $\Gamma'_k = \partial\Omega' \cap \Gamma_k$, $k = 1, 2, \ldots$, see Fig. 3. Now the composite grid solution can be approximated by the following procedure:

1. Solve the coarse grid problem in Ω_H.

2. Transfer and interpolate the computed values on Γ'_i.

3. Solve the problem in Ω'_h with Dirichlet boundary conditions on Γ'_i given by the values from the previous step, the boundary conditions on Γ'_e and the load in Ω' taken directly from the problem in Ω.

4. From the solutions u_H from the step 1 and u_h from the step 3 we can compose an approximation of the composite grid solution

$$u = \begin{cases} u_H & \text{in } \Omega \setminus \Omega' \\ u_h & \text{in } \Omega' \end{cases} \tag{1}$$

The quality of this approximation will be very good if u_H is close to the exact solution in the region $\Omega \setminus \Omega'$.

The quality of the approximation (1) can be also verified by computation of the residual

$$r = f - Au \tag{2}$$

where f is the load vector, A is the stiffness matrix of the composite grid problem.

It is important that this residual can be assembled from contribution of individual finite elements without necessity to assembly the stiffness matrix A. Moreover, if the problems 1 and 3 are solved exactly, the residual will be nonzero only along the inner boundary Γ'_i and therefore its computation is very cheap.

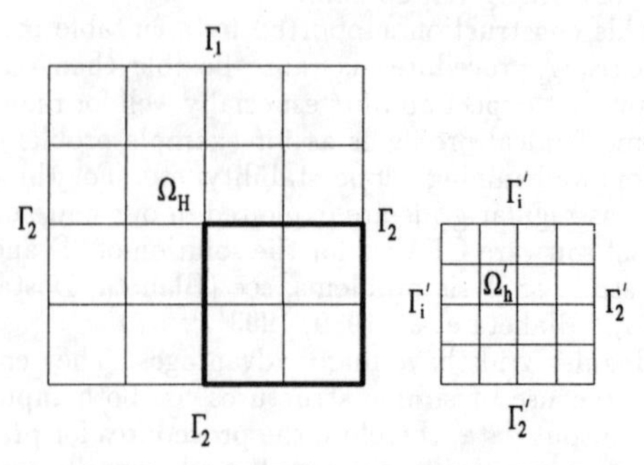

Figure 3: Composite grid problem

If the residual is not small enough, then we can approximate the correction

$$\Delta u = A^{-1}(f - Au) \tag{3}$$

again by the above described procedure (steps 1–4). We must only replace the load f by the residual r.

Note that this iterative procedure is identical to the Fast Adaptive Composite Grid Method, see (McCormick, Thomas 1986), (Heroux, Thomas 1992).

This iterative procedure constitutes the composite grid technique which brings the following advantages:

- use of completely regular data structures,
- solution of smaller and better conditioned problems from the steps 1 and 3,
- possibility of adaptive construction of the refinement.

The efficiency of iterations depends on the convergence properties, which seem to be very good (only a few iterations are sufficient, see the example from the next section).

5 IMPLEMENTATION AND EXAMPLE

The described composite grid technique was implemented in GEM22 finite element software. For the users it means that they specify the whole coarse grid problem, mark the region of refinement and specify the density of refinement. The generation of the refined grid problem and the solution on the composite grid are then performed automatically.

Fig. 4 shows a problem arising from the analysis of a longwall mining situation. Here we can see the material interfaces and the region for refinement where the grid density is doubled. Also the partly refined coarse grid is depicted in this Figure.

The convergence properties can be seen from the following Table:

Iteration	l_2-norm of the residual	Reduction factor	$\Delta\sigma_I$	$\Delta\sigma_{II}$
1	14.58	–	0	–
2	3.34	0.229	0.771	0.802
3	0.798	0.237	0.190	0.200
4	0.190	0.239	0.045	0.047
5	0.046	0.243	0.011	0.011

$\Delta\sigma_I$ – maximal difference in the computed stress outside Ω'

$\Delta\sigma_{II}$ – maximal difference in the computed stress in Ω'

The difference in computed stresses within the refined region can be seen from the Fig. 5.

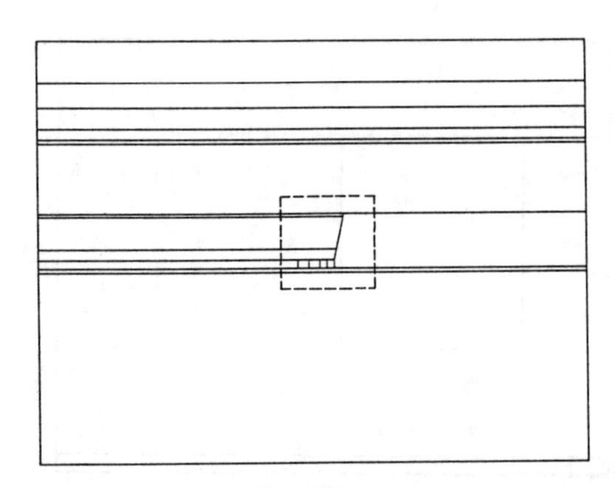

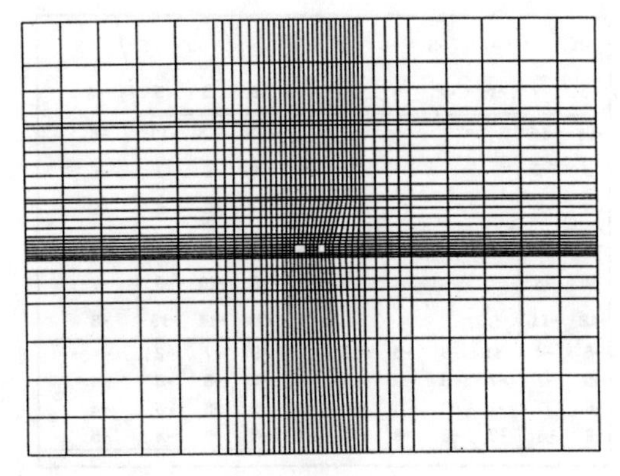

Figure 4: Longwall mining problem:
a) material interfaces. The region of refinement is marked by dashed lines,
b) the partly refined coarse grid.

With respect to the magnitude of the computed stresses, the stress differences displayed in the above Table are negligible after 2 or 3 iterations. Thus, 2 or 3 iterations are fully sufficient.

6 CONCLUSIONS

Our numerical example (see Fig. 5) confirms that it is important to use locally refined grids in finite element analysis of many geomechanical problems.

The described composite grid technique enables us to perform the local refinement and keep all the benefits provided by the regular grids.

There are also some related topics which will be discussed elsewhere, e.g. the case when the coarse grid does not fit the given problem in the region of refinement, the case of multiple refinement, the use of aposteriori error estimates, etc.

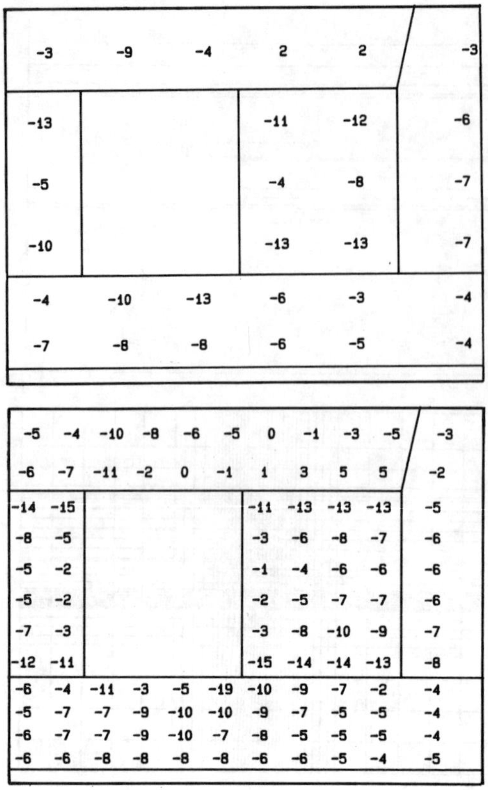

Figure 5: The horizontal stresses
a) in the coarse grid,
b) in the fine grid in the refined region.

REFERENCES

Blaheta, R., Jakl, O., Kohut, R., Kolcun, A., Slíva, J. 1989, 1993. *GEM22, User Guide.* Report of the Mining Institute Cz. Acad. Sci., Ostrava. (In Czech.)

Blaheta, R., Dostál, Z. 1988. On the solution of large 3D geomechanical problems. In Swoboda (ed.), *Numerical Methods in Geomechanics*: p. 1911–1916. Rotterdam: Balkema.

McCormick, S., Thomas, J. 1986. The Fast Adaptive Composite Grid Method for Elliptic Equations. *Math. Comp.* 46: 439–456.

Heroux, M., Thomas, J. W. 1992. A comparison of FAC and PCG methods for solving composite grid problems. *Comm. Appl. Num. Meth.* 8: 573–583.

Křížek, M. 1991. On semiregular families of triangulations and linear interpolation. *Appl. Math.* 36: 223–232.

Geomechanics 93, Rakowski (ed.) © 1994 Balkema, Rotterdam, ISBN 90 5410 354 X

Domain decomposition methods for modelling of deformable block structure

Zdeněk Dostál
Technical University Ostrava & Institute of Geonics of the CAS, Ostrava, Czech Republic

ABSTRACT: We present domain decomposition based algorithms for modelling of system of deformable blocks. We suppose that the blocks are elastic and that their relation is described by linearized contact conditions. Starting from formulation of the conditions of equilibrium by means of the variational inequality, we present the algorithm for its solution that exploits solution of subproblems for individual block with either prescribed zero displacement or boundary tractions.

1 INTRODUCTION

Since more than twenty years, a number of methods have been developed for analysis of stress and deformations of deformable block structure. We can observe two essentially different approaches. The first one is based on using various interface elements. Let us quote Goodman and John (1977) and Desai and others (1984) to give important examples in this line. The advantage of this method is that it is easy to implement and that it can be used to modelling of very complex nonlinear response of the interface. However, from the mathematical point of view, the method may be identified with the penalty method so that is suffers from the well known drawbacks of the latter method.

In our lecture, we deal with the other approach that is based on discretization of the variational inequality which describes the conditions of equilibrium of bodies in contact. Comprehensive exposition of the theory with other references may be found in Kikuchi and Oden (1988). If we restrict our attention to elastic bodies in contact without friction, the variational formulation leads to the problem of quadratic programming to minimize the energy functional

$$j(x) = \frac{1}{2}x^T Kx - f^T x \qquad (1)$$

on the set $V = \{x: Bx \le o\}$. Here K is the stiffness matrix of the system with possibly enhanced equality constraints, f is an n-vector of nodal forces, x is that of nodal displacements and $B = \{b_1,\dots,b_k\}$ is a matrix describing linearized incremental kinematical contact conditions. In general, we do not suppose the functional $j(x)$ to be coercive so that K may be positive semidefinite. Thus we can consider "floating" blocks in our analysis.

We shall concentrate on exploiting so called domain decomposition methods. The idea of these methods is to subdivide a physical domain of the problem to gain some computational advantage. Though the method has been first developed for the solution of linear problems (see Schwarz (1870)), it seems that it may be even more successful for the solution of our problem. The reason is that the physical domain of our problem consists of several subdomains (blocks) as in Figure 1, so that no additional data are necessary for description of the decomposition.Moreover, the numerical solution of such problems is usually reduced to numerical solution of a sequence of related linear problems so

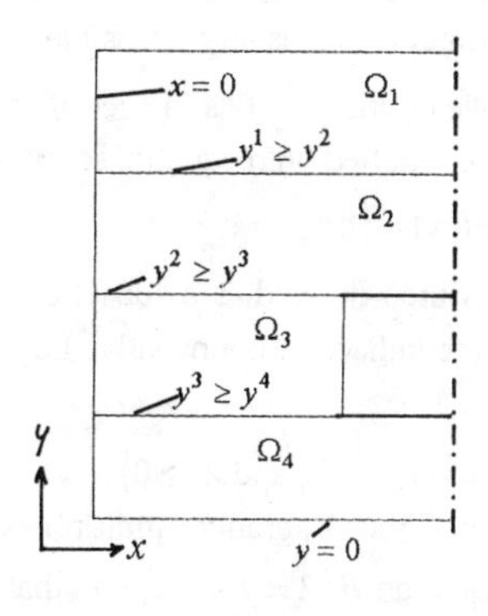

Figure 1 Contact model of a block structure.

that some more expensive preliminary computations may pay off.

2 DISCRETIZED FORMULATION

Suppose that K is the stiffness matrix of the order n resulting from a finite element discretization of the system of elastic bodies $\Omega_1,\ldots,\Omega_k$ with possibly enhanced zero normal displacement conditions. With suitable numbering of nodes, we can achieve that $K = diag(K_1,\ldots,K_k)$, where K_i is the stiffness matrix of the body Ω_i.

The linearized incremental contact conditions are supposed to be defined by the matrix B and the vector c,

$$B = (b_{*1},\ldots,b_{*n}) = \begin{pmatrix} b_{1*} \\ \vdots \\ b_{k*} \end{pmatrix}, \; c = \begin{pmatrix} c_1 \\ \vdots \\ c_n \end{pmatrix}.$$

The rows b_{i*} of B are vectors that enable us to evaluate the change of the normal distance from $c_i \geq o$ in a reference configuration of a given pair of nodes; the formula for the displacement u is $b_{i*}u$. Let use observe that the matrix B is sparse as nonzero entries of b_{i*} may be only in positions of nodal variables that correspond to the nodes involved in some constraint. For convenience, if $b_{*i} = o$, we shall call u_i a free variable. Finally, f is a vector of nodal forces.

3 POLYAK ALGORITHM

Our algorithms are built on the well known Polyak algorithm as described in Polyak (1969). It exploits so called active set strategy. If x is any feasible vector, i.e. $Bx \leq c$, then we define $\mathcal{A}(x)$ as the set of indices of relations that are satisfied with equality so that

$$\mathcal{A}(x) = \{i{:}b_{i*}x = c_i\}.$$

The binding set is then defined as a subset of $\mathcal{A}(x)$ which comprises the indices of nonnegative Lagrange multipliers λ_i, i.e.

$$\mathcal{B}(x) = \{i{:}b_{i*}x = c_i \text{ and } \lambda_i \geq 0\}.$$

Let use recall that the Lagrange multipliers are solutions of the equation $B^T\lambda = f - Bx$, and that their physical meaning is that of the contact model forces. Finally, let us denote by $y = ccg(x,I,ins)$ the procedure of constrained conjugate gradients which returns either the minimum of $j(x+\xi)$ on the face

$$W_i = \{\xi{:}b_{i*}\xi = c_i \text{ for } i \in I\}$$

and $ins = 1$, or $ins = 0$ and y which satisfies $\mathcal{A}(y) \supset \mathcal{A}(x)$, $x \neq y$. The minimization in the body of ccg is carried out by means of the conjugate gradient procedure of Hestenes and Stiefel (1952). The Polyak algorithm now reads as follows:

(Quadratic Programming by Polyak). Given the starting vector x, the algorithm returns the solution of the problem (1) in a finite number of steps.

function: $y = qpp(x)$
    $I = \mathcal{A}(x)$
    $y = x$
    while Kuhn-Tucker conditions not satisfied
        $y = ccg(y,I,ins)$
        if $ins = 0$
            $I = \mathcal{A}(y)$
        else
            $I = \mathcal{B}(y)$
        end
    end
end *ccg*

4 DECOMPOSITION INTO DIRICHLET PROBLEMS

Our first domain decomposition algorithm exploits solution of auxiliary problems of Figure 2 to preconditioning of inner iterations of *ccg* procedure. The result is that the iterations are reduced

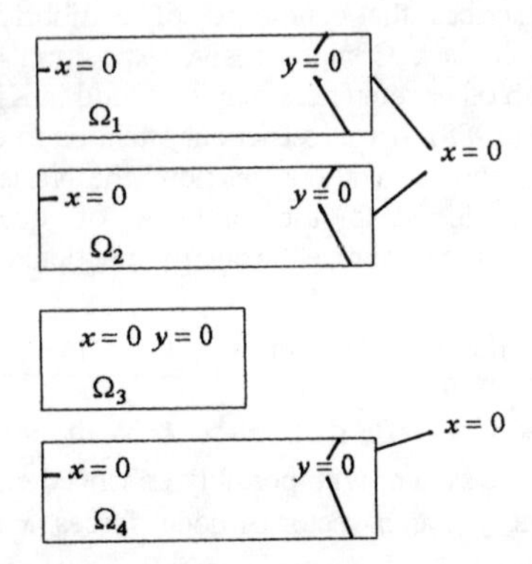

Figure 2 Auxiliary problems with Dirichlet conditions.

to interfaces between blocks with strong preconditioning effect. Instead with the matrix K, the iterations are carried out with the matrix KQ, where Q is the conjugate projector on the possible interface. It is defined by

$$Q = I - U\left(U^T K U\right)^{-1} U^T K,$$

where U is an $n \times m$ matrix that arises from the unit matrix I by crossing out the columns that do not correspond to free variables. Simple computations show that U has similar block diagonal structure as K so that

$$U^T K U = \begin{pmatrix} U_1^T K_1 U_1 & & \\ & \cdots & \\ & & U_k^T K_k U_k \end{pmatrix}.$$

The matrices in diagonal blocks may be identified with the stiffness matrices of auxiliary problems of Figure 2. Numerical experiments on a problem similar to that of Figure 1 confirmed the theoretical assumptions and reduced the cost of computations by 4 on single processor. Due to high degree of parallelism, we can suppose that parallel implementation would be even more efficient. Details may be found in Dostál (1992) or Dostál (1993a).

5 DECOMPOSITION INTO NEUMANN PROBLEMS

The obvious drawback of the approach described in Section 4 is that it does not exploit full information on the stiffness matrix K. In this section, we show that it is possible to exploit solution of singular problems, exactly those of Figure 2 without prescribed zero displacements. The difficulty with treating these floating domains may be overcome by duality. The starting point is based on a classical result which says that under conditions which are satisfied in our case, the problem to find solution of (1) is equivalent to problem to find

$$\max\{\theta(\lambda) : \lambda \geq o\},$$

where

$$\theta(\lambda) = \inf\{L(u,\lambda) : u \in R^n\}$$

with the Lagrange function

$$L(u,\lambda) = \frac{1}{2} u^T K u - f^T u + \lambda^T (Bu - c).$$

For fixed λ, the Lagrange function $L(.,\lambda)$ is convex in the first variable and the gradient argument shows that any minimizer u of $L(.,\lambda)$ satisfies

$$Ku - f + B^T \lambda = o. \qquad (2)$$

The equation (2) has a solution iff

$$f - B^T \lambda \in \operatorname{Im} K. \qquad (3)$$

The latter condition can be expressed more conveniently by means of the matrix R whose columns span the null space of K. The equivalent condition to (3) is then

$$R^T\left(f - B^T \lambda\right) = o. \qquad (4)$$

Simple computations show that for any λ which satisfies (4) there is u which satisfies (2) and

$$u = R\alpha + K^+\left(f - B^T \lambda\right), \qquad (5)$$

where K^+ is a pseudoinverse to K.

Substituting into the formula for $\theta(\lambda)$ yields that the vector of contact nodal forces λ is determined by the solution of

$$\frac{1}{2} \lambda^T B K^+ B^T \lambda - \lambda^T\left(B K^+ f - c\right) \rightarrow \min$$

subject to $\lambda \geq 0$ and (4). Theoretical results of Roux (1992) which concern linear problems indicate that the distribution of the spectra of $B^T K^+ B$ is much more favourable than that of the restrictions of KQ to the interface. The matrix K^+ is again block diagonal so that parallel implementation of a variant of Polyak algorithm is possible, as well as further preconditioning. An interesting choice may be $B^T K B$, which is very cheap and efficient in linear case as shown by Farhat and Roux (1992). Another possibility is to use the Schur complement. Both methods are tested at the moment (Dostál, 1993c).

6 CONCLUSION

Domain decomposition methods may be efficient for the solution of problems arising from discretization of conditions of equilibria of a deformable block structure. The tests have shown that even the basic decomposition into Dirichlet problems considerably improves the performance of the Polyak algorithm based methods of solution and that there are theoretical results that indicate that there are even more efficient domain decomposition methods. Another way of improving the efficiency of such algorithms may be based on the control of precision of auxiliary problems based on theoretical and experimental results of Dostál (1993b).

REFERENCES

Goodman, R.E. & C.St.John 1977. FE analysis for discontinuous rocks. In C.Desai & Christian (eds.). Numerical Methods in Engineering. Mc Graw-Hill, New York.

Desai, C.S. at al 1984. Thin layer element for interfaces and joints. Int. J. Num. And Analyt. Meth. in Geomechanics 8.

Kikuchi, N. & J.T.Oden 1988. Contact Problems in Elasticity. SIAM, Philadelphia.

Schwarz, H.A. 1870. Viertelsjahresschrift der Naturforschenden Gesellschaft in Zürich, Bd. 15.

Polyak, B.T. 1969. The conjugate gradient method in external problems. USSR Comput. Math and Math. Phys. 9: 94-112.

Hestenes, M.R. & E. Stiefel 1952. Method of conjugate gradients fro solving linear system. J. Res. Nat. Bur. Stand. 49: 409-436.

Dostál, Z. 1992. Conjugate projector preconditioning for solution of contact problems. Int. J. Num. Met. Eng. 34: 271-277.

Dostál, Z. 1993a. The Schur complement algorithm for the solution of contact problems. To appear in Proc. of 6th Domain decomposition conference, to be published by AMS.

Roux, F.-X. 1992. Spectral analysis of interface operator. 5th Symposium on Domain Decomposition Methods. SIAM, Philadelphia.

Farhat, C. & F.-X.Roux 1991. A method of finite element tearing and interconnecting and its parallel solution algorithm. Int. J. Num. Met. Eng. 32: 1205-1227.

Dostál, Z. 1993b. Directions of large decrease and the Polyak algorithm. Submitted to Lin. Alg. Appl.

Dostál, Z. 1993c. A Saddle-point domain decomposition for the solution of contact problems. To be presented at the 7th Domain decomposition conference, University Park, Pennsylvania.

Geomechanics 93, Rakowski (ed.) © 1994 Balkema, Rotterdam, ISBN 90 5410 354 X

Evaluation of opening stability with rocks technological heterogeneity being taken into account

Nina N. Fotieva & A. S. Sammal
Tula State Technical University, Russia

ABSRACT: Method of openings stability evaluation taking into account rocks technological heterogeneity, i.e. the rock deformation modulus decrease near the opening surface being the result of blasting operations is proposed. The method developed has been programmed for the computer. Examples of the design are given.

One of the ways to estimate the workings stability is based on determining the sizes of so-called conditional zones of nonelastic deformations (Bulychev, Fotieva 1977). Those zones are understood as regions surrounding the working in which the stresses determined by solving a plane problem of elasicity theory do not satisfy the Kolon-More Strength condition

$$(\sigma_\theta - \sigma_\rho)^2 + 4\tau_{\rho\theta}^2 \leq \sin^2\varphi(\sigma_\rho + \sigma_\theta + 2C\,ctg\,\varphi)^2, \quad (1)$$

where σ_ρ, σ_θ, $\tau_{\rho\theta}$ are the components of stress tenzor being determined in the curvilinear coordinates connected with a conformal transformation of a circular opening exterior upon the exterior of the pregiven shape opening; C is rock cohesion coefficient; φ is the angle of internal rock friction.

The technique generalising the approach mentioned for determining the boundaries of conditional zones of nonelastic rocks deformations with regard to the rock technological heterogeneity, i.e. the rock deformation modulus and cohesion coefficient decrease in the working proximity being the result of blasting operations is offered in the paper presented.

The solution of the problem determining the rock massif stressed state taking the rock technological heterogeneity into account is based on a representation of any deformation modulus change law by a discrete change law and on solving the elasticity plane contact problem for a multi-layer non-circular ring hardening the opening in a linearly deformable medium. The design scheme is given in Figure 1.

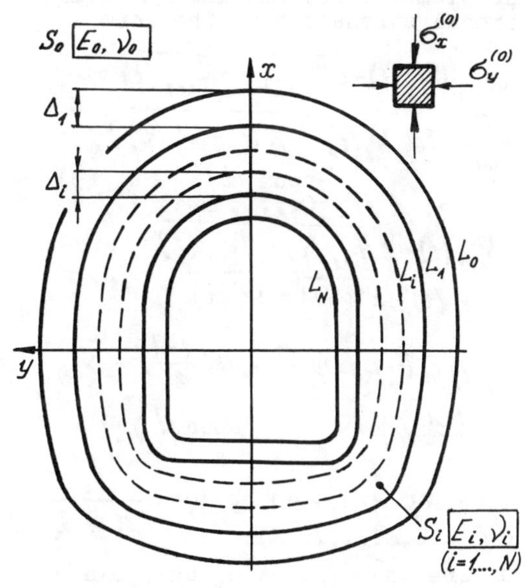

Fig. 1 Design Scheme

The S_0 medium and S_i (i = 1, ..., N) ring layers are of different materials with E_i, ν_i (i = 0,..., N) deformation modulii and the Poisson ratios and have an initial stresses field called forth by the rock's own weight

$$\sigma_x^{(i)(0)} = -\gamma H, \quad \sigma_y^{(i)(0)} = -\lambda \gamma H \qquad (2)$$

$$(i = 0, 1, \ldots, N)$$

where γ is the rock specific weight, being the same both for the layers and the medium, H is the working depth, λ is the lateral pressure coefficient in an intact massif.

The problem has been solved with the application of the complex variable analytic functions theory (Muskhelishvili 1966), the conformal representations apparatus and the complex series.

On total stresses having been represented as the sums of initial stresses (2) and additional stresses appearing due to presence of the working and after the introduction of the $\varphi_i(z)$, $\psi_i(z)$ (i = 0, ..., N) complex potentials regullar in corresponding S_i spheres and being turned to zero upon infinity, the boundary conditions of complex variable functions theory problem for determining the additional stresses have the form

$$\varkappa_{i+1}\varphi_{i+1}(t) - t\overline{\varphi'_{i+1}(t)} - \overline{\psi_{i+1}(t)} = \frac{\mu_{i+1}}{\mu_i}\left[\varkappa_i\varphi_i(t) - t\overline{\varphi'_i(t)} - \overline{\psi_i(t)}\right],$$

$$\text{upon } L_i \qquad (3)$$

$$(i = 0, \ldots, N-1)$$

$$\varphi_{i+1}(t) + t\overline{\varphi'_{i+1}(t)} + \overline{\psi_{i+1}(t)} = \varphi_i(t) + t\overline{\varphi'_i(t)} + \overline{\psi_i(t)};$$

$$\varphi_N(t) + t\overline{\varphi'_N(t)} + \overline{\psi_N(t)} = \gamma H\left(\frac{1+\lambda}{2}t - \frac{1-\lambda}{2}\bar{t}\right) \qquad (4)$$

$$\text{upon } L_N$$

Here $\varkappa_i = 3 - 4\nu_i$, $\mu_i = \dfrac{E_i}{2(1+\nu_i)}$ (i = 0, 1,..., N);

t are the affixes of the points in the corresponding outlines.

The solution of this problem described in the paper by Fotieva, Sammal (1988) allows to determine as a result the total $\tilde{\sigma}_\rho = \sigma_\rho/\gamma H$, $\tilde{\sigma}_\theta = \sigma_\theta/\gamma H$, $\tilde{\tau}_{\rho\theta} = \tau_{\rho\theta}/\gamma H$ stresses at all points of the area under investigation around the working.

Method of opening stability evaluation taking into account rock technological heterogeneity has been programmed for the computer. The program developed may be used for determining the boundaries of conditional zones of nonelastic deformations at any deformation modulus, rock cohesion coefficient and the angle of internal rock friction change law, since it is possible to consider the quentity of layers up to 20 ($N \leq 20$) and to obtain admissible results even for continuous change of rock characteristics.

The example of determining the conditional zones of nonelastic rock deformations around the arch form working with a 3.6 m span and a 2.75 m height is given below.

The deformation modulus changes under the law (Rukin, Ruppeneit 1968)

$$E(r) = E_\infty\left(1 - \frac{\kappa R^m}{r^m}\right), \qquad (5)$$

where E_∞ is the intact massif deformation modulus, R = 1.58 m is the mean working radius, r is a distance of the working centre; parametres κ = 0.8; m = 2.

Ten layers in the working's surrounding were singled out for the purpose of calculation, the deformation modulus remaining stationary in each of them (Figure 2).

The same law with the κ = 0.95, m = 3 of the rock cohesion coefficient change was applied, i.e.

$$C(r) = C_\infty\left(1 - \frac{0.95 R^3}{r^3}\right), \qquad (6)$$

where $C_\infty = 0.1\gamma H$.

The corresponding values of the layers thickness Δ_i/R (i = 1,..,10), deformation modulii E_i/E_0 (i = 1,...,10) and cohesion coefficients C_i/C_0 (i = 1,...,10) for

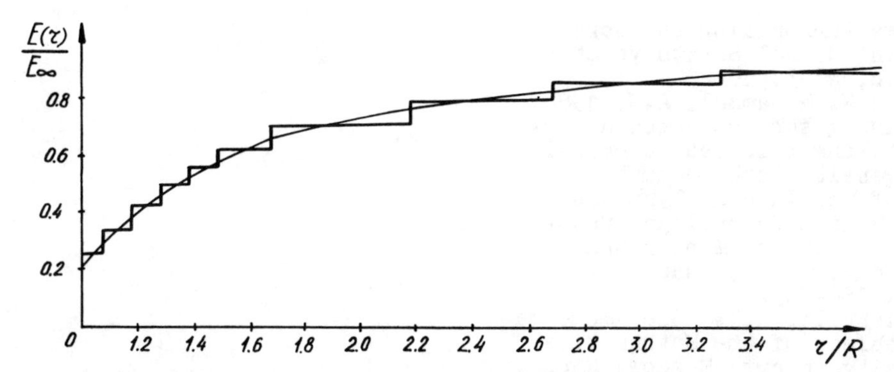

Fig. 2 Dependence of the value $E(r)/E_\infty$ upon the relative distance r/R

Table 1. The Δ_i/R thicknesses, E_i/E_0 deformation modulii, C_i/C_0 cohesion coefficients for each layer

Parametres	Number of layer									
	1	2	3	4	5	6	7	8	9	10
Δ_i/R	0.60	0.60	0.50	0.50	0.20	0.10	0.10	0.10	0.10	0.09
E_i/E_0	0.90	0.85	0.78	0.70	0.61	0.55	0.49	0.43	0.34	0.25
C_i/C_0	0.96	0.92	0.86	0.77	0.66	0.55	0.48	0.40	0.26	0.13

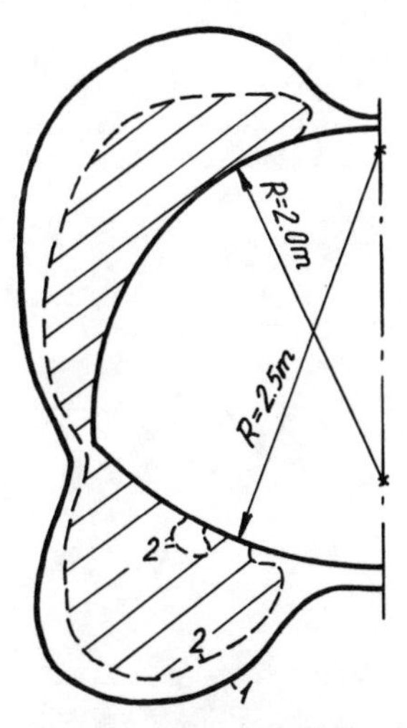

Fig. 3 The boundaries of the calculated conditional zones of nonelastic rocks deformation

each layer are given in Table 1.

The angle of rock internal friction is $\varphi = 40°$.

The boundaries of the calculated conditional zones of nonelastic rocks deformations are shown in Figure 3 by curve 1. For comparison the boundaries of the same zones in a homogeneous massif at $E_i = E_0 = E_\infty$ and $C_i = C_0 = C_\infty = 0.1\gamma H$ are shown in Figure 3 by curve 2.

In conclusion we can mark that the technological rock heterogeneity has negativ influence on rock stability and should be taken into account at the underground opening designing.

REFERENCES

Bulychev, N.S. & Fotieva, N.N. 1977. Evaluation the stability of rocks surrounding the workings. In Journal Shaxtnoye Stroitelstvo, N° 3: 16-22.

Fotieva, N.N. & Sammal, A.S. 1988. Determining stresses around workings taking rock technological heterogeneity into account. Proc. of the Intern. Symp. on Modern Mining Technology: Shandong Institute of Mining and Technology, Taian, Shandong, P.R.C.: 286-293.

Musckhelishvili, N.I. 1966. Some basic problems of the Mathematical Elasticity Theory. Moscow: Nauka.

Rukin, V.V. & Ruppeneit, K.V. 1968. Mechanism of pressure tunnels designing interaction with massif. Moscow: Nauka.

Geomechanics 93, Rakowski (ed.) © 1994 Balkema, Rotterdam, ISBN 90 5410 354 X

Non-linearities in soil mechanics – Case histories

H. Konietzky & D. Billaux
ITASCA Consultants, Bochum, Germany & Ecully, France

P. Meney & R. Kastner
INSA Lyon, France

ABSTRACT: The paper describes the importance of non-linearities regarding the material law for soil and soft rock in the engineering practice. Numerical modelling results regarding the ground behaviour are shown on two examples: an excavation, supported by a sheet pile, in an analog granular material and a near-surface tunnel with shotcrete lining in soft rock.

1 INTRODUCTION

Soil and soft rock are often characterised by non-linearities regarding the material behaviour. To what extent and in what way these non-linearities have to be considered in soil- and rock-mechanical calculations depends essentially on the objective. This is demonstrated on two examples of numerical modelling from the engineering practice.

The first one shows the analysis of a sheet-pile supported excavation in a granular material, where the classical Mohr-Coulomb law has brought reasonable results, whereas in the second example regarding subsidence calculations for a tunnel project this law fails and only a more complicated non-linear law, with a second yield-surface including strain-hardening of the cap-pressure and different loading and unloading moduli has led to satisfying results.

All the numerical modelling, described within this paper was performed with the explicit 2D-Finite-Difference-Code FLAC, Vers. 3.22 (Itasca 1992), configured for plane-strain and large displacement conditions.

2 EXAMPLE 1

2.1 Problem

A physical model was used by INSA (Institute National des Sciences Appliquees) to experiment on the behaviour of a trench in frictional material for the Lyon subway system (Masrouri 1992). A 2D analogical material made of steel rods was used (Kastner 1982) to reproduce first a 'triaxial test' in order to determine equivalent soil parameters, then the trench excavation, supported by a sheet pile with or without struts (Masrouri 1986). The excavation was performed by steps. Displacements and movements in the sheet pile were recorded at each step. A general layout is given in Figure 1.

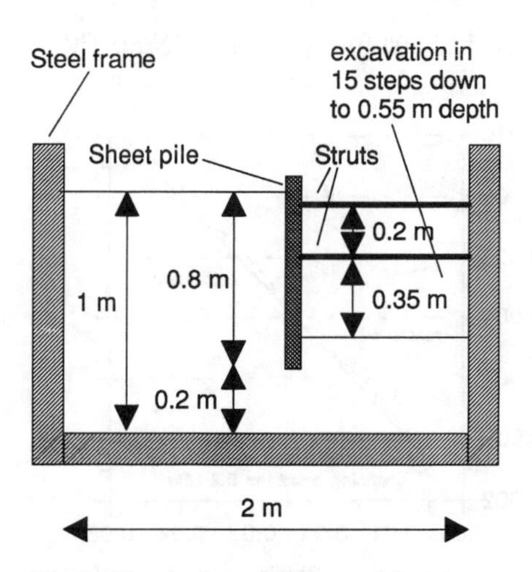

Fig.1 Physical model general layout

2.2 'Triaxial' experiment

A rectangular (0.4 by 0.2 m) analog sample was loaded to shearing failure, with confining pressures from 0.2 to 0.5 MPa. This was modelled using FLAC with a Mohr-Coulomb non-associated material. The material parameters were fitted to reproduce the stress-strain curves and the 'volume changes versus strain' curves obtained in the experiment. A good fit was easily found, using the following values:

Young's modulus: 28 MPa
Poisson's ratio: 0.48
Dilation angle: 6°
Friction angle: 20°
Density: 6500 kg/m^3

A typical volumetric change curve is given in Figure 2, together with the fitted curve (confining pressure: 0.2 MPa). These results were used directly for the excavation modelling.

2.3 Excavation modelling

The 0.8 m long sheet pile had the following characteristics:

Cross section: 0.016 m^2
Moment of inertia: $1.92*10^{-7}$ m^4
Young's modulus: 72.5 GPa,
intended to represent the Lyon subway support.

The struts had a stiffness of 83.3 MN/m. These supports were modelled by beam elements with the above characteristics. A frictional interface (friction angle: 14°) was placed between the sheet pile and the adjoining zones, on both sides.

Figure 3 represents the displacements of the sheet pile for two excavation depths, when no struts are used. This was the most difficult case, since very large plastic deformations were allowed to occur. Displacements for an excavation depth of 35 cm are underestimated, while moments in the sheet pile were close to the measurements.

Figure 4 represents the moments computed in the sheet pile when restrained by two struts, for the final excavation depth of 55 cm. These compare well to the measured ones. Displacements of the beam for this case are within 10% of the measured ones.

Although maybe a better fit could be achieved using a more complex material model, the Mohr-Coulomb model seems to be quite satisfactory for representing the steel-rod experiments.

3 EXAMPLE 2

3.1 Problem

The observed deformation of the soft-rock above and around a tunnel was modelled by back-analysis of results obtained over a certain section

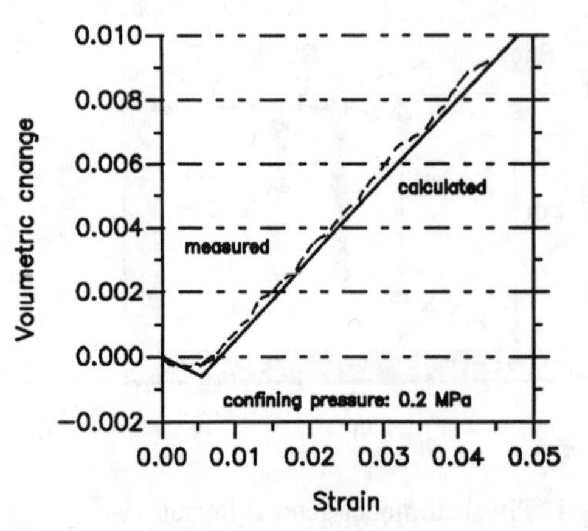

Fig.2 Volumetric change versus strain (confining pressure: 0.2 MPa)

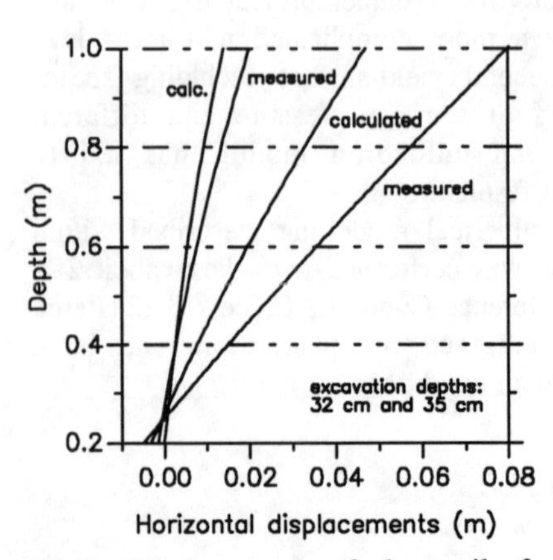

Fig.3 Displacements of sheet pile for several excavation depths - no struts

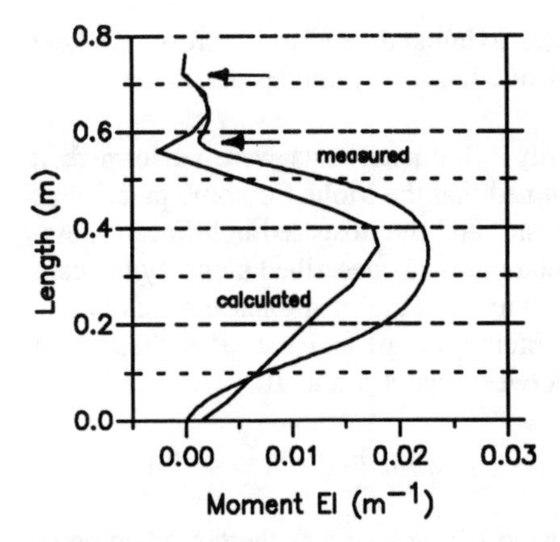

Fig.4 Moments in sheet pile with two struts, excavation depth: 0.55 m

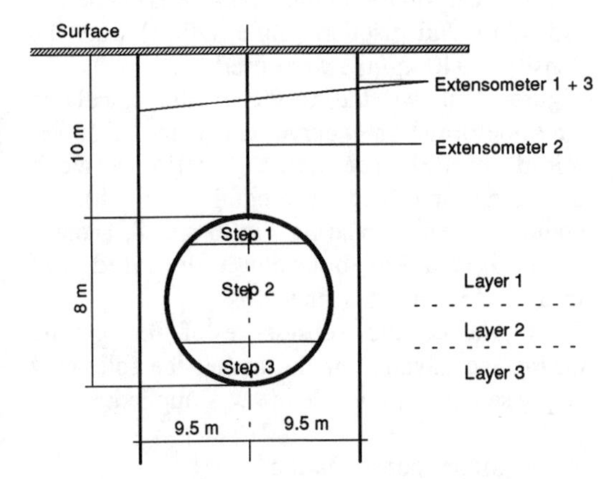

Fig.5 Details of problem modelled

during advance of the tunnel. The geological formation, excavation steps during the tunnel advance, and the measures are shown schematically in Figure 5 (Estermann 1991).

3.2 *Lab-Testing and material model*

Three Odometer-tests with cyclic loading and unloading were performed on soil samples of layer 1. A resulting stress-deformation curve is shown examplary in Figure 6.

The tests show features departing from the classical Mohr-Coulomb model:

- Strong non-linearity in the stress-strain-behaviour.

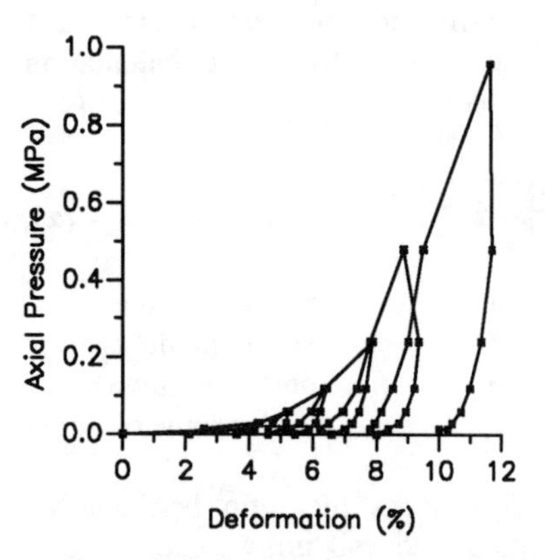

Fig.6 Stress-deformation curve (Odometer-test)

- Different loading and unloading moduli, with strain hardening.
- Plastic volumetric strain.

On the basis of the above mentioned characteristics a 'Double-Yield' model was chosen to reproduce the observed material behaviour.

The implementation of the Double-Yield-Model is based (in addition to the classical Mohr-Coulomb law) on the following three assumptions:

(1) Two independent yield-functions are used for yielding under shear (f_s) and volumetric strain (f_v):

$$f_s = S1 - S3 * \frac{1+\sin\phi}{1-\sin\phi} + 2c * \sqrt{\frac{1+\sin\phi}{1-\sin\phi}} \tag{1}$$

$$f_v = \frac{1}{3} * (S1+S2+S3) + p_c \tag{2}$$

where:	S1,S2,S3	principal stresses
	Φ	frictional angle
	c	cohesion
	p_c	cap-pressure

(2) The strain hardening was modelled by an increasing stiffness for both the elastic and plastic part of the moduli, assuming a cap-pressure-table, witch relates the actual cap-pressure to the plastic volumetric strain.

(3) For the difference between loading and unloading moduli the following calculation scheme was used:

$$\frac{dp}{de_v} = \frac{h*K}{h+K} \tag{3}$$

where:

h	plastic modulus
K	elastic modulus
p	pressure (stress)
e_v	total volumetric strain

If we introduce R as the ratio between the elastic and the plastic moduli, we get:

$$\frac{dp}{de_v} = \frac{R*h}{1+R} \tag{4}$$

Figure 7 illustrates the yield-surfaces in the S1-S3 diagramme.

The 'Double-Yield' model has the capability to describe the strain-hardening or -softening not only on the basis of a cap-pressure-function, but also on the basis of functions regarding the cohesion, the friction angle and dilation.

The most convenient way to determine the role of the individual parameters in the hardening/softening behaviour is to perform two independent lab-tests, one triaxial test done at constant mean stress (deduction of Mohr-Coulomb parameters) and an other triaxial test in which the axial and confining stress are kept equal (deduction of cap-model parameters).

Since only Odometer-tests were performed, it was assumed that the Mohr-Coulomb parameters are constant, and the observed non-linear stress-strain behaviour was described alone by the cap-pressure table (actual cap-pressure related to plastic volume strain) and the R-value, which varies between about 4 and 10.

3.3 Numerical Modelling

The first modelling step was the reproduction of the Odometer-tests (axisymmetric modelling). Additionally to the parameters mentioned above the classical Mohr-Coulomb parameters cohesion (15 kPa) and friction angle (27.5°) and the density (1940 kg/m^3) were used.

Figure 8 shows the result of the calculated stress-deformation curve using the 'Double-Yield' model (the actual ratio between unloading and loading moduli is R= 8). As could be seen comparing Fig. 6 and 8, a fairly good agreement between calculated and measured values is achieved.

To reproduce the subsidence of the ground during the advance of the tunnel the following stepwise calculation scheme was applied:

- Model consolidation
- Built-in of the shotcrete lining

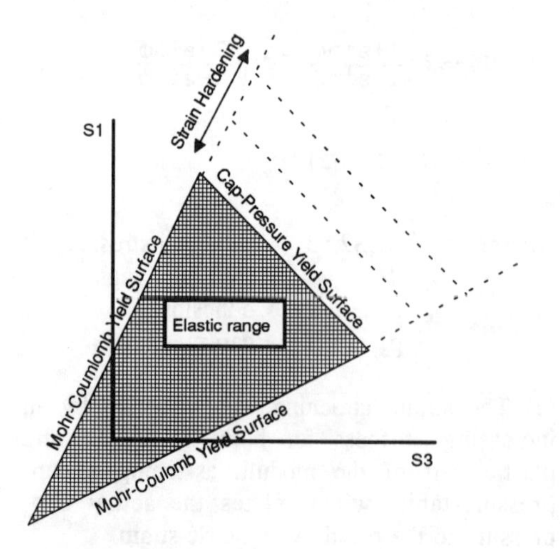

Fig.7 Yield-surfaces of the 'Double-Yield' model

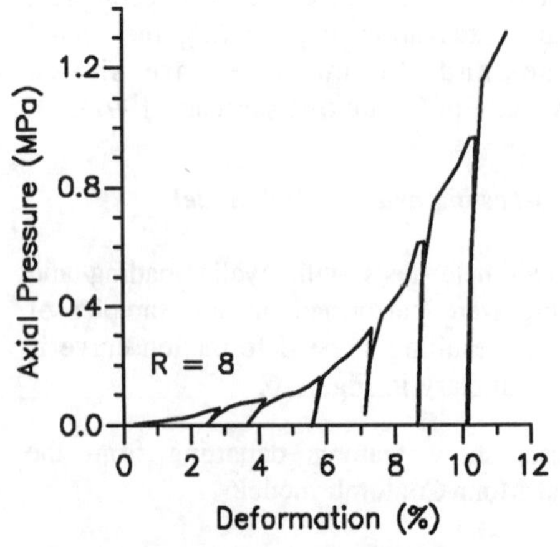

Fig. 8 Calculated stress-deformation curve for the 'Double-Yield' model (R=8)

3-setp procedure to duplicate the tunnel advance

For the tunnel lining the following parameters were chosen:

Thickness:	180 mm
Young's modulus:	15 GPa
Moment of inertia:	$4.86*10^{-4}$ m^4

For layer 1 the material law, deduced from the lab-tests was directly implemented, whereas for the layers 2 and 3 the 'cap-pressure' table was multiplied by factors of 1.5 and 3, respectively, to take into account that these layers are considerably stiffer. Table 1 summarizes the other parameters for all 3 layers (the bulk- and shear moduli are only valid for a model run with a pure Mohr-Coulomb model, which is discussed later):

Table 1. Numerical model parameters

	Density [kg/m^3]	Bulk Mod. [MPa]	Shear Mod. [MPa]	Cohes. [kPa]	Frict. Angle [°]
Layer 1	1940	13.33	5.00	15	27.5
Layer 2	2140	18.56	8.65	20	25
Layer 3	2140	27.77	16.66	20	25

Figure 9 shows the measured and calculated subsidence values for the three measurement profiles: surface subsidence and subsidence according to the two extensometer measurement lines (Extensometer 2 and Extensometers 1+3). Two mean features have been observed during all performed model variants of the Mohr-Coulomb model (see Table 1), which are contradictionary to the observations:

1. Uplifts were calculated (see Fig.9), which were not observed.

2. A detailed investigation of the development of the subsidences with ongoing excavation steps has shown a strong reduction instead of an increase as observed.

Therefore the Mohr-Coulomb model is considered as an inadequate material law, the chosen 'Double-Yield' model instead has shown in all these features a satisfying agreement between measured and calculated subsidence values.

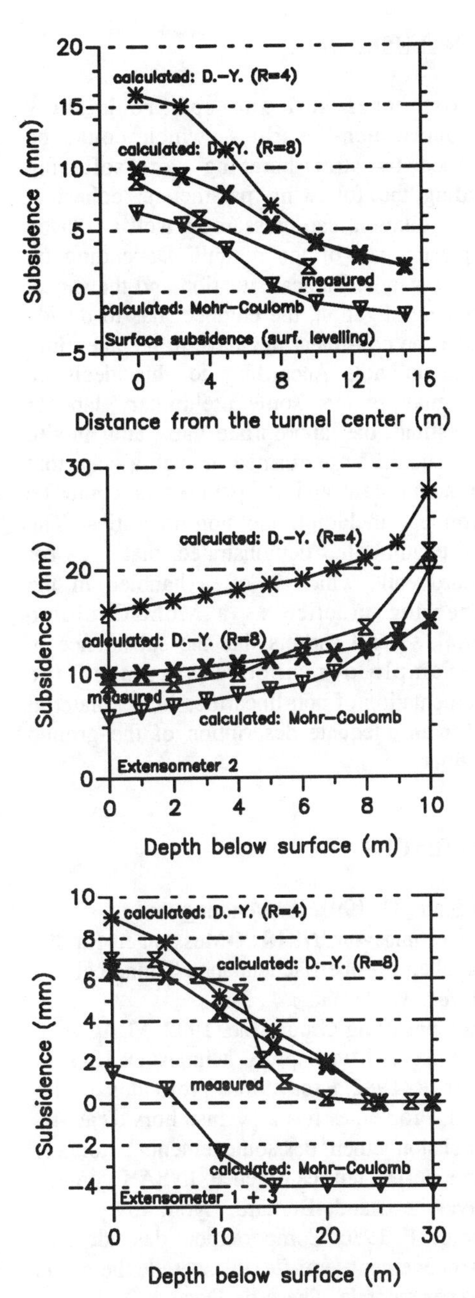

Fig.9 Calculated and measured subsidence values

4 CONCLUSIONS

In most cases soil and soft rock show pronounced non-linearities, which could be described by strain-harening or -softening regarding the following parameters: cohesion, friction, dilation, cap-pressure as well as elastic and plastic part of the moduli. Depending on both, the geomechanical situation and the aspect under investigation, the engineer has to decide whether the non-linearities have to be taken into account or not. According to this decision, which may require some preliminary lab- or field-testing, the appropriate lab- and in-situ testing should be performed in such a way, that the material law and its parameters could be determined, including the non-linearities. The tunnel example has demonstrated, that even for 'standard soil', which is often handled in the engineering practice as a Mohr-Coulomb material, several aspects like the subsidence in the example described above, need the implementation of non-linearities in the material law for an adequate description of the ground behaviour.

REFERENCES

Estermann, U. 1991. Anwendung und Auswertung von INKREX-Messungen für Aufgaben im oberflächennahen Tunnelbau. *Interfels Nachrichten 4:* 3-9.

Itasca Consulting Group, Inc. 1992. FLAC Version 3.2. Minneapolis, Minnesota: ICG.

Kastner, R. 1982. Excavations profondes en site urban. Problemes lies a la mise hors d'eau. Dimensionnement des soutenements butonnes. These de Docteur es Science, INSA Lyon et Universite Claude-Bernard, Lyon, 409 p.

Masrouri, F. 1986. Comportement des rideaux de soutenement semi-flexibles: etude theorique et experimentale. These de Doctorat, INSA Lyon, 247 p.

Masrouri, F. & R. Kastner 1992. Anchored flexible retaining walls calculated by the the reaction modulus method. *International Conference on Retaining Structures, I.C.E., Cambridge,* July 1992

Geomechanics 93, Rakowski (ed.) © 1994 Balkema, Rotterdam, ISBN 90 5410 354 X

Contact problem for modelling of bolt action in rock

J. Malík
Institute of Geonics of the CAS, Ostrava, Czech Republic

ABSTRACT: In recent years numerical models have become an integral part of tunnel design. In this paper a new model of bolt system is described. This model describes the behaviour of bolt system by a special continuous procedure. For the sake of simplicity first we study the elastic behaviour of rock and then the non - linear model is formulated.

1 INTRODUCTION

In the last twenty years the New Austrian Tunelling Method has become one of the most popular method for stabilization tunnels in rock [1], [2]. In this paper two models of bolt action in rock mass are described. These models describe two kinds of bolt reinforce.

- Every bolt is reinforced only in two points (Fig.1)
- Every bolt is reinforced along all its lenght (Fig.2)

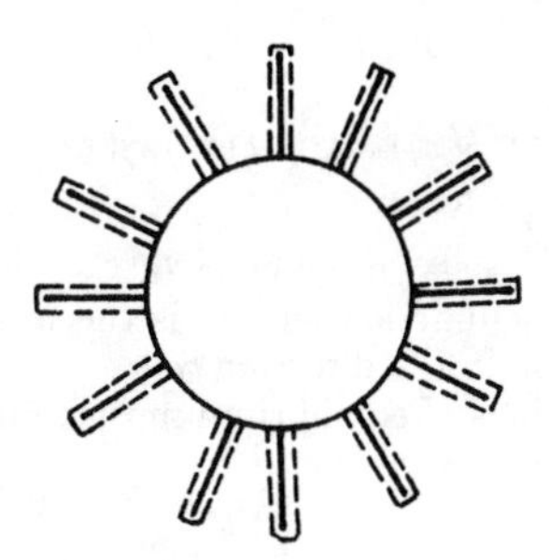

Fig. 2

2 THE FIRST MODEL

Let us imagine the situation which is described on the Fig. 1. The instalation of bolts is described by the continuous transformation $\xi : S_1 \longrightarrow S_2$ (Fig. 3).

We will describe the behaviour of bolts as a continuous system. Supposing that there is not any contact between the bolts and the rock except the points in which the bolts are reinforced and that

Fig. 1

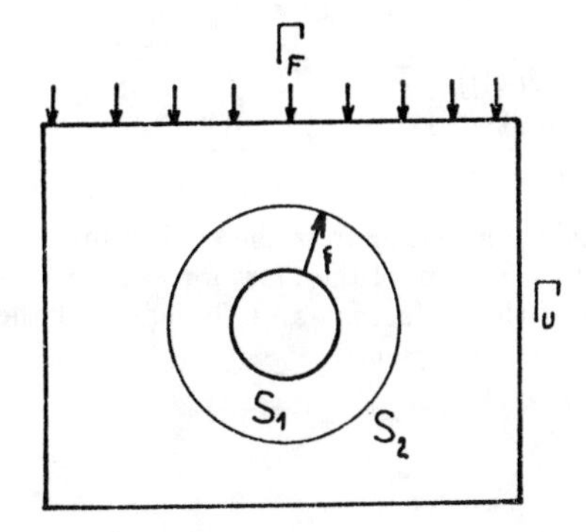

Fig. 3

there are only tensil forces in the bolts. For the sake of simplicity we will suppose that the behaviour of rock is linear elastic everywhere except the surface S_1 and that the equations of equilibrium are fulfilled.

$$-\frac{\partial}{\partial x_j} c_{ijkl} e_{kl}(u) = f_i$$

$$e_{ij}(u) = \frac{1}{2}(\frac{\partial u_i}{\partial x_j} + \frac{\partial u_j}{\partial x_i})$$

$$u_i = g_i \quad on \quad \Gamma_U \tag{1}$$

$$T_n^i(u) = c_{ijkl}e_{kl}(u)n_j = hi \quad on \quad \Gamma_F$$

n_j – components of external normal

Let us define the function

$$\gamma(x) = \frac{\xi(x) - x}{\mid \xi(x) - x \mid}$$

This vector function is defined on the surface S_1 and determines the bolt direction in the point x. Let us denote,

$$f\gamma(x) = < f(x), \gamma(x) >$$

$$f\gamma(\xi(x)) = < f(\xi(x)), \gamma(x) >$$

$<,>$ – scalar product in R^n $\quad n = 2,3$

f is a vector function and f_γ is the projection of this function in the direction γ.

The conditions of equilibrium must be fulfilled also on surfaces S_1, S_2, i.e.

$$T_n(u(x)) + B(x)(u_\gamma(\xi(x)) - u_\gamma(x) - \\ -l(x))\gamma = 0 \quad on \quad S_1$$

$$T_n(u(\xi(x))) + T_{-n}(u(\xi(x))) - \\ -B(\xi(x))(u_\gamma(\xi(x)) - u_\gamma(x) - l(x))\gamma = 0 \quad on \quad S_2$$

These conditions express the equilibrium of forces which is the sum of the mass forces from the both sides of the surface S_2 and the effect of the bolt system deformation.

$$u_\gamma(\xi(x)) - u_\gamma(x)$$

can be interpreted as a deformation of the bolt which comes from point x. The equilibrium of forces on S_1 is the sum of the mass forces from the inner side of the surface S_1 and the effect of the bolt system deformation. (Fig. 4).

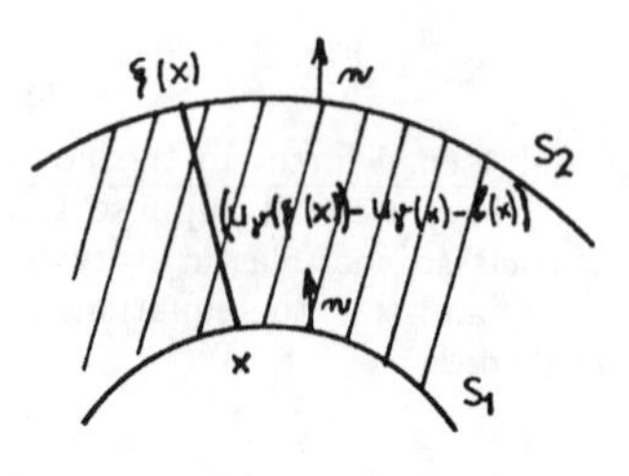

Fig. 4

$B(x) = K\varrho(x), B(\xi(x)) = K\varrho(\xi(x)), K$ is the stiffness contact of bolt,$\varrho(x)$ is the bolt density on S_1 and $\varrho(\xi(x))$ is the bolt density on S_2. We can say that $B(x), B(\xi(x))$ are the stiffness functions of the bolt system. The following conditions between $B(x), B(\xi(x))$ are fulfilled,

$$B(x)dS_1 = B(\xi(x))dS_2, \qquad dS_1 \longrightarrow^{\xi} dS_2.$$

Values of the function $l(x)$ describe that fact the bolt system is applied after some initial deformation. We will illustrate it on concrete geomechanical problem later. Applying all these equations and conditions and Betti's formula [3] we obtain the following variational formulation: to look for a minimum of the functional

$$\Phi(u) = \frac{1}{2}\int c_{ijkl}e_{ij}(u)e_{kl}(u)dx + \\ +\frac{1}{2}\int_{S_1} B(x)(u_\gamma(\xi(x)) - u_\gamma(x) - l(x))^2 dS_1 - \\ -\int_\Omega fudx - \int_{\Gamma_F} hud\Gamma.$$

The minimum is looked for on the set of functions satisfying the condition $u = g$ on Γ_u.

Now we describe a simple geomechanical problem to demonstrate the meaning of the function $l(x)$. We want to excavate the space on Fig. 5 in two steps. First we excavate one part Fig. 5a) then apply bolts Fig. 5b) and than excavate the rest of the space Fig. 5c).

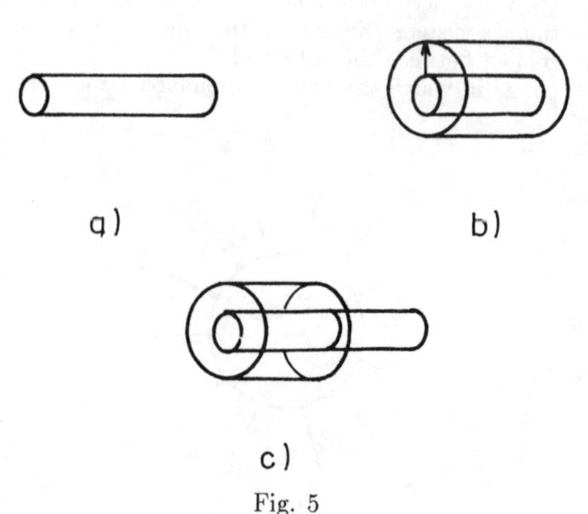

Fig. 5

Modelling this problem first we solve normal boundary problem of the Fig. 5a) to obtain the solution v. The values of the function $l(x)$ are expressed by the formula $l(x) = v_\gamma(\xi(x)) - v_\gamma(x)$ and than we can solve the above introduced variational problem.

3 THE SECOND MODEL

The situation is similar to the first one (Fig. 6) The instalation of bolts described by the continuous transformations $\xi : S_1 \longrightarrow S_2$. We define $\gamma(x) = \frac{\xi(x')-x'}{|\xi(x')-x'|}$ where $\xi(x') - x'$ is the abscissa which goes through the point x. We can say that $\gamma(x)$ represents the direction of the bolt that goes through the point x. If we take a part $\mathcal{O}$ of Ω' we can formulate the equilibrium condition in the following way.

$$\int_{\partial\sigma} T_n(u)dS+$$

$$+\int_{\partial\sigma} B(x)D_\gamma(u_\gamma - l(x)) < \gamma(x), n(x) > \gamma dS+$$

$$+\int_\sigma f(x) = 0$$

$$u_\gamma =< u(x), \gamma(x) >$$

$$D_\gamma(f(x)) = \frac{d}{dt}f(x + t\gamma(x))/_{t=0}$$

$$B(x) = E\delta\varrho(x)$$

E is Young's modulus of material of bolt, δ is the area of bolt cross-section and ϱ is bolt density in the point x. On the rest of Ω equations of elasticity are fulfiled (1). Applying similar mathematical operations we obtain the following functional,

$$\Phi(u) = \frac{1}{2}\int_\Omega c_{ijkl}e_{ij}(u)e_{kl}(u)dx+$$

$$+\frac{1}{2}\int_{\Omega'} B(x)(D_\gamma(u_\gamma(x) - l(x)))^2 dx-$$

$$-\int_\Omega fudx - \int_{\Gamma_F} hud\Gamma.$$

Our problem is equivalent to look for the minimum of this functional on the set of functions satisfying $u = g \quad \text{on } \Gamma_U$

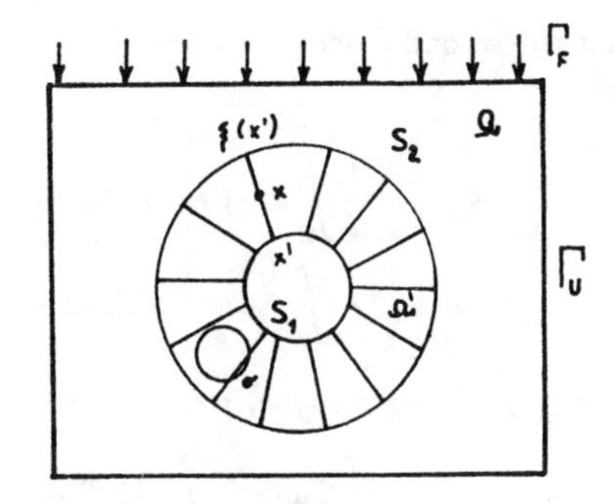

Fig. 6

4 NON - LINEAR MODELS

We supposed the linear elastic behaviour of rock mass in the previous models. Now we consider the material which behaves according to deformation theory of plasticity described by the following conditions.

$$0 < k_0 \leq k(x) \leq k_1 < \infty$$

$$0 < \mu_0 \leq (x, s) \leq \frac{3}{2}k(x)$$

$$0 < \kappa_0 \leq \mu(x, s) + 2\frac{\partial\mu(x, s)}{\partial s} \leq \kappa_1$$

The functions k, μ are continuous differentiable [3]. Let us introduce the function

$$\Gamma(u) = -\frac{3}{2}(e_{ii}(u))^2 + 2e_{ij}(u)e_ij(u)$$

Applying similar considerations and mathematical formulas we obtain the followig two variational formulations. We look for a minimum of the functional Φ , which is for the first model of bolt system

$$\Phi(u) = \frac{1}{2}\int_\Omega k(x)(e_{ii}(u))^2 dx+$$

$$+\frac{1}{2}\int_\Omega\int_0^{\Gamma(u)} \mu(x, s)dsdx+$$

$$+\frac{1}{2}\int_{S_1} B(x)(u_\gamma(\xi(x)) - u_\gamma(x) - l(x))^2 dS_1-$$

$$-\int_\Omega fudx - \int_{\Gamma_F} hud\Gamma_F,$$

and for the second model

$$\Phi(u) = \frac{1}{2}\int_{\Omega} k(x)(e_{ii}(u))^2 dx + \\ +\frac{1}{2}\int_{\Omega}\int_{0}^{\Gamma(u)} \mu(x,s)\,ds\,dx + \\ +\frac{1}{2}\int_{\Omega'} B(x)(D_{\gamma}(u_{\gamma}(x) - l(x)))^2 dx - \\ -\int_{\Omega} fu\,dx - \int_{\Gamma_F} hu\,d\Gamma_F.$$

These functionals are defined on the set of functions which satisfy $u = g$ on Γ_U.

REFERENCES

[1] Swoboda, G., Marenče, M. 1992. *Numerical modelling of rock bolts in interesection with fault system*, NUMEG IV, Swansea, 24-27 August 1992, edited by G.M.Paude, Pietruszczak, S.

[2] Swoboda, G., Marenče, M. 1992. *FEM modelling of rock bolts*, Proceeding of the Seventh International Conference on Computer Methods and Advances in Geomechanics, Cairus, 6-10 May 1991, edited by G. Beer, J.R. Booker, J.P. Carter.

[3] Nečas, J., Hlaváček, J. 1981. *Mathematical Theory of Elastic a Elasto-Plastic Bodies*, Elsevier, Amsterdam.

Geomechanics 93, Rakowski (ed.) © 1994 Balkema, Rotterdam, ISBN 90 5410 354 X

Coupled elastoplastic and viscoplastic analysis of a potash mine

P. Ramírez Oyanguren & L. R. Alejano Monge
Department of Mining Engineering, Politechnic University of Madrid, Spain

ABSTRACT: The modelisation of a potash mine is carried out. From convergence measurements in rooms and an initial set of creep parametres obtained studying the behaviour of drifts, new creep parametres have been adjusted to fit the new data and the numerical method has been calibrated by means of a sensibility analysis.

1 INTRODUCTION

The Subiza potash mine is located 20 km south from Pamplona, north Spain. Currently the operation produces about 2 millions tons of potash ore per year, with a mean grade of 15% K_2O. A 3.6 metre-high layer of potash ore, located at depths from 200 to 800 m, is mined by a room and pillar method.

This paper deals with the modelisation of the mechanical behaviour of the mine in order to meet safety standards and to improve mineral recovering.

2 PRELIMINARY REMARKS

2.1 Geological setting

The mine lies in the northflank of a tertiarian basin. The regional average dip is about 12° towards the center of the basin.

The stratigraphic sequence from top to wall is as follows:
- Undiano Marls (I): Up to 400 m of grey and red marl layers with intercalations of gypsum and clay.
- Galar Sandstones (II): 80 m of fine grained sandstone materials with disseminated marls and clays.
- Bedded Marls (III): 60 m of heterogeneous clays and alternating red, green and brownish marls. The content of sand grows towards the top of this formation.
- Top Salt (IV): 75 m of alternating rock salt and marl layers.
- Evaporitic Deposit (V): 30.7 m of evaporitic material containing the orebody, where three different beds can be separated from top to wall. First, there is a roughly 12 m bed of carnallitite, consisting of mainly rock salt and some carnallite and clay beds. Then, there is the mineral deposit, a 3.6 m potash bed composed of coarse-grained sylvinite with some interbeded rock salt layers with minor disseminated marl and only sporadic ocurrences of carnallite and anhidrite. Finally there is a 15 m rock salt bed.
- Pamplona Marls (VI): 250 m of marls and clays situated under the evaporitic materials.

2.2 Mining method

A room and pillar method has been designed to mine the inclined potash orebody, where continuous mining techniques are used. The mine is divided into pannels, each one containing five 11 m wide rooms separated by 4 m wide yield pillars. The pannels are separated by 29 m wide strong barrier pillars. Each pannel is 79 m wide and about 200 m long. The rooms are roughly 3.6 m high, corresponding to the thickness of the mineral deposit. Up to date more than fifty pannels have already been mined.

Selective continuous mining is carried out in the direction of the slope by means of boom heading machines, which need two stoping steps of 5.5 m each to mine the whole room.

3 BACKGROUND OF THE NUMERICAL SIMULATION

3.1 Objectives

To meet the safety standards it is desirable that the roof does not lower more than 20 cm while the mine crew stays in the room. Thus, it is certainly interesting, to be able to predict how much time it will take the ceiling to achieve this position.

The main objective of this analysis is to asses the mechanical behaviour of the evaporitic formation sourrounded by elastoplastic materials, taking especial account on the cavity closing within time. To achieve this goal, the results of the simulation will be compared with in situ convergence measurements in order to check the creep parametres of the behaviour law of salt. The creep parametres initially used were obtained in a former stage of research (Alonso, Martí & Ramírez, 1988) from measurements of convergence in drifts. If this parametres do not match the in situ measurements in rooms, a new and more accurate adjustment will have to be done.

Later, a sensibility analysis will be done in order to obtain realiable numerical results, which can be used in future mine developing.

3.2 Mechanical behaviour of evaporitic materials

The salt formation is supposed to have a viscoelastic behaviour. This behaviour can be simulated, at a constant temperature and within a determined range of deviatoric stresses by means of a creep constitutive equation in time-domain (Langer,1981), which can be implemented with the Norton power law. The standard form of this law is:

$$\dot{\epsilon}_{cr} = A \cdot \sigma^n \qquad (1)$$

Where:

$\dot{\epsilon}_{cr}$ = *Strain rate.*

σ = *Stress deviator.*

A, n = *Constant values.*

The creep parametres of the whole evaporite formation which contains the orebody were calculated, from the strain rates measured in some mine drifts during earlier exploitation stages of the deposit.

3.3 Software

To carry out the simulation, the code FLAC has been used. This is a two-dimensional explicit finite difference code which simulates the mechanical behaviour of structures of rock which may undergo plastic flow. FLAC presents as well several built-in constitutive models which permit the simulation of highly non-linear, time-dependent, materials like salt (Itasca, 1989).

One of the main advantages of this code and the reason why it was selected is its capability to work with more than one material having different models of behaviour, in this case: elastic, elastoplastic and viscoplastic or creep in the same area of discretization.

3.4 Other geotechnical data

As it has been said, salt has a viscoplastic behaviour while the overlying materials have an elastioplastic one. So, they will be simulated with the Mohr-Coulomb model. In the other hand the underlying Pamplona marls will be modelised as an elastic material due to the fact that they will not plastificate .

The mechanical properties of the elastoplastic rock masses have been empirically estimated by means of laboratory standard uniaxial, triaxial and brasilian tests. The values of the main geotechnical parametres -expressed in International System Units- used for this analysis are summed up in this table:

TABLE I. GEOMECHANICAL PARAMETRES OF THE ROCK MASSES

	I & III	II	IV	V	VI
ρ	2300	2300	2300	2300	2300
K	2.50e9	5.84e9	5.00e9	19.84e9	8.34e9
G	1.15e9	2.69e9	2.31e9	9.69e9	3.84e9
C	3e6	8e6	5e6	--	--
ϕ	35	35	35	--	--
σ_T	1e7	1e7	1e7	--	--
A*	--	--	--	4e-22	--
n	--	--	--	2.7	--

* Pa^n/year ρ=density σ_T=tensile strength

4 CONVERGENCE MEASUREMENTS

In the first stages of the mine exploitation, mechanical extensometers type Mark 1 and convergence stations were installed in some galleries, which allowed us to highlight some important topics about the instrumentation and to obtain a first approximation of the creep parametres.

The direct measurement convergence stations, which have proved to be an adequate way of measuring in drifts, can only be safely used during more or less a year in rooms, because at that time, due to the stratification of salt and to the presence of fine seams of clay, blocks begin to fall from the ceiling and measurements cannot be safely taken. So, remote reading convergence stations were installed in rooms to ensure good quality measurements during a representative period of time, which have achieve at present five hundred days of measurements.

Figure 1 depicts the registers of four convergence meters installed in the mine rooms. Number 1 at 510 m and number 2 at 524 m of depth, with initial heights of 2.73 and 3.5 m respectively, were both located in room B4/P1/CO3. Number 3 at 503 m and 3.00 m high and number 4 at 520 m and 3.5 m high were in room B4/P2/CO3.

It is important to remark in order to be able to achieve a good interpretation of the results, that all the convergence meters were installed once finished the exploitation stage of the rooms, so that the elastic strain ocurred instantaneously after the excavation could not be registred as it could not also be an important part of the primary or transient creep strain. So, what it has been really measured is the stationary part of the viscoplastic behaviour or secondary creep, which generally makes the bigger part of the creep strain and is easier to modelise.

Another interesting point to take into account is that a not negligible part of the convergence in rooms is due to the strata detachment through fine clay seams. This bed separation occurs in the first metre of the ceiling. From measurements taken by means of extensometers, it can be stated that the deformation ocurred in this way makes up to 20 % of the whole convergence of the room. As far as the code cannot simulate this kind of behaviour, this effect introduces a significative error in the calculations.

The installation of remote measuring convergence meters connected with floor and roof extensometric borehole stations has already been planned in order to measure as accurately as possible the convergence of rooms and to asses the influence of bed separations in displacements.

5 MATHEMATICAL MODELLING

As it has already been pointed out, in this analysis the values of the parametres A and n estimated in a former stage of research have been used. If these values do not match properly the present in situ measurements, they will have to be recalculated in a sensibility analysis. Due to the fact that the previously described convergence

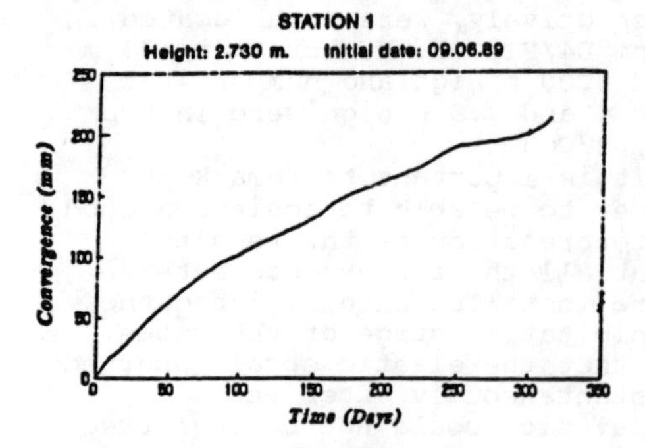

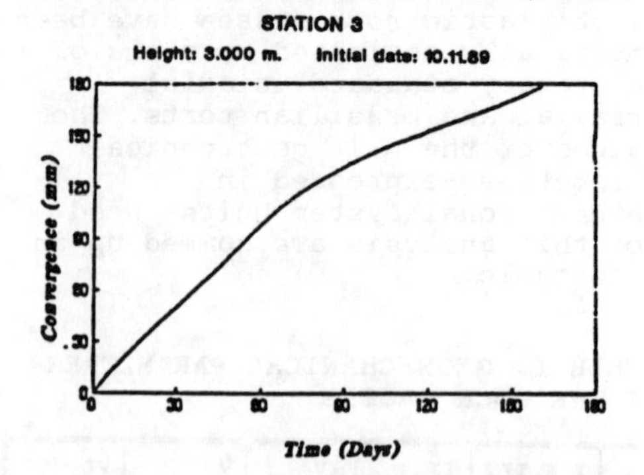

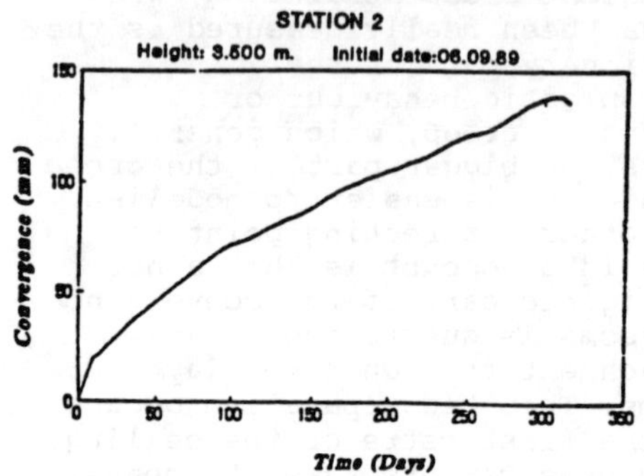

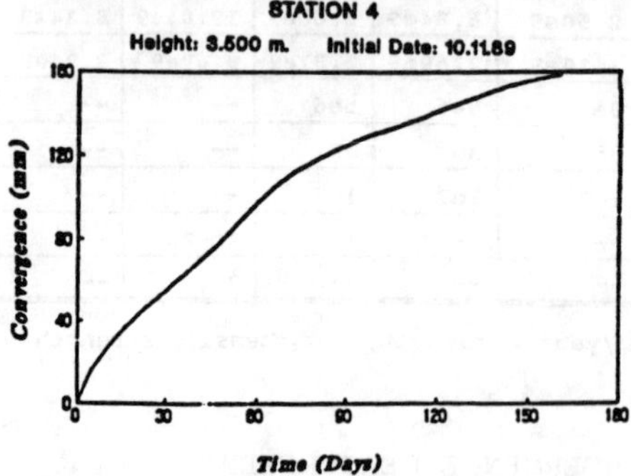

FIGURE 1: Vertical convergence in situ measurements in two different rooms.

measurements have been taken in rooms located at a depth of about 500 m, the simulations represents this situation.

5.1 Features of the simulation

The area of discretization, shown in figure 2, is big enough -500 m. wide x 300 m. high- to guaratee that the influence of the boundaries in the model behaviour is not important. Thus, the model was fixed in the horizontal direction on the simmetry line and on the left edge, located roughly 250 metres away from the first room. The lower boundary, whose movement has been restrained in the y-direction, lies one hundred metres underneath the rooms, where its influence has been tested to be negligible.

In the upper boundary an evenly distributed load equal to the weight of 200 m of overlying materials has been applied. This can be done because the stress state in this edge is not affected by the cavities excavated 300 metres downwards.

Due to the difficulties to measure the in situ stresses, an isotropic and hydrostatic stress state has been assummed; with the vertical stresses as well as the horizontal ones equal in every point to the weight of the overlying materials.

Starting from the before mentioned set of geological and geometrical data and the defined boundary and initial conditions, the simulation has been performed in plane-strain. It has been decided to stop the study once roughly 1.5 years from the opening of the cavities have been elapsed, cause it's that the time during which measurements have been taken in representative rooms.

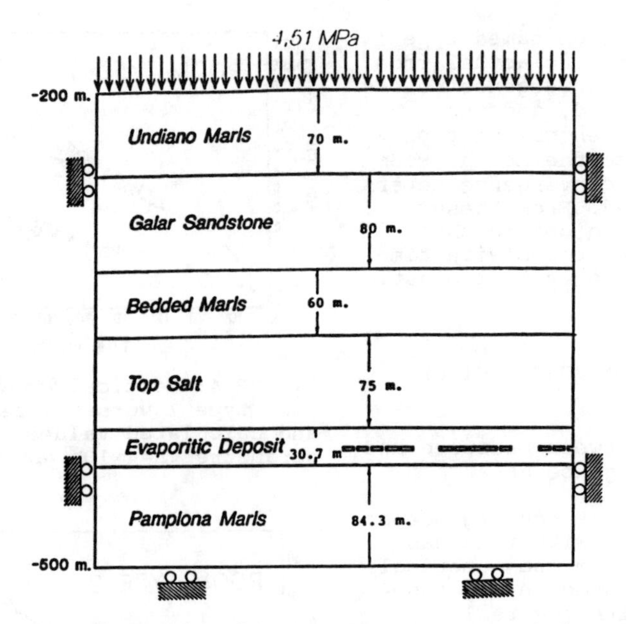

FIGURE 2: Area of discretization and boundary conditions. The defined local stratigraphy can be observed.

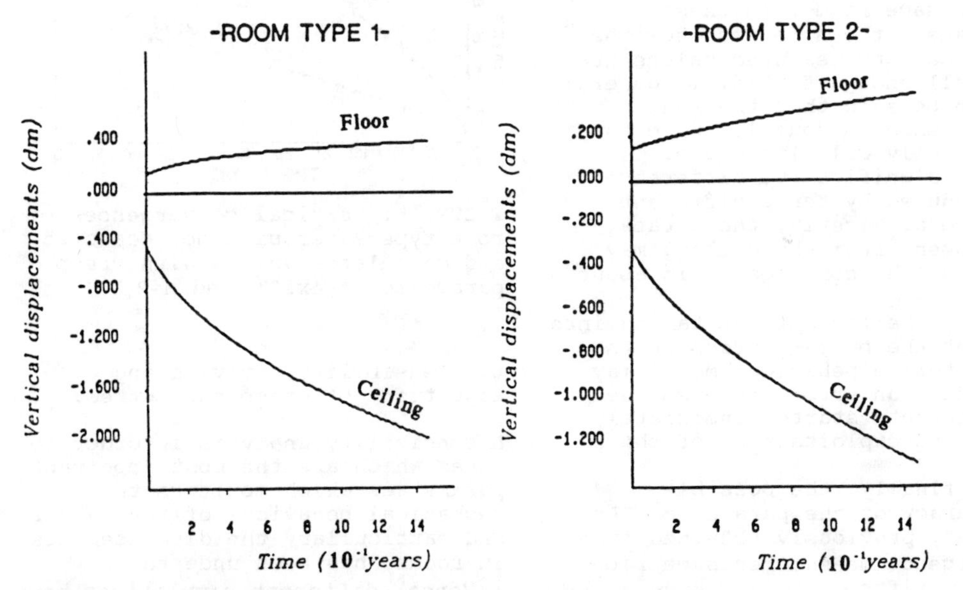

FIGURE 3: Calculated vertical displacements in the floor an ceiling of rooms type 1 & 2.

5.2 Stress-strain general behaviour

The main goal of this simulation is to obtain the convergence in rooms. Thus, in figure 3, obtained from the numerical simulation of 500 m deep panels, the evolution with time of the vertical displacement of the central points of floor and ceiling of two different rooms - lateral and central room of the second pannel- are plotted. The lateral room has been named type 1 and corresponds to the actual room B4/P1/CO3, where convergence meters 1 & 2 were installed, and the central one has been called type 2 and corresponds to the actual room B4/P2/CO3, where convergence meters 3 & 4 were located. From these plots it is easy to obtain the convergence of the rooms with time by substracting the initial elastic strain.

6 COMPARISON AND ADJUSTEMENT OF CREEP PARAMETRES

6.1 Comparison between measured and calculated convergence data.

The evolution of the convergence at rooms type 1 and 2 with time has been calculated from the numerical model and represented in figures 4 and 5 together with the real measured values in equivalent rooms -the average of the results from both convergence meters installed in the same room-. In these diagrams, it can be seen how the numerical and measured values are not well enough fitted. In general, it can be said that the real convergence is roughly twice the numerically calculated one.

This significative difference can be produced by three different phenomena. Firstly, the strata detachment from the ceiling may increase the convergence in about 20%.

Then, the fact of not taking into account the primary creep in the constitutive behaviour model may introduce an error, because the measurements started inmediately after the exploitaition of the rooms.

And finally, the possible inaccuracy of the parametres "A" and "n", previously obtained from convergence measurements carried out in drifts.

In our oppinion, this last topic could be of major importance, involving the greatest part of the error. That is why further research has been done in order to obtain better values of these parametres.

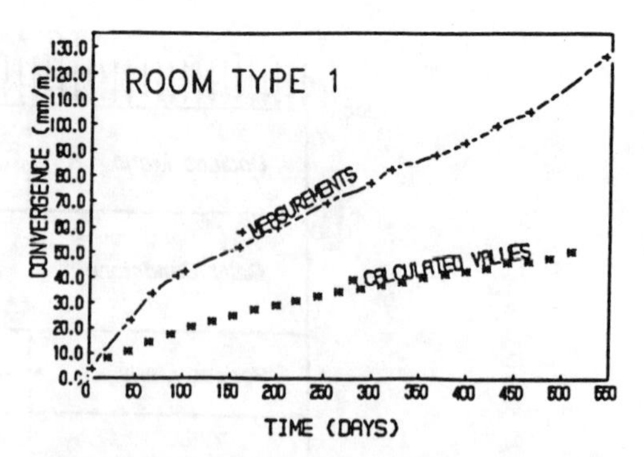

FIGURE 4: Vertical convergence of room type 1 versus time. Measured and calculated values with creep parametres $A=4x10^{-22}$ an n=2,7.

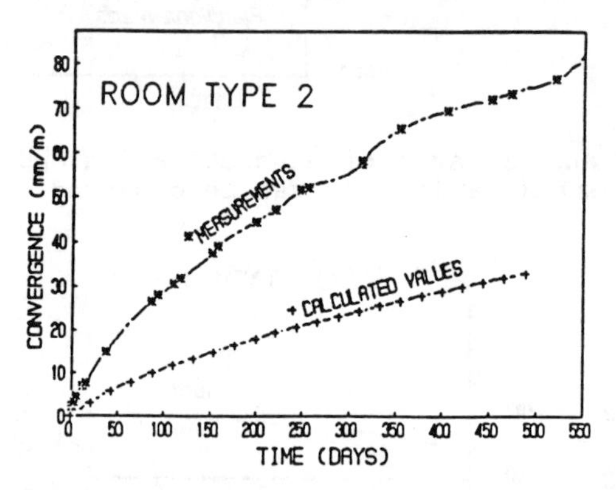

FIGURE 5: Vertical convergences of room type 2 versus time. Measured and calculated values with creep parametres $A=4x10^{-22}$ and n=2,7.

6.2 Sensibility analysis and adjustment of creep parametres.

A sensibility analysis in order to asses which are the most important parametres which monitors the mechanical behaviour of the massif and particullary the displacements in rooms, has been undertaken.

Hence, different simulations have been done varying the bulk and shear modulus, the initial stress state, the creep law -it has been checked that a two component power law das not alter significatively the results (Frayne & Mraz, 1991)- and of course the creep parametres "A" and "n". From these study, it must be formerly highlighted than the influence of the three first topics is almost negligible with differences in the convergence of rooms after one and a half years not bigger than 2 or 3 %. For instance, the influence of the initial stress state in the final displacements is not bigger than 3 % of the total movement, in cases such as:

$$\sigma_H = 0.8 \cdot \sigma_V$$
$$\sigma_H = 1.2 \cdot \sigma_V$$

So, as the primitive stresses are not known and the error produced by this parametre is fairly negligible, a hydrostatic stress state was adopted.

However, the influence of the creep parametres has shown to be terribly important, specially the stress exponent "n". This clear, it has been tried to adjust this constant as well as possible (figure 6), and after a few runs it has been decided that the best value of "n" is 2.8 .

Once fixed that value and knowing that the numerical results should give convergence values about 80% of those measured, due to the bed separation in the ceiling, the parametre "A" has been adjusted (figure 7). A value of 2 x 10^{-22} Pa^{-n}/year was obtained.

With these new creep parametres, and adding to the curve of convergence numerically produced a 20 % of displacement due to bed separation, the adjustments between measured and calculated data seems to be accurate enough in room type 1 (figure 8) as well as in room type 2 (figure 9).

Therefore a new power law is proposed:

$$\dot{\epsilon}\ (year^{-1}) = 2x10^{-22} \cdot \sigma^{2.8}$$

With this new parametres it is possible to simulate with a good degree of accuracy the behaviour of

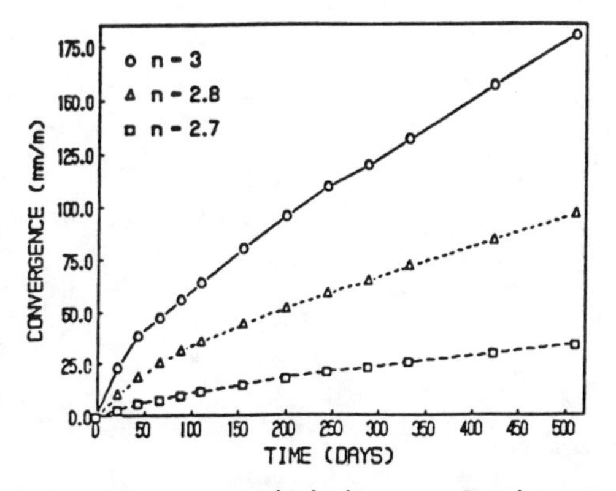

FIGURE 6: Sensibility analysis of the creep parametre "n". Calculated vertical convergences in a room for different "n" values.

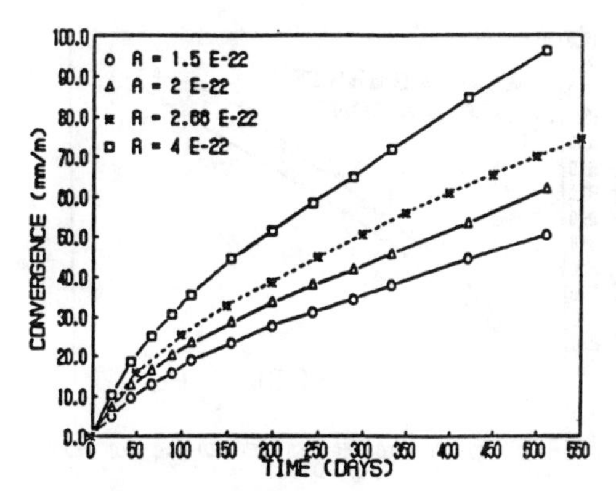

FIGURE 7: Sensibility analysis of the creep parametre "A". Calculated vertical convergences in a room for different "A" values.

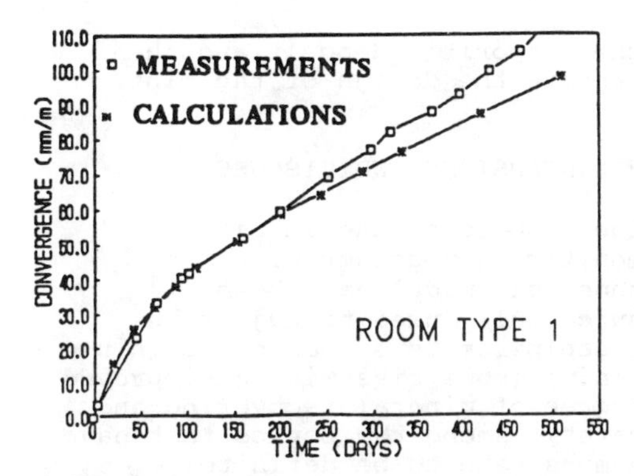

FIGURE 8: Vertical convergence of room type 2 versus time. Measured and calculated values, with creep parametres A=2x10^{-22} and n=2,8, and taking into account bed separation.

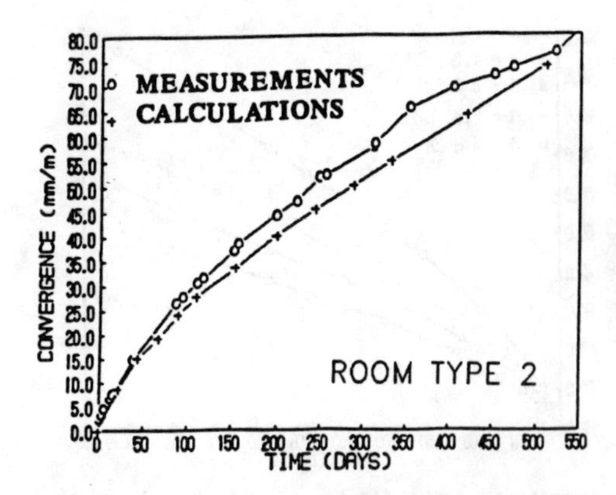

FIGURE 9: Vertical convergence of room type 2 versus time. Measured and calculated values, with creep parametres $A=2x10^{-22}$ and n=2,8, and taking into account bed separation.

the evaporitic deposit and then improve the design of the mine.

7 CONCLUSIONS AND DISCUSSION

The results of the in situ monitoring together with a numerical model have been succesfully used to adjust the creep parameters and to design new mining geometries with an improved degree of mineral recovering and safety. Among the topics that have demonstrate to be definitely important in the behaviour of a room and pillar potash mine, we highlight the constants "A" and "n" of the constituve equation of the evaporites.

Whilst other topics have proved to be less important, for example:
- Type of viscoelastic law (one or two component) at not so deep locations.
- Elastic Parametres (Young's modulus and Poisson's ratio)
- Natural stresses, if not very far from an hydrostatic state.

Some open issues related to this study can be summed up:
- Influence of primary creep. May be parametres "A" and "n" are overestimated and the role played by primary creep is quite more important than it's supposed.
- Influence of the moment of instalation of convergence meters related to the sequence of excavation of rooms.
- Due to the stratification of the salt deposit,should its behaviour be represented by means of an anisotropic creep law?

REFERENCES

Alonso,J., Marti,J. & Ramírez, P., 1988. Analysis of cavities in salt formations .Int. Symp. of Rock Mechanics and Power Plants. Madrid, Spain.

Frayne, M.A. & Mraz, D.Z., 1991. Calibration of a numerical model for different potash ores. Int. Congr. on Rock Mechanics. Aachen, Germany.

Itasca, 1989. User manual for FLAC - Fast Lagrangian Analysis of Continua, Version 3.03. Itasca Consulting Group Inc., Minneapolis, U.S.A.

Langer, M., 1981. The rheological behaviour of rock salt. First Conf. of the mechanical behaviour of salt. Pennsylvania, U.S.A.

Munson, D.E. & Dawson, P.R., 1981. Salt constitutive modeling using mechanism maps. Conf. of the mechanical behaviour of salt. Pennsylvania, U.S.A.

Geomechanics 93, Rakowski (ed.) © 1994 Balkema, Rotterdam, ISBN 90 5410 354 X

Effect of joints and adit to bearing capacity and failure mechanism of rock mass pillar

J.Vacek
Klokner Institute, Czech Technical University, Prague, Czech Republic

P.Procházka
Institute of Geotechnics ASCR, Prague, Czech Republic

ABSTRACT: The paper compares lab data obtained from a special experimental models from physically equivalent material and results from a mathematical model which enclude nonlinear behaviour along the dislocations.

The coupled modelling (physical and mathematical models) gives a possibility to remove disadvantageous of both the models and bring new views into the problematics. As an example of our investigation we present the case of failure mechanism of rock mass pillar with adit and joints.

INTRODUCTION

Failure of rock mass is not a static process, but has its own history. Its duration varies from several seconds to several years. Movements of rock mass significantly change its original shape. On inhomogeneous models we can observe the onset of failure (during the failure the first cracks appear), the chronology of various stages of failure (cavings, slides), and the final shape of rock mass, after the discontinuation of the failure. We can also observe an influence of modelled joints on the development of deformations and lines along which the dislocations occur.

Similar problems are encountered when the bearing capacity of a pillar originated by block caving of ore deposits is assessed. The aim of the present paper is to describe the behaviour of a pillar at Měděnec in North Bodemia in connection with some local heterogeneities in rock mass. Selected cases have been studied experimentally (physical models) and mathematically. The study is so extensive that it cannot be described in detail in the present paper. It involved also the pillars in mines Miková, Cínovec, Kutná Hora, etc.

EXPERIMENTAL PART

As an example of our investigations we shall present the case of a rock mass

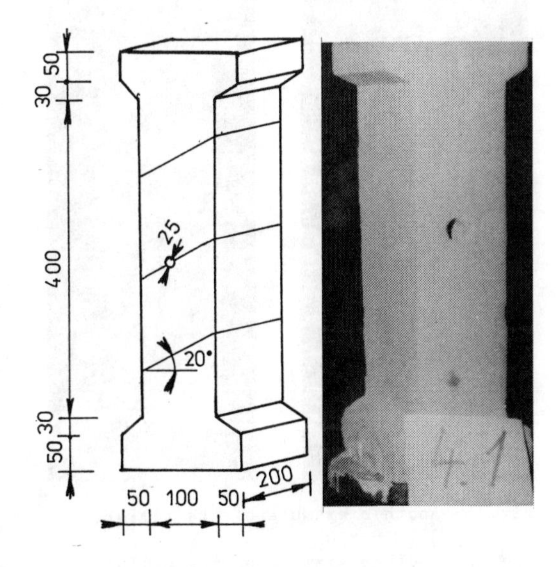

Fig. 1 Schema of pillar 4.1
Fig. 2 Pillar 4.1 before test

pillar with cracks and an adit.

The figures show the failure history of a pillar with an adit in its centre and three joints (Fig. 1). The dimensions of the actual pillar are height 50 m, width 10 m and length 20 m. The width of the adit is 2.4 m. The rock mass is magnesite. Joints with sand filling were made in the

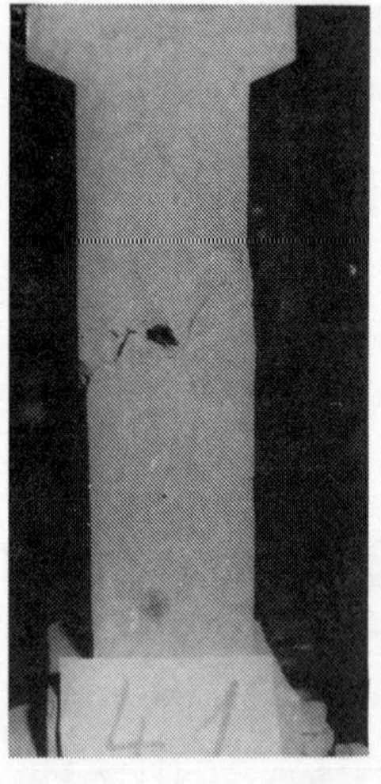

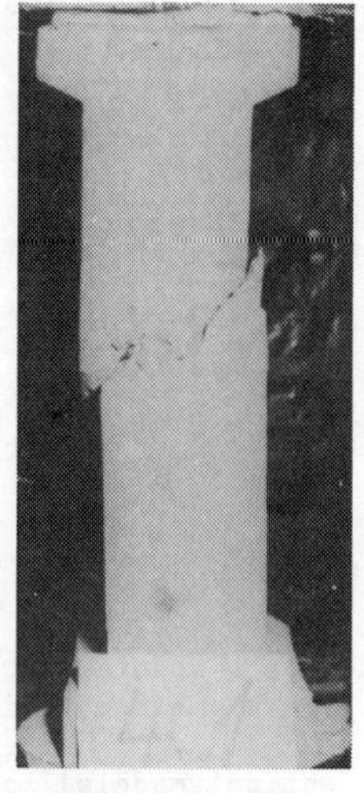

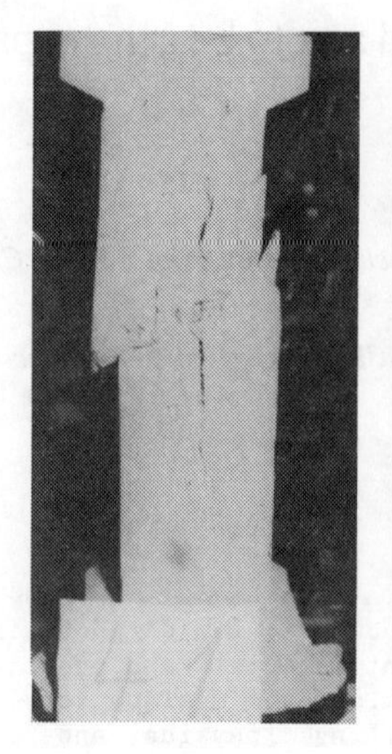

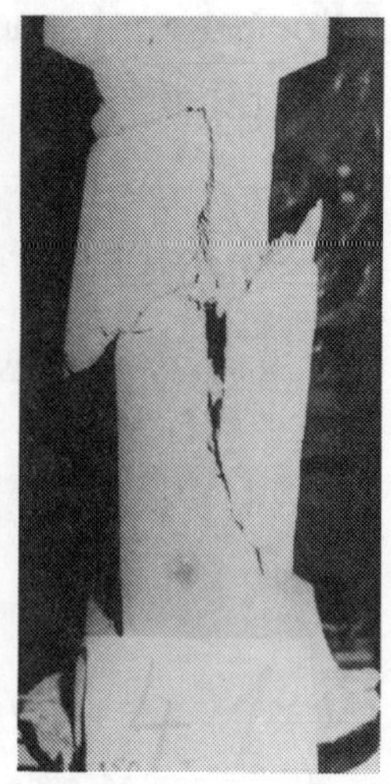

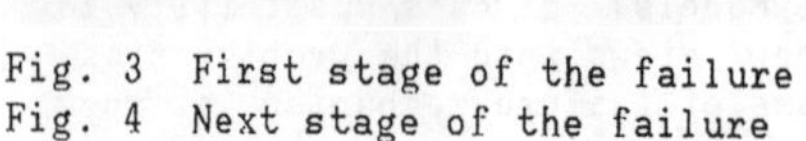

Fig. 3 First stage of the failure
Fig. 4 Next stage of the failure

Fig. 5 Next stage of the failure
Fig. 6 Final stage of the failure

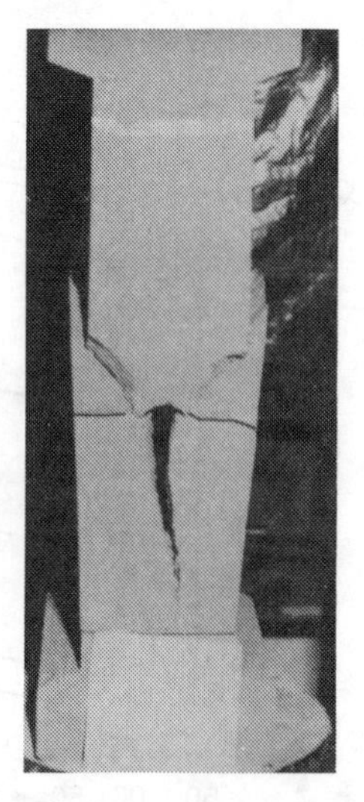

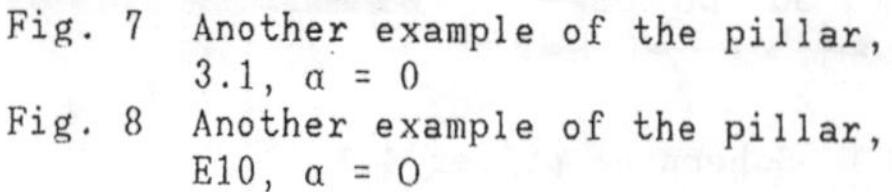

Fig. 7 Another example of the pillar, 3.1, $\alpha = 0$
Fig. 8 Another example of the pillar, E10, $\alpha = 0$

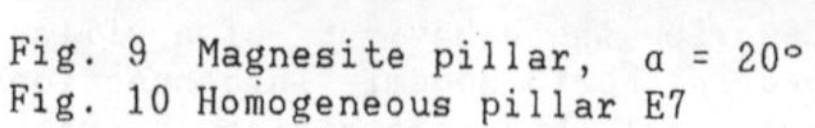

Fig. 9 Magnesite pillar, $\alpha = 20°$
Fig. 10 Homogeneous pillar E7

model. Fig. 2 shows the model before testing. Fig. 3-6 show successive stages of the failure. Deformations of the pillar are depicted in Fig. 12. A similar situation on the model from magnesite after testing can be seen in Fig. 9. Differences between the homogeneous (Fig. 10) and inhomogeneous pillars are clearly visible. The failure of the homogeneous pillar was sudden, with one or two shear surfaces. The vertical deformation before the failure was about 1.5 %. The failure of the inhomogeneous pillar was more complex. It was not instantaneous (the duration of failure depends on the loading rate; in the model presented it was 390 s) and the axial deformation amounted to 3-10 % (in the model presented it is 8 %). It follows that, in spite of a lower bearing capacity (54 %), the pillar with

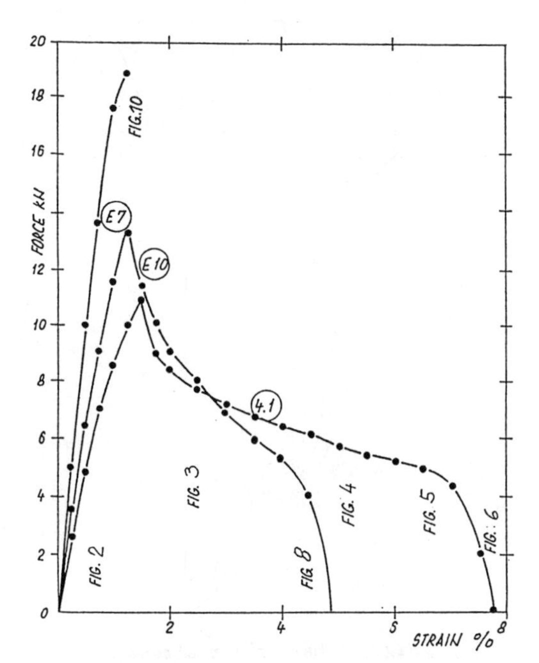

Fig. 11 Load strain graphs for pillar 4.1, E7, E10

an adit was, from the point of view of mining, more advantageous. The model displays an extremely complicated failure. Usually the failure is simpler (Figs. 7 and 8), but it is never so simple as in an homogeneous model. If the homogeneous model has a bearing capacity around 20 kN is it hardly believable that the model, at the stage shown in Fig. 5, can still bear 5 kN, and 1.5 kN at the stage in Fig. 6. The load-deformation curves of Pillars 4.1 (Figs. 2-6), E7 (Fig. 10) and E10 (Fig. 8) are shown in Fig. 11.

It should be pointed out that the more heterogeneous the structure of the model, the more pronounced the symptoms preceding the failure. Moreover, in heterogeneous models, the failure was more complicated and, as a rule, it did not occur instantaneously but proceeded gradually in a series of partial cavings (Figs. 2-6). On the contrary, homogeneous models only exhibit such phenomena rarely. Their failure usually occurs suddenly and definitely (Fig. 10).

To record model movements we used either the method of close-range photogrammetry or videorecording. The first technique is very precise and able to measure model deformations of 0.01 mm. The accuracy of the second method is not high but it does yield a continuous recording. The best

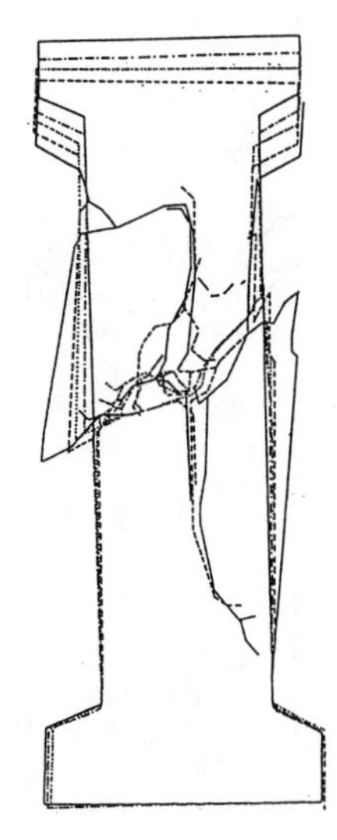

Fig. 12 Deformations of the pillar 4.1 during failure, fig. 2-6

approach is to combine the two methods. Computer analysis of movements is widely useld. Programs for adjusting the measured values on surfaces are available.

MATHEMATICAL PART

Numerous procedures have recently been developed to describe the behaviour of rock mass by means of a mathematical model. Some of them use contact elements, other describe dislocations with the aid of complex valued function, etc. In this paper we concern ourselves with a variational formulation of the non-linear problem of Mohr-Coulomb type which is quite rigorous from both the experimeental and analytical points of view. For the purpose of numerical analysis, the finite element method is applied to the problem of propagation of fractures (dislocations) in rock mass. The mathematical model is adapted to the results of the experimental model in order to obtain more accurate distribution of internal parameters of the constitutive law.

Comparison of the laboratory data, obtained by means of special experimental models from physically equivalent materials, and the results, obtained from a mathematical model, facilitates the inclusion of non-linear behaviour along the dislocations in our deliberations. the coupled modelling (physical and mathematical models) makes it possible to eliminate disadvantages of the two models

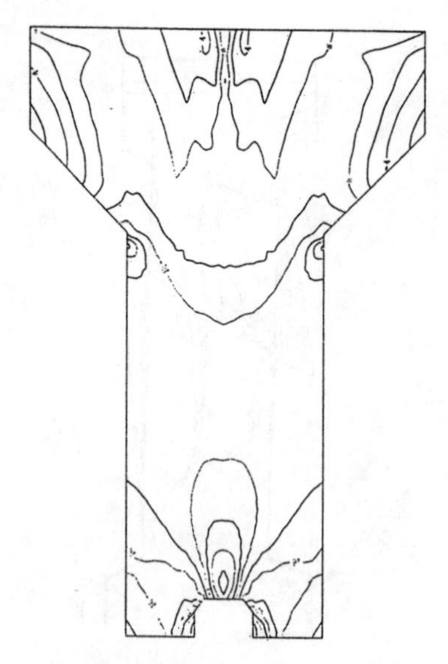

Fig. 13 Principal normal stress

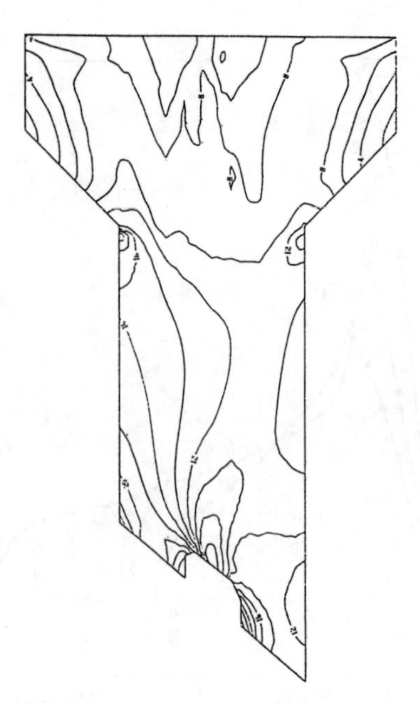

Fig. 14 Main shear stress

and throws new light on the problems involved.

In Figs. 13-14 some results of the mathematical solution are presented. Figs. in isohypse style show the principal normal and shear stresses in a pillar with an adit and the central layer inclined at 0°, and 40°. Concentration of tensile and shear stresses can be seen on the top of the adit. It can provoke tenstile cracks, typical of a pillar with addit failure.

CONCLUSIONS

Knowledge of the failure mechanism and its history in jointed rock mass is of great importance to the development of science. An experimental study on physical models makes it possible to investigate a problem continuously in time. The figures showing the failure model demonstrate differences between the types of failures in homogeneous and structured materials. The differences are not only in the geometry of failure but also in its duration and in the final shape of the model after the failure. Continuous videorecording makes it possible to identify failure symptoms and thus to estimate more precisely the deformations that signal hazard conditions. Analogous investigations can be carried out of surface subsidence due to underground mining, and of tunnel head stability, etc., in structural (jointed) rock mas. If the behaviour of the model is verified on tentative models before the final test, agreement of its results with practical conditions is very good.

Mathematical modelling yields the stress-strain field before failure and it has thus become a convenient supplement to experimental investigations.

REFERENCES

Procházka, P., Vacek, J. 1992." Determination of Internal Parameters of Open-Cast Slopes by Coupled Modelling", NUMEG'92, Prague.

Vacek, J. 1991. "Similarity in geotechnics and calibration of models from equivalent materials", 32. Symp. on Rock Mechanics. Oklahoma. Balkema: 745-754.

Numerical methods
Technical notes

Geomechanics 93, Rakowski (ed.) © 1994 Balkema, Rotterdam, ISBN 90 5410 354 X

Numerical method of analysing influence of underground coal mining on mine structure

Uroš Bajželj
University of Ljubljana, Slovenia

Jakob Likar & Franc Žigman
Mining Institute Ljubljana, Slovenia

Janez Mayer
Velenje Lignite Mine, Slovenia

ABSTRACT: The efficiency of analysing the changing stress-deformation conditions in rocks around progressing underground works depends on the application of suitable numerical method. Although the changed geotechnical conditions due to excavation are known for hanging walls, for planning the preparation and winning works with downward excavation the conditions of footwall are also indispensable. The paper describes the system and results of simulated coal excavation in different phases, using the finite element method. With these results we analysed the corresponding stability of mine roadway in changed stress-deformation conditions with the finite difference method. The comparison of measured and calculated wall movements of mine roadway showed an acceptable difference. This is the proof that the applied method is suitable for solving of similar problems.

1 INTRODUCTION

Numerical methods implemented for analysing the changes of stress-deformation states are nowadays generally known in mining and used in common practice. There we meet some definite problems regarding their use. Due to large dimensions of openings, the changes of stress-deformation states spread far away of excavation works. Therefore the analysis of changed stress-deformation state includes rather large area, in which a bigger number of excavations could be included. Thus the continuum is discretised to certain number of finite elements used in applied numerical method. Because of some limitations such as: the number of elements, their size and ratio of two adjacent elements, excavation of smaller rooms in such a net is rather difficult. The solution described in this paper includes:

- Simulation of progressing excavation and caving using finite elements method.
- Transfer of stress state of the minor area into a new net of smaller elements. This net is built again concerning dimensions and shapes of smaller excavated room.
- Simulation of excavation and supporting of smaller room in a new net of elements and determination of wall movements using finite difference method.

The solution described in this paper is applied in the Velenje Lignite Mine. As a proof that the described method is suitable, there is a comparison between measured and calculated wall movements of pit objects. The differences are acceptable.

2 MINING ACTIVITIES IN THE VELENJE LIGNITE MINE (VLM)

In the VLM the thickness of coal bed is over 100 m. The excavation is mechanised applying longwall face mining method with undercutting and vertically concentrated. The area is mined in strips up to 1000 m long, 80 m wide and 10 m high. The self advancing hydraulic support is used. The hanging wall behind the support is continuously crushed and falling down. The excavation of strips is moving from hanging wall to footwall. Generally, two kinds of mining works are carried out:

- Preparation works consisting of two longwall galleries and some cross-roads to prepare the lower strip for excavation
- Excavation of strip, winning of lignite

The program UTAH II (W.G. Pariseau, Metallurgical and Fuels Engineering - Dept. of Mining, Salt Lake City, 1973), developed for mining and capable to concern the crushing of caved spaces, proved to be suitable for simulation of winning works using bigger number of finite elements.

For simulation of preparation works the program FLAC (Itasca Consulting Group, Minneapolis, 1989) concerning the supporting elements is more suitable. It enables the simulation of roadway supporting and quick analysis of interaction between rock and support.

3 SIMULATION OF WINNING WORKS

The simulation of roadway preparation for excavation of next strip requires the knowledge of stress state before the roadway was made. This stress state is a consequence of winning works. While the changes of stress-deformation states due to large dimensions of excavated strips spread rather far, the simulation should also concern more distant works. We supposed that in our case the winning works more than 400 m away have no influence on the stress-deformation state of the roadway. The supposition was confirmed by the results of simulation. Thus the field 800 x 500 m large, with the roadway in the middle, were used for calculating the changed stress-deformation states. Figure 1 shows field discretised to finite elements with indicated excavations.

The applied program package UTAH was completed with programs for preparing the net of finite elements with digital indication and programs for graphic indication of results on screen, plotter and colour printer.

The simulation was performed in the following stages:

- Calculation of primary stress state
- Calculation of stress states in steps, where one step was the simulation of excavating a strip, or roadway. The next step was its crushing. The simulation was performed in the sequence corresponding the actual mining works.

The result of simulation is the stress-deformation state as the consequence of all mining works within the observed field. The progressing excavation and crushing of pit spaces is taken into account. Figure 2 indicates the directions and sizes of main stresses.

Safety factors were calculated as ratio between the calculated stress and strength. The safety factor iso - lines shown in Figure 3 indicate that the observed preparation roadway was not located within the most safe site. There the safety factors are only about 1.2. It would have been better to locate it about 40 m right from the existing site (seen in the profile), where the safety factor is 3.3 -3.4.

With determined stress state in the profile observed we can plan and locate the excavation and supporting of all kinds of pit objects. The first criterion for siting is the safety factor calculated for the field concerned. The criterion for dimensioning the supports is the stress state around the site selected. This stress state should be calculated with interaction between rock and supports, when the pit object is to be supported. We can not know beforehand the micro location of higher safety factors, but we must generally know the site before the net of finite elements is formed along the limit of excavated space. There the net should be correspondingly denser. Therefore we transferred the calculated stress state (σ_x, σ_y, and σ_{xy}) into a new net of finite elements adequately shaped. In this new net we simulated the excavation of objects, supports of different dimensions and checked the safety factors of the supports, etc. Then we calculated the new stress state as the consequence of the pit object as determined and accomplished.

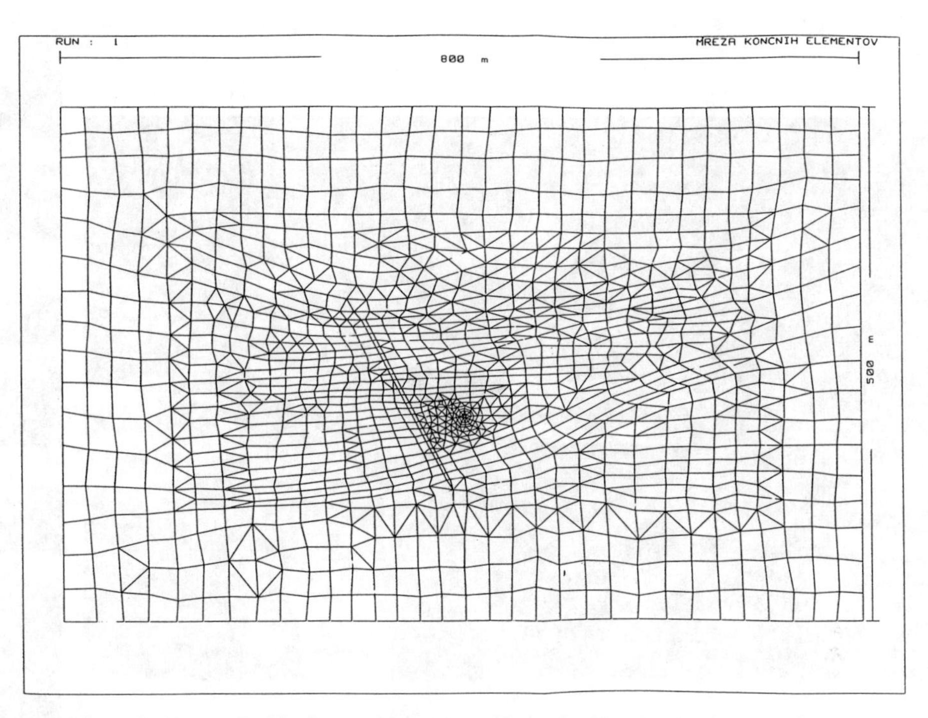

Fig. 1 Net of finite elements with indicated excavations.

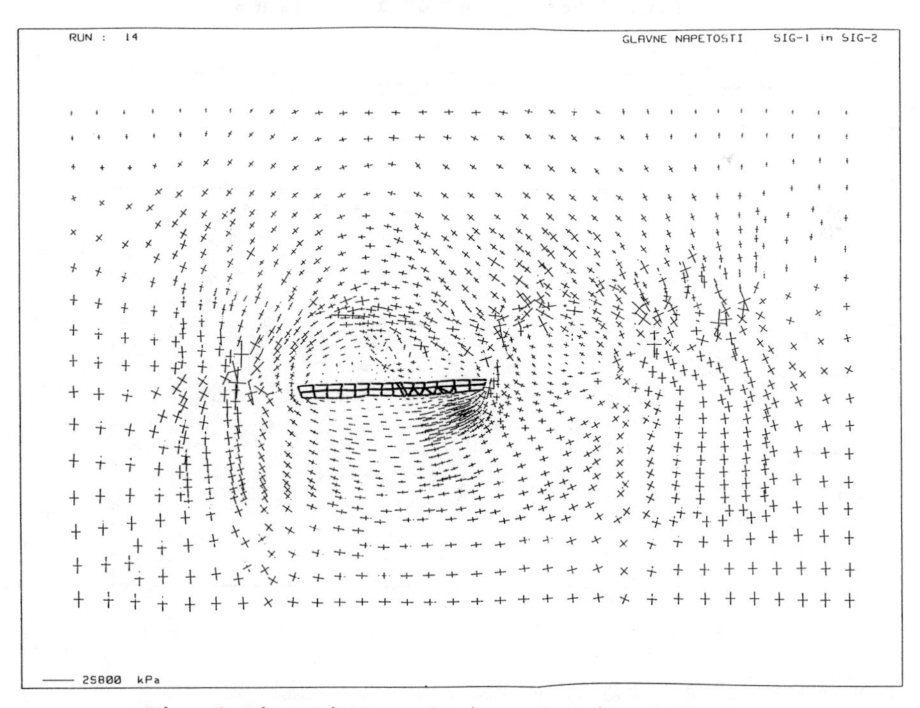

Fig. 2 Directions and size of main stresses

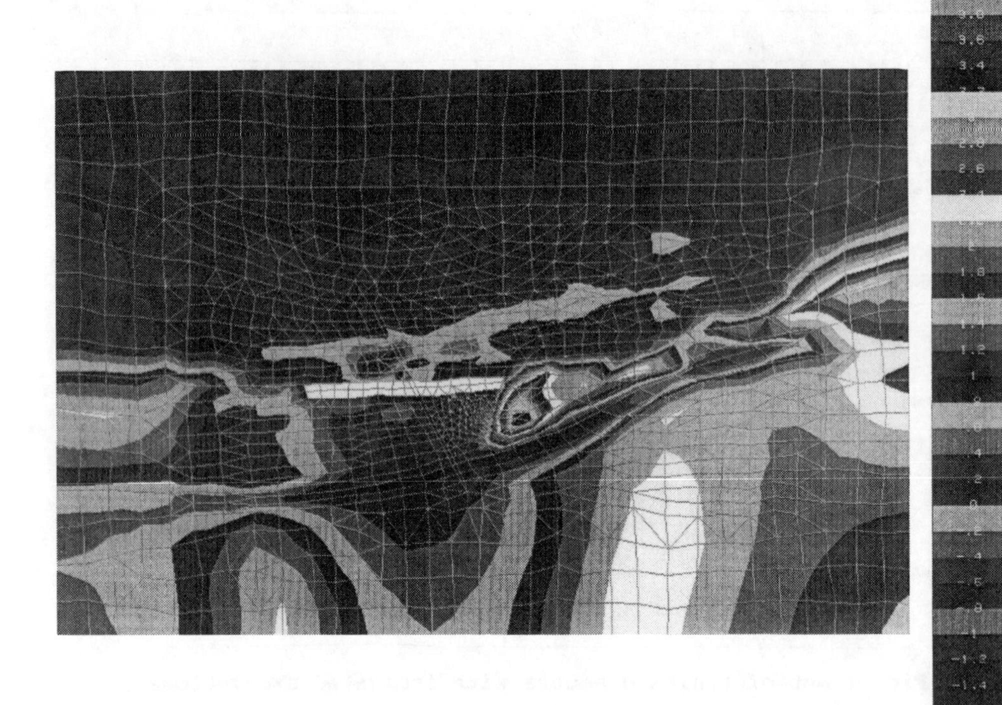

Fig. 3 Safety factor iso - lines

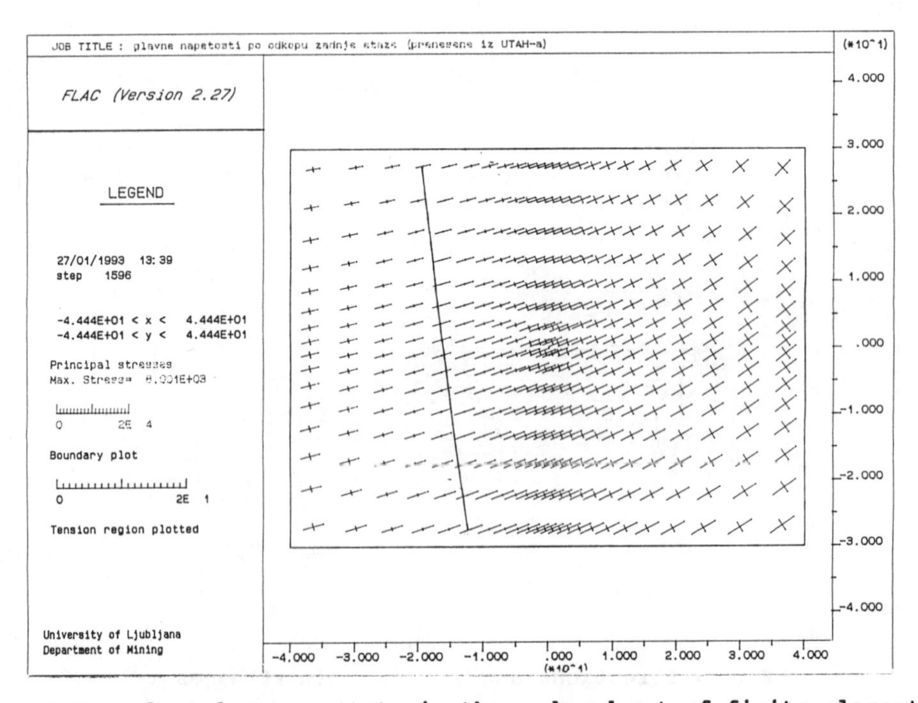

Fig. 4 Transferred stress state in the reduced net of finite elements.

4 TRANSFER OF STRESS STATE INTO THE NEW NET

With the equations for circular borings in elastic medium (Hirsch) we calculated the stress state. For the planned diameter of roadway, the radial and tangential stress components differ only 4 per cent on 20 m distance. Thus along this length there should be no influence of roadway construction upon the changes of stress - deformation state. Since the rock is inelastic and due to boundary conditions in calculations we chose the 80 x 60 m field. The analysed object is in its middle.

The initial stress state is the picture of stress state related to the minor part of field concerned. This was calculated as the result of excavating. Because of this, the transfer of stress state into a new net of finite elements was necessary. The transferred stress state in the new net of finite elements is shown in Figure 4.

The transfer of stress state, relevant for the new calculation after the roadway is completed, requires to determine the functions:

$$\sigma_x = f(x,y) \quad (1)$$

$$\sigma_y = g(x,y) \quad (2)$$

$$\sigma_{xy} = h(x,y) \quad (3)$$

where x, y are the coordinates around the roadway.

The above functions are linear, since in the program for simulation of roadway preparation the linear change of stress along the boundary of the new net can be taken.

On disposal are the calculated stresses σ_x, σ_y, and σ_{xy} related to individual points with known coordinates, that is in center points of finite elements. The linear function (equation of the plane) is to be found with which the stress component could be calculated for each point.

The coordinates of center points and related stresses can be obtained from the picture of net with finite elements. Therefrom we can obtain the equations of planes according the method of smallest quadrates as shown graphically for σ_x, σ_y, and $\tau = \sigma_{xy}$.

Using these equations we can determine the initial stress state before the simulation of roadway excavation.

The equations are as follows:

for σ_x:

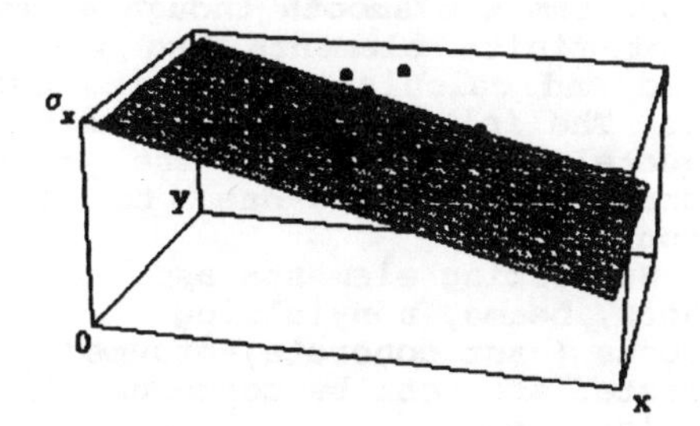

$$\sigma_x = 1594 - 23{,}98x - 0{,}2722y \quad (4)$$

for σ_y:

$$\sigma_y = -5354 - 4{,}660x + 8{,}572y \quad (5)$$

for σ_{xy}:

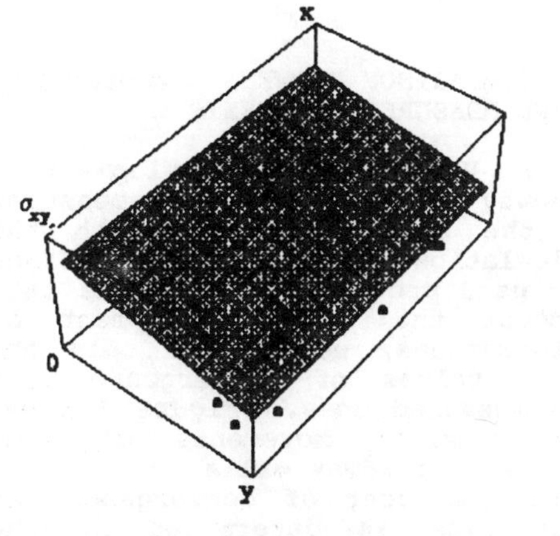

$$\tau = \sigma_{xy} = 2014 + 0{,}8299x - 6{,}063y \quad (6)$$

5 SIMULATION OF EXCAVATION AND SUPPORTING OF GALLERIES

The calculation of stress state and prediction of deformations, which were also checked with measurements, were performed with computer program FLAC. The use of new program was smooth though a new net of finite elements had to be shaped and calculation carried out anew. The initial stress state was adequately prescribed. The FLAC program was used due to two reasons:

- Supporting elements as anchors, beams, nonyielding supports (cast concrete) grouted concrete, etc. can be concerned in calculations.
- This program is substantially quicker than the program used before.

The program can be used on some IBM PC compatible computers which correspond to the requirements. Stress-deformation states are calculated with the explicit finite difference method.

The result of simulation is the stress-deformation state after the roadway was made. According the described method we can predict movements of walls and supporting of all kinds of galleries within the influence field, where excavations change the stress-deformation states. A comparison between measured and calculated movements within the roadway gives a proof that the described method is a correct one.

6 A COMPARISON BETWEEN CALCULATED AND MEASURED MOVEMENTS

In the VLM, the time development of roadway wall movements was measured in the points through which the calculation profile passes. Since the used programs do not take into account the time development of deformations, we compare only the final values of convergence with the measured ones. Figure 5 shows the measured movements of some points of roadway walls.

The per cent of convergence for each side is determined by the equation:

$$k = \frac{a_2 - a_k}{a_2} \tag{7}$$

where k, percent of convergence; a_2, initial side length; a_k, final side length

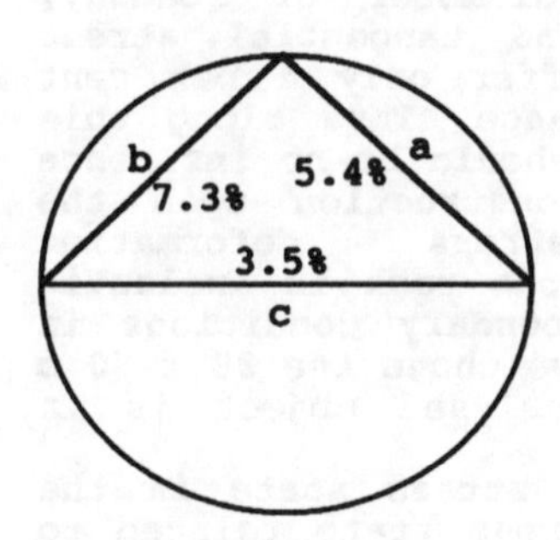

Fig.5 Measured convergence of a roadway profile

The results of simulating the excavation and preparation of roadway were the basis to determine the expected convergence from the calculated movements of points on sides and ceiling of the roadway. See Figure 6.

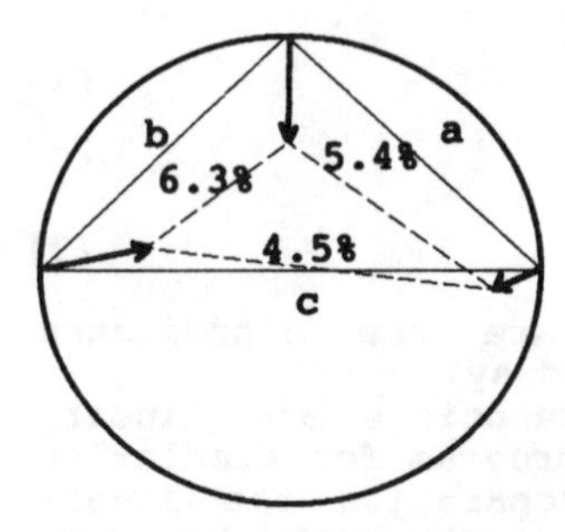

Fig.6 Expected convergence based on similar and calculated movements

The comparison between the calculated and measured decrease of side lengths is given in Table I.

Table 1. Decrease of side lengths.

Side	Measured [%]	Measured [cm]	Calculated [%]	Calculated [cm]
a	5.4	16.0	5.4	16.1
b	7.3	21.4	6.3	18.9
c	3.5	13.4	4.5	19.0

The comparison between the calculated and measured convergence, using the results of simulation, shows good coherence.

7 CONCLUSIONS

The influences of excavation reflect on changing stress-deformation state around the excavations. The changes are most discernible above excavations though the changes underneath are not negligible. The simulation of excavation makes the prediction of changed stress-deformation state possible. This is a tool for locating and dimensioning of supporting for any pit object. The criterion for locating is the safety factor within the surface analysed. The criterion for dimensioning of supporting is the stress-deformation state as the result of excavation and supporting of the particular object. This state should be calculated with analysis of interaction between rock and supports. Generally, it requires the transfer of stress state before the object is made into a new net of finite elements and the calculation of changed stress-deformation state. The UTAH program for simulation of excavations proved to be suitable for such operations. It was specially developed as well as the FLAC program which enables quick analyses of interaction between rock and supports. Thus the parameters of supports can be varied. Through quick analyses the minimal dimension of supports can be determined.

The comparison between the calculated and measured convergence, using the results of simulation, shows good coherence. This is a proof that the approach is suitable. It also means that this method can be used as a tool for planning works, having a definite degree of credibility to give correct results.

REFERENCES

Pariseau, W.G. 1973. Interpretation of Rock Mechanics Data. Dept of Mining, Metallurgical and Fuels Engineering, Salt Lake City.

Itasca Consulting 1989. Fast Lagrangian Analysis of Continua. Itasca Consulting Group, Minneapolis.

Brady & Brown 1985. Rock Mechanics for Underground Mining, George Allen & Unwin, London.

Stacey & Page 1986. Practical Handbook for Underground Rock Mechanics, Trans Tech Publications, Clausthal-Zellerfeld.

Geomechanics 93, Rakowski (ed.) © 1994 Balkema, Rotterdam, ISBN 90 5410 354 X

Mathematical analysis of inverse problems of electrostatics

Nina Častová
Technical University Ostrava, Czech Republic

ABSTRACT: The paper concerns with formulation and solution of one of the most frequent inverse problems in analysis of resistivity method in applied geophysics. In particular, the choice of optimization criterium, convergence and the stability of numerical solution of a problem that arises from diskretization of unbounded operator are discussed.

1 INTRODUCTION

The procedures of investigation into electric field (the vertical electric sounding and profiling) are widely applied in the geophysical practice. The task of finding out geoelectrical parameters of the geological environment on the basis of electric current and potential between electrodes on the earth surface belongs to inverse problems of mathematical physics. Although first results we obtained almost 50 years ago, the interest in their solution is permanent.

In general both the direct and inverse problems can be expressed by the operator equation

$$\mathbf{A}\boldsymbol{v} = f, \qquad (1)$$

where $\mathbf{A}$ is the linear operator of the problem, $\boldsymbol{v}$ denotes a geoelectric parameters and f denotes the potential function. Usually, the restriction $\bar{f}$ of f to a discrete set of points is given instead of f.

2 DIRECT TASK

Solution of the direct problem of electrosounding is closely related to the evaluation of integral

$$\frac{\rho_z}{\rho_1} - 1 = r^2 \int_0^\infty R(m) m J_1(mr)\, dm. \qquad (2)$$

which, unfortunately, in this form is divergent and standard application of an integral transformation is impossible.

In (2) $\rho_z = \frac{KU(z)}{I}$ is the apparent resistivity of the environment, K is a coefficient depending on the arrangement of electrodes, $R(m)$ is a function varying with parameters of cross section and with the variable m, i.e. $R(m, \bar{h}, \bar{\rho}, N)$, and J_1 is the Bessel function of the 1st order.

If we express (2) as a limit

$$\frac{\rho_z - \rho_1}{\rho_1} = \lim_{\alpha \to 0} \int_0^\infty m r^2 R(m) J_1(mr) e^{-\alpha m}\, dm, \qquad (3)$$

then it may be proved that the last integral is convergent (absolutely) for $m, r \in \langle 0, \infty)$. Then, applying the Hankel's transformation, we get

$$R(m) = \lim_{\alpha \to 0} \int_0^\infty \left(\frac{\rho_z}{\rho_1} - 1 \right) J_1(mr) e^{\alpha m} \frac{dr}{r}, \quad m \in \langle 0, \infty).$$

The numerical evaluation of the last inegral can be written in the form of

$$g(x_i) = \sum_{j=-\infty}^{+\infty} t(x_i - y_j) \int_{y_j}^{y_{j+1}} b(y)\, dy = \sum_{-\infty}^{+\infty} B_j . T\left(x_i - y_j\right).$$

The filtration coefficients B_i can be determiden for instance by the least-square method. Then the stabilization coefficient, the optimum step as well as the filtration window can be determined. The coefficient B_i strongly depends on the selected formula (e.g. Vaňjan, Keller, Koefoed) for calculation of theoretical values of $R(m)$ (see Table 1). The formula of the latter two authors are valid for mildly varying parameters of layers only and yield the same results with an error not exceeding 10^{-5}.

Table 1. The filtration coefficients B_i.

Author	Gosh	Abramova	Častová
n	9	8	8
Shift	1.05	-	.9481
Number	3	5	5
K	5	2	2
i \ j	$B_j = B_i \quad j = 1,\ldots,n$		
	B_j	B_j	B_j
1. $-K$	.0148	-.1003	.0279
2. $-K+1$	-.0814	1.1306	-.1830
3. $-K+2$	.4018	-3.8623	.5431
4. :	-1.5716	3.9230	-.9317
5. :	1.9720	-.9641	1.0042
6.	.1854	1.0591	-.6995
7.	.1064	-.2750	.3053
8.	-.0499	.0890	-.0671
9.	.0255	-	-

In Table 1, n denotes a number of filtration coefficients evaluated,
Shift is a constant part of linear filtration,
Number denotes a number of points in any interval $\langle 10^k, 10^{k+1}\rangle$, and $-K$ is an initial point of the filter.

3 INVERSE PROBLEM

The inverse problem is known to belong to the class of conditionally correct problems. In general conditionaly correct problems can be transformed to correct problems by means of restriction of the set on which the operator $\mathbf{A}$ is defined to

$$M \equiv D_0(\mathbf{A}) \subset D(\mathbf{A}).$$

For example, we can define a compact set for solutions M on the basis of an a **priori** information. In our problem, we prescribe intervals for the parameters of a geoelectric crosssection

$$\rho_i \in \langle \rho_{i,\min}, \rho_{i,\max} \rangle, i = 1,2,\ldots,N$$
$$h_i \in \langle h_{i,\min}, h_{i,\max} \rangle, i = 1,2,\ldots,N-1; h_N \to \infty,$$
$$N \in \langle n_{\min}, n_{\max} \rangle.$$

The solution of inverse problems is usually decomposed in to the solution of a sequence of direct problems in order to minimize the deviation between the actual and approximate solution.

From the viewpoint of available a **priori** information on the set of solutions M, we can encounter three possibilities:

1. Neither qualitative nor quantitative information is available on the set M. In principle there is not possible to derive a criterion for evaluation of accuracy of a solution.
2. A qualitative information is available only in terms of the smoothness of solution v in compact normed space M. In such case, there is the criterion for evaluation of the accuracy of the solution in the norm of the space M with reasonably small δ.
3. Qualitative and quantitative information is available on the basis of which we can define a weak compact $M \subset L_2$. In such case, the criterion for evaluation of the accuracy of the solution can be compiled in a special norm for any $\delta > 0$, where δ characterizes the accuracy.

Provided there is enough a **priori** information, we can introduce the criterion for minimization in form of the Tichonov parametric functional

$$F_\alpha = \|v - \tilde{v}\|^2 + \alpha s^2(v). \qquad (4)$$

Here $\alpha(\delta) = \alpha > 0$ is the parameter of regularization, which depends on the error of input data (representing some 3 to 5 % of the measured value).

Table 2. Results of interpretation

	Set-up intervals		Value $\alpha = 0$		Value $\alpha = .1$	
i	ρ_i	h_i	ρ_i	h_i	ρ_i	h_i
1.	30-50	.1 - .5	40	.1	43	.2
2.	40-50	.5 - 1	45	.7	48	.8
3.	30-40	1 - 3	35	1.6	38	1.8
4.	45-55	6 - 9	50	7.5	53	8.5
5.	30-40	25-35	35	27.5	38	29.2
6.	90-150		120		90	
N	4	10	N=6		N=6	
	$\varepsilon_\rho = 1$	$\varepsilon_h = .1$	$\min F_\alpha = 1.068$		$\min F_\alpha = 0.526$	

The example given in Table 2 is based upon functional referring to the function of maximum fidelity

$$F_\alpha = \sum_{j=1}^{K} \left(\frac{R_j^T - R_j^E}{R_j^T} \right)^2 + \alpha s^2(v). \qquad (5)$$

Here K is the number of values measured on a single

curve with various span of electrodes, R_j^E, R_j^T are the figures of the function $R(m)$ calculated from experimental data and with the help of analytic formulae for the given parameters. $\alpha(\delta)$ is the parameter of regularization,

$$\alpha s^2(v) = \sum_{j=1}^{2N} q_j.\log\left[\left(v_{j,\max} - v_j\right)\left(v_j - v_{j,\min}\right)\right],$$

$2N$ is the number of parameters, q_j are the weight coefficients of the individual parameters, v_j is the value of the jth parameter, and $v_{j,\max}, v_{j,\min}$ are the upper and lower bounds on the parameters.
If a restriction for longitudinal

$$\left(g_j = \frac{h_j}{\rho_j}\right)$$

or transversal conductivity

$$\left(g_j = h_j.\rho_j\right): g_j \in \left\langle g_{j,\min}, g_{j,\max}\right\rangle$$

is available, then

$$s^2(v) = \sum_{j=1}^{2N} q_j.\log\left[\left(g_{j,\max} - g_j(v)\right)\left(g_j(v) - g_{j,\min}\right)\right]$$

where q_j are the weight coefficients.

4 COMMENTS AND CONCLUSIONS

To propose the algorithm of multidimensional interpretation of the vertical electric sounding, we recommend the following procedure:

1. On the basis of interpretation of geophysical data with application of the cluster analysis and of the correlation analysis of multidimensional objects, the quassi-homogeneous blocks (having the same geoelectric properties) can be separated.
2. For the solution of a direct problem with exact operator $\mathbf{A}$ and a continuos function f, the deviation of the approximate solution $\tilde{f}$ from the exact solution given in the norm $\left\|\mathbf{A}v - \tilde{f}\right\| = \varepsilon, \varepsilon \to 0$ may be chosen as the criterion of minimization. If, however, the operator $\mathbf{A}$ is known only approximately $(\tilde{\mathbf{A}})$, then exact solution of the approximated problem $\tilde{\mathbf{A}}v = f_p$ need not converge towards the exact solution of the original problem. Thus $\left\|\mathbf{A}v - \tilde{f}\right\| = \varepsilon$ cannot be applied as the criterion of minimization, especially when the numerical stability of solution is not granted.
3. Direct application of a criterion of minimizing $\|v - \tilde{v}\| = \varepsilon$ at numerical solution for the inverse problems (incorrect problem defined on weak compacts) my yield very bad results. However, a correct criterion can be formulated with application of the regularization operator or by creation of a special norm for a weak compact.
4. On the ground of correct criterion, the aposterious errors can be calculated and the sensitivity analysis can be carried out.

REFERENCES

Morozov, V.A. 1987. Regularnye metody rešenia nekorrektno postavlennych zadach. Moskva: Nauka. In Russian.

Gaponenko, J.L. 1989. Nekorrektnye zadachi na slabych kompaktach. Moskva: MGU. In Russian.

Častová N. & Müller K. 1974. Matematicky model interpretace krzivek VES, p. 275-285. Brno: Sb. Teorie a PC v geofyzice

Geomechanics 93, Rakowski (ed.) © 1994 Balkema, Rotterdam, ISBN 90 5410 354 X

Elasto-plastic analysis of deformation soil body with 3D-finite and infinite elements

J. Králik & M.Šimonovič
Slovak Technical University, Bratislava, Slovakia

ABSTRACT: In this paper mapped 3D-infinite elements and 3D-finite elements are used for solution the structures on undeground. The static analysis of a structure-soil system supported by homogenous elasto-plastic soil is formulated. Near field is modeled by finite elements and far field by infinite elements. Stress-strain relationship is formulated for nonassociated plasticity theory. Drucker-Prager, Mohr-Coulomb and Hoek-Brown yield criterion describe well the elasto-plastic soil.

1 INTRODUCTION

Solution of contact problems by 3D-finite element method results in large number of unknows. This becomes the problem because of limited computer memory.

That' s why the new ways of elastic foundation modelling are being searched for. The possible way are either modelling of elastic foundation by simplificated models, or reduction of space problem to 2D-model (e.g. Boundary elements method). At present, Winkler' s, Pasternak' s and Bousinesque' s models are the most used ones or the foundation is modelled by semi-infinite soil region.

Unfortunately the majority of practical design problems fall outside the reach of such closed solutions, due to complex and irregular geometric forms of the structures, complexity in loading patterns, non-linearity and inhomogeneity in properties of materials etc. In such cases the designer must resort to approximate solutions.

The method of infinite elements becomes one of the latest most progressive methods. It may be combined with semi-infinite soil region properties very well.

Marquest and Owen developed the mapped infinite 1D, 2D and 3D-elements. Modeling of infinite soil domain by 3D finite and infinite elements corresponds effectively with the real behaviour of undeground structures as tunnels, pipes, piles, etc. due to non-linearity and inhomogeneity in soil properties.

2 INFINITE 3D-ELEMENT

The very first models of infinite elements were based on a finite element that had been transformed to the element with one point at infinity.

Taking the assumption of zero value of searched guantity in the consideration, Bettess (1977) derived the infinite element with dumping of basic field variable consisting of Lagrange polynomial approximation and exponential or reciprocal dumping function.

G.Beer and J.L.Meek (1981) derived special infinite elements for problems of tubular gape in 2D and 3D infinite region.

They obtained the mapping functions by combination of both basic finite element and dumping functions.

This is the most effective access because of using the combination of finite and infinite elements and full utilizing the finite elements benefits, first of all for solving of nonlinear problems.

The mapping function of 3D elements is given by boundary condition, the value of the function at infinity is zero.

The mapping function is a product of three functions

$$M_i(\xi,\eta,\zeta) = M_i'(\xi).M_k''(\eta).M_l'''(\zeta), \qquad (1)$$

where $M_i'(\xi), M_k''(\eta), M_l'''(\zeta)$ are the mapping functions for one-dimensional problem.

The interpolation from local to global position for

one-dimensional element with two nodes and one point at infinity is defined by

$$x(\xi)=\sum_{i=1}^{2}M_i(\xi)x_i, \qquad M_1(\xi)=-2\xi/(1-\xi),$$

$$M_2(\xi)=(1+\xi)/(1-\xi), \qquad (2)$$

This mapping function is chosen so that $\xi=(-1,0,1)$ correspond respectively with the global coordinate $x=(x_1,x_2,\infty)$. The requirement that the mapping must be independent of the coordinate system necessitates that $M_1+M_2=1$.

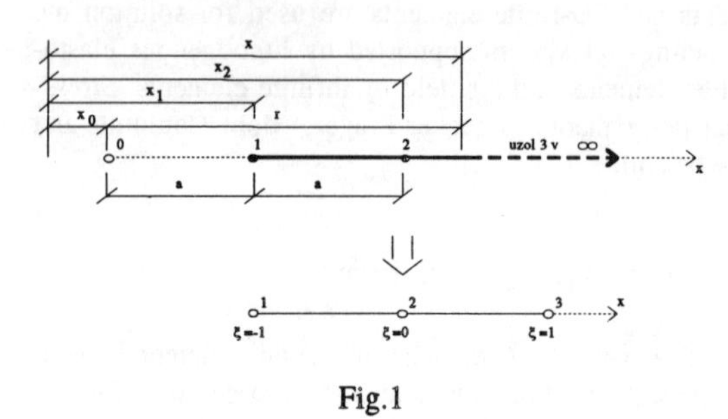

Fig.1

The basic function f is interpolated by standard shape function

$$f=\sum_{i=1}^{3}N_i f_i=\frac{1}{2}(\xi^2-\xi)f_1+(1-\xi^2)f_2+\frac{1}{2}(\xi^2+\xi)f_3 \qquad (3)$$

Solving (2) for ξ yields

$$\xi=1-2a/r \qquad (4)$$

where r is distance from the pole 0 to a general point within the element and $a=x_2-x_1$. Substituting (4) in (3) gives the unknown f dependent on global coordinate r

$$f=f_3+(-f_1+4f_2-3f_3)a/r+(2f_1-4f_2+2f_3)a^2/r^2 \qquad (5)$$

If r tends to infinity, f at infinity is zero, this boundary condition is automatically satisfied by simply summing only over the finite nodes 1 a 2 in expression (3). Including the higher order polynom leads to the higher accuracy.

In this paper we use 3D infinite elements as presented in [2]

4-node superparametric element (Fig.2a) is obtained from 8-node isoparametric element due to

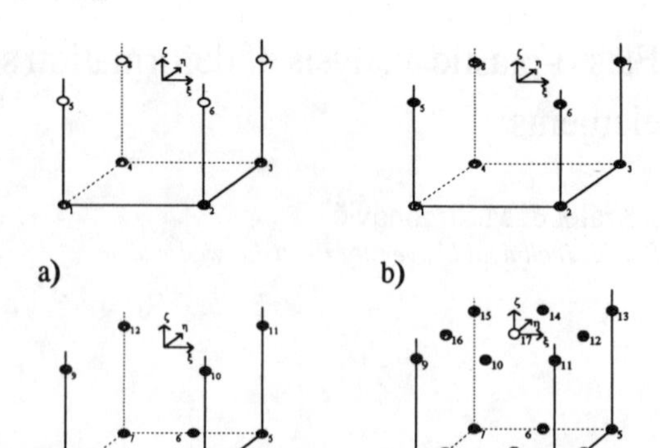

Fig.2

translation of 4 nodes at ζ=1 to infinity. There are 4 mapping joints.

8-node Lagrange isoparametric element (Fig.2b) is derived from 12-node isoparametric element also due to translation 4 nodes at ζ=1 to infinity. The mapping and shape functions are defined at 8 nodes.

12-node serendipity isoparametric element (Fig.2c) is derived from 20-node element. There are 8 nodes at ζ=1 translated to infinity.

16-node superparametric element (Fig.2d) is derived from 27-node finite element.

3 ELASTO-PLASTIC FORMULATION

For any increment of stress, the changes of strain are assumed to be divisible into elastic and plastic components, so that

$$d\varepsilon=d\varepsilon^e+d\varepsilon^p, \qquad (6)$$

where $d\varepsilon^e$ is elastic and $d\varepsilon^p$ is plastic strain increment.

The elastic strain increment is related by

$$d\varepsilon^e=D_e^{-1}d\sigma, \qquad (7)$$

where D_e is elastic matrix and $d\sigma$ is stress vector increment. The plastic strain increment is proportional to the stress gradient of a quantity termed the plastic potential Q, so that

$$d\varepsilon^p=d\lambda\frac{\partial Q}{\partial\sigma} \qquad (8)$$

where $d\lambda$ is a proportionaly constant.

The yield function is defined as follows

$$F(\sigma,\kappa,\alpha)=0 \qquad (9)$$

where κ is the hardening parameter and α is the yield surface translation vector, so that

$$\kappa = \int_{\varepsilon} \sigma^{T} d\varepsilon^{p}, \qquad \alpha = \int_{\varepsilon} C\, d\varepsilon^{p} \qquad (10)$$

where C is the material parameter. The translation of the yield is dependent on yielding history. The stress-strain relationship may be obtained from (7) in following form

$$d\sigma = \boldsymbol{D}_e(d\varepsilon - d\varepsilon^p) = \boldsymbol{D}_e(d\varepsilon - d\lambda \frac{\partial Q}{\partial \sigma}) = \boldsymbol{D}_{ep} d\varepsilon \qquad (11)$$

where $\boldsymbol{D}_{ep}$ is elasto-plastic matrix

$$\boldsymbol{D}_{ep} = \boldsymbol{D}_e - \frac{\boldsymbol{D}_e \left\{\frac{\partial Q}{\partial \sigma}\right\}\left\{\frac{\partial F}{\partial \sigma}\right\}^T \boldsymbol{D}_e}{-\frac{\partial F}{\partial \kappa}\sigma^T \left\{\frac{\partial Q}{\partial \sigma}\right\} - C\left\{\frac{\partial F}{\partial \alpha}\right\}^T \left\{\frac{\partial Q}{\partial \sigma}\right\} + \left\{\frac{\partial F}{\partial \sigma}\right\}^T \boldsymbol{D}_e \left\{\frac{\partial Q}{\partial \sigma}\right\}} \qquad (12)$$

The relantioship (12) is valid for nonassociated theory of plasticity if $F \neq Q$. In this case elasto-plastic matrix $\boldsymbol{D}_{ep}$ is non symmetric. For practical solution it is better to operate with a symmetric matrix. To preserve the symmetry of the matrix, for analyses with a nonassociative flow rule, the second term of the right-hand side of equation (12) is evaluated using F only and again with Q only and the two matrices averaged. For izotropic hardening the stress-strain relationship (12) changes to

$$\boldsymbol{D}_{ep} = \boldsymbol{D}_e - \frac{\boldsymbol{D}_e \left\{\frac{\partial Q}{\partial \sigma}\right\}\left\{\frac{\partial F}{\partial \sigma}\right\}^T \boldsymbol{D}_e}{A + \left\{\frac{\partial F}{\partial \sigma}\right\}^T \boldsymbol{D}_e \left\{\frac{\partial Q}{\partial \sigma}\right\}} \qquad (13)$$

where the value of parameter A obtained from Huber-Mises-Hencky' s yield criterion form

$$A = -\frac{1}{d\lambda}\frac{d\sigma_T}{d\kappa} d\kappa = -\frac{1}{d\varepsilon^p_{eq}}\frac{d\sigma_T}{d\varepsilon^p_{eq}} d\varepsilon^p_{eq}\sigma_T = -\frac{d\sigma_T}{d\varepsilon^p_{eq}} = -H' \qquad (14)$$

Drucker-Prager, Mohr-Coulomb and Hoek-Brown yield criterion [2,5] describe well the elasto-plastic behaviour of soil. Coefficient of cohesion c and angle of friction φ are expressed with respect to the yield stress in compresion σ_c and in tension σ_t of material

$$s = \frac{\sigma_t}{\sigma_c} = \frac{1-\sin\varphi}{1+\sin\varphi}, \quad \varphi = \arcsin\left(\frac{1-s}{1+s}\right), \quad c = \sigma_c \sqrt{s}/2 \qquad (15)$$

Plastic potential $Q(.)$ is obtained from yield criterion including the dilatation instead of angle of friction.

Nonlinear equations have been solved on the base of Newton-Raphson method with the aim to attain equilibrium of external and internal system forces.

4 NUMERICAL RESULTS

Based on the previous theory the authors developed program MEP3D for solution the geomechanics problems in space. The program is written in FORTRAN 77. There are five finite and four infinite elements as result of this program.

Example 1

For example the cylindrical cavity in an infinite medium was used. The loading is applied in one step and consists of an internal pressure $p = 100 kNm^{-2}$. The parameters of cylindrical cavity were used as follows :

$h = 1m, a = 1m, E = 10^5 kPa, \mu = 0{,}33$.

We used the analytical results by [3]

$$u = pa^2 \frac{(1+\mu)}{Er}, \quad \sigma_r = -\frac{pa^2}{r^2}, \quad \sigma_\varphi = \frac{pa^2}{r^2} \qquad (13)$$

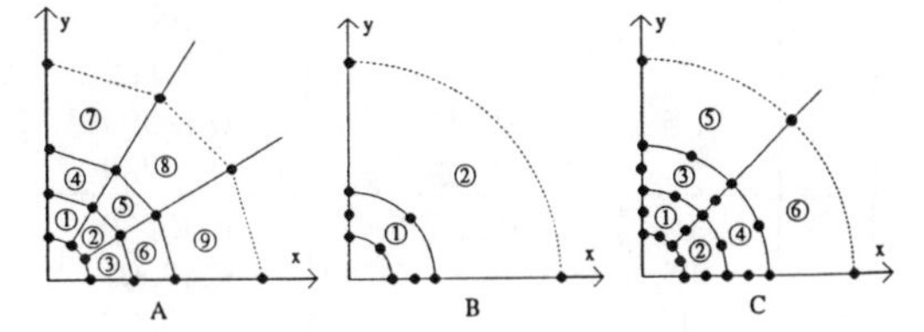

Fig. 3 Model of cylindrical cavity

A) 6 finite 8-node elements and 3 infinite 8-node elements,

B) 1 finite 20-node element and 1 infinite 12-node element,

C) 4 finite 20-node elements and 2 infinite 12-node elements

In the Tab.1 there is a comparison between analytical and numerical results.

Tab.1

r	u_{ANAL}[mm]	u_A[mm]	u_B[mm]	u_C[mm]
1	1,33	1,16154	1,29426	1,31086
2	0,665	0,60878	0,66174	0,66070
3	0,443	0,40213	-	0,43976
5	0,266	0,23930	0,26264	0,26147

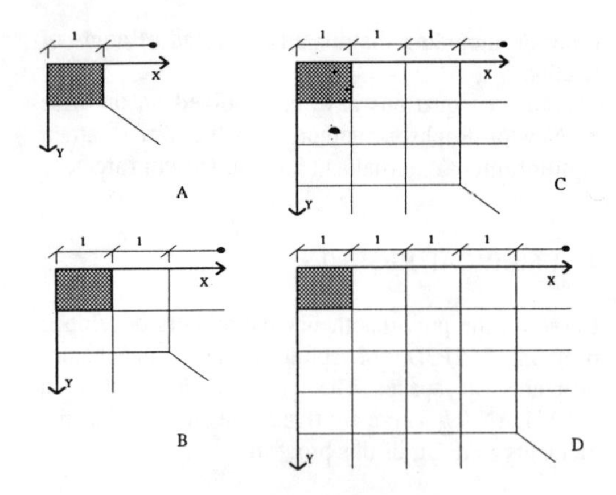

Fig..4 Finite/infinite 3D-model
A) 1 finite and 3 infinite elements, B) 8 finite and 12 infinite elements, C) 27 finite and 27 infinite elements, D) 64 finite a 48 infinite elements

Tab.2

Mo-del	x [m]	w_{anal} [m] 10^{-5}	w_{num} [m] 10^{-5}
A	0	0.52595	0.44703
	0.5	-	0.44265
	1	-	0.27399
B	0	0.52595	0.508414
	0.5	-	0.48929
	1	-	0.34862
	1.5	0.21176	0.19899
	2	0.15488	0.14789
C	0	0.52595	0.50950
	0.5	-	0.49005
	1	-	0.35039
	1.5	0.21176	0.20181
	2	0.15488	0.15003
	2.5	0.12336	0.11401
	3	0.10124	0.09303
D	0	0.52595	0.49220
	0.5	-	0.47275
	1	-	0.33326
	1.5	0.21176	0.18503
	2	0.15488	0.13375
	2.5	0.12336	0.09808
	3	0.10124	0.07675
	3.5	0.08638	0.06193
	4	0.07536	0.05124

Maximum error of solution in the displacements is 2,68%. In the case C we obtained the best results.

Example 2

In this example we presented the behaviour of body soil loading by continuous load $p = 0{,}25 kNm^{-2}$. The parameters of this model were used as follows : $a = b = 2m$, $E = 10^5 kPa$, $\mu = 0{,}25$. The pole of infinite elements was $x = 5$. We used the analytical results by [3] .

The results in Tab.2 show the accuracy of solution these problems by finite/infinite element method.

5 DISCUSSION AND CONCLUSION

This is particularly important in elasto-plastic 3D-problems where the nonlinear solution involved is often extremaly costly.

Since the formulation of mapped infinite elements based on standard finite element procedure is very effective.

REFERENCES

Desai,C.S.: Constitutive Modeling of Geotechnical Materials. Tucson Arizona, 1984.

Marquest,J.M.C., Owen.D.R.J.: Infinite elements in quasi-static materially nonlinear problems. Computers & Structures, Vol.18, No.4, pp. 739-751, 1977.,

Králik,J. & others: Interaction problems of soil and beam, plate structures respectively by elasto-plastic and elasto-viscoplastic material properties. VU A/5/14/91/III, Research work , SvF STU Bratislava, 1992.,

Nowacki,W.: Teorija uprugosti, Moskva, 1977.,

Osvald,J.: Riešenie interakcie konštrukcií s vrstevnatým podložím špeciálnymi postupmi MKP. Kand. dizert. práca, SAV, Bratislava, 1990.

Owen,D.R.J. & Hinton,E.: Finite Elements in Plasticity, Pineridge Press, Swansea, 1980.

Geomechanics 93, Rakowski (ed.) © 1994 Balkema, Rotterdam, ISBN 90 5410 354 X

Contact problem of reinforced concrete girder and non-linear winkler foundation

J. Králik & N. Jendželovský
Slovak Technical University, Bratislava, Slovakia

ABSTRACT: This paper deals with bending of reinforcement concrete girder rested on non-linear elastic foundation. The beam is discretized by layered isoparametric finite elements. The concrete material is modelled with stiffening in compression and softening in tension. The properties of the steel are exhibited by a bilinear stress-strain diagram. Soil structure interaction is idealised by using non-linear Winkler modules in the axial and transversal direction.

1 PREFACE

The non-linear solution of foun-dation' s construction is being applicated more a more at present time. The aim of such solution is to obtain exact image of their real impact.

In this paper we solve inte-raction of reinforced concret gir-der and non-linear Winkler foundation. We analyse influence of plastic deformations of concrete, its break, stiffening of rein-forcement, non-linear properties of foundation and one-way bond on load capacity of girder. Discretization of girder on finite element and di-vision the cross-section on layers enables us to describe these phenomena.

On the basisis of Lagrange' s principle of virtual work, we express the variation of potential energy of system girder - foundation in form

$$\delta\Pi = \int_{\Omega} \delta\boldsymbol{\varepsilon}^T \boldsymbol{\sigma}\, d\Omega - \int_{\Gamma} \delta\mathbf{w}^T \mathbf{p}\, d\Gamma - \int_{\Gamma_c} (\delta\mathbf{w}_n - \boldsymbol{\varphi})^T \mathbf{p}_n d\Gamma = 0 \tag{1}$$

where Ω (Γ) is volume (surface) of girder, Γ_c is contact surface of girder and founadation and φ is function of form of the foundation.

The vertical contact stress (on contact surface betwen girder and foundation) we are modelling by Winkler' s model in form

$$p_n = C\,w \tag{2}$$

2 FINITE ELEMENT MODEL

We are taking Thimosenko' s theory of bending beam int con-sideration which is more suitable for deformation of beams with massive cross-section, because it includes the work of shear deformation.

Vector of wertical displacement w, horizontal displacement u and rotation υ we approximate on the isoparametric element in form

$$\begin{aligned} w &= N_1.w_1 + N_2.w_2 \\ u &= N_1.u_1 + N_2.u_2 \\ \upsilon &= N_1.\upsilon_1 + N_2.\upsilon_2 \end{aligned} \tag{3}$$

where

$$N_1 = 1 - \xi,\ N_2 = \xi,\ \xi = (x - x_1)/L \tag{4}$$

We define the vector of deformation parameters on element

$$\{r\} = \{u_1, w_1, \upsilon_1, u_2, w_2, \upsilon_2\}^T \tag{5}$$

and

$$\begin{Bmatrix} u \\ w \\ \upsilon \end{Bmatrix} = N.r \quad , N = \begin{bmatrix} N_1 & 0 & 0 & N_2 & 0 & 0 \\ 0 & N_1 & 0 & 0 & N_2 & 0 \\ 0 & 0 & N_1 & 0 & 0 & N_2 \end{bmatrix} \tag{6}$$

We get the equilibrium equation in form

$$(\boldsymbol{K}_b + \boldsymbol{K}_c^*)\ \boldsymbol{r} = \boldsymbol{f} \tag{7}$$

where $\boldsymbol{K}_b$, is stiffeness matrix of beam in bend and shear, $\boldsymbol{K}_c^*$ is stiffeness matrix of founadation.

$$\boldsymbol{K}_b = \frac{EI}{l^3\kappa}\begin{bmatrix} k_{11} & k_{1i} & k_{16} \\ & k_{ii} & \\ sym & & k_{66} \end{bmatrix} \tag{8}$$

where

$$k_{11} = k_{44} = \frac{Al^2\kappa}{I},\ k_{14} = \frac{-Al^2\kappa}{I},\ k_{23} = k_{26} = -k_{35} = 3l^2,$$
$$k_{22} = -k_{25} = k_{55} = -k_{56} = 6,\ k_{33} = k_{66} = l^2(\kappa + 1{,}5),$$
$$k_{36} = l^2(1{,}5 - \kappa),$$
$$k_{12} = k_{13} = k_{15} = k_{16} = k_{24} = k_{34} = k_{45} = k_{46} = 0, \tag{9}$$

where κ is in form $\quad \kappa = \dfrac{6EI}{G\hat{A}l^2}$

Matrix $\boldsymbol{K}_c^*$ depends of magnitude of contact surface. We will deal with its form later in the article.

3 PHYSICAL PROPERTIES OF MATERIAL

Reinforced concrete is non-homogeneous and anizotropic material. Its properties we model as the properties of composition material, that compounds of concrete and steel.

Behaviour of concrete and steel we consider by stress - strain diagram by Eurocod. For calcu-lation of deformations and forces we are taking norm value of physical quantity.

Concrete is characterized by values $R_{bt}, R_b, \varepsilon_{b1}, \varepsilon_{bu}, \varepsilon_{bt}$ a E_{bn}.

In pressure territory we are taking parabolical course of σ - ε linear per parts (Fig.1).

By Eurocod we are taking

$\varepsilon_{b1} = -0{,}0022, \qquad \varepsilon_{bu} = -0{,}0035, \qquad \varepsilon_{bt} = 0{,}004,$

$\varepsilon_{bt_0} = 0{,}0002.$

If $\varepsilon_b > \varepsilon_{bt_0}$, the stifening of concrete arises in tension as late as crack originates. e.g. $\varepsilon_b = \varepsilon_{bt}$ and $\sigma_b = 0$.

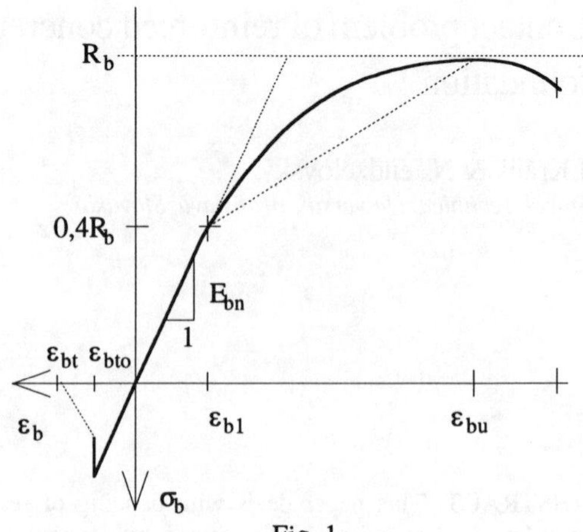

Fig. 1

Steel is characterized by value $R_s, R_{sc}, E, H_s', \varepsilon_{su}, \varepsilon_{scu}$ (Fig.2). By Eurocod we are taking $\varepsilon_{su} = 0{,}01$ a $\varepsilon_{scu} = -0{,}0025$ in case when the steel is not hardened $H_s' = 0$.

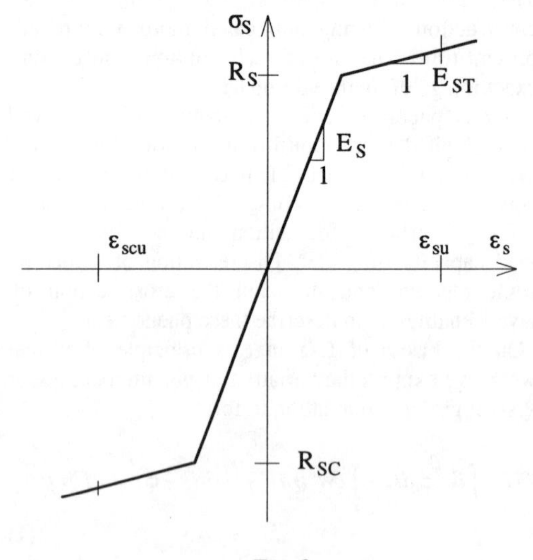

Fig. 2

In process of origin and development of micro fractures in concrete in tension load is shear stiffness of cross-section de-creasing. This phenomena explains reduced shear module by Cervenka

$$\overline{G} = \beta.G, \qquad \beta = 1 - \left(\frac{\varepsilon}{\varepsilon_{bt}}\right)^{\alpha} \tag{10}$$

when by with $\varepsilon_{bt} \cong 0{,}004 \div 0{,}005$ the crack failure of concrete ari-ses($\alpha = 0{,}3 \div 1{,}0$).

We solve the bending stiffens for elastic-plastic cross-section of beam from assumption of layered cross-section. Cross-section of beam has finite number of layers while each layer might have different physical properties. Each layer has constant course of stress. We get stiffness of cross-section in form

$$EJ = \sum_l E_l b_l z_l^2 h_l \quad a \quad \overline{G}\hat{A} = \sum_l k_s \overline{G}_l b_l h_l \qquad (11)$$

Value of bending moment and shear force we get by summation at the cross-section in form

$$M = \sum_l b_l \sigma_{x_l}(\varepsilon) z_l h_l \ , \quad Q = \sum_l b_l \tau_{xz_l} h_l \qquad (12)$$

4 NON-LINEAR MODEL OF FOUNDATION

Winkler' s model for foundation models is applied which is charac-terized by coefficient of foun-dation $C_w [kNm^{-3}]$, or by rate of spring $k_z = C_w.b\ [kNm^{-2}]$ (b - is width of beam on contact surface). Non-linear properties of foundation we consider by Yankelovsky, on Fig. 3 shows three-linear polynom.

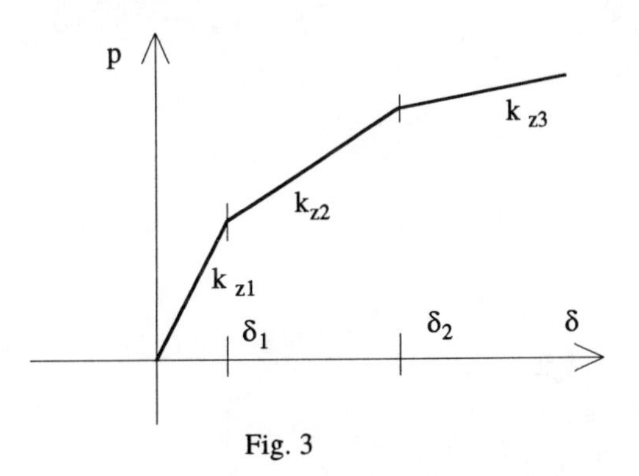

Fig. 3

For δ < 0 s contact stress p = 0, for δ > 0 is p > 0. Top points of polygon path δ_1 and δ_2 are limit value of vertical displacement w.

Stiffness matrix of foundation under beam element of length $L = x_2 - x_1$ we obtain from variation work of contact stress on foundation in form

$$\delta \Pi_C = \delta \Delta \mathbf{r}^{\mathbf{T}} \left(\int_{x_1}^{x_2} N^T N k_z dx \right) \Delta \mathbf{r} \qquad (13)$$

Expression in parentheses is stiffness matrix of foundation. If we have a beam with rate of spring k_{z1} on length $\langle x_1, x_c \rangle$ and with rate of spring k_{z2} on $\langle x_c, x_2 \rangle$ we obtain stiffness matrix by integration per partes in form

$$[K^*_{Cx}] = \int_{x_1}^{x_c} C_{x1} \begin{bmatrix} N_1^2 & N_1 N_2 \\ N_1 N_2 & N_2^2 \end{bmatrix} dx + \int_{x_c}^{x_2} C_{x2} \begin{bmatrix} N_1^2 & N_1 N_2 \\ N_1 N_2 & N_2^2 \end{bmatrix} dx \qquad (14)$$

After integration we obtain stiff-ness matrix $\boldsymbol{K}^*_C$.

5 NON-LINEAR EQUATIONS

By the solution of non-linear equations (7) the Newton-Raphson method is applied, with the aim to attain equilibrium of external and internal system forces. Vector of pseudo forces $\{\psi(r_{i+1})\}$ for (i+1)-step we formulate trough first two members of Taylor series in form

$$\{\psi(r_{i+1})\} = \{\psi(r_i + \Delta r)\} \cong$$

$$\cong \{\psi(r_i)\} + \frac{\partial\{\psi(r_i)\}}{\partial\{r\}}(\{r_{i+1}\} - \{r_i\}) = 0 \qquad (15)$$

after separation we get

$$\{r_{i+1}\} = \{r_i\} - \left[\frac{\partial \psi(r_i)}{\partial\{r\}} \right]^{-1} \{\psi(r_i)\} \qquad (16)$$

and

$$\{r_{i+1}\} = \{r_i\} - \left[\boldsymbol{K}_{ot} + \boldsymbol{K}_{st} + \boldsymbol{K}^*_{Ct} \right]^{-1} \{\psi(r_i)\} \qquad (17)$$

where subscript t indicates tan-gential stiffness matrix in i step. Stiffness matrix can by constant or variable in process of iteration. When norm of nonequlibrium nodal force is less then norm of nodal forces, iteration process finishes.

$$100(\sum_i \Psi_i^2) / \sum_i f_i^2 \le toler(\%), \qquad (18)$$

(i=1,n - number of nodal)
where toler is admissible error of solution in %.

6 EXAMPLE

We solution three - span continuous beam on

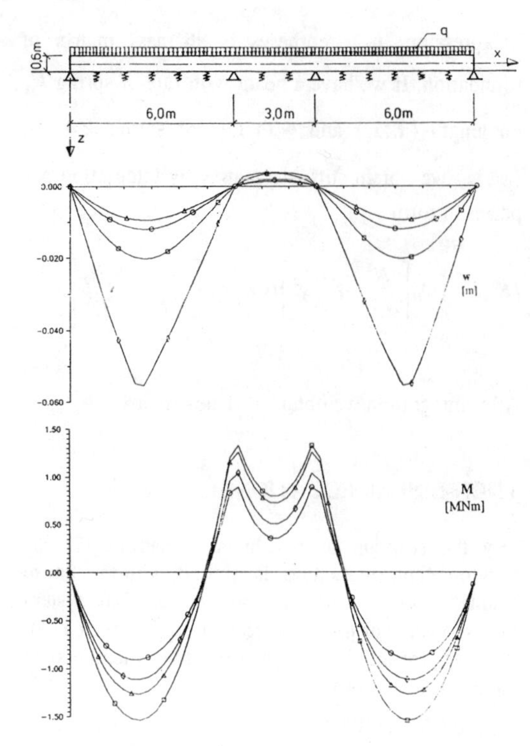

Fig. 4. Elastoplastic beam on non-linear foundation, w - vertical displacement, M - bending moment, □ e/ne, ✧ ep/ne, ○ e/e, Δ ep/e

foundation Fig. 4. The loading is aplplied continuous uniform load q.

We solution four case: - elastic beam and elastic foundation (e/e), - elastoplastic beam and elastic foundation (ep/e), - elastic beam and non-linear elastic fondation (e/ne), - elastoplastic beam and non-linaer elastic foundation (ep/ne).

Cross-section of beam is 0,6/0,6 m, it has 22 layers.

Concrete: E=23000MPa, G = 9583 Mpa, $R_{bt} = 0{,}75 MPa$, $H_{T1} = 0$, $R_b = 8{,}5$, $H_{C1} = 0$, $\rho = 2300\ kgm^{-3}$.

Steel E = 210000 MPa, $R_s = 190 MPa$, $H_{T1} = 0$, $R_{sc} = 190 MPa$, $H_{C1} = 0$. Foundation $C_1 = 50\ MNm^{-3}$, $C_2 = 5\ MNm^{-3}$, and $\delta^I = 2$ mm

Limit of elasticity is for q=0,077 MN/m and limit of plasticity is for q=3,89 MN/m.

REFERENCES

Bazant P.,Z., and the others : Finite Element Analysis of Reinforced Concrete, ASCE, New York, 1982.

Jendželovský,N.,Králik,J.: Design of contact of elastic-plastic reinforced concrete beam on non-linear Winkler' s subgrade. Inžinierske stavby 41, 2-3, 1993, 53-60, (Slovak)

Owen D.R.J.,Hinton E.: Finite Elements in Plasticity. Theory and Practice. Swansea, U.K.,1980.

Papadopulos P.G.: Nonlinear Static Analysis of Reinforced Concrete Frames by Network Models., Adv. Eng.Software, 1988, vol10, No 3.

Yankelovsky,D.Z.,Adin,M.A.,Eisenberger,M.: Analysis of Beams on Nonlinear Winkler Foundation. Computer and Structures 1989, Vol 31, No 21, pp 287-292.

Eurocod No.2, Design of oncrete Structures, Part 1, 1990, Bruxelles.

Geomechanics 93, Rakowski (ed.) © 1994 Balkema, Rotterdam, ISBN 90 5410 354 X

Shear asymmetries of layered rocks

Z. Sobotka
Mathematical Institute of the CAS, Prague, Czech Republic

ABSTRACT: The paper deals with two kinds of shear asymmetry of layered rocks. The first kind is characterized by different mechanical behaviour and defined by different shear moduli and shear resistance of rocks when acted on by shear stresses in two opposite directions. It is shown that the rocks exhibit the maximum resistance to the pure shear when the principal compression is perpendicular to the layers and the principal tension is parallel to them. The second kind of shear asymmetry occurs at the boundaries of individual layers and is manifested by different conjugate perpendicular shear strains and stresses. This asymmetry is to some measure analogous to Cosserat phenomena. It is also shown how the orientation of layers with respect to the slope exerts the influence upon the stability.

1 POSITIVE AND NEGATIVE SHEAR MODULI

The layered rocks exhibit different mechanical properties under the opposite shear stresses which Sobotka (1984) has called the positive and negative shear. The positive shear accompanied by an increase in volume is characterized by higher shear moduli and strength than the negative shear causing the decrease in volume.

The fundamentals for the definitions of moduli of the positive and negative shear are represented by the relations between the normal and shear strains in the state of of pure shear. In order to obtain these relations, let us consider the pure shear of a prismatic element with the square cross-section consisting of horizontal layers as shown in Fig. 1. This element is bounded by the planes whichare parallel to the z-axis and forming the angles 45° with the x and y-axes. In the case of positive pure shear with higher resistance to deformation and failure, the horizontal layers are pressed against each other and the element is acted on by the stresses $\sigma_x = -\sigma_y$ while the stress σ_z in the direction of perpendicular z-axis is equal to zero (Fig. 1a).

Summing up the forces along and perpendicular to the sides of the cross-section, we can see that the normal stresses on these sides are zero and the shear stress is

$$\tau = \frac{1}{2}(\sigma_x - \sigma_y) = \sigma . \tag{1}$$

In the course of deformation, the right angles of the initially square cross-section change for $\pi/2 - \gamma$ or $\pi/2 + \gamma$, respectively. From the geometrical relations in Fig. 1a representing the positive pure shear, we obtain:

$$\tan\left(\frac{\pi}{4} + \frac{\gamma}{2}\right) = \frac{1 - \varepsilon_y}{1 + \varepsilon_x}, \quad \tan\left(\frac{\pi}{4} + \frac{\gamma}{2}\right) = \frac{1 + \varepsilon_x}{1 - \varepsilon_y} . \tag{2}$$

The foregoing relations yield:

$$\tan\frac{\gamma}{2} = \frac{\varepsilon_x + \varepsilon_y}{1 + \frac{1}{2}(\varepsilon_x - \varepsilon_y)} . \tag{3}$$

This equation represents the exact relation between the normal and shear strains; limiting our considerations to the first-order effects, we can for small deformations approximately write:

$$\gamma = \varepsilon_x + \varepsilon_y . \tag{4}$$

The modulus of positive shear is then given by

$$G_P = \frac{\tau_P}{\gamma_P} = \frac{\tau_P}{\varepsilon_{xP} + \varepsilon_{yP}} . \tag{5}$$

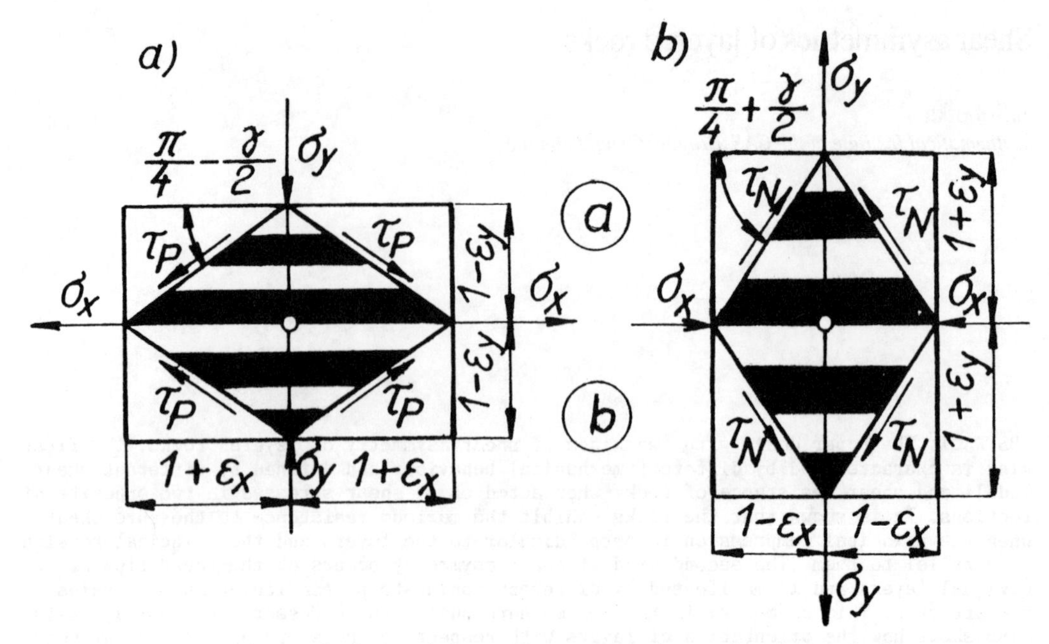

Fig. 1 The scheme of the positive and negative shear of a layered element

In view of Eq.(1), the Hooke law yields for the positive pure shear the following strains

$$\varepsilon_x = \tau_P\left(\frac{1}{E_{Tx}} + \frac{\mu_{Tx}}{E_{Cy}}\right), \tag{6}$$

$$\varepsilon_y \quad \tau_P\left(\frac{1}{E_{Cy}} + \frac{\mu_{Cy}}{E_{Tx}}\right), \tag{7}$$

where E_{Tx} is the modulus of elasticity in tension in the direction of the x- axis, E_{Cy} is that in compression in the direction of the y-axis and μ_{Tx} and μ_{Cy} are the corresponding Poisson ratios.

Substituting Eqs. (6) and (7) into Eq. (5), we obtain the resultant expression for the modulus of positive shear:

$$G_P = \frac{E_{Tx}E_{Cy}}{E_{Tx}(1 + \mu_{Tx}) + E_{Cy}(1 + \mu_{Cy})}, \tag{8}$$

On the other hand, in the case shown in Fig. 1b. we get in an analogous manner the modulus of negative shear:

$$G_N = \frac{E_{Cx}E_{Ty}}{E_{Cx}(1 + \mu_{Cx}) + E_{Ty}(1 + \mu_{Ty})}, \tag{9}$$

Since the resistance to the compressive deformation in the direction perpendicular to the layers and that of layers to tensile deformation attains relatively considerable values, the moduli E_{Tx} and E_{Cy} are much higher than E_{Cx} and E_{Ty} so that the modulus G_P prevails over G_N. The expressions (8) and (9) then define the maximum and minimum value of shear modulus.

2 LIMITING CRITERIA

The limiting criteria for the first-order equilibrium of layered rocks can be formuin the following general form

$$P_{kl}\sigma_{kl} + P_{klmn}\sigma_{kl}\sigma_{mn} = 1. \tag{10}$$

Writing this relation for the state of simple tension, compresion, and uniform biaxial compression, successively, we obtain for the two-dimensional state of layered rocks the special limiting criterion:

$$\frac{\sigma_p^2}{\sigma_{Cp}\sigma_{Tp}} + \frac{\sigma_r^2}{\sigma_{Cr}\sigma_{Tr}} - \left(\frac{1}{\sigma_{Cp}\sigma_{Tp}} + \frac{1}{\sigma_{Cr}\sigma_{Tr}} - \frac{1}{\sigma_{CCpr}} - \frac{\sigma_{Cp} - \sigma_{Tp}}{\sigma_{Cp}\sigma_{Tp}\sigma_{CCpr}} - \frac{\sigma_{Cr} - \sigma_{Tr}}{\sigma_{Cr}\sigma_{Tr}\sigma_{CCpr}}\right)\sigma_p\sigma_r - \frac{\sigma_{Cp} - \sigma_{Tp}}{\sigma_{Cp}\sigma_{Tp}}\sigma_p - \frac{\sigma_{Cr} - \sigma_{Tr}}{\sigma_{Cr}\sigma_{Tr}}\sigma_r = 1, \tag{11}$$

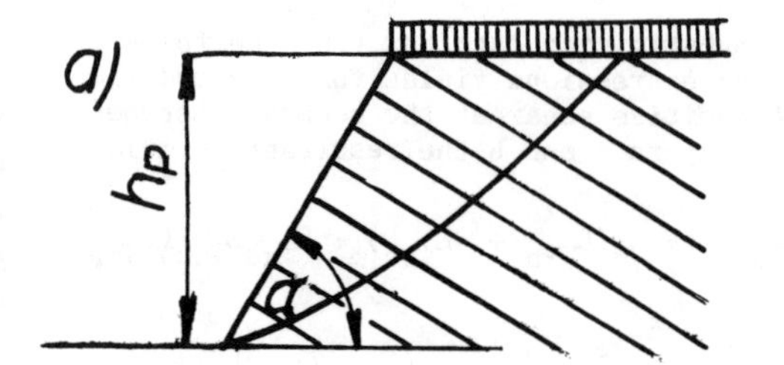

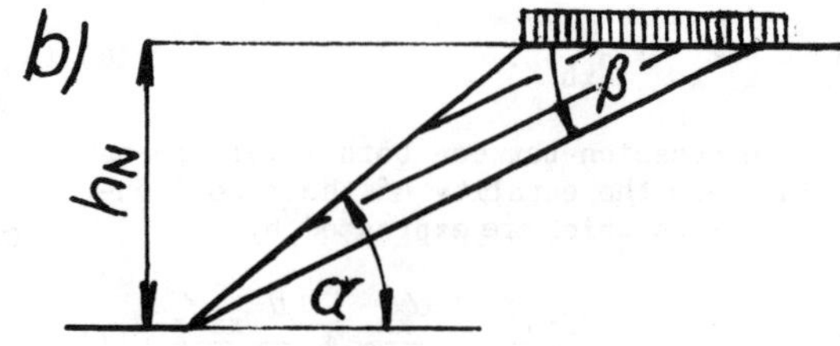

Fig. 2 Two kinds of slip in slope regions of layered rocks

where the stress σ_p which is perpendicular to the layers and σ_r acting in the direction of layers represent the principal stresses. The limiting stresses in simple compression and tension are respectively designated by the indices C and T and σ_{CCpr} stands for the limiting uniform biaxial compression in the principal directions p and r of the orientated layered structure.

Substituting into Eq. (11) the positive shear stress $\tau_P = \sigma_r = -\sigma_p$ and the negative shear stress $\tau_N = \sigma_p = -\sigma_r$, we get the quadratic equation which can be presented in the following abbreviated form

$$A\tau^2_{P,N} - B\tau_{P,N} = 1, \tag{12}$$

where A represents the sum of coefficients of the second degree in stresses and B of those at linear terms. The above equation has two roots:

$$\tau_{P,N} = \frac{B}{2A} \pm \sqrt{\frac{B^2}{4A^2} + \frac{1}{A}}, \tag{13}$$

which define the positive and negative limiting shear stresses.

3 TWO KINDS OF SLOPE STABILITY

According to the orientation of layers with respect to the slope surfaces, there exist two main kinds of principal slip lines in the slope regions as shown in Fig. 2. In the case represented by Fig. 2a, the principal slip line runs across the layers and the full shear resistance τ_0 with the angle of internal friction φ_0 of the rock material is in action against the slip. The approximate limiting height h_P of such a slope with the angle of inclination has been defined by Sobotka (1956) by the formula:

$$h_P = \frac{2\tau_0 \tan(\frac{\pi}{4} + \frac{\varphi_0}{2})}{\varrho(1 - 2\tan\varphi_0 \cot\alpha)} - \frac{p}{\varrho}, \tag{14}$$

where is the unit weight of the rock and p is the uniform load of the horizontal surface above the slope.

In the second case shown in Fig. 2a, the slip follows the interstitial plane between the layers with the angle of inclination β. From the limit equilibrium of the slope region, another limiting height can be obtained:

$$h_N = \frac{\tau_i \cos\varphi_i}{\varrho \cos\beta \sin(\beta - \varphi_i)(1 - \sqrt{\cot\alpha \tan\beta})} - \frac{p}{\varrho}, \tag{15}$$

where τ_i and φ_i are respectively the shear cohesion and angle of friction between the layers.

From the foregoing relations, it can be seen how the shear stability depends on the angle between the rock layers and slope surface. These stability phenomena correspond in macroscale to the positive and negative shear effects.

4 TWO-DIMENSIONAL SHEAR ASYMMETRY

The two-dimensional shear asymmetry manifested by different perpendicular conjugate shear strains and stresses arises at the boundaries of two layers with different mechanical properties. For deriving the corresponding shear moduli, the starting configuration can be represented by the schemes in Fig. 1 with different properties of the upper part a and lower part b of the element which is again acted on by the stresses $\sigma_x = -\sigma_y$ at the positive pure shear. However, the stress σ_x is distributed into the components σ_{xa} and σ_{xb} acting on the parts a and b according to the relation:

$$o_x = \frac{1}{2}(\sigma_{xa} + \sigma_{xb}) \ . \tag{16}$$

The cohesion between both layers a and b involves the equality of their horizontal strains which are expressed by

$$\varepsilon_{xa} = \frac{\sigma_{xa}}{E_{Txa}} - \frac{\mu_{Txa}\sigma_y}{E_{Cya}} = \frac{\sigma_{xa}}{E_{Txa}} + \frac{\mu_{Txa}\sigma_x}{E_{Cya}}, \tag{17}$$

$$\varepsilon_{xb} = \frac{\sigma_{xb}}{E_{Txb}} - \frac{\mu_{Txb}}{E_{Cyb}} = \frac{\sigma_{xb}}{E_{Txb}} + \frac{\mu_{Txb}\sigma_x}{E_{Cyb}}. \tag{18}$$

From the equality of these strains, we obtain in view of Eq. (16) the epressions for the horizontal stress components:

$$\sigma_{xa} = \frac{2E_{Txa} + E_{Txa}E_{Txb}\left(\frac{\mu_{Txa}}{E_{Cya}} - \frac{\mu_{Txb}}{E_{Cyb}}\right)}{E_{Txa} + E_{Txb}}\sigma_x, \tag{19}$$

$$\sigma_{xb} = \frac{2E_{Txb} - E_{Txa}E_{Txb}\left(\frac{\mu_{Txa}}{E_{Cya}} - \frac{\mu_{Txb}}{E_{Cyb}}\right)}{E_{Txa} + E_{Txb}}\sigma_x. \tag{20}$$

The unit vertical shortenings of both layers are different:

$$\varepsilon_{ya} = \frac{\sigma_y}{E_{Cya}} + \frac{\mu_{Cya}\sigma_{xa}}{E_{Txa}} \quad , \tag{21}$$

$$\varepsilon_{yb} = \frac{\sigma_y}{E_{Cyb}} + \frac{\mu_{Cyb}\sigma_{xb}}{E_{Txb}} \quad . \tag{22}$$

Summing up the forces along and perpendicular to the sides of the cross-section of the prismatic element with unit length as shown in Fig. 1a, we can see that for different mechanical properties of the phases a and b the normal and shear stresses are given by

$$\sigma_a = \frac{1}{2}(\sigma_y - \sigma_{xa}), \ \sigma_b = \frac{1}{2}(\sigma_{xb} - \sigma_y), \tag{23}$$

$$\tau_{Pa} = \frac{1}{2}(\sigma_{xa} + \sigma_y), \ \tau_{Pb} = \frac{1}{2}(\sigma_{xb} + \sigma_y). \tag{24}$$

According to Eq. (4), we obtain two different conjugate shear strains:

$$\gamma_{Pa} = \varepsilon_x + \varepsilon_{ya}, \ \gamma_{Pb} = \varepsilon_x + \varepsilon_{yb} \ . \tag{25}$$

The layers a and b then exhibit two different moduli of positive shear:

$$G_{Pa} = \frac{\tau_{Pa}}{\gamma_{Pa}}, \qquad G_{Pb} = \frac{\tau_{Pb}}{\gamma_{Pb}} \ . \tag{26}$$

Substituting into Eq. (26) the foregoing expressions yields for both moduli of positive shear at the boundary between the layers a and b the resultant formulae:

$$G_{Pa} = \left[E_{Cya}(E_{Txa} + 3E_{Txb}) + E_{Txa}E_{Txb}\left(\mu_{Txa} - \mu_{Txb}\frac{E_{Cya}}{E_{Cyb}}\right)\right] / 2\left\{E_{Cya}(1 + \mu_{Cya})\left[2 + E_{Txa}\left(\frac{\mu_{Txa}}{E_{Cya}} - \frac{\mu_{Txb}}{E_{Cyb}}\right)\right] + (E_{Txa} + E_{Txb})(1 + \mu_{Txa})\right\} , \tag{27}$$

$$G_{Pb} = \left[E_{Cyb}(E_{Txa} + 3E_{Txb}) + E_{Txa}E_{Txb}\left(\mu_{Txb} - Txa\frac{E_{Cyb}}{E_{Cya}}\right)\right] / 2\left\{E_{Cyb}(1 + \mu_{Cyb})\left[2 - E_{Txb}\left(\frac{\mu_{Txa}}{E_{Cya}} - \frac{\mu_{Txb}}{E_{Cyb}}\right)\right] + (E_{Txa} + E_{Txb})(1 + \mu_{Txb})\right\} . \tag{28}$$

The moduli of negative shear can be obtained in an analogous manner.

5 CONCLUSIONS

Because of the orientated structure, the layered rocks exhibit different shear moduli and shear resistance when acted on by the shear stresses in two opposite directions. This phenomena correspond in macroscale to different stability of slopes when the layers are inclined against or down to slope surface, respectively. At the boundary between two layers, there exist different conjugate shear strains and stresses and different shear moduli.

REFERENCES

Sobotka, Z. 1956. Mezní stavy rovnováhy zemin (Limiting states of equilibrium of soils, in Czech). Praha: SNTL.

Sobotka, Z. 1984. Rheology of materials and engineering structures. Amsterdam, Oxford, New York, Tokyo: Elsevier.

Sobotka, Z. 1988. Shear asymmetries of structured viscoelastic bodies.. Xth International Congress on Rheology, Vol. 2: 285-287.Sydney: Australian Society of Rheology.

Geomechanics 93, Rakowski (ed.) © 1994 Balkema, Rotterdam, ISBN 90 5410 354 X

Dynamic of steel arch support – Nonlinear solution

J. Zapomĕl & P. Horyl
University of Mining and Metallurgy, Ostrava, Czech Republic

R. Šňupárek
Research Mining Institute, Ostrava, Czech Republic

INTRODUCTION

Steel arch yielding support, the most widespread support of coal mine galleries, has been designed and calculated mainly for the static loading.

In real mine conditions the arch support is ifteb danaged by dynamic loading first of all in burst-prone areas.

Methods of design and calculation of the steel arch yielding support under conditions of the static loading are well known and its results can be compared with results of experimental methods.

On the contraty the behaviour of the steel arch support under conditions of dynamic loading and its dynamic loading capacity have been much less searched.

As for as any experimental research of support under dynamic loading is extremaly expensive and proper experimental equipments are not accessible, there is only one available method - the matematic modelling.

Even from the point of view of this method the task is very complicated. Arch supports content yielding clamp joints, which cause nonlinear characteristics of strain-stress and deformation behaviour of the construction.

Due to the long-time systematic collaboration of the University of Mining Metallurgy and Departement of geomechanics, Research Mining Institute in Ostrava Radvanice original and very interesting results of the arch support behaviour under dynamic loading were obtained.

FORMULATION OF THE TECHNICAL PROBLEM

A steel arch consists of several segments coupled with special clamping joints. Exceeding of a certain limit load results in their loosening and relative slip of connected parts. The engineering problem consists of the investigation of arch behaviour during shock bumps first of all with respect to the influence of the friction connections and their possible loosening.

SETTING UP OF THE COMPUTER MODEL

For the solution of the engineering problem a computer modelling method was chosen.

The proposed model is discrete. To represent arch segments and the prop bar linear bodies are used. For the purpose of the analysis they are discretized by finite elements. Shock load is modelled by concentrated forces acting in nodes and varying in time. Beddings of the construction footings in the floor are represented by rotational kinematic pairs and the interaction between the arch and the rock by the spring assembly with stiffness and damping properties in one direction.

The clamping joints are modelled by couples of coincident nodes (each belongs to the different model segment). If the limit value of the normal force acting between them is not reached, both move together. Its exceeding leads to the constraint loosening and to the relative slip of the nodes in direction tangential to the center line.

Two kinds of damping are incorporated into the model: viscous considered as a proportional one and hysteretic having its origin in friction forces in clamping joints.

EQUATION OF MOTION OF THE MODEL SYSTEM

Motion of the model system is described by the matrix equation:

$$M \ddot{X} + B \dot{X} + K X = f_p + f_V + f_R$$

M - mass matrix

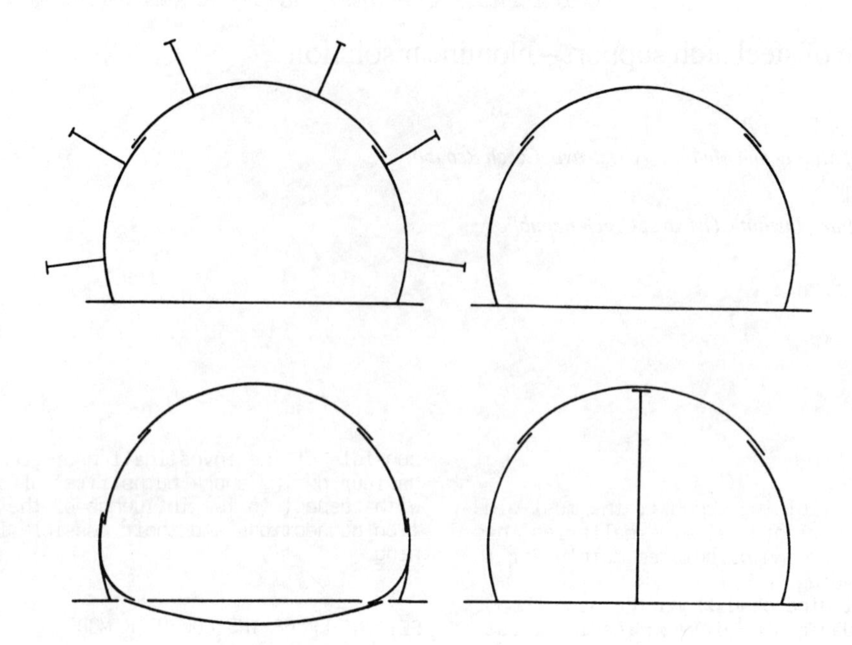

Fig. 1 Schemes of the investigated steel arches

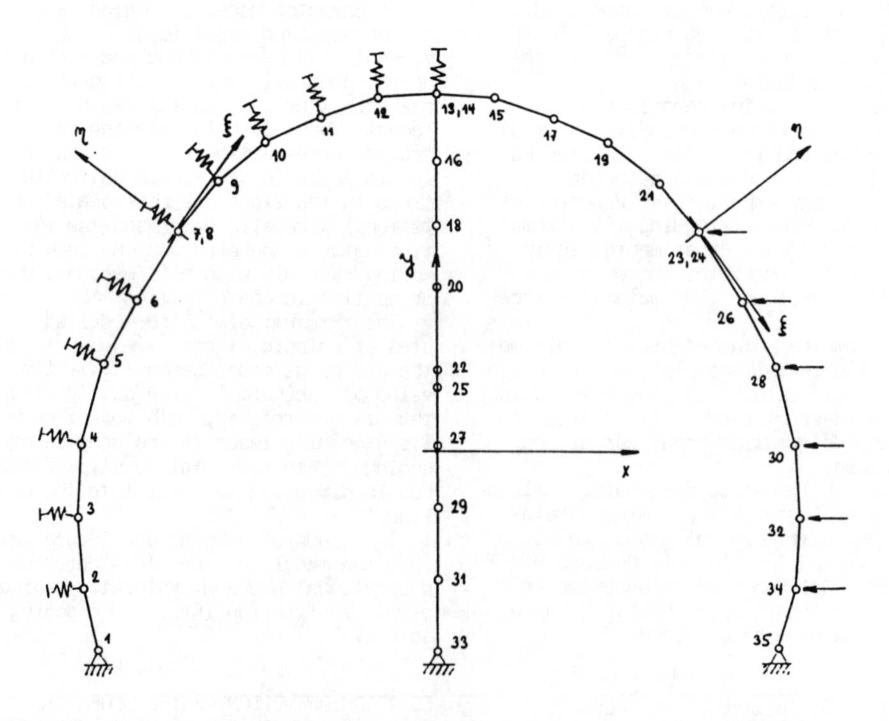

Fig. 2 Scheme of the model system (00 - 0 - 12 with a prop)

B - viscous damping matrix
K - stiffness matrix
X - generalized displacement vector
$\dot{X}$ - generalized velocity vector
$\ddot{X}$ - generalized acceleration vector
f_p - primary generalized forces vector
f_v^p - vector of internal generalized constraint forces acting in friction connections
f_R - vector of all other generalized constraint forces

The equation of motion is non-linear because the magnitude of f_v vector elements are limited from above by the limit value of the friction force.

In every moment the mechanical system can be divided into several parts that behave as flexible bodies which can move one to another. The entire process can be considered as a sequence of initial problems. During the time interval between two subsequent changes ot the system movability (slip in passive friction constraints can occur) its stiffness and damping characteristics can be considered linear and the solution may be carried out in parts always for new initial conditions.

With respect to the non-linear character of the equation of motion the direct Newmark integration method was chosen for its solution.

If a passive constraint with dry friction is released, the unknown quantities are kinematic parameters of both coincident nodes and the magnitude of the internal force acting between them is known (limit value of the friction force). In case the constraint is not released kinematic parameters of both nodes are identical and the magnitude of the internal force becomes unknown.

As for this particular problem the numerical values of both quantities differ up several orders the modification of general algorithm of Newmark method that was carried out. The aim is to eliminate unknown values of internal force and to avoid the contingent instability of the solution or inaccuracy of results.

EXAMPLES AND RESULTS EVALUATION

The described algorithm was worked out to the stage of software facilities (computer packages D1OD1, D1NA1, postprocessing programs) which were used for the investigation of several types of steel arches (fig.1). Some of the results are plotted in figs.2 through 4.

The scheme of the model system related to arch 00-0-12 with a prop can be seen in fig.2. Two couples of coincident nodes (7-8, 23-24) are adjoined to the friction connections. The behaviour of the clamping joint on the gob side is evident from fig. 3. Overcoming a certain limit state results in joint loosening and total slip (about 2,5 cm) occurs. In fig.4 energy ba-

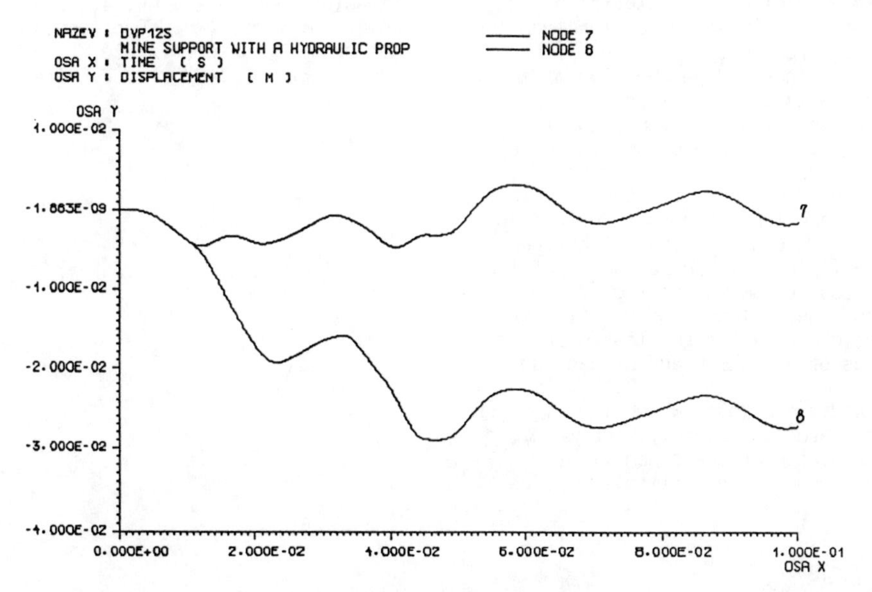

Fig. 3 Time-history of displacements of nodes 7,8 in the direction tangential to the centre line

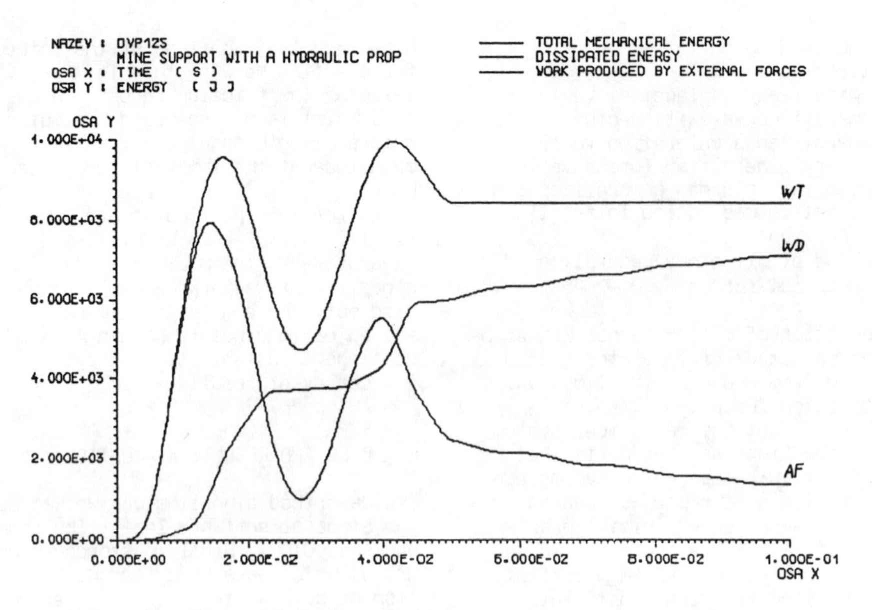

Fig. 4 Time-history of the process energy balance

lance of the whole process is plotted.From figs. 3 and 4 it is clear: in addition to viscous damping more energy is dissipated because of the dry friction after releasing of the clamping joint.

Results of the computer modelling of the steel arch behaviour show:

- clamping joints do not reduce the maximum stress value under the yield point
- the yield point of the material of the construction is reached earlier than the joints are released
- joints are loosened on the side of the arch opposite to shock load.

No substantial stress reduction in the arch with friction connections can have several causes. Stress is influenced especially by bending moment. Clamping joints limit the magnitude of normal force transmitted by the construction but not bending moment. In addition releasing of the joints leads to the increasing of the deformation. The system is given a larger amount of energy which results in larger amplitudes of vibration and stress magnitudes.

The conditions under which the steel arches occur are considerably uncertain. For better knowledge of their behaviour more additional computer experiments will be needed.

REFERENCES

Bittnar, Z., Řeřicha, P.: Metoda konečných prvků v dynamice konstrukcí, SNTL, Praha 1981.

Horyl, P., Šňupárek, R.: Parametrical study of roadway support under dynamic loading due to rock bursts.Proceedings of the International symposium on Rock support. Sudbury, Canada 1992.

Vinogradov, O.G.: Dynamics of Frictionally Coustrained Finite Rod, Computers and Structures, vol. 29, no. 4, 1989.

Zapoměl, J., Horyl, P., Ondrouch, J.: Výzkum dynamiky důlní výztuže, final research report HS 340/92, VŠB Ostrava 1992.

Water jet cutting
Main lectures

Geomechanics 93, Rakowski (ed.) © 1994 Balkema, Rotterdam, ISBN 90 5410 354 X

Water jet assisted coal and rock cutting

R.J. Fowell & J.A. Martin
Department of Mining and Mineral Engineering, The University of Leeds, UK

ABSTRACT: The application of water jet assisted (WJA) cutting of coal and rock is reviewed and laboratory instrumented cutting tests to investigate the applicability of WJA coal cutting to reduce the amount of fine coal produced are described.

1. INTRODUCTION

High pressure water jet assisted (WJA) cutting has been available on boom tunnelling machines as a commercial option for nearly a decade but has failed to find wide application for a number of reasons.

The principal perceived disadvantages of WJA cutting are: The water from the jets builds up on the floor of the tunnel and causes debris to slurry or weak floor rock to break up.

To obtain high pressures large quantities of installed power are required, which do not always compensate for the advantages gained.

Perhaps the most important reason for the few potential high pressure WJA applications in the United Kingdom is the requirement in collieries when cutting rocks containing quartz or pyrites for the track produced by each tool to be cooled by water sprays and not jets behind each tool to avoid a possible methane ignition.

The advantages of WJA cutting are now well documented. Lower tool component forces (Barham and Buchanan 1987); decreased vibration (Hurt et al, 1988); enhanced tool life (Morris and MacAndrew, 1986) and reduced dust make (Haslett et al., 1986).

Laboratory and field trials have proved the benefits but for normal mine production tunnelling WJA cutting has not found favour. Some of the bias against WJA is due to its misapplication in rocks that breakup on contact with water, the nozzles supplied easily block due to strainers not being fitted behind each nozzle, the stand off distance is too large with the jet incorrectly positioned relative to the cutting tool tip.

The principal use for water jet assistance in the authors view is for the excavation of the strong and abrasive rocks at the upper limit of drag tool application. Here the benefits of extended tool life due to cooling and the removal of debris away from the cutting area without it being forced under the cutting tool are of paramount importance.

In terms of coal excavation, the above benefits are also desirable but more interest lies in the potential of reducing the quantity of fine coal produced with coalface shearer loaders. Fine coal is defined as the minus 0.5mm size fraction: the size usually recovered from the run-of-mine by froth flotation.

The significance of fines in coal production is that 20%-30% of mine waste is estimated to consist of coal fines which are lost. Additionally, froth flotation as a washing technique costs approximately 2½ times more than the processing of larger sized coal (HQTD Bulletin, 1987). These two factors indicate that fines result in either a lost product or, where fines reclamation is attempted, an expensive recovery cost.

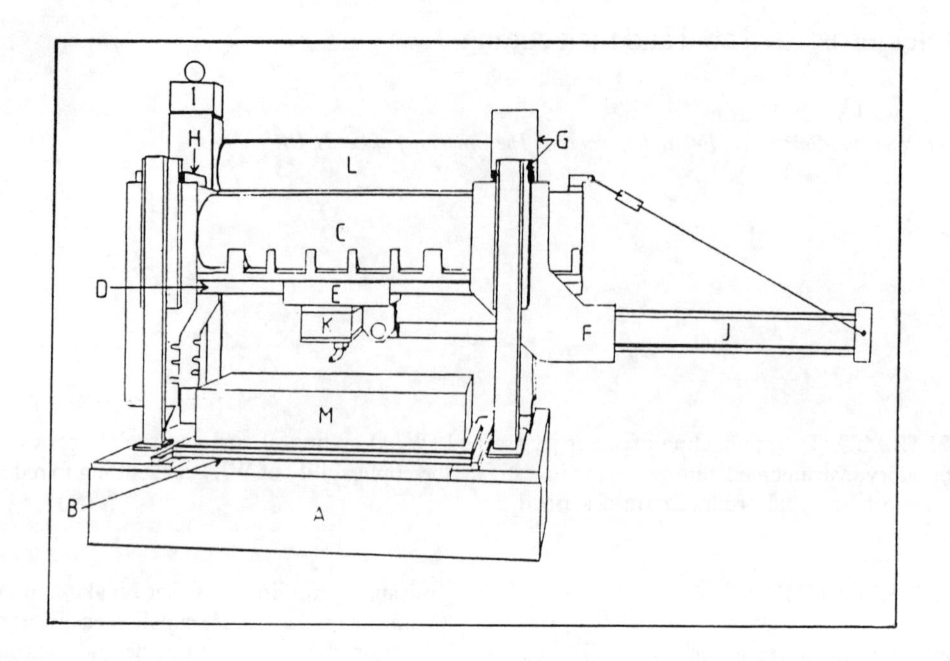

Key:

A.	Structure base.	H.	Vertical stiffeners (clamps).
B.	Specimen table.	I.	Clamp hydraulic power pack.
C.	Cylindrical strengthening beam of slide assembly.	J.	Hydraulic cylinder (pipework not shown).
D.	Slide bed.	K.	Dynamometer, toolholder and cutting tool.
E.	Cutter slider.	L.	Cylindrical strengthening beam for the main structure.
F.	Support frame for the hydraulic ram.	M.	Coal specimen.
G.	Guides for the whole slide assembly.		

Figure 1. 50 tonne Rock Planer

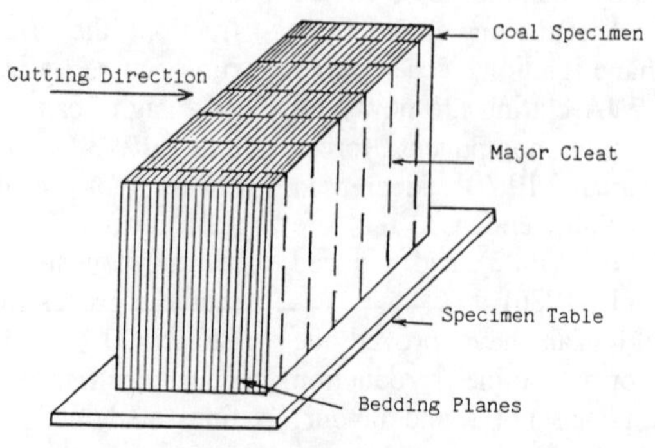

Figure 2. Orientation of major Cleat with respect to the cutting direction and coal bedding planes.

These points prompted the Rock Mechanics Branch, HQTD of the British Coal Corporation, to support a research programme into the application of high pressure water jet assistance to coal cutting.

2. RESEARCH APPARATUS AND METHOD

All the tests were performed on an instrumented 50 tonne capacity rock planing machine as illustrated in Fig.1, and described in detail by Martin (1991). The coal specimens were mounted on a specimen table which can traverse beneath the cutting tool and thereby allow the distance between succeeding cuts or line spacing to be set. The depth of cut was set by activating vertical screw jacks which raise and lower the cutting tool and associated equipment.

A hydraulic cylinder traverses the cutting tool up to a current maximum speed of 1.1m/s. The cutting tool and toolholder are rigidly attached to a four post strain gauged force dynamometer which resolves the cutting force into three orthogonal directions: parallel to the direction of cut (cutting force); perpendicular (normal force); and transverse (sideways force). A Presswell "Hydraflo" intensifier type water pump provided the high pressure water.

The jet pressure and nozzle diameter were not treated as independent factors but combined such that at the jet pressures of 35, 70 and 105 MPa, flow rates of 4, 8 and 12 l/min were obtained respectively. Therefore, a wide range of jet powers have been explored viz. 2.33kW, 9.33kW and 21kW respectively.

Wimet 75mm reach radial picks with HW tips, commonly utilized on shearer loaders, were used both in a sharp (pristine) and blunt state. The blunt tool was artificially worn to have a 10mm wide wearflat presenting a -2° clearance angle to the coal. All the coal blocks were cut through the bedding planes and parallel to the major cleat as illustrated in Fig.2.

To prevent the blocks from splitting during the course of testing, each block was cast and surrounded in pitch as a supporting medium. The other parameters were: 30mm depth of cut; 70mm line spacing; the jet was directed in front of the pick; with a 1-2mm lead distance; 1.1m/s cutting speed; and a relieved cutting pattern.

After each test had been conducted the coal debris was collected and sieved into the following size fractions (in mm):

0.5, 0.5-3.15, 3.15-6.3, 6.3-12.5, 12.5-28, 28-50, 50-100

3. COAL SPECIMENS

3.1 Bituminous Coal

This was supplied by the British Coal Corporation and obtained from the Erin Remainder Opencast site near Chesterfield, England. It is a dull dark grey colour suggesting a high durain content but also has quite frequent shiny, black bands. The coal was heavily bedded and cleated. It was friable, dusty to handle and contained a thin band of iron pyrites.

An average uniaxial compressive strength of 2.6MPa ± 3.9MPa was recorded for this coal from cubes (78mmx78mmx78mm) and its density is 1212kg/m3.

3.2 Anthracite Coal

The anthracite was also supplied by the British Coal Corporation, from the Ffos Las Opencast site near Kidwelly, Pontypridd, South Wales. It is deep black, shiny and comprised mainly of clarain and vitrain. It is very coherent, clean to handle, lightly cleated and one block contained a 40mm wide band of mixed iron pyrites and coal.

The compressive strength, determined from cubes, is 19.4MPa ± 6.0MPa with a density of 1496kg/m^3.

4. RESULTS

Figs.3 and 4, illustrate the mean forces: MNF and MCF represent the average normal and cutting forces respectively. The forces are plotted against jet power since the jet flow was not maintained at a constant rate but increased in proportion to the jet pressure as detailed above.

The coal size distributions are graphed in Figs.5 and 6, and only those results for the blunt tool

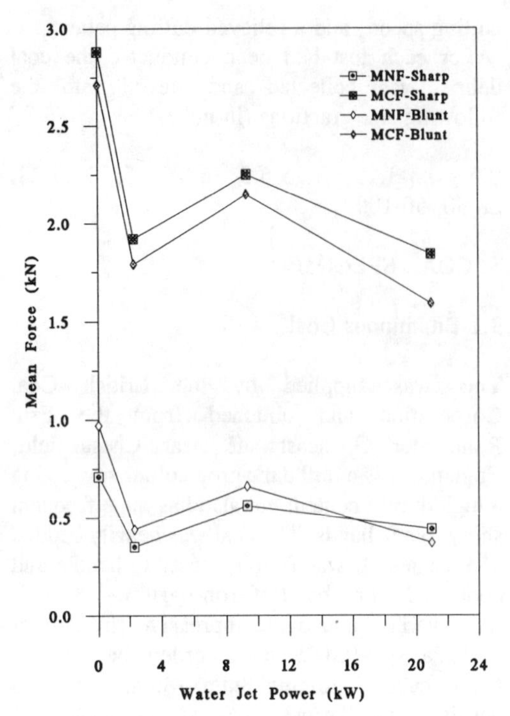

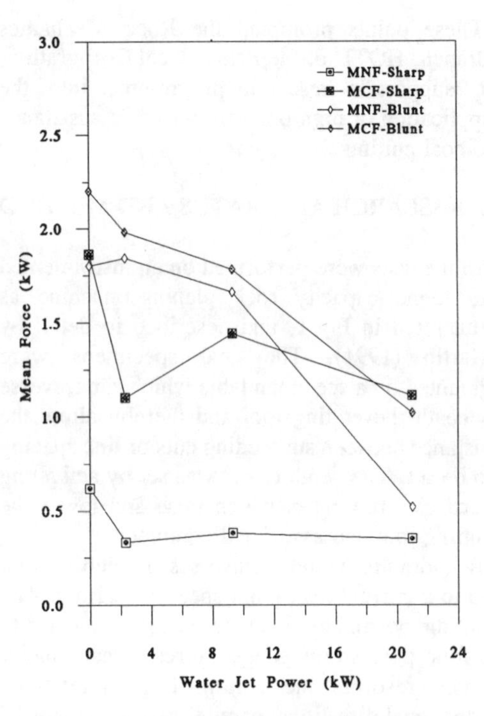

Fig.3 Cutting forces vs. jet power. Bituminous coal.

Fig.4 Cutting forces vs. jet power. Anthracite coal.

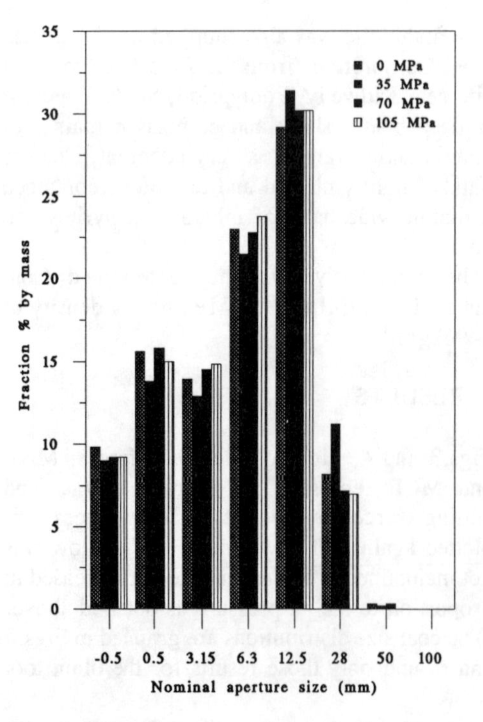

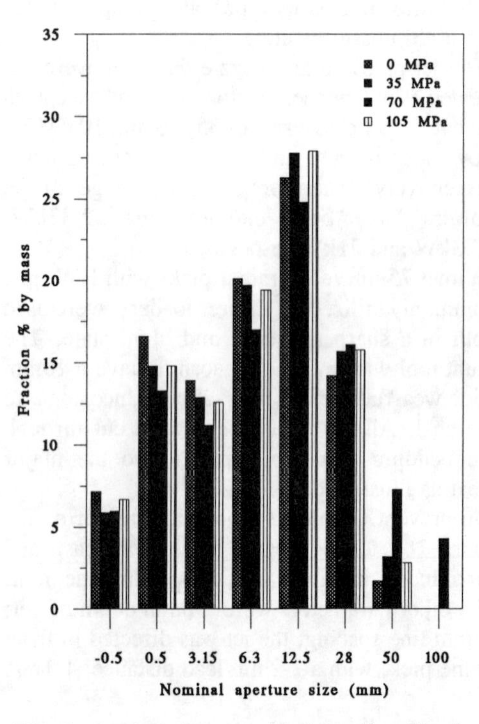

Fig.5 Bituminous coal size distribution

Fig.6 Anthracite coal size distribution

are given since these would be most representative of the field condition.

4.1 Bituminous Coal

4.1.1 Tool Wear

Surprisingly, the blunt tool condition made little impact on the force levels in this coal. This is attributed to the weak character of the coal and thus the wear flat was not yet large enough to increase the forces significantly at the depth of cut taken.

4.1.2 Jet Power

Both the sharp and blunt tool experienced force reductions with water jet assistance. On average, the percentage force reductions are 25.4% and 33.3% for the sharp pick mean normal and mean cutting forces respectively and 54.6% and 33.9% for the blunt tool. Therefore, as the cutting tool wears more proportionate benefit from jet assistance is afforded to the normal force.

Fig.3 also suggests that there is little justification in employing a jet power greater than 2.33kW (35MPa at 4 l/min) at this cutting speed, because a proportionate force decrease is not obtained.

4.1.3 Size Distribution

Each set of four columns in Fig.5 indicates the percentage of the total quantity of coal retained on that sieve size after size analysis. Cutting dry yields 9.8% fines whereas with jet assistance at 35MPa 8.2% fines was obtained: this is a 16.3% reduction. Fig.5 also suggests that the 35MPa pressure water jet returns the greatest reductions in the other size fractions, that is the 0.5mm, 3.15mm and 6.3mm fractions, and the largest increases in the 12.5mm and 28mm fractions.

4.2 Anthracite

4.2.1 Tool Wear

The harder anthracite produced a more typical result for rock as there was a sharp jump in mean normal force when cutting with a blunt tool. For the dry condition the blunt tool MNF was 2.9 times the sharp tool MNF: the advantage of maintaining a sharp tool is clearly evident here.

4.2.2 Jet Power

The effect of tool bluntness is evident again. The trace in Fig.4 indicates a similar trend to that of the bituminous coal for the sharp pick condition but an approximately linear decrease in cutting force is obtained with a blunt pick with jet power. The normal force experiences little reduction until the jet pressure and power exceeds 70MPa and 9.33kW respectively - in practical terms this somewhat prohibits the use of HPWJA cutting in this coal.

4.2.3 Size Distribution

Fig.6, illustrates that with a "dry" blunt cutting tool less fines are produced compared to the bituminous coal: 7.2% to 9.8%. This is attributed to the brittle nature of the anthracite.

Fines reduction was achieved at all the jet pressures. The 35MPa pressure jet produced the greatest reduction in fines with 7.2% to 5.9% of the total, this is an 18.1% improvement. Reductions in the other small size fractions have been obtained together with an increase in the larger size coal, hence, the median particle size has increased.

5. SPECIFIC ENERGY

An important consideration in bulk rock excavation such as coal winning is the cost to furnish the product since this determines to a large degree the econonic viability of an operation. Therefore there seems little justification in applying a novel technique if it leads to an overall increase in production costs without any compensating benefits. Consequently, a knowledge of the energy required to produce a unit volume of the coal with and without HPWJA would be of value.

The specific energies of the coals tested in this

experiment were calculated and are plotted in Fig.7 for the blunt tool results only. Specific energy for unassisted ("dry") cutting is given by the equation below:

$$\text{Specific Energy, } SE = F_c \cdot \frac{\rho}{Y} \quad \ldots (1)$$

Where F_c = Mean cutting force (N)
ρ = Coal density (kg/m^3)
Y = Yield (kg/m)

The mean cutting force is used in the calculation since this is the average force acting in the direction work is being done. When HPWJA is applied, equation (1) has to be modified to include the jet energy as follows:

$$\text{Specific Energy, } SE_j = \left\{F_{cj} + \frac{P_w}{V_t}\right\} \frac{\rho}{Y_j} \quad \ldots (2)$$

Where F_{cj} = Mean cutting force with HPWJA (N)
ρ = Coal density (kg/m^3)
Y_j = Yield with HPWJA (kg/m)
P_w = Jet power (W)
V_t = Cutting speed (m/s)

Fig.7, shows two sets of curves: The curves running almost horizontally represent specific energies without the jet energy included i.e. as calculated from equation (1). By contrast the linearly increasing curves include the jet energies and it is obvious how much the jet energy dominates the specific energy values and increases the specific energy considerably.

Ideally one would require jet assistance to reduce the specific energy from the "dry" value. In this case the following inequality would have to be satisfied:

$$SE_J < SE$$

In terms of equations (1) and (2), the inequality reduces to:

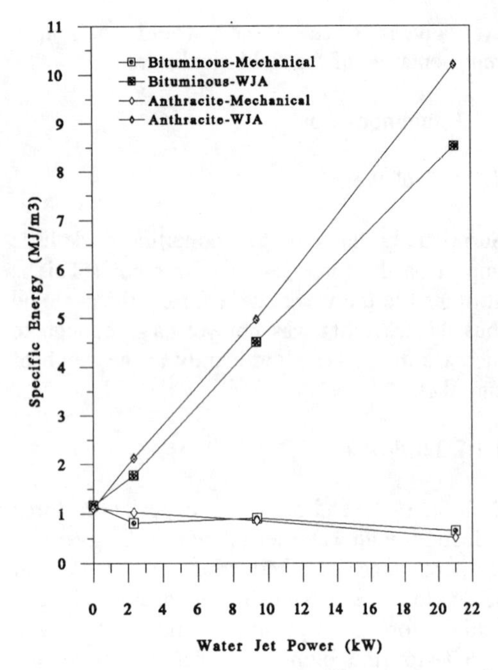

Fig.7 Correlation of specific energy and water jet power, with(WJA) and without(mechanical) the jet energy included.

$$\text{Jet Power, } P_w < \left\{F_c \cdot \frac{Y_j}{Y} - F_{cj}\right\} \cdot V_t \quad \ldots (3)$$

The ratio of yields was found to be close to unity and considering the best case possible of $F_{cj} = 0$ then.

$$\text{Jet Power, } P_w < F_c . V_t \quad \ldots (4)$$

Thus, if the jet power exceeds the power needed to cut the coal without HPWJA, then no reduction in specific energy will result no matter what force reductions are procured by the jet assistance.

The value of Fc will change as the tool wears, therefore the force level obtained from a moderately worn cutting tool should be used in the calculation.

From equation (4), jet assistance has more potential or scope in situations where the specific energy to cut the rock or coal is already high i.e. in hard minerals, providing the jets can reduce the cutting forces sufficiently.

Equation (3) may be re-arranged to express what force reduction is required at a certain jet power before a decrease in specific energy will be realised. The relationship is:

$$R > \frac{P_w}{F_c.V_t} + 1 - \frac{Y_j}{Y} \quad \text{where } R = \frac{F_c - F_{cj}}{F_c} \quad (5)$$

Taking $Y_j/Y = 1$, then (5) becomes:

$$R > \frac{P_w}{F_c.V_t} \quad \ldots (6)$$

As an example, consider the bituminous coal result for a jet pressure of 35MPa and power 2.33kW. The unassisted cutting power for the blunt tool was 2.98kW. From (6), the force reduction needed to reduce the specific energy is:

$$R > \frac{2.33}{2.98} = 0.78 \text{ or } 78\%$$

In the tests a 33% reduction was obtained, hence no reduction in specific energy could be accomplished.

Therefore, two conditions have to be met simultaneously before HPWJA cutting will reduce specific energy:

- The total jet power must not exceed the power required to cut the mineral without jet assistance i.e. $F_c.V_t$; and
- The percentage mean cutting force reduction must be greater than the ratio of gross jet power to the power required to cut the mineral unassisted.

5. DISCUSSION

It is evident from the size distribution charts that the 12.5mm fraction (12.5mm-28mm) is generally the largest and the 6.3mm fraction the second largest and so forth, despite the differences in the coal characteristics. Obviously the choice of sieve sizes affect the distribution pattern, but the authors suggest that it is also dependent upon the depth of cut, line spacing and tool geometry.

On inspection of the cutting arrangement this becomes obvious since the depth of cut and line spacing effectively determine the largest particle size that may be produced. For instance, with a depth of cut of 10mm and a line spacing of 20mm one could hardly expect any +50mm size coal at all, the basic geometry of the situation would prevent it.

These factors indicate the logic of the "deep cut principle" (Pomeroy, 1968) and the necessity of correctly spacing a cutting tool relative to the depth of cut and tool width to generate large size coal.

It is also observed that the size trends with HPWJA follow that of the unassisted case. It is infered from this that the jet assists the cutting action of the tool and does not itself bring about a fundamentally new cutting process.

The section on specific energy illustrates the high power consumption of HPWJA and the improbability that it would lead to a cheaper product - based solely on specific energy. However, other factors may mitigate the disadvantage of an increase in specific energy and these are: reduced dust make; decreased probability of frictional ignitions; reduced tool wear; less fines; and in the case of shearer loaders, a higher machine haulage speed (Mort, 1988).

6. CONCLUSIONS

The use of WJA does produce reduced cutting force components and reduce the percentage of fine coal produced but at the expense of the total energy consumed which is a most important consideration in bulk mineral excavation.

In hard rock excavation WJA does find application where alternative methods are not acceptable e.g. explosives due to induced vibrations or disc equipped machines, are too expensive.

ACKNOWLEDGEMENTS

The work on water jet assisted coal cutting was

supported by an Extra Mural Grant from the British Coal Corporation. The assistance and encouragement of Dr K Hurt, Technical Services and Research Executive, British Coal is gratefully acknowledged.
The views expressed are the authors own and not necessarily those of the British Coal Corporation.

REFERENCES

Barham, D.K and Buchanan, D.J. (1987) A review of water jet assisted cutting techniques for rock and coal cutting machines. The Mining Engineer, pp 6-14, July 1987.

Haslett, G.A., Corbett, G.R. and Young, D.A. (1986) An investigation into the effect of varying water pressure and flow rates upon the release of airborne respirable dust by a Dosco MKIIB roadheader equipped with a water jet assisted cutting head. Proceedings 8th Int. Symp. on Jet Cutting Technology, Durham, 9-11 Sept. 1986, BHRA, pp 103-111.

H.Q.T.D. Technical Bulletin, British Coal Corporation, 1987.

Hurt, K.G., MacAndrew, K.M. and Morris, C.J. (1988) Boom roadheader cutting vibration: Measurement and prediction. Proceedings, Conference on Applied Rock Engineering, Newcastle upon Tyne, pp 89-97, 6-8 Jan. 1988.

Martin, J.A. (1991) The study of high pressure water jet assisted cutting of coal samples in the laboratory. Unpublished Ph.D. thesis, Faculty of Engineering, The University of Newcastle upon Tyne, September 1991.

Mort, D. (1988) The application of high pressure water jets to longwall mining. The Mining Engineer, pp.344-350, Jan. 1988.

Pomeroy, C.D. (1968) Mining applications of the deep cut principle. The Mining Engineer, pp.506-515, June 1968.

Geomechanics 93, Rakowski (ed.) © 1994 Balkema, Rotterdam, ISBN 90 5410 354 X

Abrasive jet cutting technology: Future demands and developments

H. Louis
Institute of Material Science, University of Hannover, Germany

ABSTRACT:
Abrasive Jet Cutting is an innovating and expanding technology, however, with certain demands and developments. Main demands are the increase of knowledge and knowledge transfer, improvement of the acceptance of Jet Cutting Technology, establishment of Jet Cutting as high-tec technology.

Some of the future developments are increasing pressure level for intensifiers and plunger pump, suspension jet in production technique for quality cutting and replacement of other technologies.

1 INTRODUCTION

Jet Cutting is a very wide ranging technology. It includes cleaning, cutting and assists mechanical tools. The range of pressure goes from low pressures in the area of 10 MPa up to pressures of 400 MPa. Altogether, this technology is an increasing technology, looking for further applications, higher efficiencies and lower cost. To describe the future demands and developments in one special field of this technology is the aim of this paper. The topic is focussed on Abrasive Water Jet Cutting and here especially in the field of manufacturing. However, most of the details are also relevant for other applications. The given information is mainly based on three sources: Firstly relevant papers from the last Jet Cutting conferences (1-6), secondly discussions with colleagues at conferences and meetings, and thirdly from discussions with colleagues, manufacturers and users during our local meetings of the "Arbeitskreis Wasserstrahltechnologie" in Germany.

2 DEMANDS

The main demands related to the topic mentioned above are

- increase of knowledge and knowledge transfer
- improvement of the acceptance of Jet Cutting Technology and
- establishment of Jet Cutting as high-tec technology

Knowledge transfer is based on the knowledge itself. Even when this knowledge increases, for instance related to the process of cutting and the role of relevant parameters there are still a lot of details which are unknown. Especially on the last conferences the information regarding the process and therefore its understanding increased, for example coming up from a two-dimensional process of cutting, forming steps leading to the typical striations, figure 1. Today we can describe this process as a discontinuous process of three-dimensional step formation and leading therefore to the striation, Fig. 1. The knowledge about the process is very important to describe the influence of parameters, e.g. on the quality of the cut. However, there are still a lot of details we don't know, for example the physical sources of the jet deviation which happens periodically and repeatable. Besides this, we have a lot of problems to describe the complex interaction of the large numbers of parameters.

Even on the base of knowledge we have today,

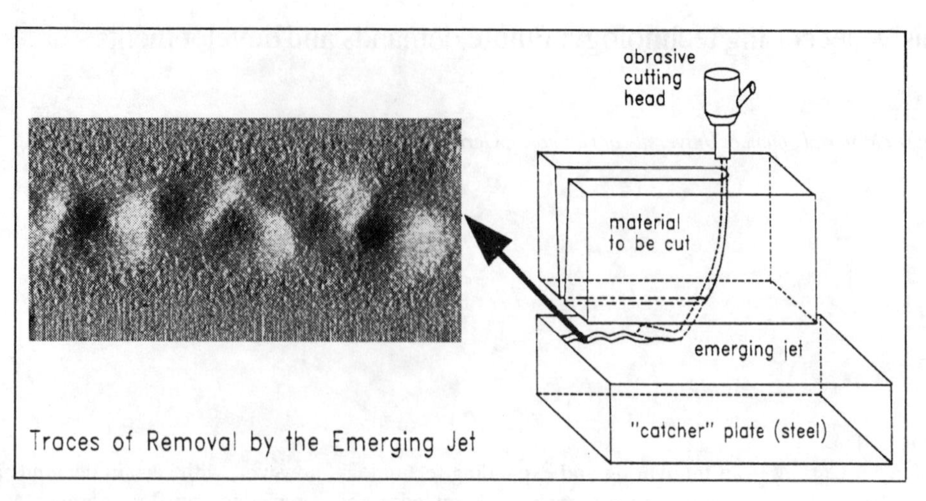

Fig. 1 3-dimensional process of step formation, leading to striation (7)

there must be the possibility of independent information of people being interested in the application of this technology. However, it is very difficult to get information not only from companies, but also from independent sources. Companies have to sell cutting units or equipment, that's their job and they often give a very positive information, especially related to their products. However, even independent research centres, e.g. laboratories at universities and technical high schools which produce a large number of publications have limitations in objectivity. Their results are based on the equipment they have. For research institutes it is not possible to change the equipment every two or three years to follow the development of this technique. Therefore it is necessary to have a forum to discuss problems as well as to give interested people a forum to get well-based information from a larger number of users, manufacturers a.s.o. The increasing number of national and local societies related to Water Jet Cutting is a very good base for this information.

The third demand is the education of people. Today it is very common that people looking for the application of Abrasive Water Jet, want to replace another technique, e.g. laser cutting or plasma cutting. The increasing number of equipment used for this purposes shows that there are very often advantages to use Abrasive Jet Cutting. However, Abrasive Water Jets have a higher potential for machining a.s.o. Figure 2 shows a part of aluminium cut by Abrasive Water Jet. This example shows that the Abrasive Water Jet Cutting Technique gives the possibility to machine very filigree parts compared to conventional machining techniques. Such a part never will be constructed, or the cost for machining such a part are very, very high. The education of students in the possibilities of Water Jet Cutting Technique is very small, especially related more to design and construction than to manufacturing.

Fig. 2 Precise cutting of aluminium Doll, Foracon (8)

3 IMPROVED ACCEPTANCE OF JET CUTTING TECHNOLOGY

In Europe the production of secondary waste is a very important factor of acceptance of Abrasive Water Jet Cutting Technology. There are only two possibilities to reduce the amount of abrasives: Firstly to minimise the use of abrasive and secondly to recycle the abrasive. Fig. 3a shows the effect of abrasive flow rate on the depth of kerf. One possibility mentioned before is to reduce the amount of abrasive, e.g. when reducing the abrasive from 10 g/s to 5.5 g/s the depth of kerfs will only be reduced to 70 %. However, even when using 5 g/s, fig. 3b, the amount of abrasive per unit a year is very high. The costs are depending on the material to be cut and on the local situation of the company, therefore the cost in Germany can vary from 20 DM to 800 DM per 1000 kg, because this is a more political price and nobody can calculate the cost of tomorrow. This is a lack of information when calculating the cost of the technology, therefore in future the abrasive material has to be recycled.

However, due to the destruction of the particles during the mixing and acceleration process and during interaction between the abrasive particles and the material to be cut, only a certain amount of the abrasive material can be reused. The breakage of the particles, however, may influence their cutting efficiency; especially very small particles show decreasing cutting efficiency. Therefore these small particles have to be removed from the reusable abrasive. This special amount of abrasive depends on the cutting tool, on the cutting parameter and on the material to be cut.

A flow chart of recycling system is given in figure 4.

The abrasive particles after feeding and cutting have to be caught very carefully to avoid an additional destruction. The separation of particles to be reused and those particles for waste disposal has to be done in wet condition. The particles which will be reused have to be dried afterwards. To dry the very small particles costs a lot of energy and causes a lot of problems. One reason is that 10 % of the small particles will have a larger surface area than the 90 % of the larger particles. There are already a few systems in tests showing excellent results.

Fig. 5 shows three different cuts: Firstly done by a synthetic particle sized distribution which will be achieved after ten time cutting with the same material. The cutting efficiency will be reduced compared to garnet material (HP 50) by about 10 %. When eliminating a certain amount of small particles by a special constructed hydro cyclone,

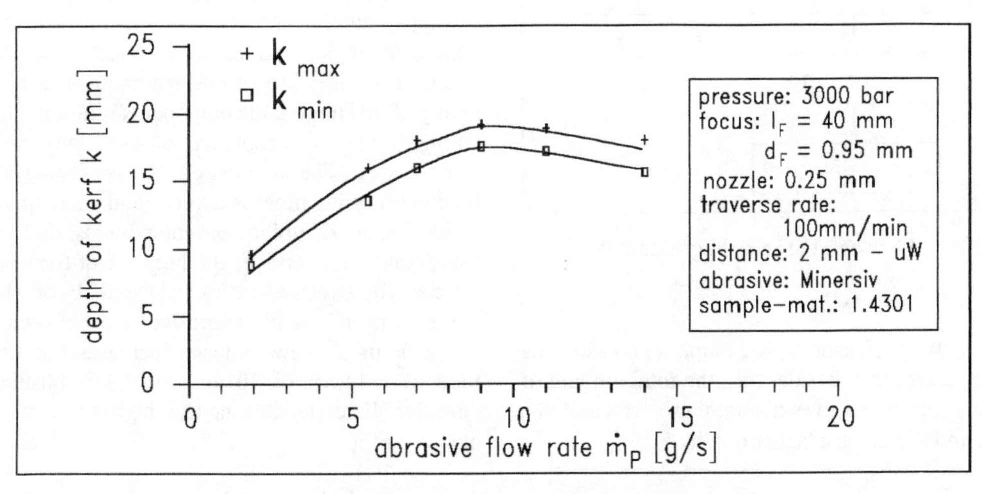

Fig. 3 Recycling potential of Abrasive Water Jets
a) depth of the kerf as a function of abrasive flow rate

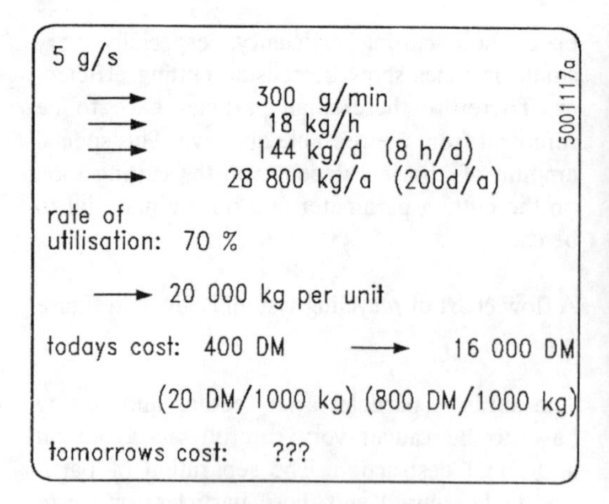

5 g/s
300 g/min
18 kg/h
144 kg/d (8 h/d)
28 800 kg/a (200d/a)
rate of utilisation: 70 %
20 000 kg per unit
todays cost: 400 DM
16 000 DM
(20 DM/1000 kg) (800 DM/1000 kg)
tomorrows cost: ???
5001117a

Fig. 3 b) calculation of consumption and cost of abrasive per unit

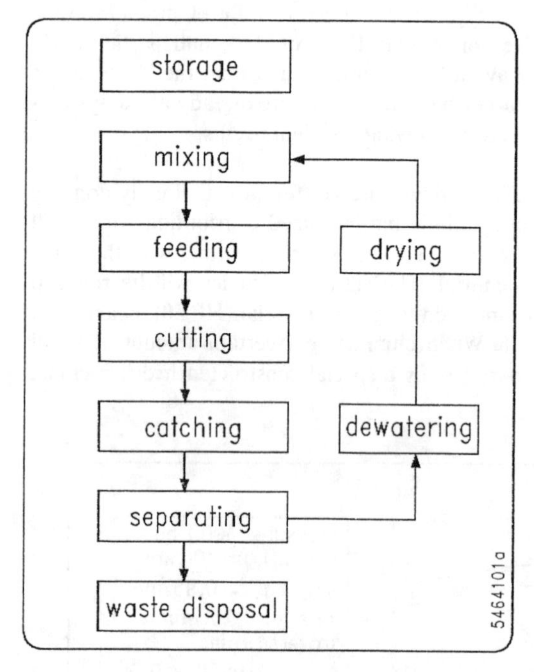

Fig. 4 Flow chart of recycling system (9)

the cutting efficiency is comparable with the unused abrasive. Besides this, the total amount of smaller particles gives a smoother surface of the cut, and therefore a higher quality.

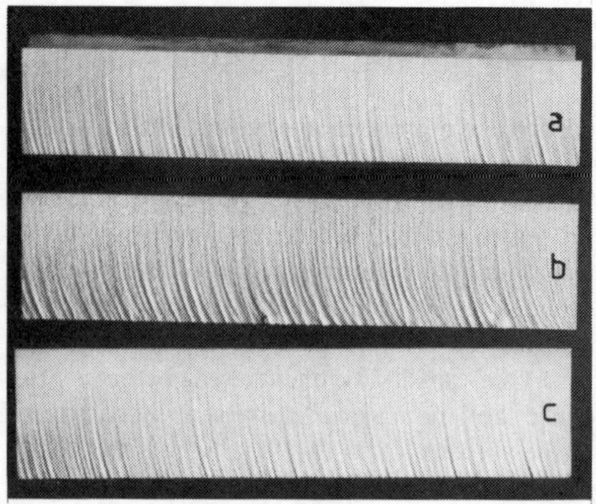

Fig. 5 Influence of particle size distribution on cutting efficiency
a) synthetic particle size distribution, simulation of used abrasive, depth 20 mm
b) HP-Garnet (HP50) unused, depth 24 mm
c) same as a), recycled and reduced to 90 % of mass, depth 24.8 mm (9)

4 JET CUTTING AS A HIGH-TEC TECHNOLOGY

Beside recycling, there are demands for process and quality control. In competitive manufacturing processes the control of the process and the results of the process, e.g. cutting efficiency, cutting quality, is nearly state of the art. An application of Jet Cutting Technique, process and quality control is just in the beginning. The reliability of machines and components is also a very strong factor in acceptance of especially new technologies. The production of new materials for the focussing tubes is a very good example of increasing in reliability of components. Before using composite carbides the life time of focusing nozzles for precise cutting (depending of the parameters and the used abrasive) was between 2 and 6 hours. A new material increases this life time by a factor of 10 and more, establishing Abrasive Water Jet Cutting as a high-tec technology.

To be accepted as a high-tec technology Water Jet Cutting Technique has to be integrated into the production processes. However, several demands have to be fulfilled to do this. Some of them are already under improved acceptance, for instance process and quality control, reliability a.s.o. However, there are some more demands, e.g. to be integrated even in the phase of construction of units. Multi-functional machining centres also show the high potential of Abrasive Water Jets.

5 DEVELOPMENTS

Besides these demands which have to be met today and in the very near future, there are already some developments which lead to an increasing of application and a spreading of this technology to other applications. One development is an increase of pressure level for intensifiers as well as for plunger pumps. At the end of the 60th when pressure intensifiers came to the market their pressure level was already mostly fixed on the level of 400 MPa. The development up to today is mostly related to their reliability, however, not to their pressure level.

Quite opposite to this is the situation in the case of plunger pumps. In the beginning of the 70th plunger pumps with an acceptable lifetime of components had a pressure level of about 100 MP. By keeping their reliability nearly constant this pressure level increased continuously now to 200 MP and more. There are developments going on to reach the range of 300 MP and even more. However, it will take about 2 to 10 more years to reach this area of pressure by plunger pumps with an acceptable reliability. By reaching pressure levels above 300 MP these plunger pumps will go into competition with pressure amplifier.

One question is whether higher pressure levels of pure Water Jets will be able to cut metal directly without using abrasive particles. A research programme is run by several American companies and a research institute involved in this question. Pressures up to /4/ 700 MPa have been reached.

Another development is related to suspension jets. Until now suspension jets are mostly being used for rough cutting on a pressure level of 350 bar and in some examples of 690 bar. Because of their higher efficiency in accelerating the abrasive particles there are developments to reach higher pressures. Parallel to this, the nozzle diameter has to go down. Besides this suspension jet technique using a by-pass flow to deliver abrasive material from a storage tank at the same pressure level there are some developments to pump the premixed suspension directly. By doing this, very small abrasive particles were used which produce a very high quality (11). This very new development will lead to further application of Abrasive Water Jet Cutting Technique. Some of these suspension jets even don't use water but synthetic liquids with high viscosity. The development of Abrasive Water Jet Cutting Technique will also lead to a larger field of application, and therefore to replacement of other technologies.

6 CONCLUSIONS

Jet Cutting Technology is an innovating and expanding technology. Information must be available and must be transferred to potential users. An exchange of experience is necessary to develop this technology. The development on Jet Cutting Technique has to follow social and industrial demands.

References

1. Miller, D.S.
   Improving the Competitiveness of Abrasive Jet Cutting
   Proc. 11th Int. Conference on Jet Cutting Techn., St. Andrews, 1992
   Jet Cutting Technology, Kluwer Academic Publishers, 1992
2. Hashish, M.
   Three-dimensional Machining with Abrasive-Waterjets
   Proc. 11th Int. Conference on Jet Cutting Techn., St. Andrews, 1992
   Jet Cutting Technology, Kluwer Academic Publishers, 1992
3. Miller, D.S.
   Expanding the Market for Abrasive Jet Cutting Systems
   Proc. of the 7th American Water Jet Conference, Seattle, 1993

4. Coleman, W.J. et al
The Next Generation of Waterjet and Abrasive-Waterjet Processing Technologies
Proc. of the 7th American Water Jet Conference, Seattle, 1993
5. Ramulu, M.
Future Research Needs, Panel Discussion
Proc. of the 7th American Water Jet Conference, Seattle, 1993
6. Guo, N.S. et al
Recycling Capacity of Abrasives in Abrasive Water Jet Cutting
Proc. 11th Int. Conference on Jet Cutting Technology, St. Andrews, 1992
Jet Cutting Technology, Kluwer Academic Publishers, 1992
7. Guo, N.S. et al
Surface Structure and Kerf Geometry in Abrasive Water Jet Cutting: Formation and Optimization
Proc. 11th Int. Conference on Jet Cutting Technology, St. Andrews, 1992
Jet Cutting Technology, Kluwer Academic Publishers, 1992
8. Mitteilung der Fa. Foracon, Bretten, Deutschland
9. Ohlsen, J.
Recycling Potential and Recycling Systeme für Wasserabrasivstrahlen
ITR-WE 92-17, Institut für Werkstoffkunde, Universität Hannover
10. Dubensky, E. et al
Hard Ceramics for Long-life Abrasive Waterjet Nozzles
Proc. 11th Int. Conference on Jet Cutting Technology, St. Andrews, 1992
11. Rhoades, L.J.
Abrasive Suspension Jet Machining
Proc. of the 7th American Water Jet Conference, Seattle, 1993

Geomechanics 93, Rakowski (ed.) © 1994 Balkema, Rotterdam, ISBN 90 5410 354 X

Materials disintegration by high pressure water jet – Review of recent development

M. Mazurkiewicz
Rock Mechanics & Explosives Research Center, University of Missouri-Rolla, Mo., USA

ABSTRACT: The advantages of using high pressure water jets (hpwj) as a new technique for material disintegration are reviewed in this paper. Some of the results of water jet properties and parameters observed in disintegration are discussed with relation to selected materials. The experience built up during numerous projects conducted in the Rock Mechanics and Explosives Research Center (RMERC) at the University of Missouri-Rolla is introduced.

1 MATERIALS DISINTEGRATION MECHANISM

The analysis of the materials disintegration must be understood in three different categories:

- soft materials having loose structure, low mechanical properties, or uniform structure,
- hard minerals with solid uniform structure, and
- compounds of soft and hard components with different mechanical properties.

In all disintegration processes by hpwj, the jet kinematics need to be carefully designed. The jet relative velocity and the jet trajectory selected should prevent as big as possible volume of material treated by jet and disintegrated to the size required. For the three cases mentioned above, the process of parameterization is entirely different.

Taking into consideration soft materials, organic or inorganic type, the disintegration effect occurrs mainly because in the area where jet collides with the material, extremely high energy concentration is developed. This happens during regular cutting of soft materials by high pressure water jet. All the material from the kerf being cut is disintegrated into very fine particles and removed from the cutting area by water jet approximately with jet's speed. The cuttings have previously not been the subject of interest.

The hard, high-strength materials like rocks or ores usually have an aggregate structure and can be disintegrated into relatively small size fractions when are under water jets attack. It has been observed that the high pressure water jet is excellent for materials disruption[1,2]. Such a capability is due to a high energy flux input to the target material. The jet's high energy density is concentrated in a very small impact zone, while intense differential pressure across the jet leads to microcrack generation and growth. Also, the jet, upon impact, creates a stagnation pressure which forces water into the cracks and microcracks. This leads to development of a hydrodynamical action in these cracks and creates an increasingly dense network of cracks in the wall of the cavity created. The water jet can more destructively act on the target if water slugs are developed, and a water-hammer effect is introduced to the collision area.[3] High velocity lateral flow of the water jet after collision can impose significant damage, when the conditions for cavitation are created.

The water jet application for waste disposal is

complicated by the presence of soft and hard components which are of value. In order to reclaim or recover these specific components it is necessary to separate out the different materials in the waste. Separation requires that the components be liberated from one another. Most of the material is not economically separable by simple mechanical disruption. Utilizing high pressure water jets moving at a high translation speed over a surface shows considerable promise for disintegrating the various components of the waste. This is particularly true in compacted waste where the bond between the two material components may be generally quite weak. The ability of the jet, by adjustment of flow rate and pressure, to preferentially cut some materials and not others, may also have particular benefit in, for example, the stripping of softer materials from other components before they are treated. The water jet can reach deep into the material, and separate the different components at their common boundary. Based on differences in the specific gravity, separation can be achieved very easy.

The picture presented is very complicated. Griffith's hypothesis[4], initially derived from an investigation of an elliptical crack on the surface of an elastic, isotropic, and homogeneous material, can be applied but must be drastically modified to describe the behavior of rock or ore, which used to be anisotropic, and varies as a function of the degree of metamorphism.

The compound of soft and hard components with different mechanical properties can be treated by hpwj. The mechanism of the jet action is very complicated, and can be understood if jet lateral flow and dispersion are taken into consideration, as well as cavitation, water-hammer, and comulative effects.

2 WOOD PULPING

The studies carried out by the authors have proven that the water jet can work as an excellent material disruption device.[5] Such a capability, indicates the potential for water jets to be introduced into the processing stages of variety of industries. One example lies in the pulp and paper industry. Paper production requires that wood fibers be prepared from solid wood. The main need is to extract long and undamaged fibers, and the length of the fibers has a significant influence on the quality of the paper. The length of fiber produced by the water jet is greater, and less damage is done to fiber integrity than during conventional grinding. The power consumption required per cublic meter of solid wood treated by the jet, when compared with conventional grinding, lies below fifty percent at the present state of this development, and can be reduced further when process optimized. There is a possibility to apply jet technology for speeding up chemical pulping. By cooking wood chips in alkali chemicals to temperatures of 170° C over many hours, the high temperature chemical jet can be applied for creating wood fibers through a combination of hydromechanical and chemical action.

The study shows that the most important parameters which influence this process are: the jet pressure, jet diameter and it's shape, stand-off distance, the jet feed rate across the wood surface and the jet attack angle with respect to the fiber orientation, as well as jet trajectory. All of these parameters have a significant influence on productivity and fiber quality.

3 WASTE PAPER PULPING

High quality wood fiber is easy to obtain from waste paper. In municipal solid waste there is as much as 41% of paper of highly recycleable quality. The numerous tests performed in the RMERC shows that waste paper can be recycleable by hpwj very efficiently without noticeable fiber quality degradation. The process of waste paper pulping by hpwj allows the treatment of high volumes in a short period of time. The specific energy is in the level of conventional methods, mainly based on a submerged propeller method. The high pressure water jet moves very fast and when the nozzle is moving with respect to the waste paper mass, the volume treated is high. The high turbulence generated during this process makes a strong rinsing effect which results in partial deinking of the pulp and significant reduction of the bleaching operation.

4 COAL AND MINERAL COMMINUTION

Comminution may be defined as a single or multi-stage process during which mineral ores or other materials are reduced from a variety of original sizes by various crushing and grinding processes, to a narrowly defined product size range required for subsequent processing. Current comminution technology is both energy intensive and inefficient. Up to 99% of the energy consumed during the operation of conventional size-reduction devices can go into nonproductive work, with only 1% of the energy input then being used to create smaller particles. Comminution is thus an appropriate target for significant energy savings, as the tonnages of materials involved in size reduction operations are so great that even small improvements in comminution efficiency would provide considerable savings in energy and mineral resources.

Recent investigations into the use of high pressure water jet[6] as a means of disintegrating organic and inorganic materials have shown that this novel tool has the potential to comminute coal and other minerals into relatively small size fractions. Additional progress in fine and ultra-fine comminution can be achieved if the compressive crushing principle commonly used at present is changed to one in which particles are fragmented by the tensile growth of pre-existing internal flaws or cracks, under internal water jet pressure. Such cracks can either be naturally present or can be developed during earlier stages of comminution. A water jet under a pressure 70 MPa is moving at a velocity of roughly 400 m/sec. This fluid will continue to penetrate a growing crack and maintain sufficient pressure in the crack to sustain growth, even at a maximum crack tip propagation velocity.

Research at the University of Missouri-Rolla has shown the possibility of using high pressure water jet in two different ways for comminution purposes. The jet can be used as a tool in its own right, or it can be used as supplementary means for supporting existing mechanical milling technologies. The second approach could be very effective if the water jet is combined with an existing mill. This concept has been proven.[6] The coal particles are pushed radially by the centrifugal force and are pressed into two tapered revolving discs. Compressive stress generated in the coal particles is condusive to the creation of cracks and microcracks, and this effect is supported by the high pressure water jet directed into the gap between the revolving discs. The high pressure water jet helps to comminute the coal particles and at the same time push them out through the defined gap. The gap size is strongly related to the product size.

5 MUNICIPAL SOLID WASTE RECYCLING

The United States faces a number of increasingly difficult problems in the disposal of the growing volumes of municipal solid waste. The average person in the United States produces some 3.7 lbs per day of disposable waste. By the end of the decade this means that the United States will face a disposal problem on the order of 100 million tons of solid waste each day.

Simplistically the waste can be divided into that which can be recycled, that which can be beneficially processed, and that which is most economically dealt with by being incinerated or buried in a landfill. It is a primary advantage of the hpwj technology that it is able to effectively achieve such a separation process, by careful tailoring of the pressure and flow rate at which a waterjet is directed at the waste materials. It is possible for a high powered jetting system to be programmed such that it preferentially attack, and break into small pieces, the organic constituencies of the waste such as paper, cardboard, and other materials while leaving the inorganic material, such as the glass and the metal, intact. Given that the organic part of the municipal solid waste can be as high as 70% of the material, this can significantly change the economics of waste separation. It also pulverizes the inorganic material into a shredded form which can be more readily treated, divided and processed. This, in turn reduces the unusable material which must necessarily be disposed of in the landfill or incinerated.

The pulverized nature of the inorganic material also provides a potential for a different method of treatment. Because of the fine nature of this material, it can be composted and through a series of treatment vats turned into either a low grade fertilizer or potentially fermented to generate methane or other resources with a practical use.

The advantage of the hpwj technology is that, as long as the container is made of an inorganic material such as glass or metal, or a heavy duty plastic (such as a battery) that the waterjet will not damage the container and breach its integrity. Thus the contents could be left inside until identified in subsequent inspection, and segregated for effective safe disposal.

High pressure water jets move at a very high speed. By controlling this speed it is possible to tailor the jet action so that it breaks up the material into very small pieces.[7] This unique capability of high pressure water jets was applied for the separation of municipal waste and the following potential applications were identified:
a) to provide an economic means of separating recycleable material such as glass and metal from the current mixture of municipal waste in a rapidly and effectivly. b) to comminute the organic material and thereby provide a sludge which can be readily disposed of by burning, or, where economical to do so, properly treated to provide a useful product. c) to reduce significantly the volume of material which cannot, at this stage be economically treated, and which must still be buried in a landfill. By reducing this volume to perhaps 10% of the current volume one can thereby significantly reduce the problems of waste disposal currently facing the municipalities.

6 CONCLUSION

The years of experimentation with the disintegration of a wide range of different materials by high pressure water jets indicates big potentials for many industries. Creation of materials with high specific surface is the main concern for many technologies, which conditioning final results of many operations. Proposed technique can also be considered for separating different materials, inducing chemicals for speeding reaction and as well as clean and wash out granular substances. Further work is necessary to conduct to implement hpwj technology in the environmental cleaning operations.

REFERENCES

1 Mazurkiewicz, M., "High Pressure Liquid Jet as a Tool for Disintegrating Organic and Nonorganic Materials", Invention Disclosure 85-UMR-009, August, 1984.
2 Mazurkiewicz, M., "Disintegration of Wood" US Patent No. 4,723,715, February, 1988.
3 Mazurkiewicz, M., "The Analysis of High Pressure Water Jet Interruption Through Ultrasonic Nozzle Vibration", Seventh International Symposium on Jet Cutting Technology, Ottawa, Canada, June 1984.
4 Griffith, A.A., "The Phenomena on Rupture and Flow in Solids", Phil. Trans. R. Soc. Series A. 221-163-198 (1920-21).
5 Mazurkiewicz, M., "Development of a Liquid Jet for Disintegrating Organic and Non-organic Materials", Final Report for Incubator Technologies, Inc., Rolla, MO.
6 Mazurkiewicz, M., "Coal Disintegation by High Pressure Water Jet", Final Report of DOE contract DOE-DE86FC91271, 1988.
7 Mazurkiewicz, M., et al. "Municipal Waste Recycling by High Pressure Liquid Jets", Invention Disclosure no. 91-UMR-005.

Geomechanics 93, Rakowski (ed.) © 1994 Balkema, Rotterdam, ISBN 90 5410 354 X

Power of pulsed liquid jets

M.M.Vijay
Institute for Engineering in the Canadian Environment, National Research Council of Canada, Ottawa, Ont., Canada

ABSTRACT: Pulsed jets can be of various types. Experimental results reported in the literature indicate that they are far superior to continuous jets and are powerful for cleaning, cutting and fragmentation of various types of materials. In this paper a brief review of the literature on pulsed jets is presented. The paper includes: (i) a brief discussion of the characteristics of pulsed jets, (ii) sample experimental results and (iii) recommendations for future work.

1 NOMENCLATURE

C_o Acoustic speed in the undisturbed liquid, m/s
D Nozzle diameter, mm
d_j Jet or pulse diameter, mm
f Frequency of interruptions or of pulsed jets, Hz
P_w Water hammer pressure, MPa
P_n Pressure at the nozzle inlet, MPa
P_s Stagnation pressure, MPa
S Standoff distance, mm
V Speed of the jet, m/s
V_{tr} Traverse speed of the sample under the jet, mm/min
ρ_L Density of liquid

2 INTRODUCTION

An extensive research project is in progress in the laboratory the purpose of which is to preweaken hard rocks with water jets ahead of continuous mining machines (Vijay & Remisz 1993). Since hard rocks of the type encountered in the Canadian mines are difficult to cut or fragment with continuous water jets, the project is concerned with developing pulsed jets for preweakening. Attention is focused on (i) low frequency (≈ 0.1 Hz) and (ii) high frequency (>5 kHz) pulsing devices. Both techniques involve modulating high speed continuous water jets. Prior to commencing this project, a thorough review of the literature was made on pulsed liquid jets. Excerpts from this review are reported here. Due to limited number of pages, no details are given on any of the pulsed jet devices reviewed. However, for those interested in pursuing the field, sufficient number of references are listed in the paper for further study.

3 BACKGROUND

The power of pulsed liquid jets can be readily seen in Fig. 1 where the mass loss (a measure of performance) of aluminum samples exposed to cavitating and continuous (in air) jets is depicted (Vijay, Zou & Tavoularis 1990). The increase in the mass loss with standoff distance is caused by droplets, termed here as "natural" pulsed jets, which are formed due to aero-dynamic drag. For example, fan jets which are used for some cleaning applications, are typically droplet laden water jets (see Houlston & Vickers 1978 and Danel & Guilloud, 1974). For this reason there is a great deal of interest in pulsed liquid jets for cleaning and fragmentation of brittle materials such as concrete and rocks.

Basically, when a high speed slug of water (liquid) impinges on a target material, the pressure at the point of impact is the water hammer pressure (Rochester & Brunton 1972; this is one of the best papers on the basics of pulsed jets). The relevant equations, which do not take into account shock wave effects (see Pritchett & Riney 1974), are:

$$\textit{Stagnation pressure}: P_s = \frac{\rho_L V^2}{2} \tag{3.1}$$

$$\textit{Water Hammer Pressure}: P_w = \rho_L V C_0 \tag{3.2}$$

$$\textit{Amplification} = \frac{P_w}{P_s} = \frac{2C_o}{V} \tag{3.3}$$

The symbols are defined in the nomenclature. Figure 2 shows the magnitude of the water hammer pressure

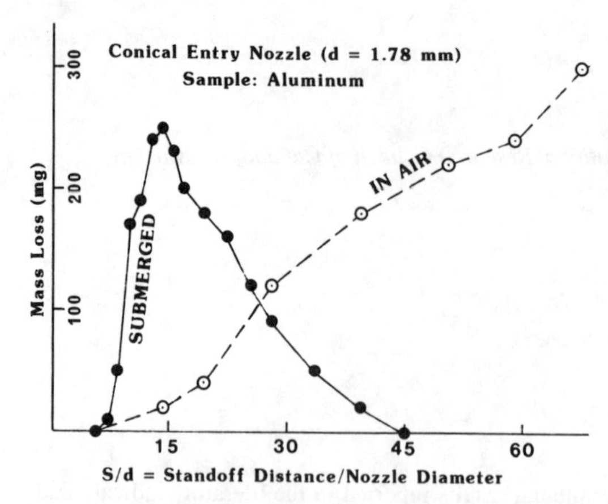

Fig. 1 Erosion of aluminum samples by three different types of jets to illustrate the performance of natural pulsed jets (Vijay, Zou & Tavoularis 1990)

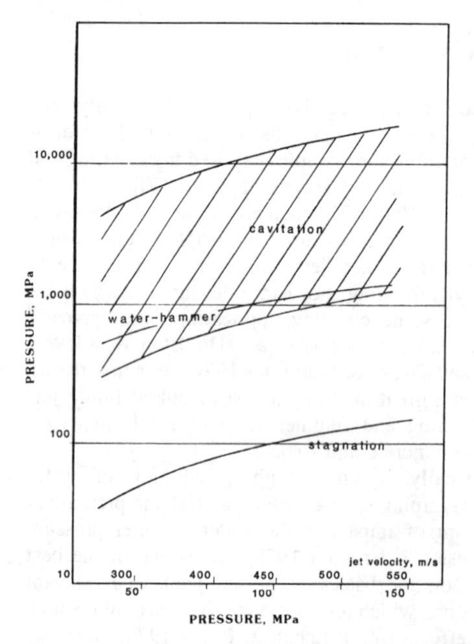

Fig. 2 A comparison of the theoretical impact pressure on a target exposed to continuous, cavitating and pulsed jets (Vijay, Remisz & Shen 1992)

with respect to the stagnation pressure on a target. Much better results can be expected when the jet is pulsed at low stagnation pressures ($\approx P_n$). If the compressibility of the target is taken into account, the magnitude of P_w is somewhat reduced (Rochester & Brunton 1972). The duration of impact depends on the shape of front of the slug. For flat faced cylindrical and spherical slugs the time of action (τ) is given respectively by the following equations:

$$\tau = \frac{d_j}{2C_o} \tag{3.4}$$

$$\tau = \frac{d_j}{2V}\left[1 - \left(1 - \frac{V^2}{C_o^2}\right)^{1/2}\right] \tag{3.5}$$

For example, for a cylindrical jet of 1.78 mm, the duration of impact is 0.6 μs. For a spherical droplet of the same diameter, the time reduces to 0.07 μs. As the performance of a pulsed jet depends not only on the magnitude of P_w, but also on the time of action, it is obvious that the shape of the pulse is quite important, although not much has been reported in the literature on this aspect (see Rochester & Brunton 1972 and Edney 1976). The main objective of the research on pulsed jet devices reported in the literature has been to improve the shape, the energy content and the duration of impact (Vijay 1991; this reference also includes a number of figures on pulsed jet devices). It should be noted in passing that pulsed jets are useful only for cleaning or fragmentation and are not suitable, at present, for precision cutting applications.

4 TYPES OF PULSED JET SYSTEMS AND NOZZLE DESIGN

Pulsed jets can be produced in several ways using different driving energy sources (Labus 1991, Vijay 1991 and Vijay, Remisz & Shen 1992). The method employed can be classified as either "natural" or "forced" (by an external source of energy). The nozzle shape and size employed depend on the technique used for producing the high speed slugs of liquid. When considering the use of a pulsed jet device, the following considerations should be borne in mind: Size and weight, practicality, ease of manufacture, cost effectiveness, ease of mobility, reproducibility of results, reliability and finally safety. For generating single pulses with high energy content ($f \leq 1$ Hz) pulse cannons (forced), which are usually of the free piston impact or pressure extrusion type, are employed. However, these are useful for massive fragmentation of brittle materials (Atanov, et. al. 1979, Atanov 1988 & 1991, Cheng, et. al. 1985, Chermensky 1976a & b, Chermensky & Davidyants 1980, Cooley 1972, Lucke & Cooley 1974, Hawrylewicz, Puchala & Vijay 1986, Kollé 1993, Mohaupt, et. al. 1978, Moodie & Artingstall 1972, Moodie & Taylor 1974, Moodie 1976, Moodie & Tomlin 1980, Pater 1984, Petrakov & Krivorotko 1978, Singh, Finlayson & Huck 1972, Vallvé, et. al. 1980, Wang, et. al. 1984, Yie, Burns & Mohaupt 1978 and Ze-sheng 1987). However, for cleaning or controlled fragmentation, pulsed jets produced by passive (Chahine, et.al. 1983, Evers & Eddingfield 1981, Evers, Eddingfield & Yuh 1983, Fang & Lin 1987, Liao & Huang 1986, Sami & Anderson 1984 and Shen & Wang 1988) or dynamic (Nebeker & Rodriguez 1976, Nebeker 1981, 1983, 1984 & 1987, Nebeker & Cramer 1983, Puchala & Vijay 1984, Sami & Ansari 1981, Vijay 1992 and Vijay, Foldyna & Remisz 1993) modulation of high speed continuous water jets have a definite advantage, the dynamic modulation being superior. Natural pulsed jets, as pointed out earlier, can be produced simply by accelerating the break-up process of a continuous jet into droplets (Danel & Guilloud 1974 and Houlston & Vickers 1978) or by using external choppers. The choppers can be lasers (Mazurkiewicz, 1983), mechanical vibration of the nozzle body or piping (Mazurkiewicz 1984 and Wylie 1972) or perforated rotating discs (Erdmann-Jesnitzer, Louis & Schikorr 1980, Kiyohashi, Kyo & Tanaka 1984, Lichtarowicz & Nwachukwu 1978 and Vijay, Remisz & Shen 1992). It should be noted, however, that these type of devices are highly inefficient and are not practical. Furthermore, although results on fragmentation of rocks reported by Vijay, Remisz and Shen (1992) were encouraging, the jets produced by this technique are not truly pulsed (to achieve a frequency of about 1 kHz, they used a bulky 0.42 m diameter rotating disc with 90 holes!).

5 PULSE CANNONS

5.1 *Free piston devices*

In these devices a piston is rapidly accelerated by using explosives (Daniel, Rowlands & Labus 1974, Daniel 1976 and Watson, Williams & Brade 1982 & 1984), bullets (Rochester & Brunton 1972) or the sudden release of the energy stored in a compressed gas (Edney, 1976 and others). In these devices, the nozzle geometry (shape, length and area ratio) is of primary importance. It determines not only the maximum exit velocity and duration of the pulse, but also the maximum static pressure generated within the nozzle (see for example, Atanov 1982 & 1988, Cooley 1985, Edney 1976, Edwards & Welsh 1978 & 1980, Edwards, Smith & Farmer 1982 and Edwards & Farmer 1984). Although several type of nozzle configurations (the most common one being the simple conical entry nozzle) have been reported, nozzles with well designed internal contours (exponential, hyperbolic and parabolic) appear to produce highly coherent jets. According to Labus (1991), free piston devices appear to suffer from several drawbacks, the most serious ones being (i) short duration of the pulse, (ii) small amount of liquid in the pulse (about 7% of the total volume in the nozzle), (iii) low efficiency (that is, conversion of piston energy to liquid pulse energy) and (iv) size and weight. Furthermore, their usefulness is limited due to high stresses in the piston and other components, fatigue failure and high level of noise.

5.2 *Pressure extrusion devices*

A pressure extrusion device (usually known as differential area intensifier "DAI") is essentially a snap acting intensifier which produces a pulsed jet by using compressed air or other gas to drive a piston to extrude the liquid through a nozzle. In contrast to the free piston type, the pulse duration is quite high and is of the order of 200 ms. The liquid content of the slug can be as high as 90-95% of the volume of the high pressure cylinder from which it is extruded. For this reason their efficiencies are also much higher (≈ 90%). Although original versions weighed as much as 1360 kg, the more recent ones appear to weigh only about 17 kg (Labus 1991).

5.3 *Hydraulic pulse generator*

In this type of device, the energy required to produce the pulse is stored in the liquid itself by compressing it in a vessel to high pressures up to 400 MPa (Kollé 1993). Release of a fast acting valve results in the formation of a high speed slug of liquid with high energy content. Kollé (1993) has developed units based on this principle with energy content in the slugs up to 250 kJ. Apart from fragmentation of rocks and concrete, other potential uses of this device are forming of metals and impact processing of materials. The operating principles of the blow-down cannon (Pater & Borst 1983 and Pater 1984 & 1986) are similar to the hydraulic pulse generator.

6 DYNAMIC MODULATION OF CONTINUOUS LIQUID JETS

This mode of generating pulsed (or cavitating) jets offers several advantages: (i) the problem of filling the nozzle with the liquid after each shot is completely eliminated, (ii) pumps which produce continuous jets up to 200 MPa are highly reliable and economical, (iii) the effects of pulsing are superimposed on those produced by the continuous jet, (iv) size and weight are not major considerations and (v) offers controlled processing of materials.

6.1 *Electro-discharge technique*

The schematics of a system currently under investigation in the author's laboratory is depicted in Fig. 3. Depending upon the nozzle-electrode

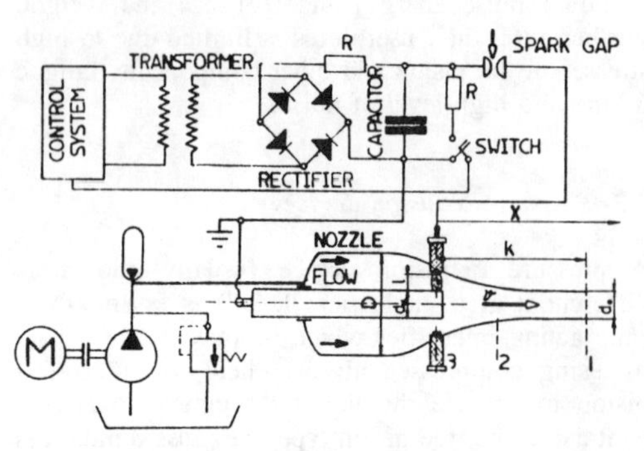

Fig. 3 Schematics of a nozzle configuration for the electro-discharge technique (Hawrylewicz, Puchala & Vijay 1986)

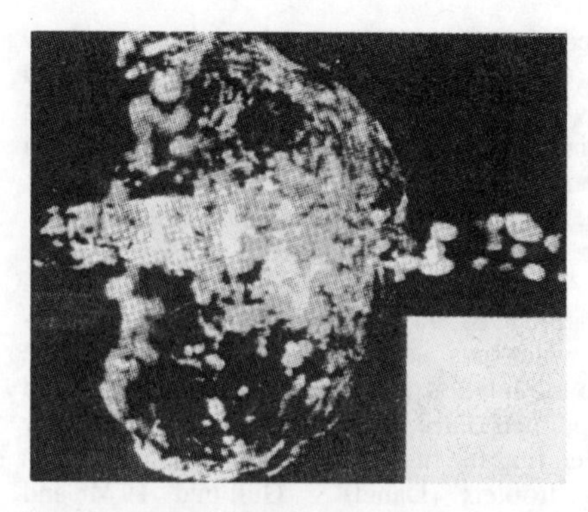

Fig. 4 Typical appearance of a percussive jet generated by dynamic modulation of a continuous jet (Nebeker 1984)

configuration this technique can be operated in several different ways (Atanov, et. al. 1979, Gustafsson 1983, Hawrylewicz, Puchala & Vijay 1986 and Zhenfang & Min 1987). For example, the nozzle can be designed to achieve all the three effects, viz., (i) the shock waves (ii) expanding plasma channel which can act as a piston to propel a slug of liquid and (iii) the plasma channel which will eventually become a powerful cavitation bubble, which accompany the high voltage discharge (energy of discharge ≈ 20 kJ) in the high speed stream of liquid. Since each phenomenon occurs at different times, it is possible to use all the three effects in succession. Although no experimental results have been reported in the literature, the combined effect of the three mechanisms could be quite effective in processing any type of material.

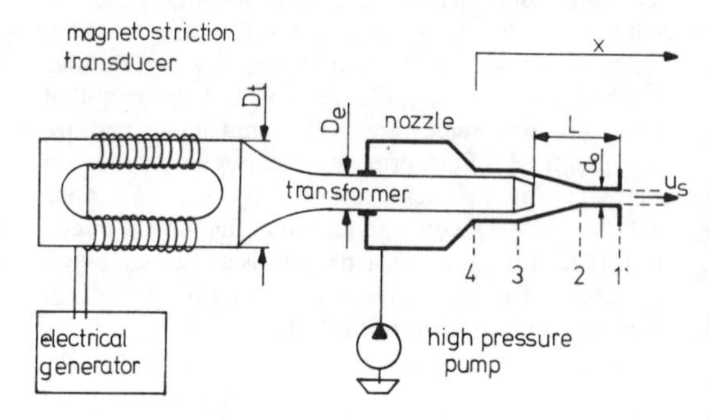

Fig. 5 Schematics of a nozzle configuration for generating ultrasonically modulated cavitating or pulsed jets (Puchala & Vijay 1984)

6.2 *Percussive jets*

A percussive jet consists of a series of large pulses which are obtained by modulating a continuous stream of liquid (Nebeker & Rodriguez 1976, Nebeker 1981, 1983, 1984 & 1987 and Nebeker & Cramer 1983). Typical appearance of a slug (water) in a percussive jet is shown in Fig. 4 (Nebeker 1984). Nebeker (1981 1983, 1984 & 1987) has reported ample results to prove the effectiveness of this technique. In the author's laboratory, a technique which employs ultrasonic waves within the nozzle (Fig. 5), is employed to modulate a continuous stream of liquid to produce either pulsed or cavitating jets (Puchala & Vijay 1984, Vijay 1992 and Vijay, Foldyna & Remisz 1993). In this mode, large pulses (resulting from the bunching effect) are produced by the velocity or flow modulation caused by the vibrating tip (Sami & Ansari 1981). Experimental results obtained to date are highly encouraging (Fig. 6, see also Vijay, Foldyna & Remisz 1993).

7 PASSIVE MODULATION OF CONTINUOUS LIQUID JETS

In this mode of operation, no external forces are required to excite the continuous stream of liquid to generate pulsed jets. The characteristics of pulsed jets produced by this technique depend strongly on the nozzle configuration. These are commonly known as self-modulated jets with Helmholtz oscillators (Sami & Anderson 1984 and Shen & Wang 1988) or self-resonating pulsed jets (Chahine, et. al. 1983 and Liao & Huang 1986). In the case of the former, two orifices are placed in tandem such that vortices are generated in the flow in the intervening chamber. The strength of the vortices, which are responsible for exciting the pulsing action, is amplified when the frequency of the pressure disturbances coincide with the natural frequency of the cavity in the chamber. The designs investigated by Chahine, et. al. (1983), essentially consist of a Helmholtz chamber followed by an organ pipe (various configurations are possible, see Fig. 7).

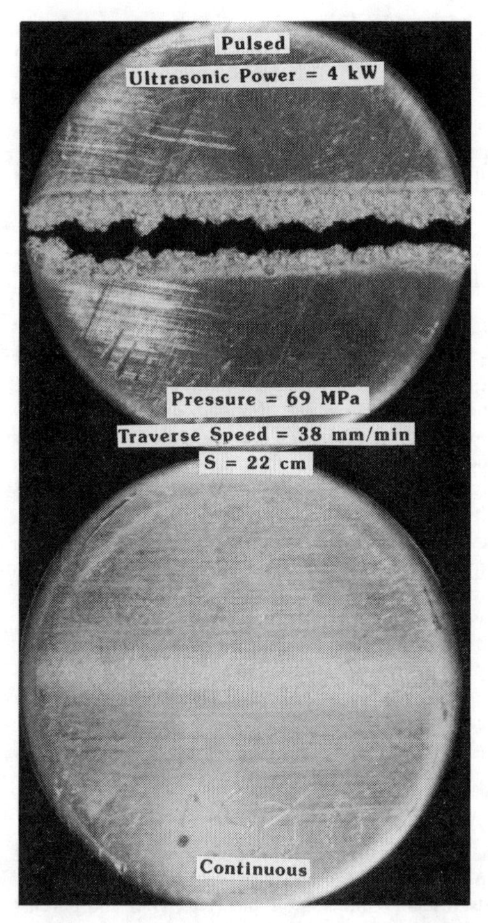

Fig. 6 Samples of copper discs cut with continuous and ultrasonically modulated pulsed jets (Vijay, Foldyna & Remisz 1993)

Several interesting cleaning applications with these devices have been reported by these authors.

8 MECHANISM OF JET MATERIAL INTERACTION

The response of the material to the high velocity transient jet impact depends on the characteristics of the pulse and the nature of the material (brittle, elastic, elastic-plastic, etc). The situation becomes even more complex when shock wave effects are taken into consideration (Pritchett & Riney 1974). Although, as yet no satisfactory theoretical analysis exists to predict the material behaviour and hence the prediction of the material removal rate, a number of investigators have made attempts to model the characteristics of the impinging pulse and the mechanism of interaction (Atanov 1982 & 1988, Beutin,

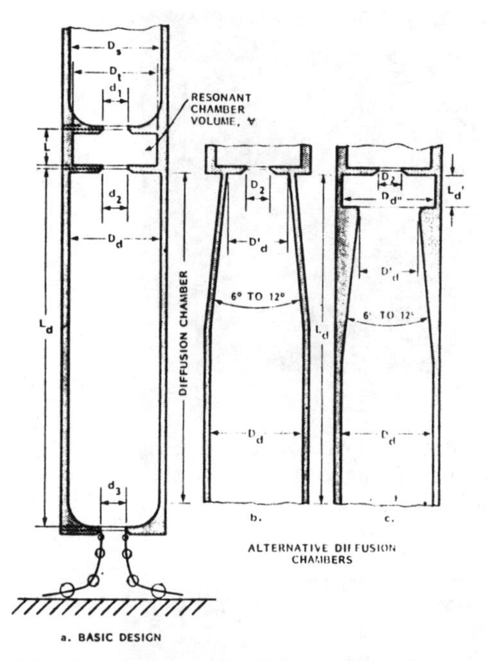

Fig. 7 Conceptual nozzle configurations for generating self-resonating pulsed or cavitating jets (Chahine, et. al. 1983)

Erdmann-Jesnitzer & Louis 1974, Cheng, et. al. 1985, Daniel, Rowlands & Labus 1974, Daniel 1976, Edwards & Welsh 1980, Edwards, Smith & Farmer 1982, Erdmann-Jesnitzer, Louis & Wiedemeier 1978, Huang, Hammitt & Yang 1972, Hwang & Hammitt 1976, Labus 1991, Lawrence 1974, Nebeker & Cramer 1983, Pritchett & Riney 1974, Rochester & Brunton 1972, Sami & Ansari 1981, Shen & Wang 1988 and Watson, Williams & Brade 1982 & 1984). The basic principles of the mechanism of interaction are explained very well by Rochester & Brunton (1972). The treatment by other investigators (for instance, Daniel, Rowlands & Labus 1974, Daniel 1976 and Pritchett & Riney 1974) is basically a refinement of the processes explained by these authors.

According to Rochester & Brunton (1972), for impact velocities below about 1000 m/s, the loading is initially compression and subsequently shear. At higher velocities, the stagnation pressure of the steady flow becomes comparable to failure stresses of most of the materials. This results in damage and penetration of the jet throughout the duration of impact. The ring type of fracture patterns observed in glass plates by these authors suggest that failure of the brittle materials is determined by the maximum tensile stresses produced during loading. In the high speed impact of a liquid jet against a plane surface, the maximum tensile stresses act radially across the

Fig. 8 A 0.9 m cube of Pennant sandstone fractured by a pulsed jet at 1000 MPa (Moodie & Taylor 1974)

Fig. 9 A block of barre granite fractured with a single pulsed jet emerging at 925 m/s. Piston energy = 45 kJ, S = 15 cm, Volume broken = 2600cm^3 (Cooley 1972)

circumference of the compressed central region which initially remains intact. For ductile materials tensile stress is of less importance, although this ultimately influences the mechanism of interaction. The plastic flow under the action of the shear stresses set up by the impact is more important. The plastic yielding develops a permanent indentation or pit on the surface. The authors also found that a layer of water on the target has a detrimental effect on the damage. This is believed to be associated with the divergence of the pressure pulse as it passes through the layer. This observation suggests that in order to retain the power of pulsed liquid jets for processing of materials, it is quite important to separate the impingement sequence of the pulses so that the surface is always free of the layer of liquid left behind by the previous pulse.

9 RESULTS

Most of the results with pulsed jets pertain to massive breaking of concrete or rocks. For this reason, it is quite difficult to find systematic or comprehensive set of results. However, abundant evidence exists to show that pulsed liquid jets are quite powerful for processing a variety of materials. Typical results on fragmentation of rocks are depicted in Fig. 8 (Moodie & Taylor 1974) and Fig. 9 (Cooley 1972). The results obtained on cutting copper samples with continuous and ultrasonically modulated pulsed jets (Fig. 6) confirms the power of pulsed liquid jets beyond any doubt. At a pressure of only 69 MPa (V_{tr} = 38 mm/min) and a standoff distance of 22 cm, the copper sample (3.2 mm thick) was completely cut through with the pulsed water jet (Vijay, Foldyna & Remisz 1993).

10 CONCLUSIONS

There are several types of pulsed water jets. These can be classified broadly as "natural" and "forced" pulsed jets. While some types (e.g., DAI and percussive jets) are highly efficient for processing different types of materials, others (external interruption or free piston) are not. The brief review and sample results presented in this paper clearly demonstrate that pulsed liquid jets are powerful and have a good potential for a variety of cleaning, cutting and fragmentation applications.

The fact that interest in the pulsed liquid jet technology has waned in recent years suggests that further work is required to: (i) improve their reliability (ii) optimize the nozzle systems and (iii) advance the knowledge of the mechanics of material removal. The latter is quite important because design of the optimal and cost effective systems depend on this knowledge.

11 ACKNOWLEDGMENTS

The author is thankful to Dr. J. Remisz and Dr. X. Shen (guest scientists at the National Research Council of Canada) for assisting in the review of the literature. Partial funding received from the HDRK Mining Research Limited, Mississauga, Ontario, Canada and the office of the IRAP (Industrial Research Program) at NRC is gratefully acknowledged.

12 REFERENCES

Atanov, G.A., N. Yu, N. Golovko and A.M. Krivoruchoko, 1979. An Electro-impulse water jet. *Izvestiya Vysshikh Uchebnykh Zavedenii - Energetika.* 7:77-81.

Atanov, G.A. 1982. Interior ballistics of impulsive water jet. Paper C5, *Proc. 6th International Symposium on Jet Cutting Technology*. pp.141-159. BHRA, Cranfield, Bedford, England.

Atanov, G.A. 1988. Powder impulsive water jetter. *Proc. 11th International Conference on Jet Cutting Technology*. pp.295-303. London, Kluwer Academic Publishers.

Atanov, G.A. 1991. The impulsive water jet device: A new machine for breaking rock. *International Journal of Water Jet Technology*. 1:85-91.

Beutin, E.F., F. Erdmann-Jesnitzer & H. Louis 1974. Material behaviour in the high-speed liquid jet attacks. Paper C1, *Proc. 2nd International Symposium on Jet Cutting Technology*. pp.1-18. BHRA, Cranfield, Bedford, England.

Bresee, J.C., G.A. Cristy & W.C. McClain 1972. Some comparisons of continuous and pulsed jets for excavation. Paper B9, *Proc. 1st International Symposium on Jet Cutting Technology*. pp.101-112. BHRA, Cranfield, Bedford, England.

Chahine, G.L., A.F. Conn, V.E. Johnson & G.S. Frederick 1983. Cleaning and cutting with self-resonating pulsed water jets. *Proc. 2nd U.S. Water Jet Conference*. pp.167-173. WJTA (Water Jet Technology Association), St. Louis, USA.

Cheng, D.Z., G.L. Liang, C.X. Lu and Y. Xu 1985. Relationship between water cannon design, pulsed water jet anatomy and rock breaking effect. *Proc. 3rd U.S. Water Jet Conference*. pp.309-326. WJTA (Water Jet Technology Association), St. Louis, USA.

Chermensky, G.P. 1976a. Breaking coal and rock with pulsed water jets. Paper D4, *Proc. 3rd International Symposium on Jet Cutting Technology*. pp.33-50. BHRA, Cranfield, Bedford, England.

Chermensky, G.P. 1976b. Experimental investigation of the reliability of impulsive water cannons. Paper H1, *Proc. 3rd International Symposium on Jet Cutting Technology*. pp.1-14. BHRA, Cranfield, Bedford, England.

Chermensky, G.P. & G.P. Davidyants 1980. Pulse water jet pressures in rock breaking. Paper C6, *Proc. 5th International Symposium on Jet Cutting Technology*. pp.155-163. BHRA, Cranfield, Bedford, England.

Cooley, W.C. 1972. Rock breakage by pulsed high pressure water jets. Paper B7, *Proc. 1st International Symposium on Jet Cutting Technology*. pp.101-112. BHRA, Cranfield, Bedford, England.

Cooley, W.C. 1985. Computer aided engineering and design of cumulation nozzles for pulsed liquid jets. *Proc. 3rd U.S. Water Jet Conference*. pp.327-335. WJTA (Water Jet Technology Association), St. Louis, USA.

Danel, F. & J.C. Guilloud 1974. A high speed concentrated drop stream generator. Paper A3, *Proc. 2nd International Symposium on Jet Cutting Technology*. pp.33-38. BHRA, Cranfield, Bedford, England.

Daniel, I.M., R.E. Rowlands & T.J. Labus 1974. Photoelastic study of water jet impact. Paper A1, *Proc. 2nd International Symposium on Jet Cutting Technology*. pp.1-18. BHRA, Cranfield, Bedford, England.

Daniel, I.M. 1976. Experimental studies of water jet impact on rock and rocklike materials. Paper B3, *Proc. 3rd International Symposium on Jet Cutting Technology*. pp.27-46. BHRA, Cranfield, Bedford, England.

Edney B.E. 1976. Experimental studies of pulsed water jets. Paper B2, *Proc. 3rd International Symposium on Jet Cutting Technology*. pp.11-26. BHRA, Cranfield, Bedford, England.

Edwards, D.G. & D.J. Welsh 1978. A numerical study of nozzle design for pulsed water jets. Paper B1, *Proc. 4th International Symposium on Jet Cutting Technology*. pp.1-12. BHRA, Cranfield, Bedford, England.

Edwards, D.G. & D.J. Welsh 1980. The influence of design and material properties on water cannon performance. Paper G3, *Proc. 5th International Symposium on Jet Cutting Technology*. pp.353-368. BHRA, Cranfield, Bedford, England.

Edwards, D.G., R.M. Smith & G. Farmer 1982. The coherence of impulsive water jets. Paper C4, *Proc. 6th International Symposium on Jet Cutting Technology*. pp.123-140. BHRA, Cranfield, Bedford, England.

Edwards, D.G. & G.P. Farmer 1984. A study of piston-water impact in an impulsive water cannon. Paper D1, *Proc. 7th International Symposium on Jet Cutting Technology*. pp.163-178. BHRA, Cranfield, Bedford, England.

Erdmann-Jesnitzer, F., H. Louis & J. Wiedemeier 1978. Material behaviour, material stressing, principle aspects in the application of high speed water jets. Paper E3, *Proc. 4th International Symposium on Jet Cutting Technology*. pp.29-44. BHRA, Cranfield, Bedford, England.

Erdmann-Jesnitzer, F., H. Louis & A.W. Schikorr 1980. Cleaning, drilling and cutting by interrupted jets. Paper B1, *Proc. 5th International Symposium on Jet Cutting Technology*. pp.45-55. BHRA, Cranfield, Bedford, England.

Evers, J.L. & D.L. Eddingfield 1981. The effect of the piping system on liquid jet modulation. *Proc. 1st U.S. Water Jet Symposium*. pp.I-1.1-13.

WJTA (Water Jet Technology Association), St. Louis, USA.

Evers, J.L., D.L. Eddingfield & J.Y. Yuh 1983. Dimensionless pipe length analysis for modulation systems. *Proc. 2nd U.S. Water Jet Conference*. pp.1-12. WJTA (Water Jet Technology Association), St. Louis, USA.

Fang, L.Z. & T.C. Lin 1987. Theoretical analysis and experimental study of the self-excited oscillation pulsed jet device. *Proc. 4th U.S. Water Jet Conference*. pp.27-34. WJTA (Water Jet Technology Association), St. Louis, USA.

Gustafsson, G. 1983. The focused shock technique for producing transient water jets. *Proc. 2nd U.S. Water Jet Conference*. pp.39-42. WJTA (Water Jet Technology Association), St. Louis, USA.

Hawrylewicz, B.M., Puchala, R.J. & M.M. Vijay 1986. Generation of pulsed or cavitating jets by electric discharges in high speed continuous water jets. Paper 36, *Proc. 8th International Symposium on Jet Cutting Technology*. pp.345-352. BHRA, Cranfield, Bedford, England.

Houlston, R. & G.W. Vickers 1978. Surface cleaning using water-jet cavitation and droplet erosion. Paper H1, *Proc. 4th International Symposium on Jet Cutting Technology*. pp.1-18. BHRA, Cranfield, Bedford, England.

Huang, Y.C., F.G. Hammitt & W.J. Yang 1972. Mathematical modelling on normal impact between a finite cylindrical liquid jet and non-slip, flat rigid surface. Paper A4, *Proc. 1st International Symposium on Jet Cutting Technology*. pp.57-68. BHRA, Cranfield, Bedford, England.

Hwang J.B. & F.G. Hammitt 1976. Transient distribution of the stress produced by the impact between a liquid drop and an aluminum body. Paper A1, *Proc. 3rd International Symposium on Jet Cutting Technology*. pp.1-15. BHRA, Cranfield, Bedford, England.

Kiyohashi, H., M. Kyo & S. Tanaka 1984. Hot dry rock drilling by interrupted water jets. Paper H1, *Proc. 7th International Symposium on Jet Cutting Technology*. pp.395-418. BHRA, Cranfild, Bedford, England.

Kollé, J.J. 1993. Development and applications of a hydraulic pulse generator. Paper 32, *Proc. 7th American Water Jet Conference*. pp.459-471. WJTA (Water Jet Technology Association), St. Louis, USA.

Labus, T.J. 1982. A comparison of pulsed jets versus mechanical breakers. Paper F1, *Proc. 6th International Symposium on Jet Cutting Technology*. pp.229-240. BHRA, Cranfield, Bedford, England.

Labus, T.J. 1991. Pulsed fluid jet technology. *Proc. 1st Asian Conference on Recent Advances in Jetting Technology*. pp.136-143. CI-Premier Pte. Ltd., 150 Orchard Road #07-14, Orchard Plaza, Singapore.

Lawrence, R.J. 1974. Elastic-plastic target deformation due to a high speed pulsed water jet impact. Paper X, *Proc. 2nd International Symposium on Jet Cutting Technology*. pp.21-29. BHRA, Cranfield, Bedford, England.

Liao, Z.F. & D.S. Huang 1986. Nozzle device for the self-excited oscillation of a jet. Paper 19, *Proc. 8th International Symposium on Jet Cutting Technology*. pp.195-201. BHRA, Cranfield, Bedford, England.

Lichtarowicz, A. & G. Nwachukwu 1978. Erosion by an interrupted jet. Paper B2, *Proc. 4th International Symposium on Jet Cutting Technology*. pp.13-18. BHRA, Cranfield, Bedford, England.

Lucke, W.N. & W.C. Cooley 1974. Development and testing of a water cannon for tunnelling. Paper J3, *Proc. 2nd International Symposium on Jet Cutting Technology*. pp.27-43. BHRA, Cranfield, Bedford, England.

Mazurkiewicz, M. 1983. An analysis of one possibility for pulsating a high pressure water jet. *Proc. 2nd U.S. Water Jet Conference*. pp.15-22. WJTA (Water Jet Technology Association), St. Louis, USA.

Mazurkiewicz, M. 1984. The analysis of high pressure water jet interruption through ultrasonic nozzle vibration. Paper P5, *Proc. 7th International Symposium on Jet Cutting Technology*. pp.531-536. BHRA, Cranfield, Bedford, England.

Mellors, W., U.H. Mohaupt & D.J. Burns 1976. Dynamic response and optimization of a pulsed water jet machine of the pressure extrusion type. Paper B4, *Proc. 3rd International Symposium on Jet Cutting Technology*. pp.47-58. BHRA, Cranfield, Bedford, England.

Mohaupt, U.H., D.J. Burns, G.G. Yie and W. Mellors 1978. Design and dynamic response of a pulse-jet pavement breaker. Paper D2, *Proc. 4th International Symposium on Jet Cutting Technology*. pp.17-28. BHRA, Cranfield, Bedford, England.

Moodie, K. & G. Artingstall 1972. Some experiments on the application of high pressure water jets for mineral excavation. Paper E3, *Proc. 1st International Symposium on Jet Cutting Technology*. pp.25-44. BHRA, Cranfield, Bedford, England.

Moodie, K. & G. Taylor 1974. The fracturing of rocks by pulsed water jets. Paper H7, *Proc. 2nd International Symposium on Jet Cutting Technology*. pp.77-88. BHRA, Cranfield, Bedford, England.

Moodie, K. 1976. Coal ploughing assisted with high-pressure water jets. Paper D6, *Proc. 3rd International Symposium on Jet Cutting Technology*. pp.65-79. BHRA, Cranfield, Bedford, England.

Moodie, K. & M.G. Tomlin 1980. The application of high pressure water jets for roadway drivage through hard rock. Paper C5, *Proc. 5th International Symposium on Jet Cutting Technology*.

pp.141-154. BHRA, Cranfield, Bedford, England.
Nebeker, E.B. & S.E. Rodriguez 1976. Percussive water jets for rock cutting. Paper B1, *Proc. 3rd International Symposium on Jet Cutting Technology*. pp.1-9. BHRA, Cranfield, Bedford, England.
Nebeker, E.B. 1981. Development of large diameter percussive jets. *Proc. 1st U.S. Water Jet Symposium*. pp.IV-5.1-11. WJTA (Water Jet Technology Association), St. Louis, USA.
Nebeker, E.B. 1983. Standoff distance improvement using percussive jets. *Proc. 2nd U.S. Water Jet Conference*. pp.25-34. WJTA (Water Jet Technology Association), St. Louis, USA.
Nebeker, E.B. & J.B. Cramer 1983. Visualization of the central core of high-speed water jets - an infrared technique. *Proc. 2nd U.S. Water Jet Conference*. pp.75-80. WJTA (Water Jet Technology Association), St. Louis, USA.
Nebeker, E.B. 1984. Potential and problems of rapidly pulsing water jets. Paper B1, *Proc. 7th International Symposium on Jet Cutting Technology*. pp.51-68. BHRA, Cranfield, Bedford, England.
Nebeker, E.B. 1987. Percussive Jets - State of the art. *Proc. 4th U.S. Water Jet Conference*. pp.19-25. WJTA (Water Jet Technology Association), St. Louis, USA.
Pater, L.L. & P.H. Borst 1983. An extrusion type pulsed jet device. *Proc. 2nd U.S. Water Jet Conference*. pp.83-98. WJTA (Water Jet Technology Association), St. Louis, USA.
Pater, L.L. 1984. Experiments with a cumulation pulsed jet device. Paper B3, *Proc. 7th International Symposium on Jet Cutting Technology*. pp.83-90. BHRA, Cranfield, Bedford, England.
Pater. L.L. 1986. The blowdown water cannon: A novel method for powering the cumulation nozzle. Paper 20, *Proc. 8th International Symposium on Jet Cutting Technology*. pp.203-210. BHRA, Cranfield, Bedford, England.
Petrakov, A.I. & O.D. Krivorotko 1978. Some experience in developing mining roadways using the experimental heading machine with pulsed water jets. Paper J3, *Proc. 4th International Symposium on Jet Cutting Technology*. pp.23-36. BHRA, Cranfield, Bedford, England.
Pritchett, J.W. & T.D. Riney 1974. Analysis of dynamic stresses imposed on rocks by water jet impact. Paper B2, *Proc. 2nd International Symposium on Jet Cutting Technology*. pp.15-36. BHRA, Cranfield, Bedford, England.
Puchala, R.J. & M.M. Vijay 1984. Study of an ultrasonically generated cavitating or interrupted jet: Aspects of design. Paper B2, *Proc. 7th International Symposium on Jet Cutting Technology*. pp.69-82. BHRA, Cranfield, Bedford, England.
Rochester, M.C. & J.H. Brunton 1972. High speed impact of liquid jets on solids. Paper A1, *Proc. 1st International Symposium on Jet Cutting Technology*. pp.1-24. BHRA, Cranfield, Bedford, England.
Sami, S. & H. Ansari 1981. Governing equations in a modulated liquid jet. *Proc. 1st U.S. Water Jet Symposium*. pp.I-2.1-9. WJTA (Water Jet Technology Association), St. Louis, USA.
Sami, S. & C. Anderson 1984. Helmholtz oscillator for the self-modulation of a jet. Paper B4, *Proc. 7th International Symposium on Jet Cutting Technology*. pp.91-98. BHRA, Cranfield, Bedford, England.
Shen, S.Z. & Z.M. Wang 1988. Theoretical analysis of a jet-driven Helmholtz resonator and effect of its configuration on the water jet cutting property. Paper D4, *Proc. 9th International Symposium on Jet Cutting Technology*. pp.189-201. BHRA, Cranfield, Bedford, England.
Singh, M.M., L.A. Finlayson & P.J. Huck 1972. Rock breakage by high pressure water jets. Paper B8, *Proc. 1st International Symposium on Jet Cutting Technology*. pp.113-124. BHRA, Cranfield, Bedford, England.
Vallvé, F.X., U.H. Mohaupt, J.G. Kalbfleisch & D.J. Burns 1980. Relationship between jet penetration in concrete and design parameters of a pulsed water-jet machine. Paper D2, *Proc. 5th International Symposium on Jet Cutting Technology*. pp.215-228. BHRA, Cranfield, Bedford, England.
Vijay, M.M., C. Zou & S. Tavoularis 1990. A study of the characteristics of cavitating water jets by photography and erosion. Chapter 3, *Proc. 10th International Symposium on Jet Cutting Technology*. pp.37-67. London, Elsevier Applied Science.
Vijay, M.M. 1991. Properties and parameters of water jets. *Proc. International Conference Geomechanics 91*. pp.207-222. Rotterdam, Balkema.
Vijay, M. M. 1992. Ultrasonically generated cavitating or interrupted jet. *U. S. Patent No. 5,154,347.*
Vijay, M.M., J. Remisz & X. Shen 1992. Fragmentation of hard rocks with discontinuous water jets. *Proc. 3rd Pacific Rim International Conference on Water Jet Technology*. pp.201-224. Water Jet Technology Society of Japan, Tokyo, Japan.
Vijay, M.M. & J. Remisz 1993. Preweakening of hard rocks with water jets. *Proc. 7th American Water Jet Conference*. pp.405-425. WJTA (Water Jet Technology Association), St. Louis, USA.
Vijay, M.M., J. Foldyna & J. Remisz 1993. Ultrasonic Modulation of High-Speed Water Jets. *Proc. International Conference Geomechanics 93.* Rotterdam, Balkema.
Wang, F.D., Z. Mou, H. Bao & D. Cheng 1984. Field applications of pulse and swing-oscillating jets for coal mine entry driving. Paper F3, *Proc. 7th International Symposium on Jet Cutting Technology*. pp.331-336. BHRA, Cranfield, Bedford, England.
Watson, A.J., F.T. Williams & R.G. Brade 1982.

Impact pressure characteristics of a water jet. Paper C2, *Proc. 6th International Symposium on Jet Cutting Technology*. pp.93-106. BHRA, Cranfield, Bedford, England.
Watson, A.J., F.T. Williams & R.G. Brade 1984. The relationship between impact pressure and anatomical variations in a water jet. Paper D3, *Proc. 7th International Symposium on Jet Cutting Technology*. pp.193-209. BHRA, Cranfield, Bedford, England.
Wu, W.Z.B., D.A. Summers & M.J. Tzeng 1987. Dynamic characteristics of water jets generated from oscillating systems. *Proc. 4th U.S. Water Jet Conference*. pp.35-41. WJTA (Water Jet Technology Association), St. Louis, USA.
Wylie, E.B. 1972. Pipeline dynamics and the pulsed jet. Paper A5, *Proc. 1st International Symposium on Jet Cutting Technology*. pp.69-80. BHRA, Cranfield, Bedford, England.
Yie, G.G., D.J. Burns & U.H. Mohaupt 1978. Performance of a high-pressure pulsed water-jet device for fracturing concrete pavement. Paper H6, *Proc. 4th International Symposium on Jet Cutting Technology*. pp.67-86. BHRA, Cranfield, Bedford, England.
Ze-sheng, M. 1987. An experimental investigation of water cannon for driving coal mine entry. *Proc. International Water Jet Symposium*. pp.5-46-50F. F.D. Wang (ed), Colorado School of Mines, Golden, USA.
Zhenfang, L. & Z. Min 1987. Design factors on electrohydraulic pulsed focus water jet generator. *Proc. International Water Jet Symposium*. pp.2-44-58. F.D. Wang (ed), Colorado School of Mines, Golden, USA.

Water jet cutting
Lectures

Geomechanics 93, Rakowski (ed.) © 1994 Balkema, Rotterdam, ISBN 90 5410 354 X

A waterjet mill for coal grinding

A. Bortolussi, R. Ciccu & W. M. Kim
Department of Mining and Minerals Engineering, Mineral Science Study Centre – CNR, University of Cagliari, Italy

ABSTRACT: The paper deals with the coal comminution results obtained with a waterjet mill designed and built at the DIMM Laboratories. Grinding efficiency appears very interesting and susceptible to substantial improvement with better design, optimization of operational variables and equipment scale-up.
Despite being a energy-intensive technology, waterjet represents a very attractive solution for fine grinding, especially in the case of coal and ores having favourable textural characteristics.

1 INTRODUCTION

Comminution is a very important step in mineral processing from both the technical and economic points of view. Indeed, size reduction often involves large energy consumption and considerable wear of the grinding media, resulting in a relatively high operational cost. Moreover traditional mills entail very important capital expenses.

Conventional mechanical methods of crushing and grinding have a very low efficiency, so that a search for advanced grinding techniques is of great importance. A variety of novel techniques of comminution is currently being explored. Among them, high pressure waterjet is now being considered for its disintegration potential.

Waterjet can be used either for assisting mechanical grinding in conventional mills (Mazurkiewicz et al. 1983) or as the only fracturing agent in specially designed apparatus (Bortolussi et al. 1993).

Very intersting micronization results have been obtained by applying a water jet to a disk mill (Mazurkiewicz et al. 1985).

The energy required to achieve a given size reduction increases as the product size decreases. This is due to many factors and is the consequence not only of the type of mill and the micromechanical features, but also of the mechanisms of failure that operate at the particulate level.

Since coal is generally anisotropic, heterogeneous, and extensively pre-cracked as a function of the degree metamorphism, an analitical approach to coal fragmentation is very complex.

In waterjet comminution, differential coal porosity and permeability are also important as these properties control the degree of water absorption which directly influences the rate of disintegration.

Due to high velocity collisions of the jet with the coal particles shock waves are generated which affect microcrack formation and induce extension of existing cracks. Water penetration into pre-exixting and created cracks supports this effect and develops the internal pressure necessary to keep cracks growing (Mazurkiewicz et al. 1993 and Ciccu et al. 1993).

2 BACKGROUND

Although the mechanism of high pressure waterjet milling is not yet completely understood, it appears that the finest product can be generated when the jet is able to attack very small and free particles.

3 HIGH PRESSURE WATER JET TECHNIQUES

The main features of a water jet are that, due to its small diameter and very high velocity, a powerful beam can be directed to the target. The jet's high energy density is concentrated in a very small impact zone, while intense differential pressure across the jet leads to generation and growth of microcracks.

The fragmentation mechanism consists in the development of an hydromechanical action by which water is forced into the existing cracks under the stagnation pressure generated by the impact of the jet, producing an increasingly dense fissuration. Only high pressure waterjet can follow the propagation velocity of the microcrack tip.

It is a well known fact that the force required for fracture propagation is reduced by the presence of the water at the crack tip, thus resulting in improved breaking efficiency.

Moreover, material is dynamically stressed by the wave energy generated by the impinging jet.

Contact time is a very important factor. If the relative velocity between waterjet and a solid particle is low, energy consumption required to produce fine product is rather high. For high feed rates the jet acts on the individual particles for a short time, thereby permitting only shallow fractures and requiring much more time for the product to form. Low feed rate permits deep cutting inside the material and much longer fractures thereby form, consuming much more energy (Ciccu et al. 1993).

Waterjet technology can be profitably used for ore beneficiation if the mechanical strength of the r.o.m. components is well differentiated. The separation of coal from the accompanying mineral matter or the recovery of the valuable minerals from a crude ore can be achieved by use of selective waterjet disintegration followed by size classification.

A crushing machine based on the use of a high pressure water jet is expected to be an extremely simple and relatively cheap-to-operate. Mechanical wear or contamination of the product is also likely to be reduced.

In the preparation of coal/water slurry the required bimodal frequency distribution of particle size can be obtained with two parallel mills by properly tuning the jet pressure or by adjusting the discharge slot width.

4 BASIC EXPERIMENTS

4.1 Material

Grinding tests have been carried out on a sample of a low-rank coal coming from the Gardanne Mine (France), after washing. The ash content of the coal air-dried is 7.51. Bond index according to the standard procedure resulted to be 11.23, for a reference size of 250 micrometres.

The sample was crushed and screened into three size classes: a coarser fraction (-15 +10 mm), an intermediate class (-10 +5 mm) and a finer fraction (-5 +1 mm). Each class was treated separately according to the grinding test plan.

4.2 Apparatus

The first prototype of a waterjet grinding mill designed and built at the Mining and Mineral Engineering Department of the University of Cagliari for studying the basic aspects of hydraulic comminution of mineral matter is the fixed jet mill shown in Figure 1.

This waterjet mill is divided in two distinguished parts:

-the mixing chamber and the collimator tube;

-the slurry vessel hosting the jet catcher device, where a rough size classification of the ground material also occurs.

The jet is directed along the axis of the collimator tube and crosses vertically the mixing chamber over a stand-off distance of about 10 cm to the tube inlet. The feed material enters at the top of the mixing chamber and falls towards the collimator tube at the bottom of the funnel-shaped section. Material flow is helped by the air draft produced by the jet itself.

Individual grains are hit by the water jet as they approach the collimator and are broken into smaller fragments to a top size allowing them to pass through the circular opening.

A further comminution takes place into the slurry vessel, as the fragments are projected at high velocity against the jet catcher consisting of a hard metal plate.

Fragmentation is completed as a result of collision between the incoming slurry and the particles bouncing away from the target.

4.3 Experimental plan

With the goal of disclosing the role of the different jet parameters and experimental conditions, the following variables have been included in the experimental plan:

- jet pressure: 40, 80 and 120 MPa;
- nozzle diameter: 0.80 and 1.25 mm;
- diameter of the collimator tube: 4.5 and 7.5 mm;

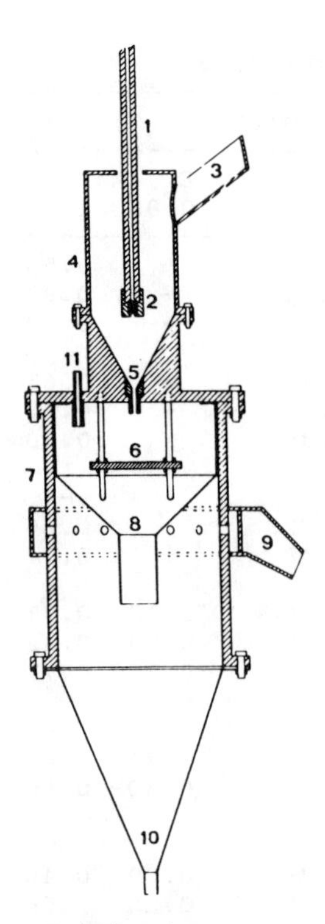

1 - Waterjet lance
2 - Nozzle holder
3 - Feed inlet
4 - Mixing chamber
5 - Collimator tube
6 - Jet catcher target
7 - Slurry chamber
8 - Slurry collector device
9 - Fine product outlet
10 - Coarse product outlet
11 - Air pressure relieve

Figure 1. Sketch of the laboratory waterjet mill

- feed rate: 50 and 80 g/s;

Each test consisted of a single pass through the waterjet mill without recirculation of the coarse fraction. Moreover the effect of stand-off distance was explored. Individual size classes were treated separately.

Water flowrate was that corresponding to the jet velocity generated through the contracting nozzles under the different pressures (40, 80 and 120 MPa), i.e. 0.21, 0.31 and 0.38 l/s for the 1.25 nozzle and 0.08, 0.13 and 0.15 l/s for the 0.8 mm nozzle, respectively.

4.4 Results

Results of wet screening of the comminution product are summarized in Table 1 in terms of d_{80}. Specific energy is calculated as the ratio between the input power of the jet and the feed rate. However data have been corrected by a factor:

$$K=(d_{80}^{-0.5}-D_{80}^{-0.5})/(d_{80(m)}^{-0.5}-D_{80}^{-0.5})$$

for comparison at equal product fineness, in agreement with the most credited theory of rock comminution (Bond's model). Size $d_{80(m)}$ is the minimum value obtained for each class, underlined in the Table and D_{80} is the reference size of the feed.

4.5 Discussion

A - Size distribution of the comminution product

Size distributions are characterized by a relatively low dispersion, being mostly concentrated within the range between 0.1 and 1.0 mm (Bortolussi et al. 1993).

Reduction ratio D_{80}/d_{80} as a function of pressure is reported in Figure 2, a) and b) for the different experimental conditions.

For a given nozzle diameter, R increases considerably with pressure, especially for the smaller collimator. As the discharge aperture increases, reduction ratio gradually decreases, especially for the finer particles which subtract themselves to the action of the jet as they freely enter the collimator tube: in this case an increase in pressure from 80 to 120 MPa has a very little effect.

Of course, since product size is controlled by the gap clearance, the largest R values are obtained for the coarser size class of the feed and for the smaller collimator diameter; the smallest values for the finer class and for the larger collimator diameter.

All the curves tend to converge toward a R value of 3 - 5, irrespective of the experimental conditions, as pressure decreases down to 40 MPa.

Except for few instances, reduction ratio for the 1.25 nozzle is 1.5 to 2 times larger than that for the 0.8 mm nozzle, at equal pressure and collimator diameter.

Discrepancies are due to experimental errors probably caused by the short duration of the tests.

The influence of the various experimental conditions is summarized here below:

Table 1. Experimental results of waterjet grinding

Collim. mm	Feed r. g/s	Size cl. mm	Press. MPa	d_{80} [mm] 0.80	d_{80} [mm] 1.25	E_s [MJ/kg] 0.80	E_s [MJ/kg] 1.25
4.5	50	- 5 + 1	40	-	0.99	-	0.56
			80	-	0.50	-	0.87
			120	-	0.22	-	0.90
		-10 + 5	40	-	1.86	-	0.61
			80	-	0.51	-	0.62
			120	-	0.35	-	0.90
		-15 +10	40	-	-	-	-
			80	-	0.72	-	0.66
			120	-	0.48	-	0.94
	80	- 5 + 1	40	1.37	0.86	0.19	0.31
			80	0.57	0.35	0.25	0.42
			120	0.41	0.35	0.35	0.77
		-10 + 5	40	-	-	-	-
			80	0.92	0.65	0.24	0.46
			120	0.52	0.42	0.30	0.63
		-15 +10	40	-	-	-	-
			80	0.51	0.78	0.29	0.40
			120	0.61	0.45	0.28	0.56
7.5	50	- 5 + 1	40	-	0.95	-	0.54
			80	-	0.75	-	1.22
			120	-	0.68	-	2.06
		-10 + 5	40	-	2.26	-	0.75
			80	-	1.00	-	1.09
			120	-	0.78	-	1.55
		-15 +10	40	-	-	-	-
			80	-	1.37	-	1.04
			120	-	0.94	-	1.46
	80	- 5 + 1	40	1.50	1.26	0.21	0.45
			80	0.79	0.86	0.33	0.86
			120	0.46	0.72	0.39	1.35
		-10 + 5	40	-	2.42	-	0.50
			80	1.68	1.00	0.40	0.63
			120	0.96	0.75	0.46	0.94
		-15 +10	40	-	-	-	-
			80	1.98	1.68	0.36	0.76
			120	1.11	1.54	0.43	1.31

-Feed size

The reference size d_{80} of the comminution product is affected by the size of the feed material: the coarser the feed, the coarser the product, experimental conditions being the same.

-Waterjet nozzle diameter

Using the 0.80 mm nozzle the d_{80} of the product is on average about 40 % (from 20 to 60 % according to cases) coarser than for the 1.25 mm nozzle. At a given pressure, it seems that grinding fineness is roughly inversely proportional to nozzle

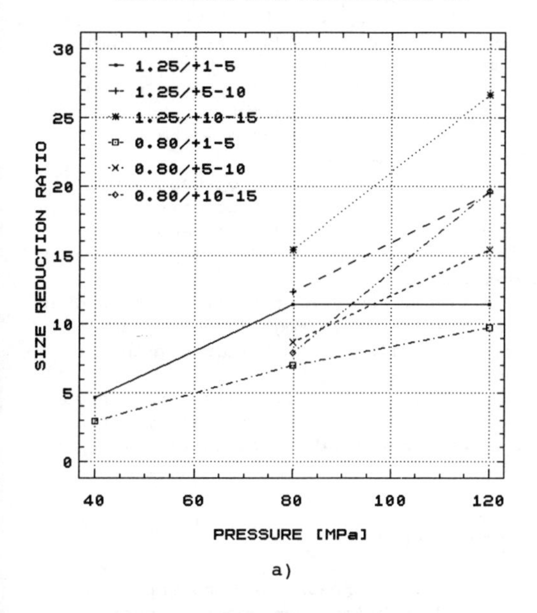

a)

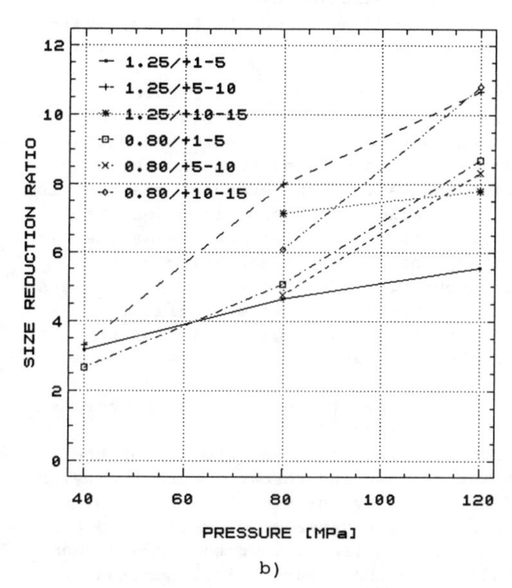

b)

Figure 2. Size reduction ratio R as a function of pressure. Feed rate: 80 g/s. Collimator diameter: 4.5 mm (a) and 7.5 mm (b).

diameter and not to flowrate (and thence to jet power).

-Diameter of the collimator tube
The larger the collimator tube, the coarser the product. For the size class -5 +1 mm the effect of pressure above 80 MPa is negligible, irrespective of feed rate.

-Feed rate
Generally an increase in feed rate from 50 to 80 g/s produces a little significant effect on product fineness: maybe the real capacity of the mill is potentially much higher than the explored range.
Indeed, grains are rapidly evacuated as soon as they are broken due to the very high velocity of the jet (several hundred of meters per second) so that the permanence time into the mixing chamber should be very small.

-Pressure
The higher the pressure, the finer the product. In fact reduction ratio increases from about 4 up to about 27 for the 4.5 mm collimator and from 3 up to 11 for the 7.5 mm collimator as pressure increases from 40 to 120 MPa.
However, the effect becomes progressively negligible, as the cross section of the collimator tube increases (the size distribution curves at 80 and 120 MPa are almost overlaying for the 7.5 mm opening).
The pressure of 40 MPa is not sufficient for the coarser size classes with the smaller collimator tube, working at high feed rate, since the material tends to accumulate into the mixing chamber until clogging occurs.
In the log plot, the gradient of the rectilinear portion of the size distribution curves is generally greater as pressure increases, meaning a reduced statistical dispersion.

B - Specific energy

Specific energy is a measure of the efficiency of a grinding operation and depends on the material properties, the fragmentation mechanism, the setting of operational variables and the reduction ratio.
As already pointed out, experimental data have been corrected for comparing the technical results as a function of experimental conditions at equal reduction ratio, obtaining the results shown in Figure 3.
Minimum specific energy is always encountered for the pressure of 40 MPa. Increasing feed rate positively affects energy efficiency until optimum is achieved, seemingly well beyond the range explored, provided that the drawbacks connected with particle crowding are eliminated through improved design.
With the 0.8 mm nozzle, specific energy is considerably lower than that achieved with the 1.25 mm nozzle, meaning that a

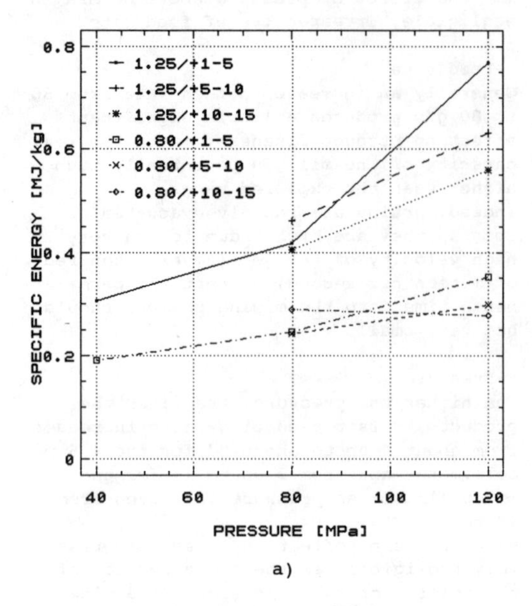

a)

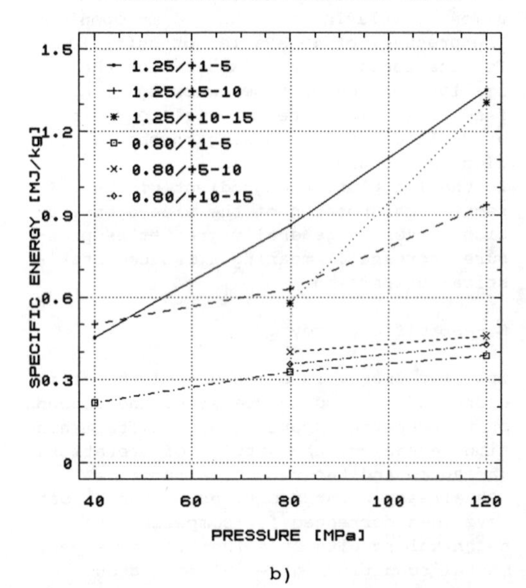

b)

Figure 3. Specific energy at equal reduction ratio for each class, as a function of pressure.
Collimator diameter: 4.5 mm (a) and 7.5 mm (b).

relatively low jet power is needed for a ready comminution of coal.

For a given nozzle diameter, specific energy calculated at equal reduction ratio R is not greatly affected by the size of the discharge gap. Actually, for the 7.5 mm collimator E_s is only about 30 % higher than in the case of the 4.5 mm collimator.

The above results suggest that coal can be more efficiently ground, from the point of view of energy consumption, by using a suitably small discharge opening than by increasing the pressure.

In fact a reduction ratio around 20 could be achieved with a specific energy around 0.19 MJ/kg (corresponding to about 50 kWh/t), working at 40 MPa with a nozzle diameter of 0.8.

The figure however is still quite higher than that normally required by traditional grinding mills for a similar comminution task.

Actually, a ball mill would give a greater size reduction with an energy consumption of about 15 kWh/t (about 3 times lower).

5 DEVELOPMENTS

In order to increase the throughput capacity and to bring down specific energy to a more competitive level, the rotating jet mill has been designed and tested, availing of the results of the above investigation.

The equipment is shown in Figure 4.
The concept followed consists in rotating a twin-jet lance over a circular slot between a cone-shaped core and the surrounding wall of the crushing chamber. The slot is slightly tapered in order that coal particles are held steady as they are hit by the jet, thus achieving better fragmentation. The angle between the jets is 30°, whereas the taper angle is 2° resulting from an angle of 14° of the chamber wall and 16° of the bell-shaped core with respect to the vertical.

A close view of the crushing chamber is shown in Figure 5.

The material is fed at the top of the tower chamber and thrown toward the cylindrical wall by the centrifugal force imparted by a spreader disk fastened to the rotating lance. Then the feed stream falls into the lower conical portion of the chamber down to the crushing zone where it is hit by the water jets and forced through the slot, undergoing intense shattering action; fragmentation is completed upon impingement against a target plate below.

Substantially, the principle of the rotating jet mill is similar to that of the fixed jet. Therefore the effect of pressure, nozzle diameter and slot opening can be assumed to be roughly the same.

Here a new variable appears, i.e. the

Figure 4. Laboratory set up of the rotating jet mill

Figure 5. Crushing chamber of the rotating jet mill.

spinning speed of the lance which affects the time of waterjet action on individual particles and thence the grinding size obtainable for a given slot opening.

Accordingly, tests have been carried out aiming at putting into light the combined effect of feed rate and spinning speed as a function of discharge gap.

To this end, coal was crushed to below 15 mm, without preliminary size classification.

Nozzle diameter was 0.8 mm for all the tests and pressure 80 and 100 MPa, giving an overall flowrate of 0.25 and 0.28 l/s, respectively (from the two angled nozzles).

Spinning speed was 100, 300 and 600 rpm and feed rate 150, 200 and 250 g/s.

Results are shown in Table 3. Specific energy has been corrected for comparison at equal reduction ratio (R = 10.43), as indicated in paragraph 2.3.

From the tests it emerged that maximum allowable mill capacity was around 250 g/s at 100 MPa, i.e. three to five times higher than with the fixed jet mill.

Beyond those limits, material started to build up into the tower chamber and feed rate could not be sustained regularly.

However, larger capacity is expected with further increase in pressure and with wider opening gap.

Pressure also improves the product fineness but its effect is considerably less important than in the case of the fixed jet trough a circular opening and sometimes even controversial.

Larger opening gap gives coarser products. Actually it appears that, at pressures around 80-100 MPa, while the top size is roughly equal to the slot clearance, the reference size d_{80} is about 60% and d_{50} about 20% of that figure.

Compared with the fixed jet mill, product is a bit coarser at equal experimental conditions (pressure, jet diameter, discharge opening). Actually, particle crowding in the fixed jet mill helps fragmentation due to inter-particle friction and collision. To this end, the shape of the discharge aperture (circular hole against linear slot) is an important factor.

It appears that the influence of jet parameters diminishes for the closer gap clearance: actually the size distribution curves practically overlay each other for the 2 mm slot width.

The effect of spinning speed seems rather important, the more as the gap opening decreases: faster speed reduces the impinging time on individual particles, increasing the grinding size.

The gain in specific energy was lower than expected, because of the fact that the twin-jet power was not taken full advantage of. Seemingly, the size of the crushing chamber must be increased, allowing the material to expose itself to the action of the jets at a higher rate.

For a reduction ratio close to 10.5, the lowest specific energy was around 0.08 MJ/kg, corresponding to about 20 kWh/t of fresh feed, 2.5 times lower than with the fixed jet mill.

In order to draw some preliminary conclusions about the possibility of industrial

Table 2. Rotating jet mill: Results of coal comminution.
Nozzle diameter: 0.8 mm;

Experimental conditions				Results			
Slot gap mm	Feed rate g/s	Pressure MPa	Sp. speed rpm	d_{50} mm	d_{80} mm	R	E_s MJ/kg
2.0	150	80	600	0.53	1.33	9.02	0.149
	100		300	0.52	1.38	8.70	0.216
			600	0.63	1.52	7.89	0.234
	200	100	300	0.33	1.15	10.43	0.140
			600	0.49	1.30	9.23	0.154
	250	80	600	0.38	1.22	9.84	0.084
		100	300	0.35	1.20	10.01	0.116
			600	0.48	1.29	9.30	0.123
3.2	150	80	300	0.95	2.34	5.12	0.249
			600	0.65	1.96	6.12	0.209
		100	100	0.62	2.22	5.40	0.248
	250	80	600	0.99	2.25	5.33	0.144
		100	100	0.55	1.88	6.38	0.169
			300	0.65	2.00	6.00	0.179
			600	0.57	1.88	6.38	0.169

application of water jet for coal grinding, a very important aspect must be underlined: mill performance depends very much on the size of the machine.
In fact, throughput capacity is believed to increase almost proportionally with the square of the linear dimension, while the energy needed is expected to grow at a slower rate. Therefore specific energy will decrease on scaling up the equipment.
Moreover, in a larger mill the power of the jet can be imparted at lower pressure and high flowrate, with additional energy saving advantages per tonne of coal treated.
Better results can be achieved by a suitable combination of waterjet with the mechanical breaking action obtained by vibrating axially or rotating the bell-shaped core with a slight eccentricity, in order to produce a periodic variation of the slot opening, like in gyratory crushers.

6 CONCLUSIONS

Waterjet represents a very promising way for ore and coal comminution down to very fine sizes from both the technical and economic points of view, though performance data of an industrial significance are not yet available, being the research at its early stage.
Among the potential advantages of waterjet in mineral grinding, the following points are worth consideration:

-high feed rate can be achieved due to the very rapid breaking action of the jet; capacity can be increased by feeding the material through a slot over which the jet is scanned with a suitable velocity;
- the size of the mill is very small in relation to its potential capacity;
- wear is greatly reduced due to limited frictional contacts; parts subjected to wear can be adjusted or easily replaced with minor cost incidence;
- noise level is low since the jet is completely encased into the mixing chamber:
- size classification can take place in the mill itself.

On the other hand a waterjet mill requires a considerable amount of power for pumping the water to the desired pressure. However, in the case of coal, which can be ground at relatively low pressure (even <40 MPa), specific energy can be reduced, making waterjet competitive with traditional mills. In addition, the capital investment for a pumping system in that range of pressure is not very high.
Presented concepts of milling devices based on the high pressure water jets after preliminary lab scale tests can be summarized as a potential technology which requires further development.

7 ACKNOWLEDGEMENTS

Suggestions and advice given by professor

M. Mazurkiewicz of the University of Missouri-Rolla are greatly appreciated. The research is being carried on with the financial support of EMSa (Sardinian Mining Body) according to the programmes approved by CNR (National Research Council) and MURST (Italian Ministry of Scientific Research).

GENERAL BIBLIOGRAPHY

D.G. Osborne: Coal Preparation Technology, Vol 1, Chapter 3 Size Raduction, Graham & Trotmen Ed., 1988, 58-113

L.G. Austin, J. McClung: Size reduction of coal, in: Coal Preparation, J.W. Leonard (Ed.) AIME, New York (1979), Ch. 7, 1-34

A.A. Griffith: The phenomena of rupture and flow in solids, Phil. Trans. R. Soc. Series A, 221-163-198 (1920-1921)

I. Evans, C.D. Pomeroy: The strength, fracture and workability of coal, Pergamon Press Ltd. (1973)

J.A. Herbst: Energy requirements for the fine grinding of coal in an attritor, Final report, Contract No. EY-77-S-02-4560 University of Utah (1978)

Comminution and Energy Consumption, Report of the Committee and Energy Consumption, Publication NMAB-364 National Academy Press, Washington DC (1981)

P. Somasundaran and I.J. Lin: Effect on the Nature of the Environment on Comminution Processes, Ind. Eng. Chem. Process. Des. Dev., 321-331 (1972)

REFERENCES

Mazurkiewicz M., White J., Karlic K.,1988: Effect of feed rate during comminution of coal by high energy waterjet, Coal Preparation (1988), Vol.6

Bortolussi A., Carbini P., Ciccu R. and Ghiani M.,1993: Waterjet grinding of coal, Proceedings IVth Int. Conf. on High Sulphur coals, Lexington.

Mazurkiewicz M.,1985: High pressure liquid jet as a tool for disintegrating organic and non-organic materials, Invention disclusure 85-UMR-009

Mazurkiewicz M. and Galecki G.,1993: Coal and minerals comminution with high pressure waterjet assistance, Proc. XVIII Int. Min. Proc. Congr., Sydney, 1993, Vol. 1, pp. 131-138.

Ciccu R., Mazurkiewicz M.,1993: Coal disintegration and beneficiation by high pressure water jets, Anais II Congresso Italo-Brasileiro de Engenharia de Minas, Sao Paulo.

Geomechanics 93, Rakowski (ed.) © 1994 Balkema, Rotterdam, ISBN 90 5410 354 X

Jet cutting assisted restoration technique

F. Doll
FORACON, Bretten, Germany
H. Louis & G. Meier
Institute of Material Science, University of Hannover, Germany

ABSTRACT:
Conventional restoration technique means beside linear cutting technique by different kind of saws an immense quantity of hand working. By abrasive water jets the amount of hand work can be reduced drastically.

However a larger effect in minimising hand work and in automatisation can be achieved by using three dimensional jet cutting techniques. For this purpose 5-axis-machines are necessary. The most efficient process can be achieved if different machining techniques will be combined, for example cutting, drilling and material removal.

1 INTRODUCTION

The aim of this paper is the information about developments for application of abrasive water jets in the field of natural stone machining. There are three main reasons to go with this new technique in this field. Firstly for natural stone and especially for buildings out of this material it becomes more and more problematic to resist the attack of the environment. Even in the past there was wear of this material by wind and sand-containing wind. Depending of non-uniform quality of the material part of them show these damages, Fig. 1. It increases rapidly by the pollution in the air which attacks even materials with principally high resistance.

Secondly the awareness in being cultural monuments and therefore the strong demand to save them has increased in the last twenty years.
Thirdly opposite to this, it is more and more difficult to get enough money for a restoration. These are the reasons to look for new and more effective cutting and restoration techniques for natural stones.

2 CONVENTIONAL PROCEDURE

The conventional procedure to produce parts of natural stone starts with rough cutting of blocks, preshaping of the curvature
using all kinds of saws like gang saws, steel-shot gang saws,
circular saws, diamond wire cutting a.s.o. Afterwards the lot of handiwork has to be done. This part mostly becomes the main part
of time and cost consumption. Even in the very simple structure as shown in figure 2 the additional handiwork after cutting and preshaping is eminent.

3 ABRASIVE WATER JET TECHNIQUE

3.1 Two-dimensional jet cutting

As we know abrasive water jet cutting using injection or suspension jets is able to cut every material, also natural stones like sandstone, granite a.s.o.

In the following different steps and the results to

Fig. 1 Erosion of sandstone by wind and abrasives (1)

Fig. 3 Parapet work, cut by abrasive water jets

Fig. 2 Preshaping of structure (1)

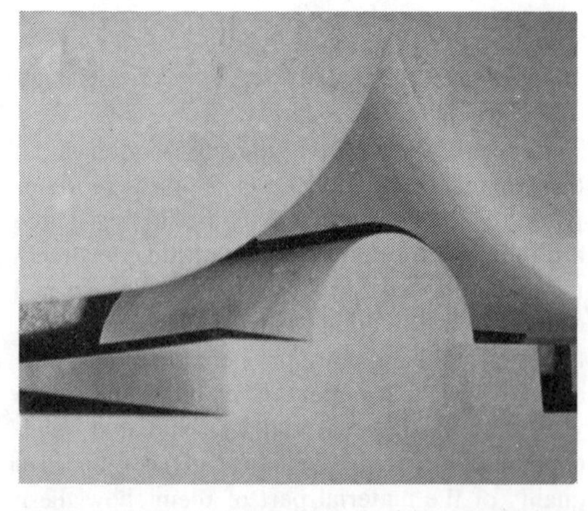

Fig. 4 Detail from fig. 3, showing the quality of the cut

use two-dimensional jet cutting technique to prepare parts of an old parapet work will be described. The material is sandstone at a thickness of 15 cm. The total shape of one element is seen in figure 3. (Conventional two-dimensional jet cutting equipment using the injection water jet has been used). The demand on accuracy was very high so that the cutting velocity has to be reduced down to 60 mm/min. This velocity makes sure that the wanted accuracy is given to all parts and sections of this element. Figure 4 demonstrates the high quality of the cut. The whole parapet work after restoration is given in figure 5, it shows some of the single elements.

Because of the high investment of jet cutting machines beside the total costs the time of cutting is of interest. Comparing three different techniques of restoration from the past, from today and with jet cutting assisted restoration is

Fig. 5 Part of the restored parapet work

given in figure 6. Typical for the restoration technique of the past was i.e. in this symbolised special contour element which is part of the section to make a central hole drilling and start contouring from this hole by hammer and chisel. The end of this procedure was a postfilling of the whole element. The calculated operating time is given with nearly 300 hours per element. A comparison with the procedure often done today, is given in the middle of figure 6. The different steps are the contour drilling, using a lot of small holes following the given geometry as close as possible. After this contour drilling the following work was done by compressed air tools, followed by manual polishing of the contour. By doing so time of operating can be reduced to half of the time, nearly 150 hours. However, when using abrasive jet cutting technique the whole contour can be cut without postfilling on manual polishing. The quality and accuracy is very high when the relevant cutting parameters, especially cutting velocity, were chosen (as to be seen in figure 5). When using the cutting velocity of about 60 mm/min to reach this quality of cut the total operating time of this element with a contour length of more than 40.8 m is less than 5 hours. On the base of normal costs of a workshop, for example 6 DM/min, the resulting costs are approx. 1.500 DM. None of the other techniques even on the base of low labour costs can reach this cost level.

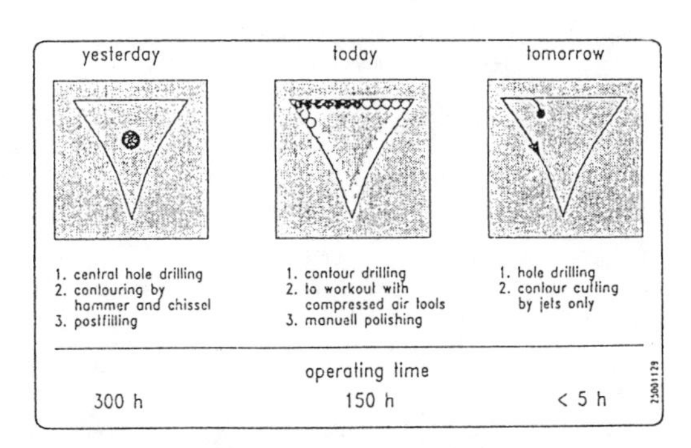

Fig. 6 Comparison of different restoration techniques, time consumption

There are several other techniques that can be used like diamond saw cutting, which are not included in this comparison. However, depending on the structure of the elements the costs are more or less close to the cost of contour drilling. Especially when a lot of small segments have to be cut off, it takes a lot of time for example to install the diamond wire saw for each element.

3.2 Three-dimensional jet cutting

Most of the elements that have to be restored are more difficult in their shape and structure, so after normal two-dimensional cutting an additional workout with compressed air tools and following by manual polishing has to be done. However, the effectivity of using abrasive water jets as shown on the example before, can be increased very intensively when using three-dimensional cutting. Indeed this will be more important in the case of complex structures. However, there are three preconditions which are necessary to fulfil. Firstly the machining possibility, secondly a software relevant to the problems and thirdly a procedure in ranking of cutting steps to get as close as possible the given contour cut. Figure 6 shows, that these machines and the software which is necessary to run this complex equipment already exist. The figure shows the cutting head and some examples in stone and in metallic material using three-dimensional machining. However the ranking of the cutting steps is still a problem which has to be solved today only for every single event. The more universal strategy to do this is not available up to now. However, it is very necessary to make a more effective process out of this technique. A

Fig. 7 Cutting head for 3-dimensional cutting

next step to the future is to combine different machining techniques such as cutting, hole drilling a.s.o. With such a process using the equipment and 3D-computing software as shown before the most effective cutting of natural stones for restoration technique can be done.

4. CONCLUSIONS

Two-dimensional jet cutting technique using abrasive injection jets can reduce time and costs in preparation of natural stone monuments at a very good quality and accuracy. Three-dimensional jet cutting technique will even be more effective, especially when using the optimal ranking of the different cutting steps. The combination of different machining processes such as cutting, drilling, milling a.s.o. and their optimal combination will be the most effective procedure to produce complex structures.

References

1. Meisel, U.
   Naturstein, Erhaltung und Restauration von Außenbauteilen
   Bauverlag GmbH Wiesbaden, Berlin
2. Blickwedel, H. et al
   Prediction of Abrasive Jet Cutting Performance and Quality
   Proc. 10th Int. Symp. on Jet Cutting Technology, BHR,
   Cranfield, UK (1990), Paper L2
3. Guo, N.S. et al
   Assistance for the Application of Abrasive Water Jets
   GEOMECHANICS 91 International Conference
   Ostrava/Czechoslovakia (1991)
4. F. Doll
   Neue Fünfachs-Abrasivstrahlschneidanlage
   Naturstein 9/93, 163

Geomechanics 93, Rakowski (ed.) © 1994 Balkema, Rotterdam, ISBN 90 5410 354 X

Abrasive water jet cutting – Methods to calculate cutting performance and cutting efficiency

N.S.Guo, H.Louis, G.Meier & J.Ohlsen
Institute of Material Science, University of Hannover, Germany

ABSTRACT: When using abrasive water jets it is impossible to quantify the influence of each parameter for all materials that can be cut. Therefore the idea of the information system is to give the dependencies for one material only in detail (reference material) for abrasive water jets.
For other materials belonging to very different groups of materials the relation to the behaviour of the reference material regarding the cutting performance can be evaluated.
The quantification of the influences of different parameters on the reachable cutting performance as well as the setting up of the relation to reference material can be realised by determining the kerfing depth or the obtainable cutting performance.
In addition investigations have been carried out regarding an increase of the cutting efficiency related to the consumption of energy, water and abrasives. Cutting parameters can be optimised to reach a better exploitation of these goods of consumption.

1 NOMENCLATURE

d_D Nozzle diameter
d_F Diameter of focusing nozzle
k Depth of kerf
l_F Length of focusing nozzle
$\dot{m}_P$ Abrasive mass-flow rate
p Pressure
P_h Hydraulic power
$P_h = p * Q_W$
P_s Cutting performance
$P_s = 0.8 * k * v$
P_s^* Specific cutting performance
$P_s^* = P_s$ / good of consumption
$(mm^2/s) / (l/s;\ g/s;\ kJ/s\ [=kW])$
$P'_{s,h}$ Specific cutting efficiency

$$P'_{s,h} = \frac{P_s}{P_h}$$

Q_W Water volume-flow rate
R Abrasive mass ratio $(\dot{m}_P/\dot{m}_W)$
(P: abrasive; W: water)
s Stand-off distance
v Traverse speed
Ω Ratio of diameters (d_F/d_D)

2 INTRODUCTION

Jet cutting technology gives the opportunity to machine a wide range of materials. But to increase the number of real applications two aspects have to be in view:

Firstly it is helpful to find a method to simplify the calculation of the cutting performance for different materials to avoid expensive parametric tests. This demand can be realised by a special type of data base.

Secondly the sensitiveness regarding environmental aspects causes an increasing number of problems when cutting with abrasive water jets. The consumption of resources like water, abrasives and energy leads to operating costs which are too high for some industrial applications especially for abrasive water jets. Especially the storage of used abrasives causes costs which are increasing very fast because of reglementations by law. Therefore the cutting performance related to the amounts of used abrasives and water and also the exploitation of energy has to be optimised keeping in mind the effect on the operating costs.

The investigation of both aspects - an easy way to calculate cutting performance for different materials as well as the possibility to reduce the consumption of abrasives, water and energy - is an important step to find an increasing number of industrial applications for abrasive water jets.

The aim of the paper is to give information about

- a special kind of data base to calculate the cutting performance for a wide range of materials and
- methods to optimise the efficiencies of water, abrasives and energy.

3 TEST CONDITIONS

All tests were carried out for abrasive water injection jets: A plain water jet produced by high pressure (up to 4000 bar) and a small nozzle (diameter up to 0.5 mm) runs through a mixing chamber (water jet pump) and aspirates abrasives transported in air. The abrasives are subsequently accelerated in the focusing nozzle (focus).

All tests are done with Garnet as abrasive material. Using different abrasives can cause a tremendous change in cutting performance. Because of the high number of abrasives available this effect can't be shown in detail. Garnet or similar abrasives are being used for most applications of abrasive water injection jets.

The reference material is AlMgSi0.5.

The quantification of the influence of different parameters on the attainable cutting performance as well as the setting up of the relation to reference material is done by kerfing tests. Samples of the reference material and of other materials were not cut through but kerfed to find an easy method to calculate the cutting performance.

The calculation of the cutting performance was realised according to figure 1. First kerfing tests were carried out by using different parameters. In order to find an easy way to use the results for a wide range of workpiece thicknesses the traverse rate as well as the kerfing depth were eliminated by setting up the cutting performance. The cutting performance is defined as the area of the shoulder of the cut related to the time being used for their generation. It can be calculated by multiplying the obtained depth of kerf with the traverse rate being used (units: mm^2/min) and the factor of 0.8 to be sure to cut through. The cutting performance is nearly independent from the traverse rate when cutting smaller kinds of thickness (e.g. aluminium less than 30 mm) (see fig 1: position 2 and 3). This dependency is valid when using a "normal" range of parameters; going for example to extremely high material thicknesses the cutting performance will be lowered because of an increasing energy loss of the jet due to wall friction inside of the deep kerf. The advantage of using the cutting performance instead of the depth of kerf is that for a given workpiece thickness the traverse rate can be calculated very easy by dividing the cutting performance by the thickness. The result gives the traverse rate for a rough cut.

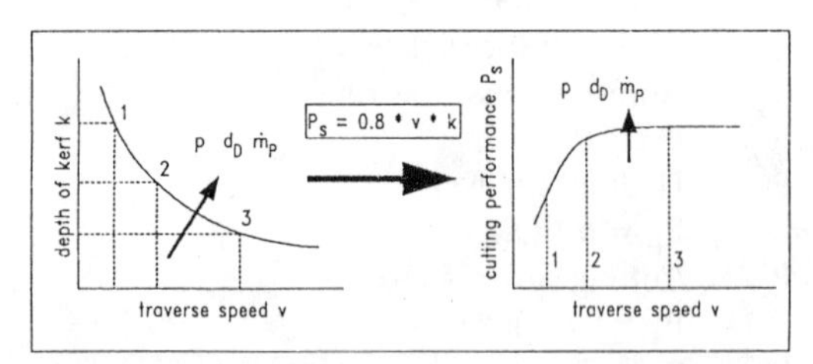

Figure 1 Calculation of cutting performance by using the depth of kerf

To quantify the efficiency of water, abrasives and energy specific cutting performances were defined: The obtained area of the shoulder of the cut can be divided by the consumed amount of water or abrasives or energy (units: mm^2/l or mm^2/g or mm^2/kJ).

4 INFORMATION SYSTEM

For industrial applications the easy calculation of the cutting capability for given materials is important to reduce the expense for preliminary testing. Therefore an information system has been created.

The cutting result for abrasive water jet cutting is depending on a large number of parameters. Some parameters of importance are given in figure 2 as well as general tendencies for the cutting performance when varying these parameters.

It is impossible to quantify the influence of each parameter in detail for any material that can be cut. Therefore the idea of the information system is to give the dependencies for one material only in detail (reference material) in figure 3.

For other materials belonging to very different groups of materials the relation to the behaviour of the reference material regarding the cutting performance can be evaluated according to figure 4.

For an evaluation of the main machining parameters for a given material the detailed diagrams of figure 3 can be used. For this purpose the relation of the cutting performances of the given material and the reference material has to be known (see figure 4).

An optimisation of the machining process can be done by using the general dependencies shown in figure 2.

Although the influences and the relations of different materials referred to the reference material are reported for abrasive water injection jets, the dependencies are right for all jet cutting methods (plain water jets, abrasive water injection jets and abrasive water suspension jets) as far as the parameters are relevant for the process. All the given dependencies are valid in the normal range of parameters only. The change of parameters to extremely high or low values can cause severe differences related to the outlined dependencies (see figure 1 for traverse rate).

4.1 General influence of parameters

The most relevant parameters involved in the process of cutting with abrasive water jets can be categorised as follows:

Hydraulic parameters
- pressure
- diameter of the water nozzle

Parameters regarding the abrasives
- abrasive flow rate
- size and shape of the abrasives
- abrasive material properties

Parameters of abrasive jet generation
- diameter of the focusing nozzle
- length of the focusing nozzle

Parameters of interaction
- traverse speed
- stand-off distance
- angle of attack

Some parameters (as underlined above) are most important for practical application of the technology. There general influence on the cutting result is shown in figure 2. The arrows give the influence of an increase of the abbreviated parameter.

Regarding the other parameters mentioned above there are some well established relationships of parameters as follows

related to the focusing nozzle
- diameter = (3 to 4) x diameter of water nozzle
- length = about 50 x diameter of focusing nozzle

related to the abrasives
- optimum abrasive flow rate in reference material to maximum cutting performance = 25 to 35% of water mass-flow rate
- influence of the kind of abrasive depends on the properties of the workpiece

related to the depth of kerf or the thickness of the workpiece
- maximum thickness of material to be cut = (0.80 - 0.85) * depth of kerf
- region of quality cut = (0.3 - 0.4) * depth of kerf
- definition of quality and rough cutting:
  * quality cut: no striation on the surface, roughness less than 10 μm (steel, titanium) to 15 μm (aluminium)
  * rough cut: striation on the downstream part of the shoulder of the cut

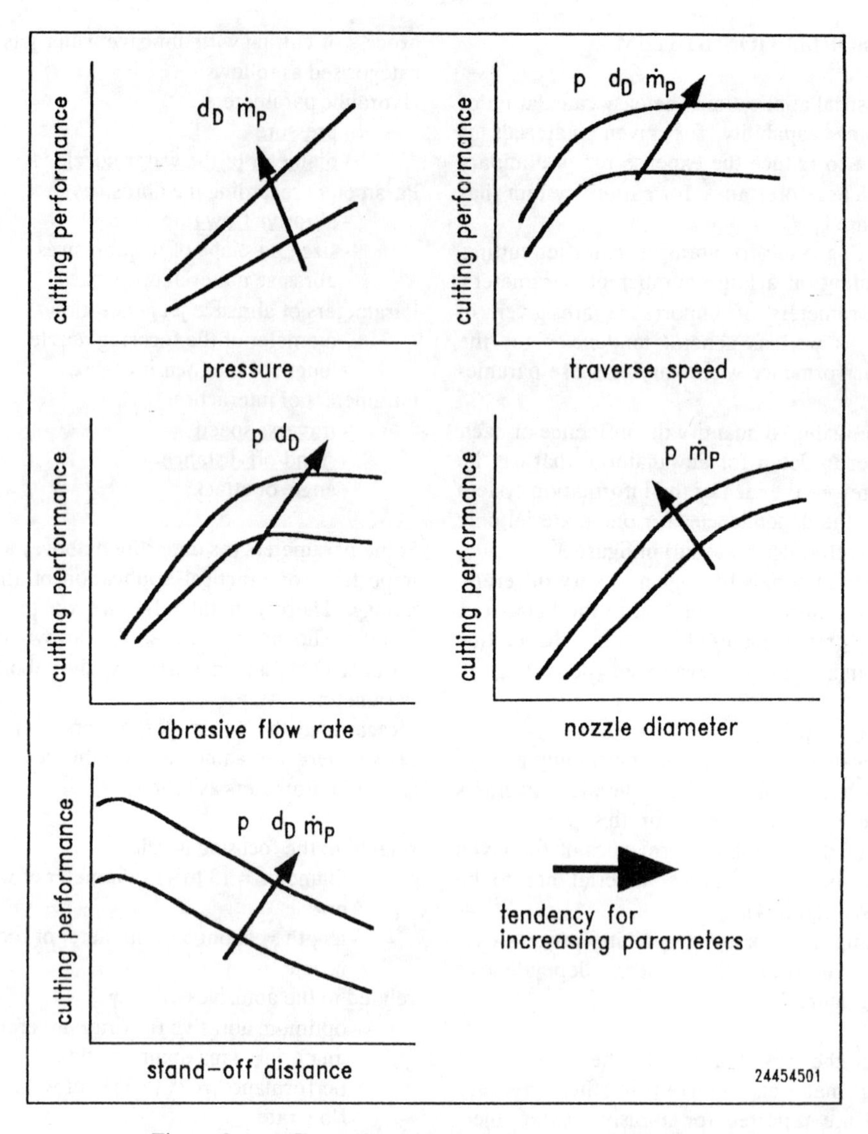

Figure 2 General tendencies of parameter influence

4.2 Influence of cutting parameters when kerfing the reference material

The influence of main cutting parameters on the cutting performance when kerfing the reference material (AlMgSi0.5) are given in figure 3. These results can be used for a calculation of attainable cutting performances for different materials when knowing the relation as given in figure 4.

4.3 Comparison of the behaviour of different materials

In the following the cutting/kerfing behaviour of different material related to the reference material are given.

The specific cutting efficiencies ($P'_{s,h}$) are used for the comparison of the cutting performances. The specific cutting efficiency is the relation of the cutting performance (P_s = k * v: produced area of the shoulder of the cut per time) and the

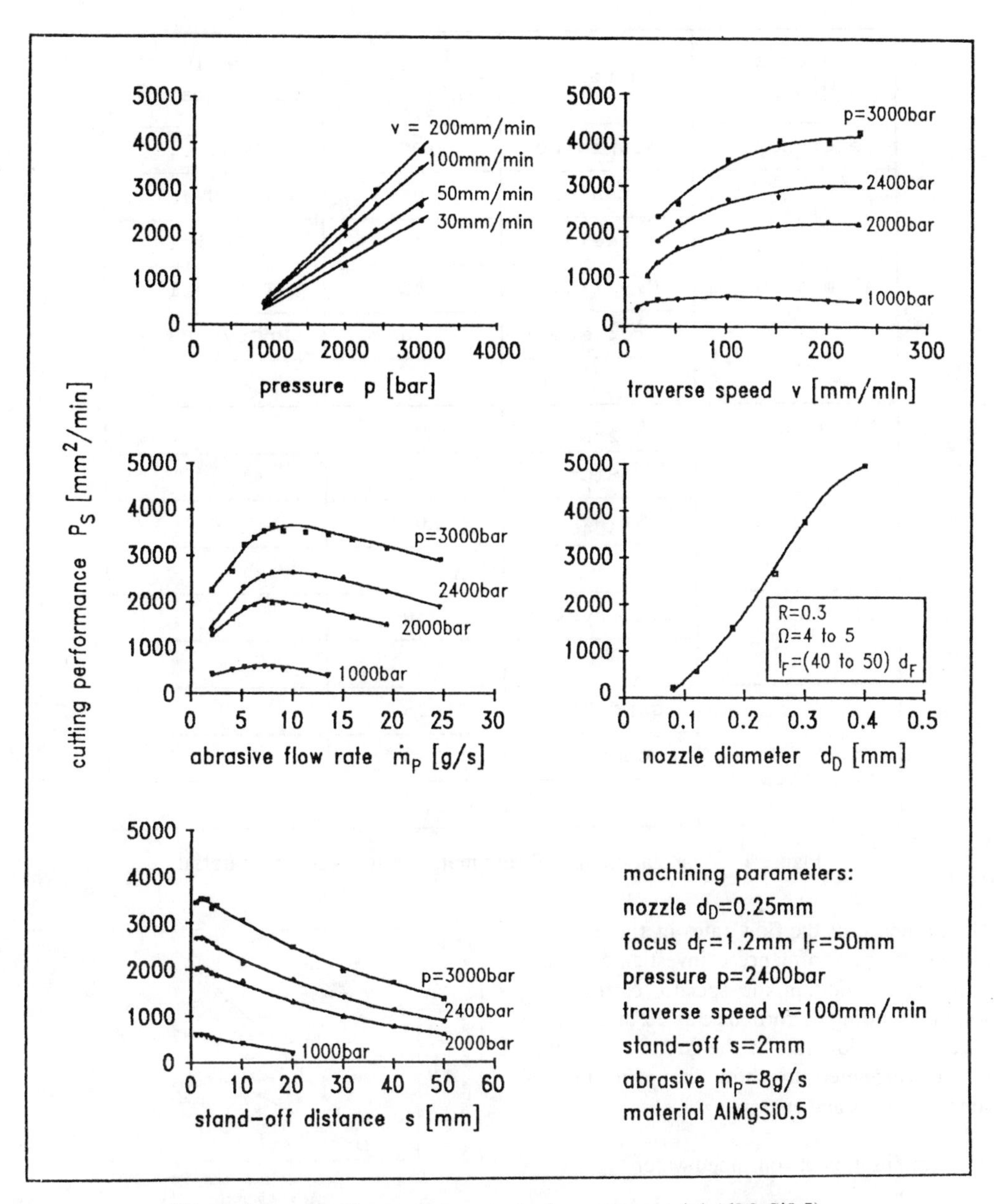

Figure 3 Cutting performance (reference material AlMgSi0.5)

hydraulic power ($P_h = p * Q_W$: product of pressure and water flow rate).

The values given in the tables are calculated as the specific cutting efficiency of the given material related to the specific cutting efficiency of the reference material in %.

5 OPTIMISATION OF THE EFFICIENCY

When using the information described above the maximum cutting performance for a given material can be evaluated quite easily. But in many applications there are additional requirements regarding the consumption of water, abrasives and/or energy to fulfil the cutting job.

material group	material	$\frac{P'_{S,h}}{(P'_{S,h})_{Al}}$ [%]	material specification
reference	aluminium	100	AlMgSi 0.5
a) metals	titanium	55	TiAl6V4
	steel	40	X5 CrNi 18 9
	Ni-alloy	42	Inconell
	hard. steel	36	90MnCrV8 63 HRC
b) ceramics	AL_2O_3	5 - 20	
	SiSiC	2	
	Si_3N_4	0.1	
c) glass	lead crystal	350	
d) others	sandstone	1435	
	concrete	310	

Figure 4 Relation of different materials to reference material

The influence of the flow rates of these goods on the cutting performance is investigated in the following. In addition the specific cutting performance (cutting performance as defined before related to the flow rate of the good of consumption) is calculated to show the efficiency of water, abrasives and energy.

5.1 Minimisation of consumed water

The water flow rate is influenced by the pressure and the nozzle diameter ($Q_W \sim \sqrt{p}$; $Q_W \sim d_D^2$). The effect of both parameters on the specific cutting performance will be described in the following figures.

Figure 5 gives the effect of the pressure on the cutting performance as well as on the specific cutting performance related to the water consumption.

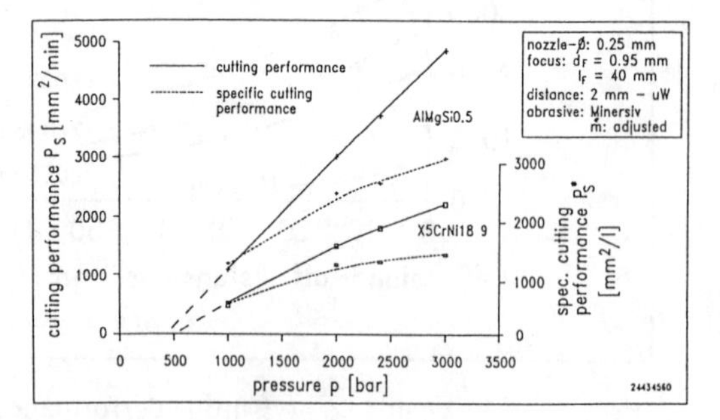

Figure 5 Effect of pressure on cutting performance

An increasing pressure causes a linear increase in cutting performance (per time) and also a better specific cutting performance (related to the water flow). For the highest pressure the exploitation of the water is the best. So it is useful to increase the pressure to values as high as possible, but in case of using more than 3000 bar there is an increasing wear of the pressure pump (sealing) and the nozzles.

For the tests the abrasive flow rate was adjusted to the water flow rate using the abrasive mass ratio R = 0.33 to reach a maximum cutting performance.

The results of a variation of the nozzle diameter are given in figure 6.

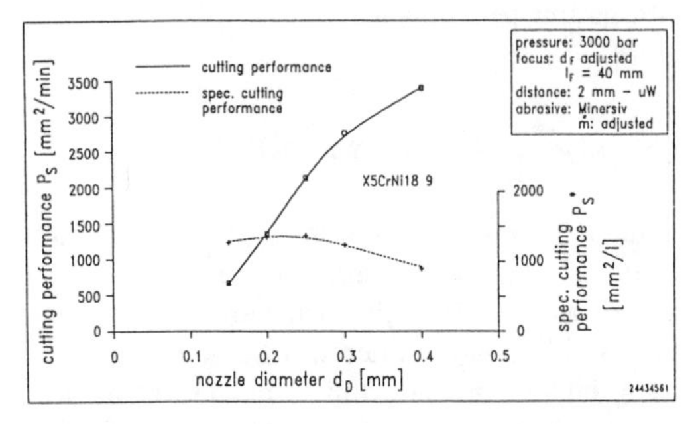

Figure 6 Effect of nozzle diameter on cutting performance

Although the cutting performance is increasing for bigger nozzles the specific performance has an optimum for medium sized nozzles. The abrasive flow rate was adjusted in the way already mentioned (R = 0.33).

In order to reach the optimal efficiency related to the water consumption a nozzle with a diameter of 0.20 - 0.25 mm should be used.

Summing up for an optimal exploitation of the water the pressure should be as high as possible using a medium sized nozzle. In addition it has to be mentioned that for these chosen parameters the machining time for a given job is not minimal, because the cutting performance (related to the time) is not maximal (see fig. 6) because of the smaller nozzle size.

5.2 Minimisation of used abrasives

Two strategies can be used to find parameters for the best exploitation of the abrasives. On one hand the best values regarding pressure and nozzle diameter can be determined for the optimal abrasive flow rate, on the other hand the abrasive flow rate can be reduced to values smaller than mentioned before. In that case the cutting performance related to the cutting time will be decreased, but the exploitation of the abrasives will be better as shown later.

Figure 7 gives the effect of the abrasive flow rate on the specific cutting performance (related to the abrasive flow rate) for different pressures.

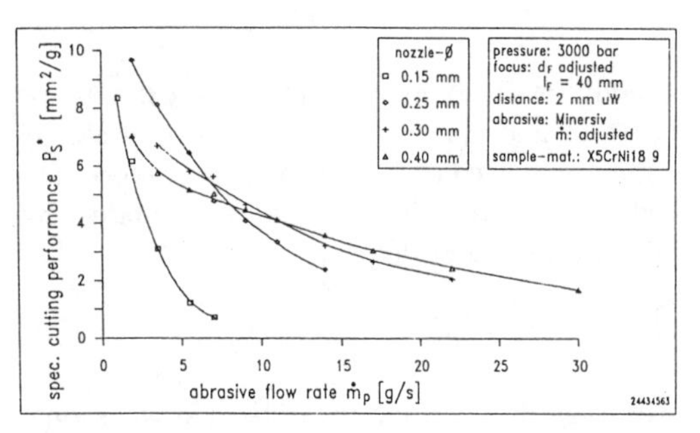

Figure 7 Effect of abrasive flow rate on specific cutting performance

The increase of pressure causes a better exploitation of the abrasives. Additionally the smaller flow rates result in a better specific cutting performance. The smallest flow rate causes the best exploitation regarding the abrasives. But, to fulfil a given cutting job, the time which is necessary to do the job will increase for smaller flow rates and so does the consumption of water and energy.

For using different nozzle sizes the effect is given in figure 8.

Smaller abrasive flow rates cause a better exploitation of the abrasives. For high flow rates bigger nozzle are being more useful, but a decrease of

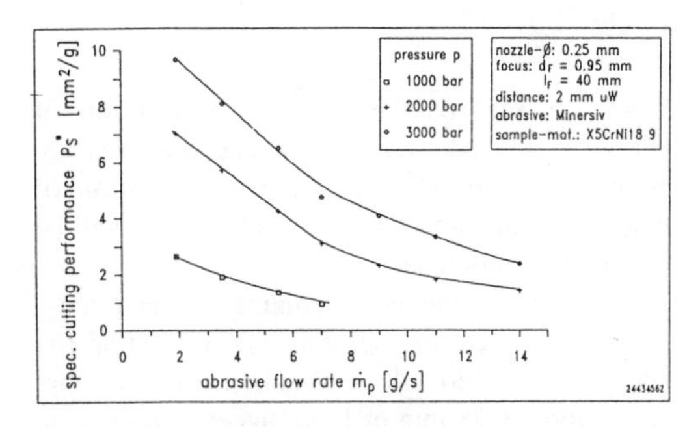

Figure 8 Effect of abrasive flow rate on specific cutting performance

the abrasive flow rate for smaller nozzle diameters leads to a faster increase of the specific cutting performance. Again it has to be mentioned that using smaller nozzles and lower abrasive flow rates help to save abrasives but leads to an increase of machining time, energy and water consumption.

5.3 Minimisation of used energy

In addition to the optimisation of the exploitation of water and abrasives it seems to be useful to minimise the consumption of energy, too. The hydraulic power results from the product of pressure and water flow rate:

$$P_h = p * Q_W \sim p^{3/2} * d_D^2.$$

With regard to figure 5 it can be said when doubling the pressure above a level of 1500 bar the cutting performance is increased by the factor of 2, too. The consumption of energy increases by the factor of $2^{3/2}$ (2.8); so the exploitation of energy is worse for an increasing pressure. Figure 9 gives the results of the tests.

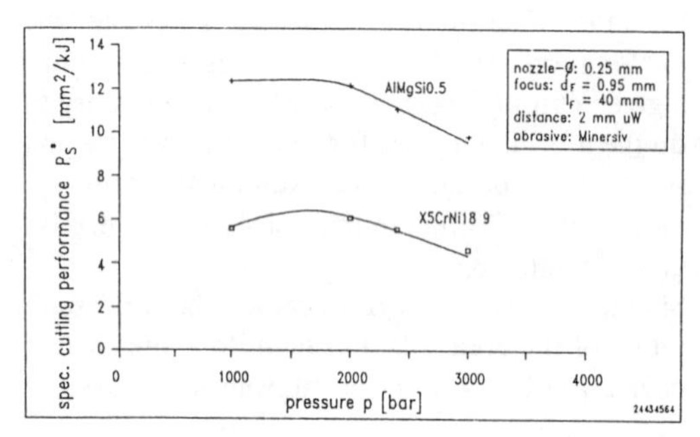

Figure 9 Effect of pressure on specific cutting performance

When looking for the optimal exploitation of energy it will be useful to reduce the pressure below 2000 bar. However, the consumption of time, water and abrasives per meter of cut will be increasing, unfortunately.

When varying the nozzle diameter the maximum specific cutting performance regarding the energy being used will be obtained with a diameter of about 0.25 mm or less. Bigger nozzles lead to a decrease in specific cutting performance. The results are given in figure 10.

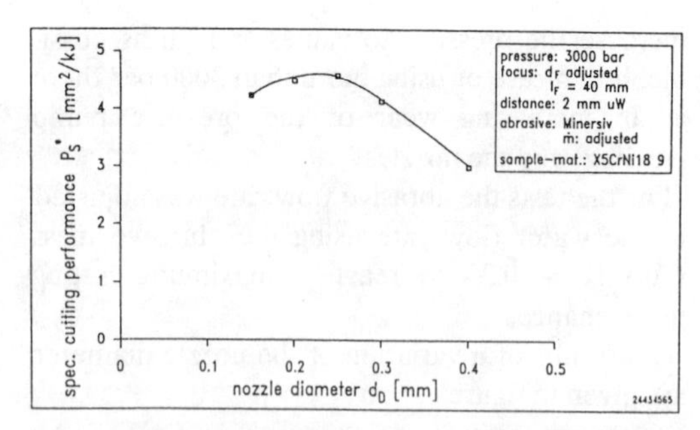

Figure 10 Effect of nozzle diameter on specific cutting performance

6 SUMMARY AND CONCLUSIONS

Summing up the optimisation of the exploitation of the goods of consumption can not be realised for all aspects with identical parameters. While the exploitation of water will be optimal when using highest pressure this is not useful to save energy. The best abrasive efficiency has been realised for smallest flow rates but in that case the cutting time and so the consumption of water and energy will be increased.

The optimisation process has to be focused on the aims of the given cutting job. Facing the job the priorities regarding the optimisation have to be defined. By using these priorities the cutting parameters can be chosen.

When using the information system for a rough calculation of the cutting performance and the know how about the possibilities to optimise the exploitation of water, abrasives and energy useful machining parameters can be defined without a high effort of preliminary studies. This will reduce the costs of abrasive water jet cutting and hopefully can help to come to an increasing number of applications.

7 REFERENCES

Haferkamp, H.; Louis, H. and G. Meier: "Weiterentwicklung des Abrasivstrahl-Schneidverfahrens zum Trennen ferritischer und austenitischer Stähle unter Wasser" Final Report Contract FI1D-0069, EUR12684DE Commission of the European Communities, Luxembourg, 1990

Meier, G.
"Unterwassereinsatz von Wasserabrasivstrahlen"
VDI-Fortschrittbericht, Reihe 2 (Fertigungstechnik), Nr. 289
VDI-Verlag, Düsseldorf, Germany, 1993
Blickwedel, H.
"Erzeugung und Wirkung von Hochdruck-Abrasivstrahlen"
VDI-Fortschrittbericht, Reihe 2 (Fertigungstechnik), Nr. 206
VDI-Verlag, Düsseldorf, Germany, 1990
N.N.
"Water Jet Technology Transfer"
Final Report SPRINT RA 156 bis
Commission of the European Communities, DG XIII-C-4, Luxembourg, 1993

Geomechanics 93, Rakowski (ed.) © 1994 Balkema, Rotterdam, ISBN 90 5410 354 X

Physical model of jet – Abrasive interaction

L. Hlaváč
Institute of Geonics of the CAS, Ostrava, Czech Republic

ABSTRACT: Since 1986 a physical modelling of various aspects of high-pressure water jet behaviour and jet - material interaction has been performed in the High pressure water jet laboratory of the Mining Institute of the Czechoslovak Academy of Sciences. After creating a new physical model of jet - material interaction, completed last year, a study of the interaction between jet and abrasive particles in the mixing chamber was begun. Several physical models describing the processes taking place during jet - abrasive interaction have been prepared and tested. Based on these models the mathematical description has been prepared to study abrasive particle deformations and failures. A theoretical base and first numerical experiments are described and discussed in this article.

1 INTRODUCTION

Physical modelling of the liquid jet interaction with brittle non-homogeneous materials is a part of the scientific activity of the Rock disintegration laboratory of the Institute of Geonics of the Academy of Sciences of the Czech Republic (formerly Mining Institute of the Czechoslovak Academy of Sciences) since 1986. The present scientific program of physical modelling forms an essential part of the grant led by ing. J. Vašek, DrSc. (Vašek 1993). The physical description of both the high-energy liquid jet (HELJ) behaviour and jet - material interaction is aimed at the intensification of the HELJ efficiency by means of abrasive particles delivery - the high-energy abrasive liquid jet (HEALJ) is created.

2 PHYSICAL DESCRIPTION

The energy accumulated in the pressed liquid passes from the static form to the dynamic form through the process of flow development in the nozzle. The efficiency of this transformation of energy is determined both by nozzle parameters and by properties of used liquid.

When the liquid - abrasive mixing processes take place after the liquid jet creation (Fig. 1a), the kinetic energy of the liquid is obtained by energy transformation in the liquid nozzle with efficiency η_1, and then it is transfered to a kinetic energy of abrasive particles with an efficiency η_2. As the efficiency η_1 can be practically the same in both the case of slurry and liquid pumping it is obvious that the condition $\eta_2 \rightarrow 1$ needs to be valid, but the efficiency of the whole cutting process is determined not only by η_1 and η_2.

3 LIQUID JET REPRESENTATIONS

For the purpose of the description of the liquid flow and the abrasive particle behaviour during their interaction in the mixing chamber and the mixing tube, several physical representations of the jet (see Fig. 1b through 1e) have been chosen: continuous flow representation, droplet representation, pressure cross-section representation. The determination of the range of their validity in respect to the input parameters is one of the objects of the experimental investigation.

The studied interactions can be divided into special problems according to the jet diameter and abrasive grain size ratio. In addition to it the particle representations can be divided as follows (Fig. 1c and 1d):

- the flow is supposed to be formed by a sequence of spherical shape drops;
- the flow is supposed to be divided into the cylindrical sectors their volumes being determined by the jet diameter, mean flow velocity and characteristic interaction time.

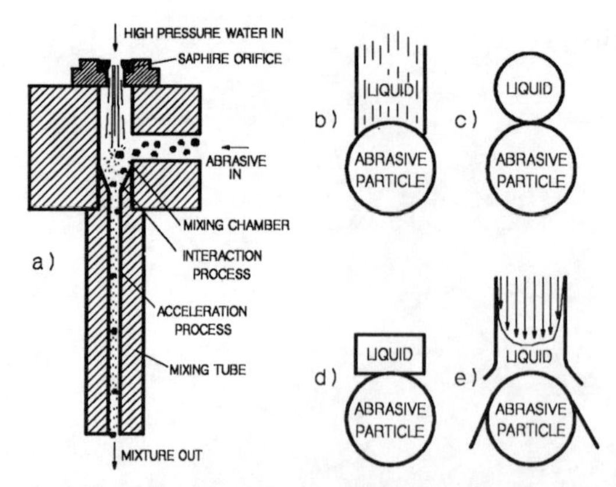

Figure 1. A schematic view of the mixing process (a) and liquid jet representations: b) continuous flow, c) spherical droplet, d) cylindrical droplet, e) pressure cross-section distribution.

4 THE LIQUID JET INTERACTION WITH AN ABRASIVE PARTICLE

The model describing the liquid jet as a continuous flow can be used in the case of relatively low velocities. This representation can be fully accepted in cases when the growth of tension inside the abrasive particle does not exceed critical value for its destruction especially even if some preliminary dispositions exist inside it.

The liquid flow has a velocity $\vec{v_o}$ and an incoming abrasive particle moves with a velocity $\vec{v_a}$. During the interaction the particle slows down in the direction transversal to the flowing and it accelerates along the jet flowing direction. The basic relation characterizing the resistance of the particle in a liquid can be written as follows:

$$F = \tfrac{1}{2}\, C_t\, \rho_o\, S\, v_r^2 \qquad [N]\ (1)$$

Both retardation of a particle in the transversal direction to the flow a_d and acceleration in the longitudinal direction a_n^* can be defined by equations (2) and (3) respectively (derived from (1)).

$$a_d = \tfrac{3}{4}\, C_t\, \rho_o\, v_a{}^2 \rho_a^{-1} a^{-1} \qquad [m.s^{-2}]\ (2)$$

$$a_n^* = \tfrac{3}{4}\, C_t\, \rho_o\, (v_o{}^* - v_{n-1})^2\, \rho_a^{-1} f^{-1} \qquad [m.s^{-2}]\ (3)$$

Characteristic interaction time t_z necessary for the evaluation of the particle velocity evolution in mixing tube by recurrent method is determined from the displacement of an abrasive grain by one tenth of its diameter with initial longitudinal acceleration. Velocity evolution of the particle is then calculated according to the equation (4):

$$v_n = v_{n-1} + a_n^*\, t_z \qquad [m.s^{-1}]\ (4)$$

For the purpose of first calculations the following assumptions have been included:
- the liquid is ideal and its behaviour is valid the laws of classical hydrodynamics;
- an abrasive grain is spherical, homogeneous and intact.

In the case of the particle representation the liquid jet is supposed to be a sequence of either spherical drops or cylindrical particles. The liquid and abrasive grain properties are the same as in the case of the continuous flow representation. For the purpose of the outlet velocity determination the theory of the inelastic impact has been used. The resulting velocity of each interacting liquid element and abrasive grain as well, is given by the relation (5) supposing that the angle θ between the velocity vectors $\vec{v_o}$ and $\vec{v_a}$ is valid the condition $\theta \leq \pi/2$. The number of accelerating impacts depends on the parameters L_a, v_o^*, v_a, m_o, m_a etc. The resulting outlet velocity is then calculated by n recurrent steps.

$$v_i = \frac{(1-k)\,.m_o\,.(v_o{}^* - v_{i-1})}{m_o + m_a} \qquad [m.s^{-1}]\ (5)$$

If the analysis of particle motion inside the mixing tube leads to a conclusion that interactions with walls cannot be neglected, friction and elastic collisions are supposed for all cases of jet representation.

For the purpose of determination and observation of the stress/strain deformation fields in the abrasive particles the liquid jet is represented by a pressure distribution across the jet diameter. The assumptions about the properties of liquid and abrasive grain are valid as well. The pressure cross-section distribution is characterized by the relation derived for this purpose sooner:

$$p_y = p_{os}\,[1 - (2\,y/d)^{\log(Re+1)}]^2 \qquad [N.m^{-2}]\ (6)$$

This jet representation enables us to decide if the infraction of the abrasive grain takes place and under which parameters of the jet (Figures 2a through 2c).

The presented physical representations will be improved according to the continuing numerical experiments (Sochor 1993) and physical measurements (Martinec 1993, Vala 1993).

5 THE ABRASIVE JET - MATERIAL INTERACTION ANALYSIS

During a physical analysis of the abrasive water

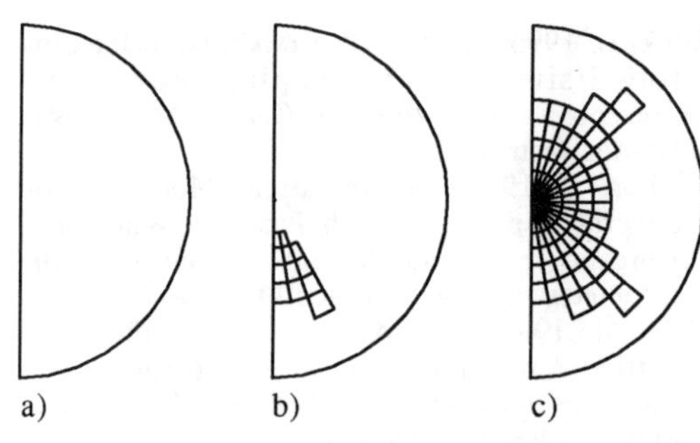

a) b) c)

Figure 2. One half of a quartz abrasive grain tested by ANSYS: a) the liquid pressure is 120 MPa, no failure appears; b) the liquid pressure is 180 MPa, the failure begins to appear; c) the liquid pressure is 240 MPa, large failure appears.

Table 1. The outlet velocity of abrasive particle according to the jet representation without (up) and with (down) friction [$m.s^{-1}$].

Jet representation	Pressure 120 MPa	Pressure 180 MPa	Pressure 240 MPa
Continuous flow	338 / 296	411 / 367	471 / 426
Spherical droplet	348 / 308	423 / 380	485 / 440
Cylindrical droplet	354 / 321	429 / 395	493 / 456

jet interaction with rocks some differences between the presented theories of abrasive water jet cutting (e.g. Hashish 1987) and new experimental results obtained by Sitek (Sitek 1993) were noticed when mentioned theory did not provide a satisfactory explanation of observed reality. It is probably caused by the fact that the model was developed for the ductile materials cutting. The behaviour of brittle non-homogeneous materials that often consist of the more-resistant brittle particles fixed in a relatively plastic less-resistant matrix, both of them including a great number of small discontinuities, can be described as follows: the abrasive particles impacting the brittle material surface crush or strike out the brittle more-resistant elements of the material while the liquid component of the jet washes off the less-resistant matrix material, the particle fragments, grains etc. This hypothesis enables us to explain the phenomena observed during stepping movement of the abrasive jet (Sitek 1993).

6 DISCUSSION

The calculation of an abrasive material grain velocity can be realized according to chosen representation of the liquid jet (see Fig. 1b through 1d). Comparison of the outlet abrasive particles velocities obtained for successive liquid jet representations is presented in Table 1. All calculations were realized for a quartz abrasive, the jet nozzle diameter 0.325 mm, the mean grain size 0.4 mm and the tube length 51.2 mm.

Continuous flow representation is supposed to be usable approximately up to 150 MPa (Fig. 2a through 2c); more exact value depends on the abrasive material characteristics and other conditions. The results of physical modelling will be correlated with the physical measurements of the abrasive jet velocity (Vala 1993).

The droplet model is in the stage of verification and further numerical experiments together with physical measurements may bring the final decision of its suitability. Comparison of the particle velocity evaluations calculated according to the liquid jet representations can be observed in Fig. 3 and Fig. 4.

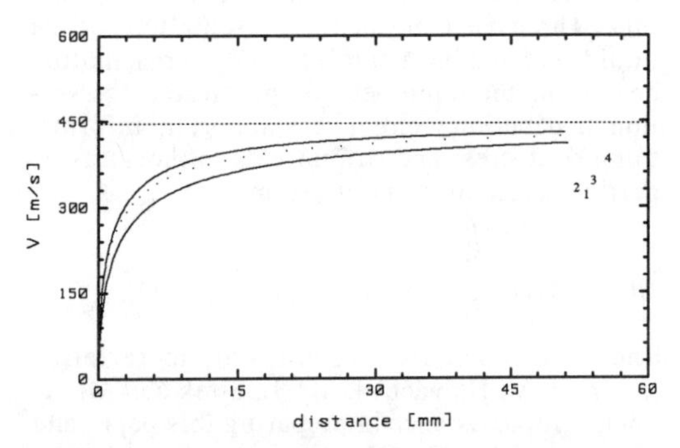

Figure 3. Evaluation of the quartz abrasive particle velocity in the mixing tube in respect to the distance from the point of its enter to the jet. Liquid pressure 180 MPa, no friction considered: 1 - continuous flow, 2 - spherical droplet, 3 - cylindrical droplet, 4 - initial liquid jet velocity.

The stress and strain deformation fields in an abrasive particle can be determined using the representation of the liquid jet by a pressure cross-section distribution. This model needs extension for dynamic processes.

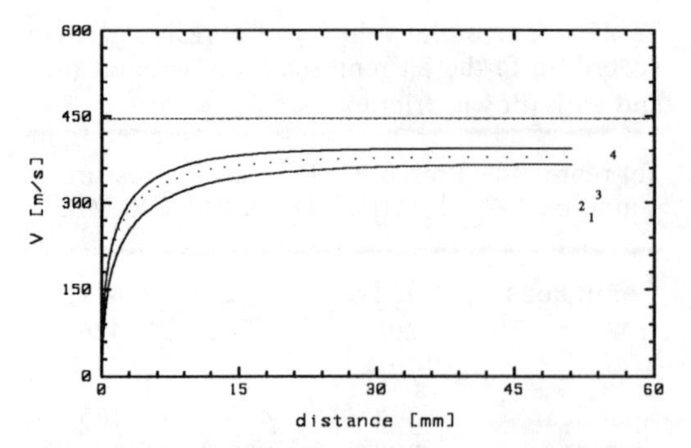

Figure 4. Evaluation of the quartz abrasive particle velocity in the mixing tube in respect to the distance from the point of its enter to the jet. Liquid pressure 180 MPa, friction considered: 1 - continuous flow, 2 - spherical droplet, 3 - cylindrical droplet, 4 - initial liquid jet velocity.

7 CONCLUSION

Several physical problems of an energy transfer from static hydraulic form to the kinetic energy of abrasive particles and of particle energy transformation to the material failure are presented in this article. A special attention has been paid to the interaction of the abrasive particles with the liquid jet in the mixing chamber and the mixing tube. Three basic physical representations of the liquid jet have been studied. The representation describing the liquid jet by a pressure cross-section distribution yields a very important information about stress and strain fields in the abrasive particle at the moment of jet impact.

ACKNOWLEDGEMENTS

The author would like to express his appreciation to Mrs. I. M. Hlaváčová, Mr. L. Sitek and Mr. T. Sochor for assistance in preparing this paper and he is grateful to the Grant Agency of the Academy of Sciences of the Czech Republic for supporting the work presented in this paper (Grant No. 31 655).

REFERENCES

Hashish, M. 1987: An Improved Model for Erosion by Solid Particle Impact. *Proceedings of the 7th Int. Conf. on Erosion by Liquid and Solid Impact*, ELSI VII, Cambridge, England.

Martinec, P. & J. Vašek 1993: Garnets for Abrasive Water Jet Cutting. Presented at *the Int. Conf. Geomechanics '93*, Hradec-Ostrava.

Sitek, L. 1993: High Pressure Abrasive Jet Cutting Using Linear Stepping Movement. Presented at *the Int. Conf. Geomechanics '93*, Hradec-Ostrava.

Sochor, T. 1993: Mathematical Modelling of Several Aspects of High Pressure Water and Abrasive Jet Action. Presented at *the Int. Conf. Geomechanics '93*, Hradec-Ostrava.

Vala, M. 1993: The Measurement of the Non-Setting Parameters of the High Pressure Water Jet. Presented at *the Int. Conf. Geomechanics '93*, Hradec-Ostrava.

Vašek, J. 1993: Basic Research of Water Jet Rock and Concrete Disintegration in Czech Republic. Presented at *the Int. Conf. Geomechanics '93*, Hradec-Ostrava.

NOMENCLATURE

a - mean abrasive particle size ...[m]
a_d - deceleration of an abrasive particle ..[$m.s^{-2}$]
a_n^* - acceleration of an abrasive particle before the n-th reccurent step ...[$m.s^{-2}$]
C_t - drag coefficient ...[$-$]
d - jet diameter ...[m]
f - surface cross-section of a drag force ...[m^2]
F - accelerating (decelerating) force ...[N]
θ - angle of the vectors $\vec{v_a}$ and $\vec{v_o}$ ...[rad]
k - coefficient of stimulation ...[$-$]
L_a - length of the mixing tube ...[m]
m_a - mass of an abrasive particle ...[kg]
m_o - mass of a liquid particle ...[kg]
p_{os} - jet axes pressure ...[$N.m^{-2}$]
p_y - liquid jet pressure in the radial distance y from the jet axis ...[$N.m^{-2}$]
ρ_a - density of an abrasive particle [$kg.m^{-3}$]
ρ_o - density of the liquid ...[$kg.m^{-3}$]
Re - Reynold's number ...[$-$]
S - abrasive particle cross-section ...[m^2]
t_z - interaction time ...[s]
v - speed of an abrasive particle ...[$m.s^{-1}$]
$\vec{v_a}$ - the vector of the abrasive particle speed
$\vec{v_o}$ - the vector of the liquid jet speed
v_a - component of the vector $\vec{v_a}$ transversely to the direction of the vector $\vec{v_o}$ ...[$m.s^{-1}$]
v_o^* - mean liquid jet velocity ...[$m.s^{-1}$]
v_i - abrasive particle velocity after the i-th impact of the liquid droplet ...[$m.s^{-1}$]
v_n - abrasive particle velocity after the n-th accelerating step ...[$m.s^{-1}$]
v_r - relative speed ...[$m.s^{-1}$]

Geomechanics 93, Rakowski (ed.) © 1994 Balkema, Rotterdam, ISBN 90 5410 354 X

Improving the surface quality by high pressure waterjet cooling assistance

R. Kovacevic, C. Cherukuthota & R. Mohan
University of Kentucky, Ky., USA

ABSTRACT: The quality of the machined product is evaluated in terms of its surface finish. A technique is developed here to improve the quality of the surfaces produced by rotary tools (typically face milling) by high pressurised waterjet cooling assistance. The effectiveness of this technique is investigated in terms of improvement in surface finish, reduction in tool wear and control of chip shapes.

1 INTRODUCTION

In the process of metal machining a tool comes in contact with the workpiece and material is removed in the form of chips due to relative motion between the tool and the workpiece to yield the final component. As an engineering product, the suitability of any machined part (in an assembly) depends upon the dimensional accuracy and the surface quality. With the advent of modern machining processes achieving the desired dimensional tolerances has become possible with relative ease. However improvement of the surface quality in the most efficient way is still a primary concern in metal machining.

The surface quality of a machined product is influenced by geometric and kinematic factors like nose radius, cutting edge angle and its shape and physical and dynamic factors like built-up-edge, friction, vibrations, heat generation, and environmental conditions existing at the cutting zone. There have been a few investigations (see Jung and Oh(1991), and Montgomery and Altintas(1991)) in the past for improving the surface finish of the machined component through prediction and control. A couple of investigations were conducted (see Wang and Zhao (1987) and Shiraishi and Sato (1990)) previously for surface finish improvement in turning operation by using non-traditional techniques.

As the thermal/frictional properties of a system are constant, the thermal/frictional conditions existing in the system can be influenced only by external means. Cutting fluids are used for cooling and lubrication (friction reduction) at the cutting zone. Rate of flow and the direction of application of the cutting fluid decide the effectiveness of the external cooling method. There have been several attempts (see Pigott and Colwell (1952), Sharma et al.(1971) and Wertheim et al.(1992)) at applying the cutting fluids in the form of high velocity jets. But they were characterised with relatively low pressures that could not sufficiently overcome contact pressures at the tool-chip interface. Recently, cooling/lubrication systems based on the high pressurised waterjets upto a pressure of 280 MPa for turning operation was proposed by Mazurkiewicz et al.(1989) and Lindeke et al.(1991). Test results showed that with the high pressurised system, a significant gain in the material removal rate, tool life and the improvement in chip shape could be achieved. A high pressurised waterjet used as a coolant/lubricant reduces secondary shear, lowers the interface temperature of the body of the insert itself and changes the shape of the chip. Until now, all attempts at using high pressurised coolant/lubricant systems focussed on the stationary, single edge cutting tools, namely in the turning operation. However there is a great need to improve the surface finish and reduce the tool wear in milling and drilling operations especially in the case of machining difficult-to-machine materials.

The objective of this work is to evaluate the effectiveness of the coolant/lubricant systems in improving the surface finish for rotary tools (tools for face milling operation) that will be based on a high pressurised waterjet (upto 120 MPa). This being a preliminary study, cutter equipped with single insert has been used. The improvement in surface finish at different cutting conditions by the application of high pressure waterjet is evaluated. This investigation also focuses on the change in chip shape and reduction in the tool wear by the use of high pressure waterjet as they provide us a better understanding of the physical phenomenon responsible for the improvement in surface finish. The performance of the high pressure waterjet with change in pressure and orifice diameter of the waterjet are evaluated.

2 EXPERIMENTAL SETUP AND PROCEDURE

In order to conduct the experimental work on the milling machine, a high pressure cooling/lubricating system suitable for rotary tools was designed (see Kovacevic,1991). This experimental setup consists of a high pressure intensifier pump which can pressurise water upto a pressure of 380 MPa and a vertical milling machine equipped with a rotary swivel, a hollow draw bar, sapphire orifice and a rotary cutter with one insert having an EDM drilled hole. High pressure water from the intensifier pump is brought to the milling machine through stainless steel tube. This setup shown in Fig. 1 is similar to that used for experiments previously (see Kovacevic et al.,1993). However the milling machine has been overhauled to eliminate chatter vibrations. The workpiece used here is stainless steel (AISI 304) of 25.4 mm thickness. Fig. 2(a) indicates the relative position of the chip with respect to the hole in the insert and Fig. 2(b) shows the orientation of the jet towards the tool-chip interface.

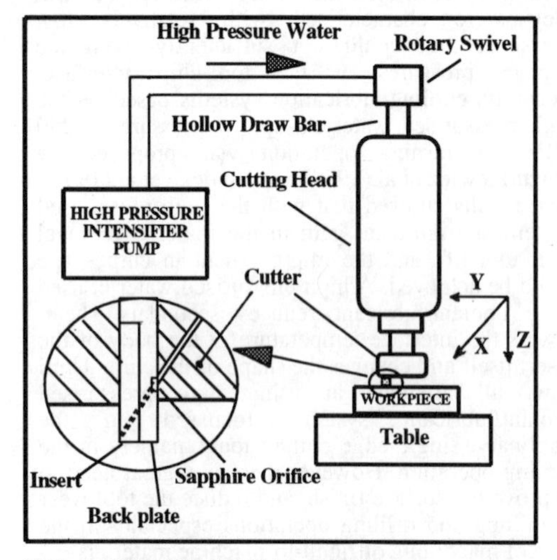

FIG. 1. EXPERIMENTAL SETUP

The process parameters for this experiment are given in Table 1. Different samples were cut by varying each cutting parameter and keeping the other parameters constant at their mean value. For each cutting condition, samples with flood cooling and high pressure waterjet cooling are analysed. The chips produced under varying cutting conditions were collected for analysis. The profile of the cut surface was measured using a profilometer. Surface finish of the profile is measured through roughness average (R_a). In order to understand the influence of waterjet on the surface finish, the above experiments were repeated for different pressures and orifice diameters under the mean cutting conditions. Scanning electron micrographs of the chips produced under different cutting conditions were analysed.

TABLE 1. PROCESS PARAMETERS

Workpiece material	- Stainless Steel AISI 304
Range of Cutting Speed	- 47.9 m/min to 95.8 m/min
Range of Cutting Feed	- 5.10 to 12.70 mm/min/tooth
Range of Depth of Cut	- 0.51 mm to 1.27 mm
Length of the Cut	- 75 mm
Diameter of the Cutter	- 50.8 mm
Type of Operation	- Face Milling
Max number of Inserts	- 8
Number of Inserts Used	- 1
Type of the Insert Used	- TPG322 (K313)
Geometry of the Insert	Rake Angle=0 Nose radius=0.8 mm Clearance angle=11
Type of Cooling Used	- Flood & High Pressure Waterjet
Range of Water Pressure	- 0 to 110.08 MPa
Range of Orifice Diameter	- 0.125 mm to 0.450 mm

3 RESULTS AND DISCUSSION

3.1 Surface finish

Typical surface profile signatures for the two types of cooling techniques are shown in Fig. 3. The change in surface roughness (R_a) with the cutting parameters are plotted in Fig. 4. It is interesting to note that for all cutting conditions, high pressure

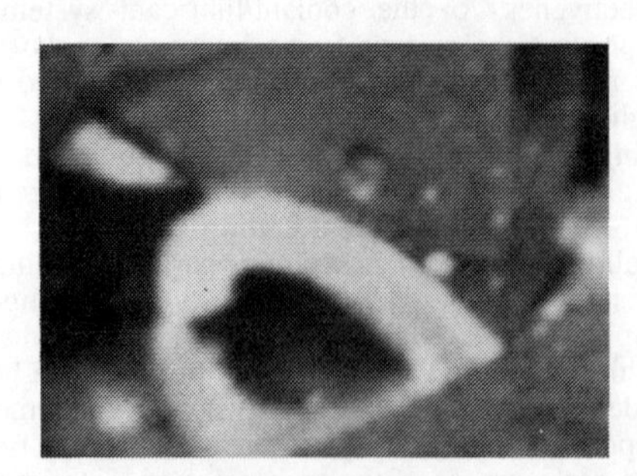

(a) Cutting without waterjet
(Feed=8.9 mm/min)

(b) Direction of flow of waterjet
(Pressure=68.86 MPa, Orifice Dia.=0.125 mm)

FIG. 2. PHOTOGRAPHS OF INSERT

waterjet cooling provides us a better surface finish compared to flood cooling. It can be noted that as the cutting speed increases, R_a steadily reduces for both flood cooling and high pressure waterjet cooling. The influence of high pressure waterjet is more pronounced at lower cutting speeds. This could be due to the fact that at higher cutting speeds, time is not sufficient for the high pressure waterjet to penetrate deep into the tool-chip interface. As expected, increase in the cutting feed tends to deteriorate the surface finish for both flood cooling and high pressure waterjet cooling. The change in the depth of cut does not indicate any clear trend in the surface roughness.

The improvement in the surface finish obtained with high pressure waterjet cooling compared to flood cooling can be attributed to several reasons. In the waterjet assisted machining process, high pressure waterjet has the dual role of a lubricant and a coolant. Due to the placement of the waterjet very close to the cutting edge, the contact area between the chip and the rake face is reduced resulting in the reduction of the frictional forces at the tool-chip interface. The waterjet is able to penetrate into the momentary "vacuum spots" that are created between the chip and the rake face during the process of machining because of its higher rate of flow, overcoming the high resistance. This phenomenon aids in its lubricating action. High specific heat and high heat of vaporisation of water assisted with higher rate of flow increases the efficiency of cooling. The heat dissipation and reduction in the frictional forces reduce the thermal stresses in the insert resulting in decelerating the tool wear. The washing action of the cutting fluid (in this case, water) is aided by the high wettability and downstream movement. These phenomena improve the surface finish considerably.

Apart from the direction in which the waterjet is injected, another important factor which determines the effectiveness of the cooling/lubrication action is the rate of flow of water. The two factors which determine the rate of flow of water are water pressure and orifice diameter. Hence separate experiments were conducted by varying the water pressure and orifice diameter (keeping all other parameters constant) to understand their influence on the surface finish. The variation of surface roughness with water pressure under the same cutting conditions is shown in Fig. 5. The point corresponding to zero pressure indicates flood cooling. From Fig. 5 it can be noted that with increase in water pressure the surface finish improves steadily. As water pressure increases the, velocity of the waterjet increases leading to a higher rate of flow. This improves the efficiency of the

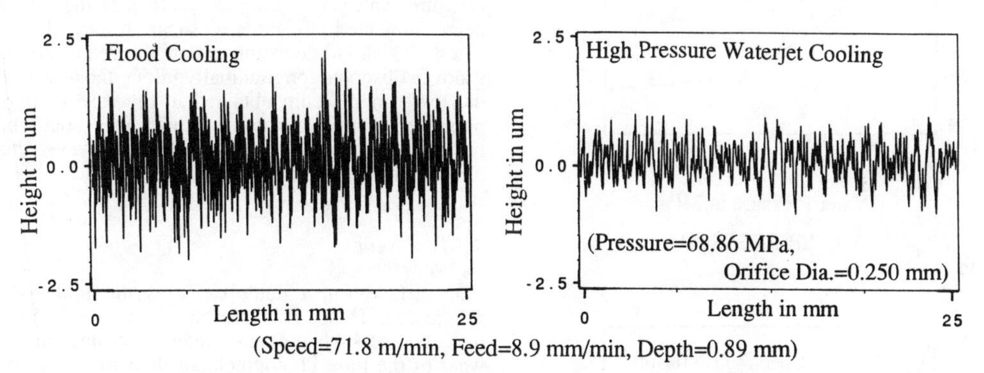

FIG. 3. TYPICAL SURFACE PROFILE SIGNATURES

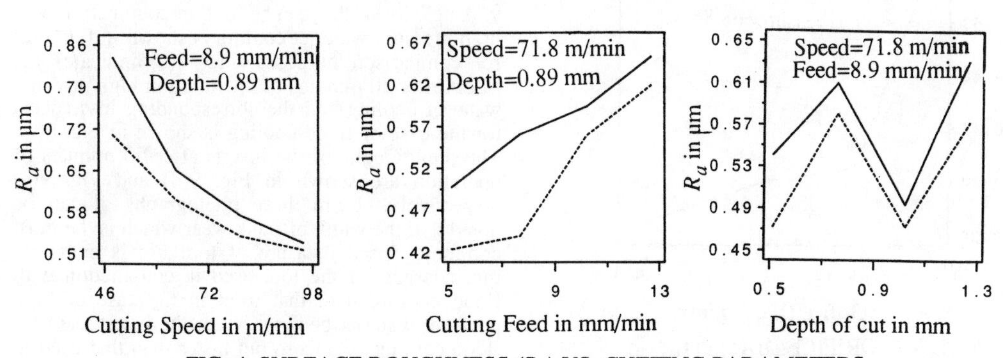

FIG. 4. SURFACE ROUGHNESS (R_a) VS. CUTTING PARAMETERS

cutting fluid as higher rate of flow aids in the effectiveness of the lubricating and cooling action.

The variation in the surface roughness with orifice diameter for the same cutting conditions is shown in Fig. 6. Surface finish improves steadily with increase in the orifice diameter. As the orifice diameter increases, the volume rate of flow increases. This improves the efficiency of the cooling/lubricating action of the cutting fluid which is the primary cause for better surface finish.

3.2 Chip Shape

A comparative study of the chip texture and geometry is done to provide more insight into the role of high pressure waterjet in the milling operation. Fig. 7 gives the chip shapes produced under similar cutting conditions with flood cooling and high pressure waterjet cooling. It was observed that the chips produced during flood cooling were blackened because of the extreme heat generated at the tool-chip interface. However the chips produced by high pressure waterjet assisted milling have bright surface. The chips produced by flood cooling are relatively bigger, irregular and more deformed whereas for high pressure waterjet cooling, the chips are folded over with smoother edges promoting the self breaking effect. As the velocity of the waterjet increases with increase in the water pressure, it penetrates deeper into the tool-chip interface. The hydraulic wedge created as a result of this, forms a cushion at the tool-chip interface reducing the tool-chip contact length. This accelerates the chip breakage and improves the surface finish.

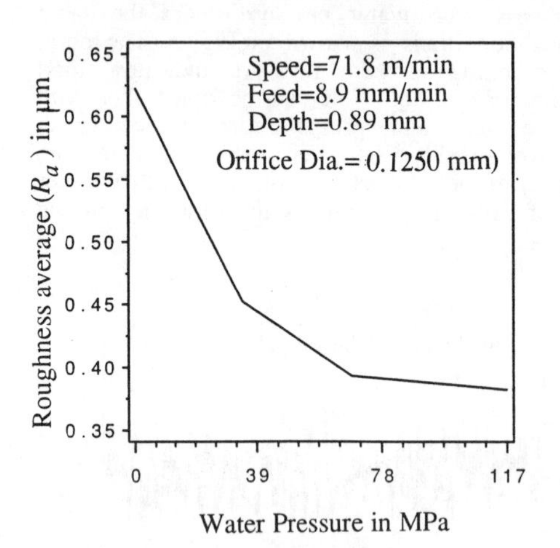

FIG. 5. R_a VS. WATER PRESSURE

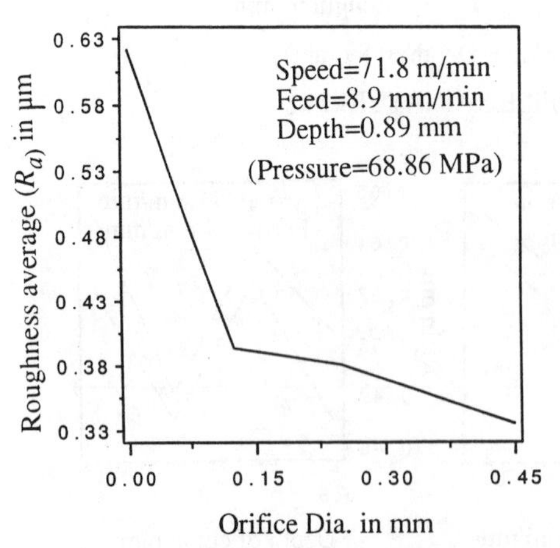

FIG. 6. R_a VS. ORIFICE DIAMETER

The SEM photographs of the chips obtained for flood cooling and high pressure waterjet cooling for maximum pressure and maximum orifice diameter are shown in Fig. 8. A comparison of the size of the chips produced indicates that the chip width ratio of the chips produced by high pressure waterjet cooling is greater than that produced in the case of flood cooling. This is due to the fact that the chips in flood cooling are subjected to intense heat resulting in more plastic deformation than those obtained by high pressure waterjet cooling. The serrations of the chips produced in flood cooling are bigger and farther apart than in the case of high pressure waterjet cooling denoting that they have been subjected to intense shear forces during machining. It is interesting to note the characteristic grooves observed longitudinally (along the direction of flow) on the chip obtained in the case of high pressure waterjet assisted milling. This could be produced by the presence of waterjet. However this phenomenon needs to be investigated further.

3.3 Tool wear

The surface finish deteriorates as the tool wear progresses. Hence it is necessary to investigate the influence of high pressure waterjet cooling on the wear of the tool. Photographs of the carbide inserts used for the high pressure waterjet cooling and in flood cooling at two stages of life are shown in Fig. 9. A new insert that can be used in conjunction with high pressure waterjet cooling is shown in Fig. 9(a) for comparison purposes. Fig. 9(b) indicates the insert after 30 minutes of operation in high pressure waterjet cooling and the corresponding insert used for the case of flood cooling is shown in Fig. 9(c). The photographs of the inserts after 50 minutes of operation are shown in Fig. 9(d) and Fig. 9(e) respectively. From these photographs, it can be noted that the width of flank wear which is the most commonly used measure of tool life is evidently much larger for the tool used in conjunction with flood cooling than that used in the case of high pressure waterjet cooling. Thus, the insert used for flood cooling gets worn out faster than that used in

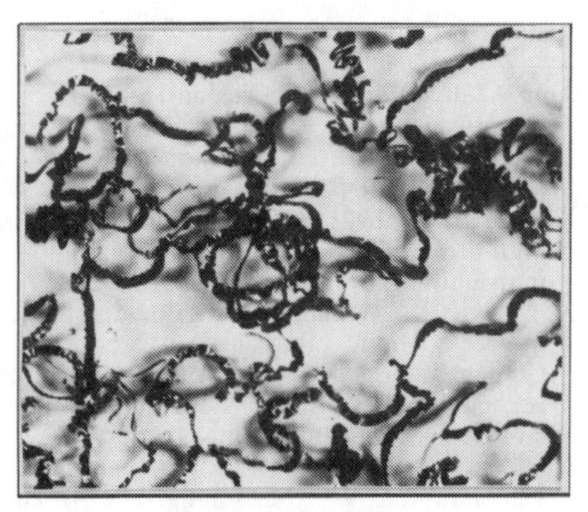

(i) Flood Cooling

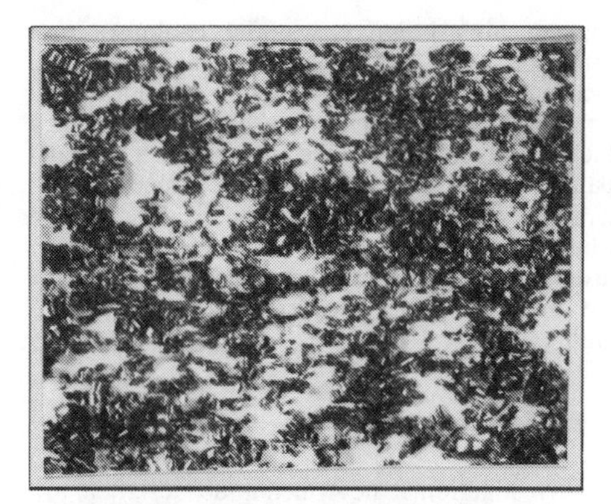

(ii) High Pressure Waterjet Cooling
(Pressure = 68.86 MPa, Orifice Dia.=0.45 mm)

(Speed = 71.8 m/min, Feed = 12.70 mm/min, Depth = 0.89 mm)

FIG. 7. CHIP SHAPES FOR DIFFERENT CUTTING CONDITIONS

Flood Cooling

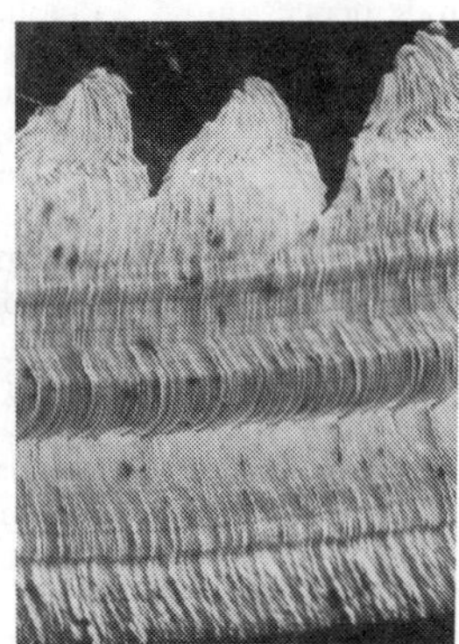

High Pressure Waterjet Cooling
(Orifice Dia.=0.450 mm)

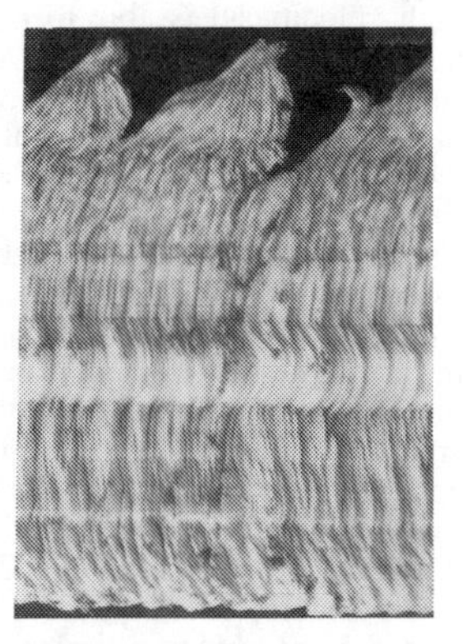

High Pressure Waterjet Cooling
(Pressure=110.08 MPa)

FIG. 8. SEM PHOTOGRAPHS OF CHIPS

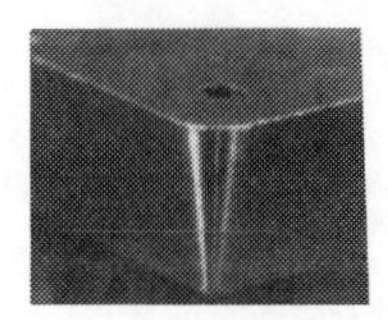

(a) New Insert

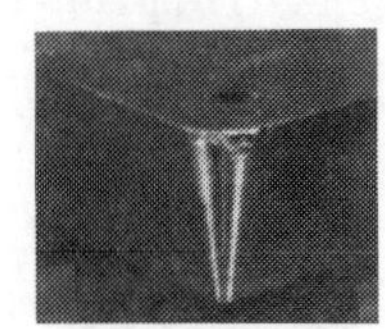

(b)High Pressure Waterjet Cooling

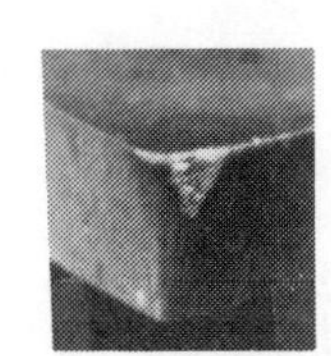

(c)Flood Cooling

Worn Insert after 30 minutes of operation

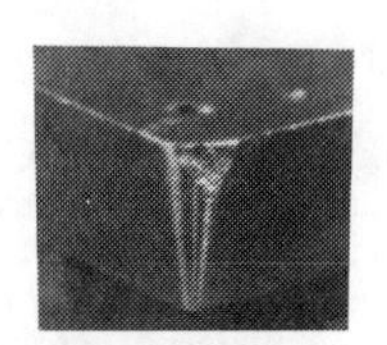

(d)High Pressure Waterjet Cooling

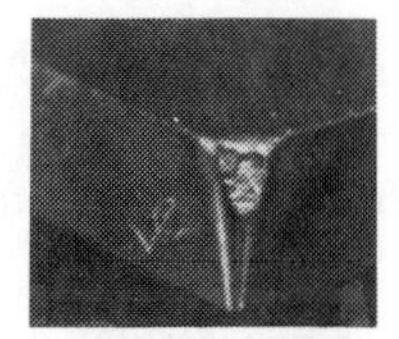

(e)Flood Cooling

Worn Insert after 50 minutes of operation

FIG. 9. PHOTOGRAPHS OF PROGRESSIVE WEAR OF INSERT

the case of high pressure waterjet cooling. The formation of hydro-wedge by the high pressure waterjet tends to reduce the tool-chip contact area. This reduces the secondary shear and lowers interface temperature. Due to higher heat dissipation, the insert is not subjected to extreme temperatures as in the case of flood cooling. This contributes to the reduction in tool wear and increase in tool life consequently leading to a better surface finish for high pressure waterjet assisted machining.

4 CONCLUSIONS

In this investigation a new technique designed to provide high pressure waterjet as the coolant/lubricant through the rake face of the carbide insert of rotary cutting tools is developed. The following conclusions can be drawn from the experimental investigations conducted:

* The surface finish produced by the high pressure waterjet assisted milling is much better than that produced by flood cooling. In the case of high pressure waterjet cooling the surface finish improves with increase in water pressure as the high velocity jet is able to penetrate deeper into the regions inaccessible in flood cooling. The surface finish also improves with increase in the orifice diameter as the volume rate of flow increases aiding the washing action of the cutting fluid.
* The chips produced by high pressure waterjet assisted milling are relatively smaller, fragmented and with smooth edges. Lower chip width ratio in the case of high pressure waterjet assisted milling leads to reduction in the tool-chip contact area promoting the chip breakage. This contributes to a better surface finish.
* Placement of the cutting fluid at the underside of the chip very close to the cutting edge of the insert enables to keep the chip away from the rake face minimising the friction at the tool tip and increasing the rate of heat dissipation from the cutting zone. This leads to the reduction in tool wear and increase in tool life resulting in a better surface finish.

ACKNOWLEDGEMENTS

The authors would like to thank the Center for Robotics and Manufacturing Systems, University of Kentucky for the financial support in executing this project, Flow International Inc., Kent, Washington, for providing us with the waterjet cutting system and rotary swivel and Kennametal Inc. for providing us with the milling machine and tools.

REFERENCES

Jung, Chung-Young and Oh, Jun-Ho, 1991, "Improvement of surface waviness by cutting force control in milling", *Int. J. of Mach. Tools Manufact.*, Vol. 31. No. 1. pp. 9-21.

Kovacevic, R., 1991, "Apparatus and method of high pressure waterjet assisted cooing/lubrication in machining", *Patent pending*, *Dec.*

Kovacevic, R., Mohan, R. and Cherukuthota, C., 1993, "High Pressure Waterjet as a Coolant/Lubricant in Milling Operation", accepted for *ASME Winter Conference*, Nov.28-Dec.3, New Orleans, LA, USA.

Lindeke, R.R., Schoenig Jr., F.C., Khan, A.K. and Haddad, J., 1991, "Machining of $\alpha-\beta$ Titanium with Ultra-High Pressure Through the Insert Lubrication/Cooling", *Transactions of NAMRI/ SME*, pp. 154-161.

Mazurkiewicz, M., Kubala, Z. and Chow, J., 1989, "Metal Machining With High-Pressure Water-jet Cooling Assistance - A New Possibility", *Journal of Engineering for Industry*, Vol. 111, pp. 7-12.

Montgomery, D. and Altintas., 1991, "Mechanism of Cutting Force and Surface Generation in Dynamic Milling', *Transactions of the ASME*, Vol. 113, pp. 160-168.

Pigott, R.J.S. and Colwell, A.T., 1952, "Hi-Jet System for Increasing Tool Life", *SAE Quarterly Transactions*, Vol. 6, No. 3, pp. 547-564.

Rasch, F.O. and Vigeland, T., 1981, "Hydraulic Chip Breaking", *Annals of the CIRP*, *Vol. 30(1)*, pp. 333-335.

Sharma, C.S., Rice, W.B., and Salmon, R., 1971, "Some Effects of injecting Cutting Fluids Directly into the Chip-Tool interface", *Journal of Engineering for Industry*, *Transactions of the ASME*, Ser. B 93, pp. 441-444.

Shiraishi, M. and Sato, S., 1990, "Dimensional and Surface Roughness controls in a Turning Operation', *Transactions of the ASME*, Vol.112, pp. 78-83.

Wang, Li-Jiang and Zhao, Ji, 1987, "Influence on Surface Roughness in Turning with Ultrasonic Vibration Tool", *Int. J. Mach. Tools Manufact.*, *Vol. 27. No.2, pp. 181-190.*

Wertheim, R., Rotberg, J. and Ber, A., 1992, "Influence of High-Pressure Flushing through the Rake Face of the Cutting Tool", *Annals of the CIRP*, *Vol. 41(1)*, pp. 101-106.

Geomechanics 93, Rakowski (ed.) © 1994 Balkema, Rotterdam, ISBN 90 5410 354 X

High pressure abrasive jet cutting using linear stepping movement

L. Sitek
Institute of Geonics of the CAS, Ostrava, Czech Republic

ABSTRACT: The problem with the realization of very small traverse rates occured during the development of equipment for the linear movement of a high pressure abrasive jet cutting head, because a high gear ratio is required. This problem can be solved by the use of a stepping motor, but it causes a linear stepping movement of the cutting head. This paper deals with an analysis and experimental modelling of the linear stepping movement of the cutting head and a comparison of the quality of the cut surface of linear stepping movement and continuous movement.

1 INTRODUCTION

High pressure abrasive jet cutting of certain materials requires very small traverse rates (order 10^{-2} mm.s^{-1}) due to either the large thickness of the disintegrated material or to a requirement of good quality of the cut surface. The problems with the realization of such speeds can occur when developing the cutting head manipulator for linear movement. Gearboxes with an extremely high gear ratio (about 1:10,000) between motor revolutions and traverse of cutting head are required.

One of the ways to solve this problem conserving electric drive is to use a stepping motor. The movement of the cutting head is not continuous then but consists of jumps of the cutting head at definite distances and certain waiting in a place. Some aspects of high pressure abrasive jet cutting using linear stepping movement were studied as a part of solution of the grant No. 31655 of the Grant Agency of ASCR in the Rock Disintegration Laboratory of the Institute of Geonics, Academy of Sciences of the Czech Republic, Ostrava (formerly the Mining Institute of the Czechoslovak Academy of Sciences). Experimental results of the disintegration of sandstone using linear stepping movement with larger steps (millimeters), where better efficiency of disintegration was reached, are presented in this paper. Results of the disintegration of the steel where some additional cuts were realized for comparison with sandstone are also presented.

2 LINEAR STEPPING MOVEMENT

The linear stepping movement realized in our laboratory on a cutting X-Y table can be divided into three periodically repeated time sections (Fig. 1). The cutting head does not move in the first section. In the second one the cutting head moves in a straight-line uniformly accelerated motion with initial speed of $v_P = 0$ and an acceleration of $a = const$. The third section is characterized with straight-line uniformly retarded motion with initial speed of $v_P = v_{max}$ and an acceleration of $a' = -a = const$.

Three cases of stepping movement can be differentiated in accordance with step distance (Vašek 1993):

1. step distance Δs is shorter than diameter of abrasive jet d_A : $\Delta s < d_A$;
2. limit case; step distance is equal to diameter of abrasive jet: $\Delta s = d_A$;
3. step distance is greater than diameter of abrasive jet: $\Delta s > d_A$.

It is useful to note that realized continuous movement is a special case of stepping movement where step distance tends to zero ($\Delta s \to 0$).

3 EXPERIMENTS

Cuts realized during experiments with an abrasive jet by continuous movement were compared with cuts realized by stepping movement. Step distances at stepping movement (sections 2 and 3 in Fig. 1) were changed, the waiting time

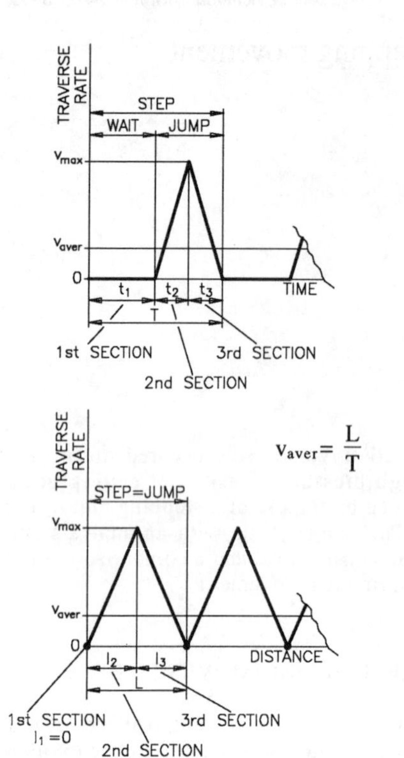

Figure 1. Sections of the linear stepping movement

of the cutting head (section 1 in Fig. 1) had been calculated for individual cuts so that the average traverse rate of stepping movement was equal to the continuous movement traverse rate. Basic information about experiments can be seen in Table 1. Properties of sandstone and steel used for tests are presented in Table 2.

4 MEASUREMENT AND ANALYSIS OF THE DEPTH OF CUT

The depth of cut as the basic parameter is often used for evaluation of the jet effects on disintegrated material.

Table 1. Parameters of the cutting process

Water pressure	250 MPa
Water nozzle diameter	0.325 mm
Abrasive nozzle diameter	1.5 mm
Abrasive flow rate	250 g/min
Cutting head	PASER II

Table 2. Properties of the cut materials

SANDSTONE	
Compression strength	121 MPa
Tensile strength	8.6 MPa
Specific weight	2655 kg/m^3
Specific volume weight	2510 kg/m^3
Porosity	5.45 %
Medium grain size	0.14 mm
Maximum grain size	0.88 mm
Young's modulus	22.6-26.1 GPa
Poisson's ratio	0.18-0.24
STEEL	
Tensile strength	300-340 MPa
Strength limit	600-720 MPa
Specific weight	8050 kg/m^3
Young's modulus	190-215 GPa
Poisson's ratio	0.3
Speed of wave propagation	5000 m/s

Experimental depths of cut have been calculated as an arithmetic mean from 5 values measured in the kerf. The depths of cut in sandstone were greater when using stepping movement, even if the step distance was greater than the diameter of the jet (Fig. 2). In addition uncut material mass between drilled holes was presupposed but it was not confirmed. On the other hand the greater the step distance of the cutting head in steel is, the lower the cutting capability of the stepping system and the smaller

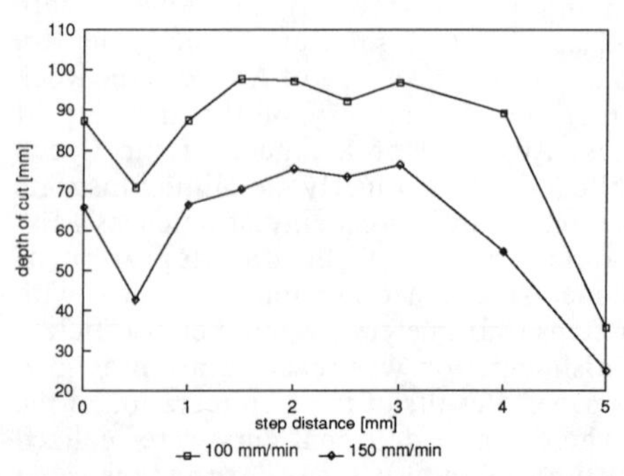

Figure 2. Jump distance in respect to depth of cut in sandstone

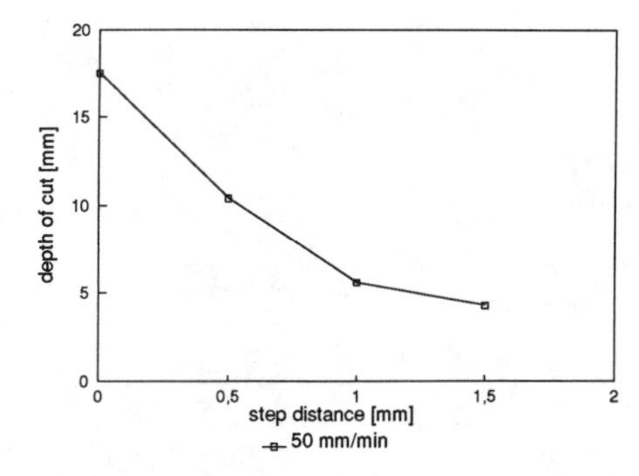

Figure 3. Jump distance in respect to depth of cut in steel

the depth of cut (see Fig. 3). Uncut material was observed between drilled holes when the step distance is greater than the jet diameter.

The study of the physical problems of the interaction of the water jet and the abrasive jet with material in our laboratory (Hlaváč 1992) leads to the conclusion that the description of the interaction process used by Dr. Hashish is especially applicable to the disintegration of ductile materials - metals and some non-metallic homogeneous plastic materials. This common theory can also be used for other materials in case of continuous movement. When using non-continuous movement, the behaviour of non-homogeneous brittle materials (rocks, concrete) differred from presumptions mentioned in theories of abrasive jet interaction mechanism (Hashish 1987).

Brittle construction elements of non-homogeneous materials can be split during interaction and the rest of these elements is probably easily washed off simultaneously with often less resistant matrix by means of the liquid part of the jet (Hlaváč 1993). That is why a higher depth of cut was reached when cutting by stepping movement in comparison to the continuous movement, in spite of longer step distance than the jet diameter. Further physical analysis will be performed and presented in subsequent publications from our workplace.

5 EVALUATION OF THE CUT SURFACE

The structure of the cut surface is studied for instance by Hashish (1992). He divided the cut surface according to the angle of particle impact into two zones. The first one is the cutting wear zone with a shallow angle of impact where the surface is relatively smooth without waviness and the second one is called the deformation wear zone where impacts at large angles cause material removal and surface waviness. He says that the external sources for waviness formations are primarily related to the dynamic process parameters i.e. traverse rate, pressure and abrasive flow rate. Chao et al. (1992) find that the waviness of the surface is caused by vibration of the nozzle during movement and can be eliminated by better leadership and nozzle fixing. According to the experience of our own experiments an additional important parameter influencing the waviness especially at deformation wear zone is local structure and the type of material (for example compare steel with sandstone), however the structure of the cut surface is the same for all materials and is inherent not only in the abrasive-water jet cutting process but also in other beam cutting techniques (plasma, laser etc.) as Singh et al. (1991) and Hashish (1992) presented. Other parameters which can considerably influence the waviness of the cut surface and even the cutting process are not-commonly measured parameters of the jet (i.e. non-symetrical speed profile of the jet, pulsation inside the jet etc.) (Vala 1993).

The cut surface when using stepping movement has rather a different structure. It can be divided into three zones: The primary waviness zone, this is caused by repeated stopping of the abrasive nozzle and is regular. The longer the step distance is, the greater the primary waviness. Commonly it is applicable only to step distances less than abrasive nozzle diameter. This first zone fluently passes to the cutting wear zone where the cutting with shallow angles of impact is applied. The last zone is the deformation wear zone with large impact angles and secondary waviness (Fig. 4 and 5). In accordance with these assumptions when using longer step distances it is possible that the waviness at the top portion of the cut surface is greater than that at the bottom portion (observed in sandstone - Fig. 5c). The factors influencing the waviness when using stepping movement are those presented above and (in systems with small toughness) vibration of the nozzle and cutting head after every stopping of the movement.

The cutting by the stepping movement causes more folded and, especially when cutting non-homogeneous material, more irregular bottom surface of the kerf comparing it with surface made when the continuous movement is applied (Fig. 4 and 5).

6 CONCLUSIONS

Experiments on cutting by stepping movement

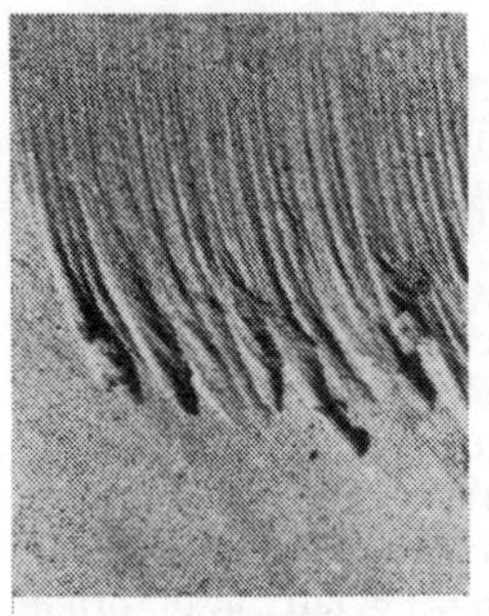

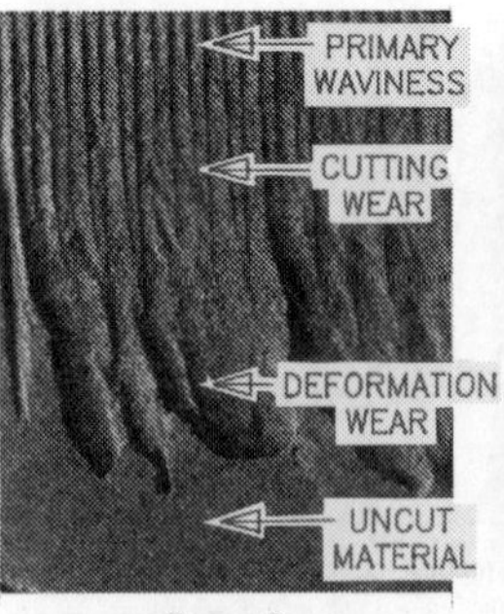

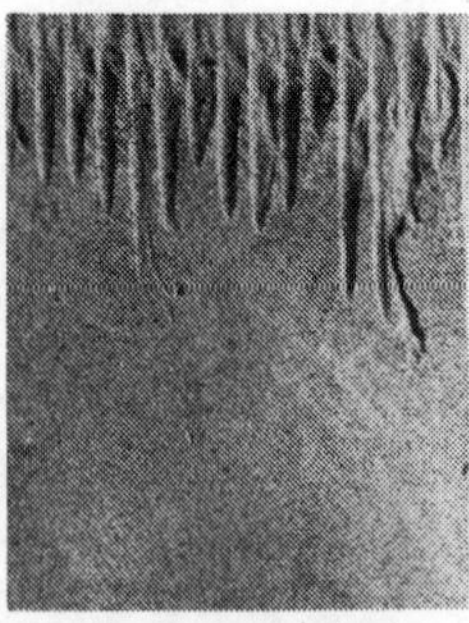

a. Continuous b. Step 0.5 mm c. Step 3 mm d. Step 5 mm

Figure 4. Cut surfaces in sandstone (average traverse rate 150 mm/min, material thickness 100 mm)

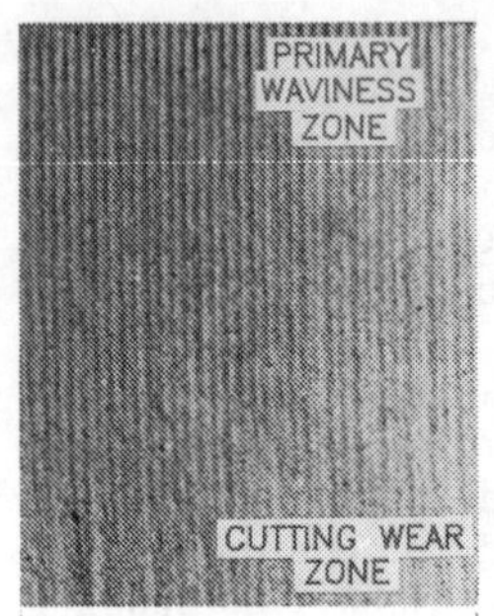

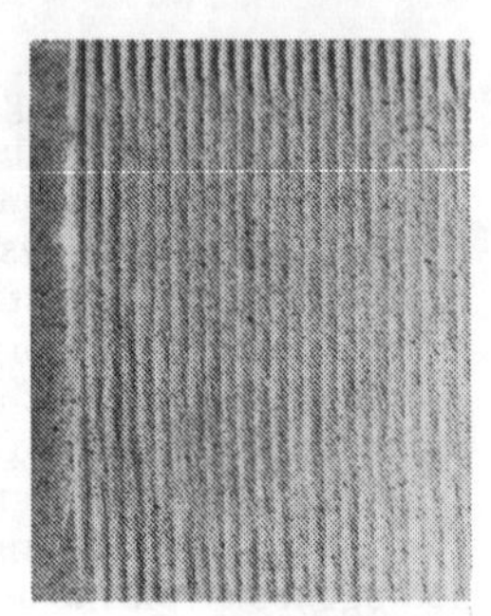

a. Continuous b. Step 1 mm c. Step 2 mm d. Step 3 mm

Figure 5. Cut surfaces in sandstone (average traverse rate 10 mm/min, material thickness 100 mm)

with larger steps imply that the higher efficiency of the cutting can be reached in non-homogeneous brittle materials (comparing it with continuous movement). Different proportions are apparently presented at the disintegration process in contrast with up to now studied disintegration of the ductile materials. The structure of the cut surface is similar to that performed by continuous movement. In addition the primary waviness zone appears in the top portion of the cut surface caused by repeated stopping of the jet movement. The bottom surface of the kerf is more irregular.

Further experimental research in this area will be directed at the disintegration of some other materials (especially rocks and concrete) by stepping movement.

REFERENCES

Chao J., E.S. Geskin & Y. Chung 1992. Investigations of the Dynamics of the Surface Topography Formation During Abrasive Waterjet Machining. *Proceedings of the 11th Int. Conf. on Jet Cutting Technology:* 593-603. BHR Group, St. Andrews, Scotland

Hashish M. 1987. An Improved Model for Erosion by Solid Particle Impact. *Proceedings of the 7th International Conference on Erosion by Liquid and Solid Impact.* ELSI VII, Cambridge, England

Hashish M. 1992. On the Modeling of Surface Waviness Produced by Abrasive-Waterjets. *Proceedings of the 11th Int. Conf. on Jet Cutting Technology:* 17-34. BHR Group, St. Andrews, Scotland

Hlaváč L. 1992. Physical description of high energy liquid jet interaction with material. *Proceedings of the Int. Conf. Geomechanics '91: 341-346.* Rotterdam: Balkema

Hlaváč L. 1993. Physical model of jet-abrasive interaction. Presented at *the Int. Conf. Geomechanics '93.* Hradec/Ostrava, Czech Republic

Singh P.J., W.L. Chen & J. Munoz 1991. Comprehensive Evaluation Of Abrasive Waterjet Cut Surface Quality. *Proceedings of the 6th American Water Jet Conference:* 139-161, Houston, Texas

Vala M. 1993. The Measurement of Non-setting Parameters of the High Pressure Water Jets.

Presented at *the Int. Conf. Geomechanics '93*. Hradec/Ostrava, Czech Republic
Vašek J. 1993. Basic Research of Water Jet Rock and Concrete Disintegration in Czech Republic. Presented at *the Int. Conf. Geomechanics '93*. Hradec/Ostrava, Czech Republic

Geomechanics 93, Rakowski (ed.) © 1994 Balkema, Rotterdam, ISBN 90 5410 354 X

Mathematical modelling of several aspects of high pressure water and abrasive jet action

Tomáš Sochor
Institute of Geonics of the CAS, Ostrava, Czech Republic

ABSTRACT: The paper is devoted to the mathematical modelling of rock disintegration using high pressure water jet. The problem analysis and one possible method for its solution are presented. Besides an important problem in abrasive jet disintegration is studied, namely behaviour of an abrasive particle during the jet generation. An example of the solution obtained using ANSYS program is presented.

1 INTRODUCTION

The experimental research of rock disintegration requires a lot of expensive results. Bearing in mind the reduction of the experimental requirements we have started utilizing applied mathematics that could enable the simulation of some experiments using mathematical modelling approach. The quality of scientific results however does not made worse.

The rock disintegration is very complicated physical process and interdisciplinary access is necessary for its exact and correct description. From physical point of view it can be represented in simplified way by failure of energy bonds in rock material by means of external power while from mathematical point of view it can be studied as a problem of optimal control and design.

The external power mentioned in previous paragraph can be generated by many different ways. The classical one is using mechanical tool. It seems however (Vašek 1993) to be reaching its physical limitations at present, especially in hard-to-cut rocks. There are several non-classical ways too. Our research work is focused especially to high pressure water and abrasive jet as a prospective disintegrating tool.

1.2 *Modelling of mechanical tool*

The research on the mathematical modelling of the rock disintegration process has been performed since 1986. The main effort has been focused to the modelling of the disintegration by mechanical cutting tool first. In spite of neglecting of certain aspects of the disintegrating process in the mathematical model the results of numerical experiments (presented e.g. in Horák (1989, 1992a)) confirm the supposition that the possibility for using of the non-assisted cutting tool for disintegration of hard rocks is limited. The reason resides in stresses generated in the tool itself during the disintegration process. The stress tensor components' values are high so far that the tool may be damaged. As a result the tool is worn quickly.

The main conclusions resulting from the mathematical modelling of the rock disintegration by mechanical tool may be summarized in the following:

- An additional improving of rock disintegration efficiency is obviously possible especially by the development of new high-quality materials for cutting tools that have to carry the higher and higher stresses. Unfortunately this way is very expensive and the results may seem to be unsatisfactory sometimes.
- The results of the mathematical modelling showed that there is one possible way to avoid excessive stress and resulting wear of the cutting tools: generating of additional stresses in the vicinity of the tool tip.

The latter conclusion is confirmed by the fact that preweakening of rock ahead of a cutter is an efficient method of reducing cutting forces in hard rocks (Morrell, Wilson 1983). Our experience (e.g. Vašek et al. 1987) shows that high pressure water jet is one of the most suitable means for doing it.

2 MODELLING OF HIGH PRESSURE WATER JET

The conclusion stated above as well as further analyses imply that our work should be oriented to the combination of the cutting tool with high pressure water jet (further HPWJ) as a tempting mean especially for hard rock disintegration.

The analysis of the disintegration process by means of the combination of the tool and HPWJ showed however that the knowledge of the mechanism of the disintegration caused by the HPWJ is not exact sufficiently to become the base for mathematical modelling of the process. It is obvious besides that the modelling of the rock disintegration by an unassisted HPWJ represents the necessary step to the mathematical modelling of the disintegration by the combination of the cutting tool with the HPWJ. The more detailed analysis is referred in Sochor (1991).

Regarding the fact mentioned above the possibilities for mathematical modelling of the unassisted HPWJ action to the rock during disintegration have been studied.

Compared with the mathematical modelling of the disintegration by mechanical cutting tool there is one substantial difference. It is very difficult to find any acceptable and correct physical representation for HPWJ unlike mechanical tool easily representable as an elastic body. The problem is studied by Hlaváč (1992). This work include a physical description of the HPWJ action however but no correct and exact physical model utilizable for mathematical modelling is developed either there or anywhere in the world. This is a main obstacle in application of mathematical methods. In fact their direct application is impossible.

Facing the need the disintegration process to be modelled I have tried nevertheless to approach the problem in the different way: based on the knowledge about HPWJ action during rock disintegration I have created mathematical model of the isolated steps which the disintegrating process can be divided into. This model should approximate the HPWJ action to rock at least at the level of individual steps. It is obvious that the method of division influences the obtained results but the freedom of choice can be reduced substantially. The reduction was made by considering of equally time-spaced divisions only.

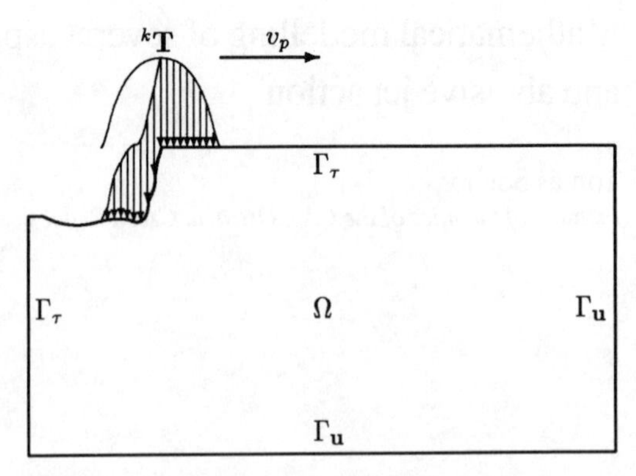

Figure 1. Modelling of the HPWJ action

2.1 *Technical description of the problem*

Linear cutting to the rock specimen with the planar upper side was selected as a basic experiment to be modelled. The jet nozzle moves along the straight line parallel to the upper rock side and the jet axis is perpendicular to it implying the constant distance between the nozzle and the rock surface. The velocity of the nozzle movement is considered constant too. The size of the rock specimen is selected so that its edges do not influence the stress distribution inside it. The modelled system layout is shown in the Fig. 1.

2.2 *Simplifying assumptions*

Several hypotheses have been assumed for the purpose of easier realization of these first steps in the mathematical modelling of the HPWJ action. The list of these simplifying assumptions is as follows:

- All dynamical effects caused by the HPWJ are neglected implying the jet action could be described as static pressure applied to the rock surface in the jet axis direction.
- The modelled problem has been formulated as planar in the vertical plane containing the straight line along which the nozzle motion is actuated. The acceptance of this assumption was motivated by the effort to build the mathematical model as simple as possible. The 3D modelling requires much more time both for computing and especially for data preparation and postprocessing.

This assumption is justified – at least for initial stage of modelling – by the plane symmetry of the problem to be modelled and by the fact that the distance of influence in the third dimension is very small both relatively and absolutely.

The magnitude of the pressure p_L acting to the rock surface was computed using the relation

$$p_L(u) = 2p_0(1-\gamma p_0)e^{-2\xi L}\mu^2 \left[1 - \left(\frac{2u}{d_L}\right)^{\log(Re+1)}\right]^2$$

where p_0 is water pressure before nozzle, L is the distance between nozzle and material, d_L is the diameter of the jet at the distance L from the nozzle and u is the perpendicular distance from jet axis. The remaining symbols γ, μ, ξ and Re indicate the parameters of the jet and their description can be found in Hlaváč (1990), where the relation is developed.

Considering the typical course of rock disintegration using mechanical cutting tool described e.g. in Morrell, Wilson (1983) or Vašek et al. (1987) which is observed to be non-continuous process consisting of the successive breaking out of the rock chips on the one hand and the observation of experimental cuts in rocks by HPWJ on the other hand we have accepted the assumption that the progress of the rock disintegration process is analogous to the mechanical tool disintegration, namely non-continuous one consisting of the successive rock particle creating and removing. Therefore the mathematical model of the rock disintegration by HPWJ has been designed in a way that the rock surface is loaded by the static pressure representing HPWJ action repeatedly; after every load cycle the rock disintegration is evaluated, broken rock is removed and the load is applied again. It can be seen that the both carrying off the disintegrated rock particles and the water draining is neglected in the model. This neglecting is justified by the fact that the model has been designed for rock cutting where the role of these aspects is small in the difference with HPWJ drilling.

2.3 *Mathematical model*

The conceptional model described in preceding paragraphs means that considering the accepted assumptions the mathematical modelling of the disintegrating process consists of the solution of the sequence of the linear plane elastic problems with simple static pressure load ${}^k\mathbf{T}$ acting to the boundary Γ_τ of the domain Ω representing the rock sample. This elementary problem is uniquely soluble when the boundary $\Gamma_\mathbf{u}$ with prescribed zero displacement is such that $\mu(\Gamma_\mathbf{u}) \neq 0$. More detailed problem analysis and exact description of the algorithm is given in Sochor (1991) as well as the results of several numerical experiments.

The results of the mentioned experiments with the mathematical modelling of the HPWJ action seem to be prospective. On the other hand the research works showed that there are some unsolved problems causing the limited applicability of the model. The most important ones are the following:

- The size of the time step influences the results very much but its proper selection is very difficult. The problem is in insufficient knowledge about the velocity of failure propagation in rock under HPWJ action. The problem is currently researched in our laboratory however.
- The type of brittle failure taking place by the HPWJ action is not set exactly. The most recent studies show that both tensile and shear failure play important role in the rock disintegration using HPWJ, but the portion of the failure types are not determined.

3 ABRASIVE JET MODELLING

Our main effort is oriented to the mathematical modelling of several aspects of abrasive jet disintegration recently. The reason is simple – the importance of the abrasive jet cutting grows substantially compared with water jet cutting. Moreover the analysis of the disintegration process (Hlaváč 1990) show that the efficiency of energy transformation in HPWJ disintegration is very low and using the abrasive jet is one possible way to improve it.

The abrasive jet application involves the dependence of the process upon several additional parameters incidental to abrasive material but on the other hand the process can be divided into partial processes easily. Such a division is very important for the purpose of mathematical modelling. Our work is focused on the mathematical modelling of the subprocess of adding of abrasive particles into the water jet.

For the sake of simplicity we have used the mathematical model in the plane elasticity so far. The

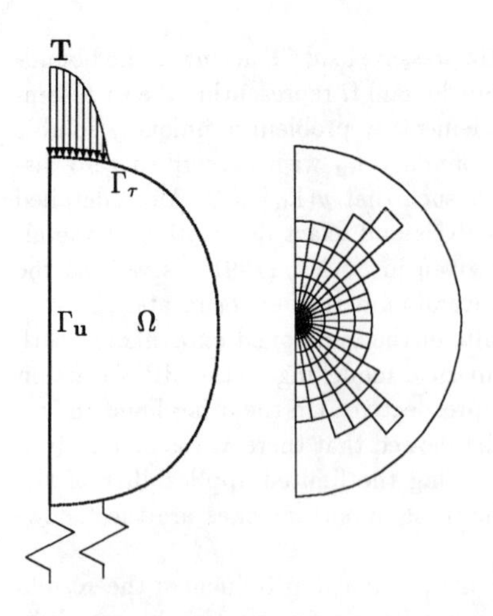

Figure 2. Modelling of the abrasive particle

abrasive particle was modelled as an elastic sphere (the symmetry of the problem along the jet axis was utilized, i.e. the hemisphere Ω with symmetrical boundary condition was modelled) supported by springs (fixed at the opposite ends) and loaded by HPWJ pressure **T** (see Fig. 2). The total stiffness of the springs was calculated from the force acting to the fixed particle due to the stream of water with the parameters taken from the appropriate HPWJ and the particle's moment of inertia. The load acting to the abrasive particle in the direction of the jet axis was determined using the relation given in par. 2.2. The failure state of the abrasive particle was studied as a part of postprocessing. For this purpose the Coulomb criterion combined with simple tension was selected.

The problem was numerically solved using finite element software ANSYS with our own extension for failure state evaluation. An example of the typical result (jet pressure 240 MPa, nozzle diameter 0.3 mm, abrasive particle diameter 0.4 mm, abrasive material properties $E = 970\ GPa, \mu = 0.085$, compression strength 250 MPa, tensile strength 40 MPa, inner friction angle 30°.) is shown at the right part of Figure 2 (failured elements are framed).

4 CONCLUSION

The mathematical modelling of the abrasive particles shows that the failure during abrasive jet generation cannot be neglected. Therefore we plan further continuing of the research and the extending of the range of our numerical experiments. It is supposed that further step in the mathematical modelling of water jet rock disintegration could be made applying the algorithm of the same type as in the case of the jet -abrasive particle interaction. Namely the rock material will be represented as a system of grains with different parameters.

As far as the HPWJ modelling is concerned, further progress depends on the improvement of the understanding the mechanism of HPWJ action to the material.

REFERENCES

Hlaváč, L. 1990: Physical modelling of the liquid action during rock disintegration. In Vašek et al. : *Mechanism of rock disintegration by high pressure water medium.* Annual report of Mining Institute Ostrava, Czechoslovakia (in Czech).

Hlaváč, L. 1992: Physical description of high energy liquid jet interaction with material. *Proc. Int. Conf. Geomechanics'91* Hradec/Ostrava, Czechoslovakia:341-346. Rotterdam:Balkema

Horák, J. 1989: Mathematical modelling of rock disintegration problem. *Proc. Mining geomechanics'89* Hradec/Ostrava, Czechoslovakia:454-464.

Horák, J. 1992a: Mathematical modelling of rock cutting. *Proc. Int. Conf. Geomechanics'91*, Hradec/Ostrava, Czechoslovakia:347-352. Rotterdam:Balkema.

Horák, J. 1992b: Numerical solution of contact problems. *Proc. Int. Conf. Geomechanics'91*, Hradec/Ostrava, Czechoslovakia:187-192. Rotterdam:Balkema.

Morrell, R. J. , Wilson, R. J. 1983: *Toward development of a hard-rock mining machine – drag cutter experiments in hard, abrasive rocks.* Bureau of Mines, USA, Report of investigation 8784.

Sochor, T. 1991: *Numerical solution of the high pressure water jet disintegration problem.* Ph. D. thesis, Mining institute of Czechosl. Acad. of Sci., Ostrava, Czechoslovakia (in Czech).

Vašek et al. 1987: *Mechanism of rock disintegration by high pressure water medium.* Annual report of Mining Institute Ostrava, Czechoslovakia (in Czech).

Vašek 1993: *Basic research of water jet rock and concrete disintegration in Czech republic.* To be presented at the Int. Conf. Geomechanics'93.

Geomechanics 93, Rakowski (ed.) © 1994 Balkema, Rotterdam, ISBN 90 5410 354 X

A fundamental test for parameter evaluation

David A. Summers & J.G. Blaine
University of Missouri-Rolla, Mo., USA

ABSTRACT: The use of high pressure water as a tool to cut and mine material is becoming more common. However, there is no good model to predict the required performance of a jet system. In order to develop a standard method of evaluating projected performance, a specialized piece of equipment has been developed. A short series of traversing tests is used to identify the operational parameters for the jets to effectively cut the material. From this experimental series it is possible to define a system for effectively removing material, and to determine a production rate for a given waterjet system.

Considerations in the operation of the equipment must be directed toward insuring that the material is removed in a controlled particle size. The reasons for this include the safety precautions required to insure that no water hammer pressures are generated during mining, as well as the simple logistical concerns for effective material removal and feed stock parameters for subsequent processing. The considerations of particle size require a proper evaluation of the path which the high pressure cutting lance must follow as it passes over the surface of the material and sequentially along the cut. Special precautions must be taken to insure complete material removal without the risk of engendering extra large fragments of debris. The operation of the system is demonstrated using an example.

1 INTRODUCTION

The growing use of high pressure waterjets for cutting and removing material has led to an increasing requirement for a method of predicting jet performance. However, while there have been a number of attempts to develop a theoretical basis on which to bse a predictive equation, these have not been effectively successful. Part of the reason for this is that the investigators have started with an assumption on the behaviour of the jet and the target which derives from the impact of a solid projectile on a continuous surface. As a result the models have looked at stress distributions within the continuum, and sought to generalize the failure mechanism within it.

A strong body of physical evidence exists to indicate that this is not the correct approach to take. Waterjets fail material by penetrating and extending surface cracks and are most effective when the pressure applied across the surface is at a strong lateral gradient. This failure mechanism requires a wide variety of inputs of it is to be effectively modelled for any one circumstance. Firstly one must be able to predict the pressure profile of the jet as it reaches the surface, and secondly one must be able to define the crack density and condition on that surface. The profile of the jet is controlled not only by the delivery capacity of the high pressure pump,. but the condition of the flow path to the nozzle, and the condition of the nozzle itself. These various parameters make a true mathematical model somewhat difficult to derive for any one situation without a considerable physical investigation of the surface. It therefore becomes, on a timely basis, more efficient to derive a physical test from which to empirically predict jet performance. It is with the development of such a simple predictive procedure that this paper is concerned.

2 SAMPLE PREPARATION

Although samples may be recieved in a variety of forms, and conditions, it is preferred that the materials be prepared to a standard shape in order to minimize un-noticed side effects from affecting the results of a test. For this reason, since many of the samples to be tested are of a rock like nature the decision was made to standardize on a cored sample shape which are typically available in the geotechnical industry. This also had the advantage that, by using a cylindrical shape, the apparatus could be easily modified to hold thinner discs of material from other sources.

Experience has shown that the most important parameters to be chosen for the cutting jet are its pressure, its diameter and the speed at which the nozzle is moved over the material surface. It is most effective if these can be evaluated within the range that can be expected to be used in the application. This will, in turn, mean that the jets will cut to the same depth as that during a washout operation. The sample size must therefore be sufficiently large to contain the cuts made during the tests, and yet small enough to be readily prepared. The most common core size is roughly 5 cm. diameter, and a test apparatus has, therefore been developed, to use samples of this size to obtain the required data. The apparatus was designed to hold samples of various lengths, although, in order to contain the cut within the sample body under all the conditions of test, it is recommended that the length be, where practical, at least 15 cm.

3 TEST APPARATUS

The device constructed was developed to gain the greatest quantity of information possible from the 5 cm. diameter samples. This equipment, known as PETE (for Parameter Evaluation Test Equipment) is used to hold the sample in a fixture ahead of the waterjet nozzle and to rotate the sample cylinder at 120 rpm. (Figure 1). While turning at this speed, a waterjet nozzle is traversed across the face of the cylinder cutting a groove in the surface. A plate is used to cover half the sample face, so that the jet will only make one cut into the material as it moves to the center of the sample.

With the plate in place the jet cuts a single spiral groove into the surface. To ensure no interference between adjacent cuts, the gearing on the lance advance drive was changed to give a feed of 3.8 mm. between adjacent passes (Figure 2). In this way, with more than 3 nozzle diameters between the two cuts, there would be no overlap of the jet cuts.

Running a test with this configuration provided a cut made at a continuing variable speed. Thus the one cut could provide several data points. To provide consistent points of measurement the center of rotation was first found for each cut. To do this the sample face was scribed with two perpendicular pencil lines which identified the center of the sample. A digital caliper was then used to identify points at which the cut centerline was at the 6.25 mm., 12.5 and 18.75 mm. radii, and the slot depth was measured at each point.

Figure 1. Parameter Evaluation Test Equipment (PETE)

Figure 2. Cut Configuration on the Face of a Sample, showing the pencil marks used for Center Location.

The experiments are normally designed to be carried out at three pressures and three nozzle diameters, but this range can be expanded, based upon the initial test results should these show that the initial range does not include the optimal conditions likely to be needed for most effective cutting.

To provide an example of the process to be followed a test series is described in which a crystalline simulant was used for the target material. For this material the test series was expanded until a series of 5 pressures, with three traverse speeds and four nozzle sizes had been evaluated.

The procedure for each test was the same, in that the same was inserted into the device, and locked in place. The sample was rotated at speed, and the jet pump switched on and brought up to pressure. A motor was then engaged which moved the nozzle over the surface at the required speed, until the sample face had been covered. The motors were then stopped, and the sample removed. Another sample could then be inserted, and, where necessary the nozzle changed before running the following test.

4 EXPERIMENTAL RESULTS

The extracted data is presented as a measured depth of cut, with the tables separated by nozzle diameter, and with the rows identified by the jet pressure, in MPa, at which the tests were run.

Averaged values for the data are presented at the end of each row and column. Rather than give the reading as defined by the radius at which it was measured, the value has been converted into the linear traverse speed at which the jet moved over the surface, in mm./second. This was felt to be a more useful representation of the data, and is so used in the subsequent data analysis.

5 DATA ANALYSIS

Because the data had been collected through an experiment using a factorial design, it was relatively easy to extract the information on the effect of change in the major parameters on the depth of cut achieved. The averaged depths of cut were plotted against each of the major variables, using the program Cricket Graph (Ref. 4), which can compute a best curve fit and plot this relative to the underlying data (Figures 3, 4, & 5).

Previous experiments at UMR and other laboratories over the years have shown that the

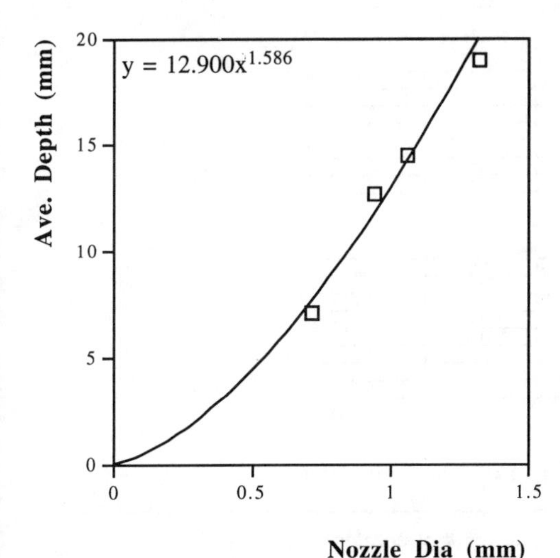

Figure 3. Depth of Cut as a Function of the Jet Nozzle Diameter.

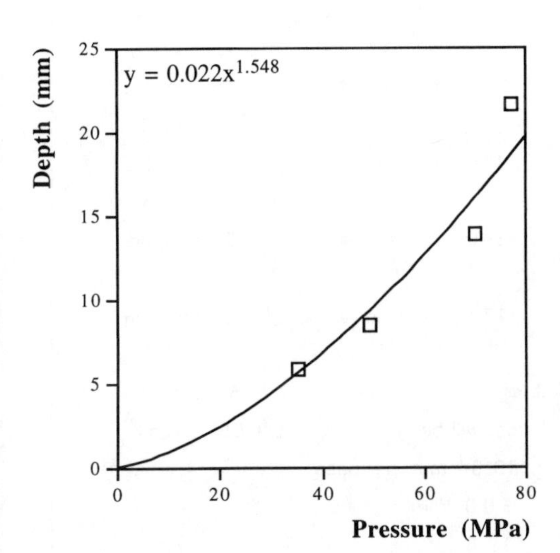

Figure 4. Depth of Cut as a Function of the Jet Pressure.

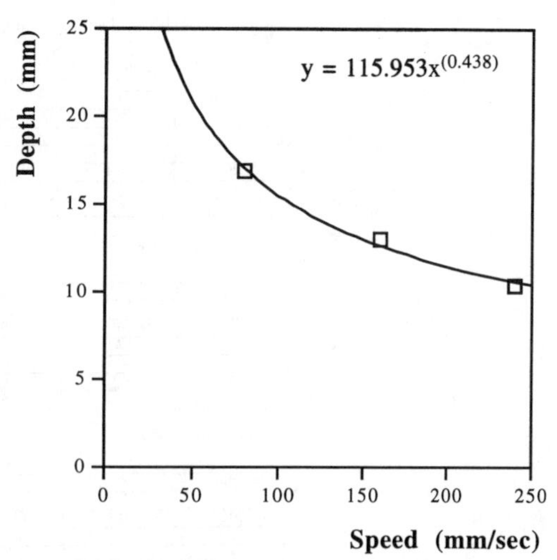

Figure 5. Depth of Cut as a Function of Traverse Speed.

relationship between these jet performance parameters and the cutting depth can be described by an equation of the form:

$$\text{Depth of Cut} = \frac{(\text{Pressure})^{1.0} .(\text{Nozzle Dia})^{1.5}}{(\text{Traverse Speed})^{0.33}}$$

This was not quite the case with the experimental relationship developed with the above simulant. However, in order to fit the predictive equation to the material under test the model is modified to use the exponential values (in this case 1.548, 1.586 and 0.438, for the pressure, diameter and traverse speed exponents respectively).

6 PROJECTIONS FROM THE DATA

The cutting path to be followed in removing material from any casing will be controlled by the casing geometry. To illustrate the next stage in the development, consider a chemical reaction vessel which is 30 cm. in diameter, and some 50 cm. long filled with the above material. To obtain a performance estimate for the cleaning of this vessel one uses the above relationship in a form which can be loaded into a spreadsheet.

The following model has been generated on a computer spread sheet using the EXCEL computer Program. A copy of the spreadsheet is available, from the author.

The program layout is, to a large degree self-explanatory. It can be broken down into several parts. The bold numbers are data which should be input by the programmer.

1) On the second line, the program operator is asked to input the nozzle diameter, the jet pressure and the number of nozzles required for the jetting operation.

Table 1. Depth of Cut for Varying Jet Parameters

Nozzle diameter 0.71 mm.

Pressure	speed (mm./sec) 80	160	240	Ave
77	13.2	10.2	8.9	10.7
70	10.9	9.9	5.3	8.6
63	11.3	9.8	8.8	9.9
49	4.1	3.3	3.8	3.8
35	2.5	2.5	2.1	2.3
Ave	8.4	7.1	5.8	7.10

Nozzle diameter 0.94 mm.

Pressure	speed (mm./sec) 80	160	240	Ave
77	20.3	20.0	13.5	18.0
70	26.3	19.8	13.5	19.0
63	17.5	19.0	11.2	16.0
49	4.5	4.3	4.1	4.3
35	6.9	6.3	5.8	6.3
Ave	14.7	14.0	9.6	12.7

Nozzle diameter 1.06 mm.

Pressure	speed (mm./sec) 80	160	240	Ave
77	37.1	25.4	22.4	28.2
70	20.1	15.0	13.2	7.6
63	21.6	15.2	8.9	15.2
49	10.7	7.6	4.6	7.6
35	6.6	6.1	3.6	5.3
Ave	19.3	14.0	10.4	14.5

Nozzle diameter 1.32 mm.

Pressure	speed (mm./sec) 80	160	240	Ave
77	41.9	23.1	23.1	29.5
70	30.0	16.5	14.7	20.3
56	29.7	23.9	*21.6*	25.1
49	23.9	15.7	14.0	17.8
42	13.5	10.7	10.9	11.7
35	11.7	8.4	8.4	9.4
Ave	25.1	16.5	15.5	19.0

Table 2. A Result derived from the Experimental Program

Jet Dia (mm) Jet Pressure (MPa) # of nozzles

1.25 **70** **2**

At **1** rev/sec @ **150** mm. diameter

traverse velocity equals 471.18 mm./sec.

If a

56 MPa jet cuts **21.60** mm. at **240** mm./sec @ **1.32** mm. dia

then a

70 MPa jet cuts 27.00 mm. at 240 mm./sec @ 1.32 mm. dia

then a

70 MPa jet cuts 24.88 mm. at 240 mm./sec @ 1.25 mm. dia

then a

70 MPa jet cuts 19.92 mm. at 471 mm./sec @ 1.25 mm. dia

If the largest piece of debris we wish to allow has a width of **7.5** mm.

then the speed of rotation of the lance around the vessel axis will be 23.56 seconds

then the advance rate down the vessel will be 39.84 mm. per pass

length of the vessel is **500** mm.

number of passes required becomes 12.5

time required to clean the vessel 5.7 minutes

2) On the next line information is needed on how fast the cutting head will rotate about its axis, and the diameter of rotation of the nozzles about the cutting head axis. (For this example the head was presumed to move along a path half way between the wall and axis of the vessel. The nozzle movement was then set to cover a circle moving from the vessel axis to the vessel wall, or half the vessel internal diameter).

3) On the following line the program calculates the lateral traverse velocity of the nozzle over the surface.

4) The operator should then input a representative data value from the experimental program which has been carried out. The data input is that marked with a double asterisk in Table 1.

5) The program goes through a series of steps, based on the above equation for depth, to determine the depth of cut achievable with the jet parameters input as being used on the lance.

6) The operator is asked to input the maximum size of particle which can be allowed to break from the solid. This is an important value since larger fragments can get stuck in any narrow entries into the vessel or can block critical flow passages, it should therefore be set at a relatively small value (perhaps one third of the width of the narrowest passage that the particle must pass through).

7) Based on the rotation speed of the head, the nozzle diameter used and the vessel size, the program calculates how long it will take for the cutting head to complete a circuit around the axis of the vessel.

8) The depth of cut achieved by the head on each pass will have been calculated. Note that because the head has multiple nozzles it is assumed that the depth of cut will be equal to the number of nozzles multiplied by the individual depth of cut achieved by each jet, and earlier calculated.

9) The operator is asked to input the depth of material within the vessel.

10) The program calculates how many passes, at the given depth of cut, it will take to reach the back of the vessel.

11) The program calculates, based on the number of passes, and the time taken per pass how long it will take to clean the vessel. Note that it adds two additional passes, one at the beginning and one at the end of the jetting program, to catch any residual material, and to flush the vessel clear of debris.

The model allows iterations, by changing jet operating parameters, to reach a desired cleaning speed. Only one such possible option is shown in the demonstration example of the model given above.

REFERENCES

Giltner, Scott G., Sitton, Oliver C., and Worsey, Paul N., "The Reaction of Class 1.1 Propellants and Explosives to Water Jet Impact," 24th International ICT Conference, Karlsruhe, Federal Republic of Germany, June 29 - July 2, 1993.

Summers D.A., Wright D., & Galecki G. "Failure Mechanisms under Waterjet Impact," 34th U.S. Rock Mechanics Symposium, U of Wisconsin, 1993 (in press).

Lesser M.B. & Field J.E. "The geometric wave theory of liquid impact," 6th Conf. on Erosion by Liquid and Solid Impact, Cavendish Laboratory, University of Cambridge, UK.

C-A Cricket Graph III, Computer Associates.

Geomechanics 93, Rakowski (ed.) © 1994 Balkema, Rotterdam, ISBN 90 5410 354 X

Ultrasonic modulation of high-speed water jets

M.M.Vijay
Institute for Engineering in the Canadian Environment, National Research Council of Canada, Ottawa, Ont., Canada

J.Foldyna
Institute of Geonics of the CAS, Ostrava, Czech Republic (Presently: National Research Council of Canada)

J.Remisz
Applied Fluid Mechanics Inc., Willowdale, Ont., Canada (Presently: National Research Council of Canada)

ABSTRACT: Improving the performance of high-speed water jets is necessary for extending their use in mining and underground applications. From the analysis of the impact of a water jet on the material surface one can show that pulsed jets provide an effective method for cutting or fragmentation of hard and brittle materials. Modulation of the continuous high-speed jet represents one of the most promising methods for generating pulsed jets. In this paper, a brief summary of theoretical models, which attempt to describe the behaviour of the modulated jet after leaving the nozzle, is presented followed by some aspects of the ultrasonic modulation of high-speed continuous water jets. Preliminary tests to evaluate the effectiveness of modulated water jets were performed using aluminum, copper and cast iron samples. The jet was modulated by the vibrations of a piezoelectric transducer and velocity transformer (horn) at 15 kHz.

1 INTRODUCTION

Despite the impressive advances made in the field of water jet technology during the last two decades, the use of water jets for hard rock mining and other underground engineering applications remains unattractive. The main reason for this apparent lack of interest is that the performance of water jet techniques is not competitive with the existing conventional mechanical systems. Therefore, extending the use of water jets to these areas requires significant improvement in their performance. Pulsed jets, as shown in this paper, offer this possibility.

From an analysis of the impact of a water jet on a target, it can be shown that the impact pressure generated by a slug of liquid is considerably higher than the corresponding stagnation pressure generated by a continuous jet. Therefore, if a continuous jet could be divided into a train of slugs, the resulting pulsed jet could significantly improve the performance in the cutting of hard rocks. Furthermore, one can assume that additional effects (for example, enhanced penetration and the subsequent crack propagation in the material) due to impact of pulsed jets may provide new methods for the cutting and fracturing of hard and brittle materials.

Since the early seventies, incessant attempts have been made for generating various types of pulsed water jets (for details, refer to the paper by Vijay, 1993). In this paper, attention is focused on a particular method of generating pulsed jets which involves modulating a continuous stream of water with an ultrasonic vibrator upstream of the nozzle exit (Puchala & Vijay, 1984 and Vijay, 1992).

Unlike single pulse and interrupted jets, a modulated jet escapes from the nozzle as a continuous stream of liquid having unsteady velocity (cyclically modulated over time). Slow and fast portions of each cycle tend to flow together, forming a train of "bunches" in the free stream, which eventually separate (Nebeker & Rodrigues, 1976 and Nebeker, 1981). It should be noted that other techniques for modulating continuous water jets are available (for instance, refer to Chahine & Conn, 1983; Chahine et al., 1983; Conn, 1989; Johnson et al., 1982; Nebeker & Rodriques, 1976; Nebeker, 1981; Sami & Anderson, 1984; Shen et al., 1987 and Shen & Wang, 1988). In this paper a brief discussion on theoretical aspects of modulated high-speed water jets is given with sample experimental data obtained on cutting metallic samples (aluminum, cast iron and copper) to illustrate the superior performance of modulated jets compared to continuous water jets.

2 THEORETICAL ASPECTS OF HIGH-SPEED WATER JET MODULATION

2.1 *Natural break-up of the jet*

When a continuous liquid jet of cylindrical form emerges from a nozzle, the interaction between the cohesive and disruptive forces on the surface perturbs the jet giving rise to oscillations. Under favourable conditions, the oscillations are amplified and the liquid body disintegrates into drops {according to Vijay (1993) this is the so-called natural pulsed jet}. The mathematical model of the mechanism of break-up of a laminar inviscid jet was first published by Rayleigh (1878). His analysis shows that all disturbances on the jet with wavelengths (jet velocity divided by frequency) greater than the jet circumference will grow. Furthermore, his results show that the jet is unstable only for symmetrical disturbances and will break up eventually. Although Rayleigh's theory is not applicable to high speed turbulent jets, it does show that it is possible to produce pulsed jets in this manner and the average diameter of the drops produced is about twice the diameter of the undisturbed jet. Following Rayleigh, many attempts have been made to analyze the break-up mechanisms of low speed liquid jets (Weber, 1931; Yuen, 1968; Nayfeh, 1970; Chaudhary & Redekopp, 1980 and Busker & Lamers, 1989). However, for high speed liquid jets, natural break-up mechanisms are not well understood (Reitz & Bracco, 1986).

2.2 *Forced break-up of modulated jets*

In the discussion given above, the surface tension induced displacements are believed to be the dominant factor for the natural break-up of continuous jets. These theories are restricted to infinitely small initiating amplitudes and capillary type instabilities of the jet. However, the mechanism of forced break-up of modulated jets is rather different. The initiating amplitude of modulation is of finite magnitude, and is the main cause for the instability of the jet. Thus, the kinematics of the jet is quite important in the analysis of forced break-up of modulated jets.

Crane et al., (1964) and McCormack et al., (1965) used the high-frequency mechanical vibration of a nozzle to study the break-up characteristics of the emerging jet, and presented their semi-quantitative theory for the "bunching" phenomenon. Their study appears to be the first account of a jet break-up due to an inertial-type mechanism. The physical nature of the initiating disturbance was identified and incorporated into the theory. Pimbley's (1976) analysis showed that drop formations obtained with the axial velocity modulation are quite different from those offered by Rayleigh's theory.

More recently, Sami & Ansari (1981) attempted to theoretically determine the surface configuration of the modulated jet r_0 (x, t) along the jet axis at any given distance (x). They considered a water jet issuing at time (t) from a nozzle of radius (R) with a velocity $V_0(0, t)$ that is uniform across the exit area, but periodic about a mean value (V_m). They formulated the equations of motion by assuming the flow to be one-dimensional and axi-symmetric. They obtained the following dimensionless expression for the radius of a modulated jet:

$$\frac{r_0}{R} = 1 + \pi m \beta \cos(2\pi\tau) \tag{1}$$

where m is the velocity modulation ratio = $\Delta V/V_m$, ΔV = amplitude of the velocity modulation, $\beta = x/\lambda$, λ = modulation wavelength, and τ = f t, f = modulation frequency.

Equation (1) shows that the diameter varies cyclically with increasing x, while the amplitude increases after each cycle. At some distance from the nozzle exit, the jet becomes a train of discrete bunches of water. This distance, the jet break-up length (l_0), is given by:

$$l_0 = \frac{V_m}{\pi m f} \tag{2}$$

Chahine & Conn (1983) presented an alternate approach to the problem of the influence of modulation amplitude on the bunching process. They argued that the distance required to form a bunch (l_0) occurs when a crest overtakes a trough after a time (T) given by:

$$T = \frac{\lambda}{4\,\Delta V} \tag{3}$$

which yields:

$$l_0 = T V_m = \frac{\lambda}{4} \frac{V_m}{\Delta V} = \frac{V_m}{4 m f} \tag{4}$$

Thus, according to Sami & Ansari (1981) and Chahine & Conn (1983), the break-up length of the modulated jet is directly proportional to the mean velocity of the jet, and inversely proportional to both the velocity modulation ratio and modulation frequency. Interestingly, both models show that the break-up length of a modulated jet is not influenced by its initial diameter. Furthermore, both models, despite the different approaches and simplifications used in their analyses, lead to the same expression for the break-up length, except for a constant factor of $\pi/4$.

Both models of the forced break-up of a modulated jet introduce the term velocity modulation (or velocity modulation ratio). Therefore, an attempt was made to determine the magnitude of velocity modulation produced by three different modes of operating the ultrasonically modulated jet.

3 MODES OF ULTRASONIC JET MODULATION

Ultrasonic modulation of a jet is produced by the vibrating tip of an ultrasonic velocity transformer located inside a nozzle. The vibration is generated by an ultrasonic transducer connected to the velocity transformer. A detailed description of the concept and configuration of the ultrasonic nozzle can be found in Puchala & Vijay (1984) and Vijay (1992). The geometric configuration of the nozzle with the vibrating tip is shown in Fig. 1. The tip vibrates axially with amplitude (A) so that both the distance (a) and the gap (b) periodically change from maximum to minimum values. As stated above, the three different modes of modulation were: (i) flow rate modulation, (ii) velocity modulation and (iii) modulation induced by acoustic pressure. Analysis of the problem was simplified by making the following assumptions:

- Ambient pressure (p) upstream of the tip is constant;
- Flow through the large space (c) between cylindrical sections of the tip and the nozzle chamber is not affected by the position of the tip;
- Pressure losses in the nozzle are neglected and hence, nozzle discharge coefficient (C_D) = 1.0;
- The flow rate is modulated by the variations in flow cross-sectional area only.

3.1 *Modulation of the flow rate*

When the cross-sectional area of the gap is equal to or greater than the cross-sectional area of the nozzle orifice, the momentary position of the tip does not affect the flow rate through the nozzle (a threshold condition). However, when the cross-sectional area of the gap is smaller than that of the nozzle orifice, the vibrating tip operates like a dump valve, which periodically changes the cross-sectional area of the gap and thus modulates the rate of flow at the constant pressure p. The maximum velocity of the jet corresponds to this pressure. The mean velocity of the modulated jet is lower than the velocity of a continuous jet under the same hydraulic parameters. By a simple analysis of the kinematics of the flow, the velocity modulation ratio (m_Q) for the flow rate modulation mode can be written as:

$$m_Q = \frac{2(d_t A + aA\sin 2\alpha)}{2d_t a + (a^2 + A^2)\sin 2\alpha} \tag{5}$$

3.2 *Modulation of the jet velocity*

When the cross-sectional area of the gap is larger than that of the nozzle orifice, the mechanical action of an axially vibrating tip imposes a sinusoidally varying velocity onto the constant velocity (V_0) of the steady jet. This velocity V_0 corresponds to the pressure p, and it constitutes the mean velocity of the modulated jet. For this case, the velocity modulation

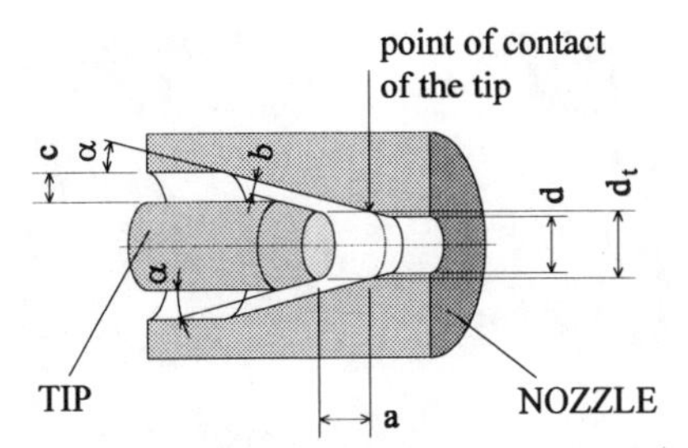

Fig. 1. Geometric configuration of the ultrasonic nozzle.

ratio (m_V) can be written as:

$$m_V = 2\pi f A (2p/\rho)^{-1/2} \tag{6}$$

where ρ is the liquid density.

3.3 *Modulation by acoustic pressure*

Regardless of the tip position inside the nozzle, vibration of the tip generates acoustic pressures which modulate the ambient pressure downstream of the tip and thus the velocity of the jet. The cyclically varying pressure downstream of the tip (P) can be written as follows:

$$P = p + 2\pi f \rho C_0 A \cos(2\pi f t) \tag{7}$$

where C_0 is the sound velocity in liquid.

Amplitude of the velocity modulation (ΔV) and the velocity modulation ratio (m_A) are:

$$\Delta V = (4\pi f C_0 A)^{1/2} \tag{8}$$

$$m_A = \left(\frac{2\pi f \rho C_0 A}{p}\right)^{1/2} \tag{9}$$

It should be emphasized that equations (5), (6) and (9) are based on a simplified description of the process of ultrasonic modulation of the jet. The actual process is much more complex and probably includes combination of all the three modes of modulation.

4 EXPERIMENTAL RESULTS

Although an extensive experimental program is in progress in the laboratory, due to limited length of this paper, only a few sample results are presented to illustrate the difference in the cutting performance of continuous and ultrasonically modulated pulsed jets. A water jet emerging from a 2.08 mm diameter nozzle at a pressure of 34.5 MPa was modulated by

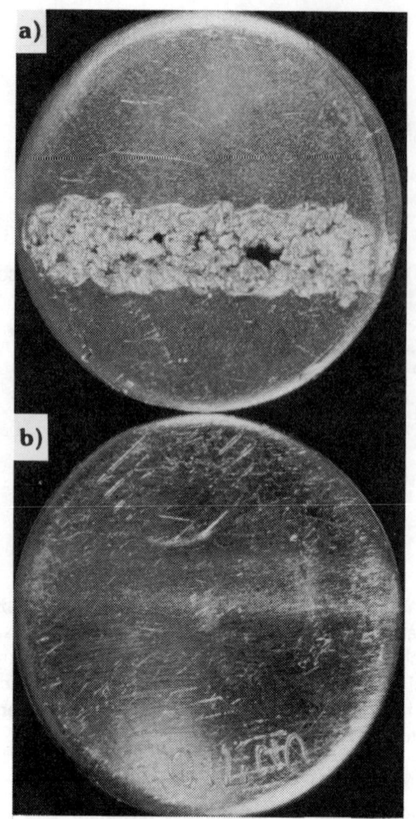

Fig. 2. Aluminum samples exposed to (a) modulated jet and (b) continuous jet. (p=34.5 MPa; v_{tr}=254 mm/min)

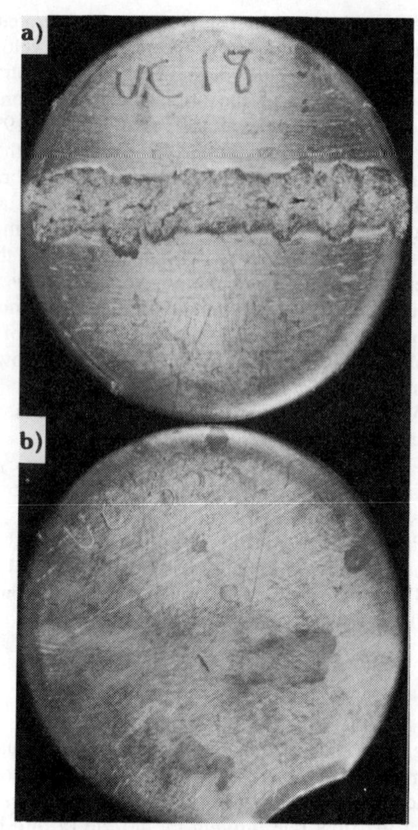

Fig. 3. Copper samples exposed to (a) modulated jet and (b) continuous jet. (p=34.5 MPa; v_{tr}=38 mm/min)

the vibration of a 2.4 mm diameter tip. This tip was situated inside the nozzle at a distance of 11.6 mm from the exit plane of the nozzle. Vibration at a frequency of 15 kHz was generated using a piezoelectric transducer. Ultrasonic power was 4 kW.

Thin discs (38 mm in diameter and 3.2 mm thick) of three different type (ductile and brittle) of materials were used as targets: aluminum 1100-H4 (>99% pure, tensile strength 124 MN/m^2), electrolytic copper (> 99.9% pure, tensile strength 207 MN/m^2), and grey cast iron ASTM A48 (tensile strength 207 MN/m^2).

Figure 2 shows the results pertaining to aluminum. The standoff distance was 140 mm and the traverse velocity was 254 mm/min. The modulated jet caused a wide and deep slot with material squeezed out to the edges of the slot (Fig. 2a). The sample was pierced at several points. The rate of mass loss was 3642 mg/min. The continuous jet operating at the same parameters, on the other hand, hardly made a dent on the surface, the rate of mass loss being 8 mg/min.

Figure 3 shows the results for the copper sample exposed to the jets at a standoff distance of 120 mm, and a traverse velocity of 38 mm/min. For the modulated jet, (Fig. 3a), the nature of the slot is similar to that of the aluminum sample, except that the slot is narrower and not as deep. Rate of mass loss was 2124 mg/min. For the continuous jet (Fig. 3b) the kerf is hardly visible, and the rate of mass loss was merely 3 mg/min.

The cast iron (brittle material) discs exposed to the jets at a standoff distance of 160 mm and a traverse velocity of 38 mm/min are illustrated in Fig. 4. The modulated jet formed a slot irregular in both shape and depth (Fig. 4a). However, the edges of the slot were clearly defined, without visible deformations. Rate of mass loss was 623 mg/min. In Fig. 4b the sample exposed to the continuous jet is depicted. Kerf is almost invisible, with the rate of mass loss being only 2 mg/min.

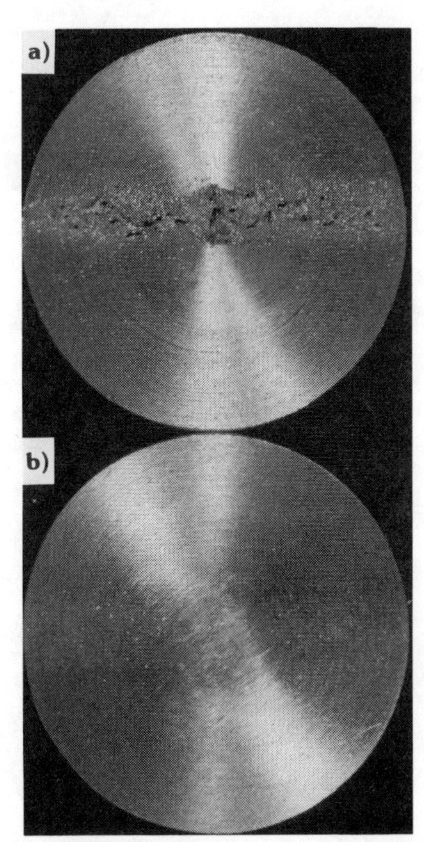

Fig. 4. Cast iron samples exposed to (a) modulated jet and (b) continuous jet. (p=34.5 MPa; v_{tr}=38 mm/min).

5 CONCLUSION

The conclusions from the simple theoretical analysis and the sample experimental results presented here are:

- The break-up length of a forced modulated jet is directly proportional to the mean velocity of the jet, inversely proportional to both the velocity modulation ratio and modulation frequency, and is not influenced by the initial diameter of the jet;
- It is possible to generate pulsed jets by modulating a continuous stream of water with ultrasonic waves;
- The performance of the acoustically (ultrasonic) modulated pulsed jets is significantly better than the continuous jet at the same parameters.

6 ACKNOWLEDGMENTS

The authors are grateful to Mr. N. Paquette, Technical Officer for the Water Jet Project, and to Mr. S. V. P. George, Faculty of Engineering, University of Alberta, Edmonton, for their technical assistance. The interest shown in this project by HDRK Mining Research Ltd., Canada is highly appreciated.

7 REFERENCES

Busker D. P. & A. P. G. G. Lamers 1989. The non-linear breakup of an inviscid liquid jet. *Fluid Dynamics Research 5*: 159-172.

Chahine, G.L. & A. F. Conn 1983. Passively-interrupted impulsive water jets. *Proc. 6th Int. Conf. on Erosion by Liquid and Solid Impact*, Cambridge: 34-1 - 34-9.

Chahine, G.L., A. F. Conn, V. E. Johnson Jr. & G. S. Frederick 1983. Cleaning and cutting with self-resonating pulsed water jets. *Proc. 2nd U.S. Water Jet Symposium*, University of Missouri-Rolla, Rolla, Missouri: 167-173.

Chaudhary, K. C. & L. C. Redekopp 1980. The non-linear capillary instability of a liquid jet. Part 1. Theory. *J. Fluid Mech. 96*: 257-274.

Conn, A.F. 1989. Asbestos removal with self-resonating water jets. *Proc. 5th U.S. Water Jet Symposium*, Toronto: 133-139.

Crane L., S. Birch & P. D. McCormack 1964. The effect of mechanical vibration on the break-up of a cylindrical water jet in air. *Brit. J. Appl. Phys. 15*: 743-750.

Johnson Jr., V.E., A. F. Conn, W. T. Lindenmuth, G. L. Chahine & G. S. Frederick 1982. Self-resonating cavitating jets. *Proc. 6th Int. Symp. on Jet Cutting Technology*, Guildford, Paper A1: 1-25.

McCormack, P. D., L. Crane & S. Birch 1965. An experimental and theoretical analysis of cylindrical liquid jets subjected to vibration. *Brit. J. Appl. Phys. 16*: 395-408.

Nayfeh, A. H. 1970. Non-linear stability of a liquid jet. *The Physics of Fluids 13 (4)*: 841-847.

Nebeker, E.B. & S. E. Rodriguez 1976. Percussive water jets for rock cutting. *Proc. 3rd Int. Symp. on Jet Cutting Technology*, Chicago: B1-1 - B1-9.

Nebeker, E.B. 1981. Development of large diameter percussive jets. *Proc. 1st U.S. Water Jet Symposium*, Golden, Colorado: IV-5.1-IV-5.11.

Pimbley W. T. 1976. Drop formation from a liquid jet: A linear one-dimensional analysis considered as a boundary value problem. *IBM J. Res. Develop., March*: 148-156.

Puchala, R.J. & M. M. Vijay 1984. Study of an ultrasonically generated cavitating or interrupted

jet: Aspects of design. *Proc. 7th Int. Symp. on Jet Cutting Technology*, Ottawa, Paper B2: 69-82.
Rayleigh, W.S. 1878. On the instability of jets. *Proc. London Mathematical Society, Vol. 10*: 4-15.
Reitz, R. D. & F. V. Bracco 1986. Mechanisms of breakup of round liquid jets. In Cheremisinoff, N.P. (ed.), *Encyclopedia of liquid mechanics, Volume 3 - Gas-liquid flows*: 233-249. Gulf Publishing Company: Houston, Texas.
Sami, S. & H. Ansari 1981. Govering equations in a modulated liquid jet. *Proc. 1st U.S. Water Jet Symposium*, Golden, Colorado: I-2.1 - I-2.9.
Sami, C. & C. Anderson 1984. Helmholtz oscillator for the self-modulation of a jet. *Proc. 7th Int. Symp. on Jet Cutting Technology*, Ottawa, Paper B4: 91-98.
Shen, Z., G. Li & C. Zeng 1987. Experimental study on rock erosion by self-resonating cavitating jets. *Proc. International Waterjet Symposium*, Beijing: 2-35 - 2-43.
Shen, Z. & Z. M. Wang 1988. Theoretical analysis of a jet-driven Helmholtz resonator and effect of its configuration on the water jet cutting property. *Proc. 9th Int. Symp. on Jet Cutting Technology*, Sendai, Paper D4: 189-201.
Vijay, M. M. 1992. Ultrasonically generated cavitating or interrupted jet. *U. S. Patent No. 5,154,347.*
Vijay, M. M. 1993. Power of pulsed liquid jets. *Proc. International Conference: Geomechanics 93*, Ostrava.
Weber, K. 1931. Zum Zerfall eines Flussigkeitsstrahles. *Z. angew. Math. Mech. 11*: 136-154.
Yuen, M. C. 1968. Non-linear capillary instability of a liquid jet. *J. Fluid Mech. 33*: 151-163.

Geomechanics 93, Rakowski (ed.) © 1994 Balkema, Rotterdam, ISBN 90 5410 354 X

The measurement of the non-setting parameters of the high pressure water jets

M.Vala
Institute of Geonics of the CAS, Ostrava, Czech Republic

ABSTRACT: The efficiency of the cutting process using high pressure water jets not only depends on the choice of the main operating parameters but as well as especially upon the static and dynamic stagnation force. An original method of providing a measure of the static and dynamic stagnation force of the inner active zone of the high pressure water jets is described. The design of the measuring device is delineated. Subsequently some results of initial experiments are presented and some initial conclusion are indicated in this paper.

1 INTRODUCTION

Since the high pressure water jet has been using for rock material cutting many research works were performed to found interrelationship between parameters of high pressure water jets and parameters of rock materials because in cutting, fragmenting or rock materials weakening by high pressure water jets is desirable to have theoretical equations to predict the result of water jets affect upon other related parameters.

In despite of considerable advances have been made in the field a knowledge of parameters which are relevant in the cutting process are still very poor. Mathematical solution of the flow field of high velocity circular jet emerging in air is highly complicated because its characteristic properties change for a given set of operating parameters.

Between operating parameters of the high pressure water jet we can place namely the pressure in front of the nozzle, the nozzle diameter and its geometry, the traverse velocity, the standoff distance and the angle of impact of the jet. Their adjustment and measurement is not problem but for many reasons these operating parameters cannot be regular correlation parameters for proper evaluation of the effect of cutting process.

The hydrodynamic effects at the inlet of the nozzle and the aerodynamic drag on the jet outside of the nozzle contribute to the variation of the velocity profiles and the dynamic behavior of the jet. These different jet structure naturally cause different cutting ability.

In generally the cutting action of the high pressure water jet is based on combined a static and dynamic loading of the workpiece by the jet. On the one hand it has to be assumed that the kind of stress of rock materials is similar to the jet velocity distribution on the other hand the stress is strongly influenced by spread of the jet upon the workpiece area.

From these facts it can be concluded that not the real jet diameter is the regular correlation parameter, but the effective jet diameter originates in an energy-rich inner zone of high pressure water jet is necessary to determine.

Therefore the know-how of these "non-setting" parameters is very important for studying of the jet nature and creates a base of knowledge for mathematical (Sochor 1993) and physical (Hlaváč 1992, Hlaváč & Vašek 1992) modeling of the high velocity water jet in our research program (Vašek 1993).

Between non-setting parameters of the high pressure water jet we can place mainly the value of the real dynamic pressure and its distribution upon the target material, respectively the static and dynamic stagnation forces as a result of this dynamic pressure action.

2 EXPERIMENTAL APPARATUS AND PROCEDURE

Due to the fact that the high pressure water jet is not only aggressive for rock materials but is

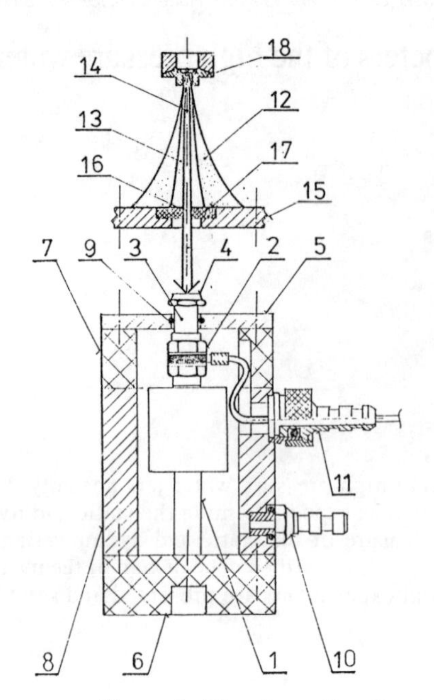

1. double quarter vave cylindrical transformer
2. crystal force transducer
3. strain bolt
4. inpingement disk
5. cover plate
6. rubber base plate
7. rubber pad
8. housing
9. rubber "O" ring seal
10. air inlet port
11. cable and air outlet
12. droplet layer
13. outer zone
14. inner zone
15. shield screen holder
16. penetration hole
17. shield screen
18. nozzle

Figure 1. The stagnation force measuring system with shield screen.

regardless of measure devices too, the direct measurement of the non-setting parameters is difficult because the jet diameter is very small and the impact pressure very high. Despite of it, the pattern of the high velocity flow can be obtained by means of the axial momentum flux measurement which consist of measurement of the thrust force of impinging jet on the flat disk and measurement of its effective diameter by an original method (Vala 1993) using the shield screen for checking non-effective parts of jet.

2.1 *Experimental apparatus*

The experimental apparatus can be described with the aid of schematic diagram in Fig. 1. The measurement of the real jet's stagnation forces especially dynamic forces features the difficulties because of its small value, high frequency range and generally the high jet's erosive ability. For these reason the measuring system employs a crystal force transducer in order to achieve high natural resonant frequency. The natural resonant

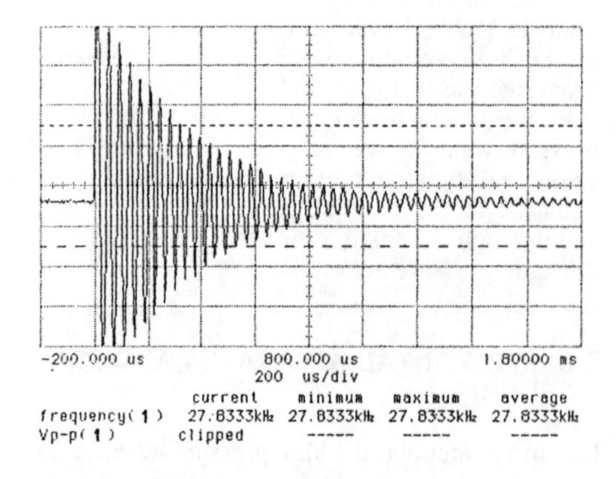

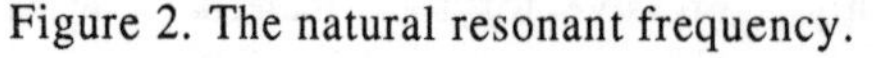

Figure 2. The natural resonant frequency.

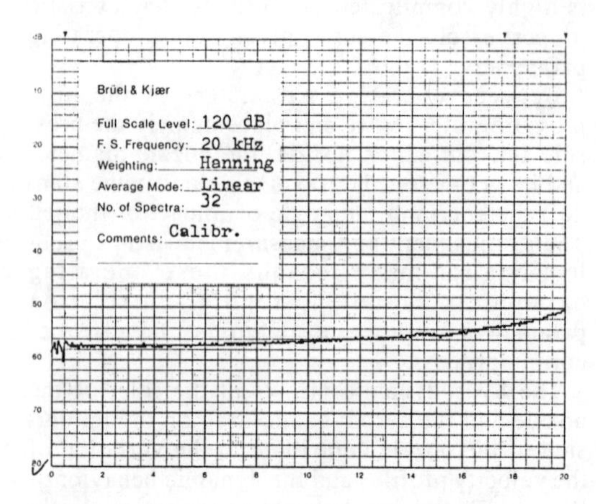

Figure 3. The frequency characteristics.

frequency of this device was determined to be over 27 kHz as is illustrated in Fig. 2.

Special care had to be taken to obtain a flat frequency characteristic of the useful measurement range, is illustrated in Fig. 3 and overcome of the interference effect by a double quarter wave cylindrical transformer (Vala 1990).

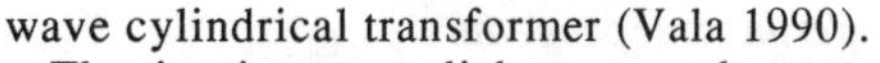

The impingment disk was made out of cemented carbide and the shield screen was made out of a hardness phosfor-bronze tin. During preliminary tests the thickness of the shield screen was 0.4 mm.

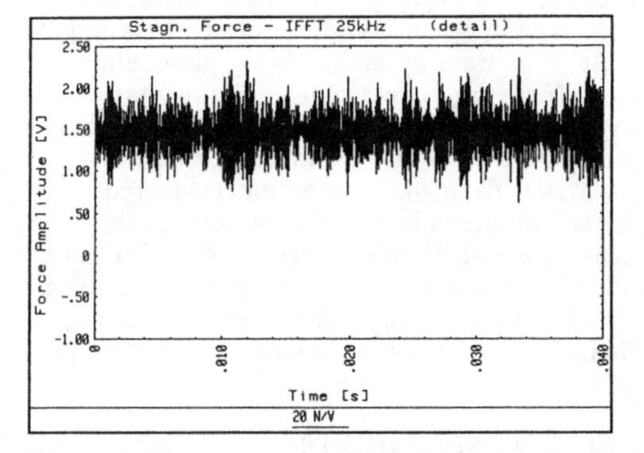

Figure 4. The restored record of the waveform of stagnation forces by the IFFT 25 kHz.

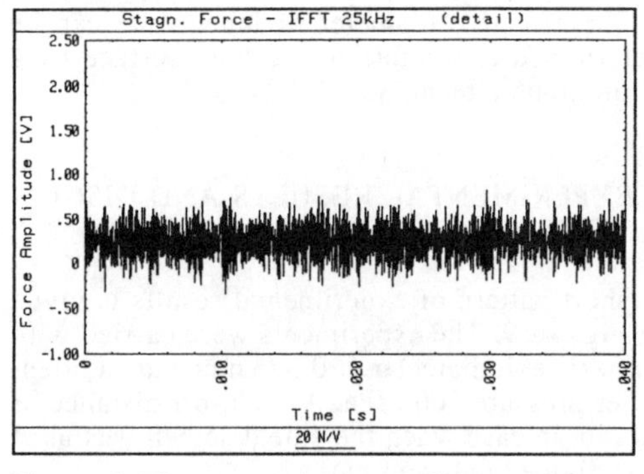

Figure 5. The restored record of the waveform of stagnation forces by the IFFT 25 kHz - shielded.

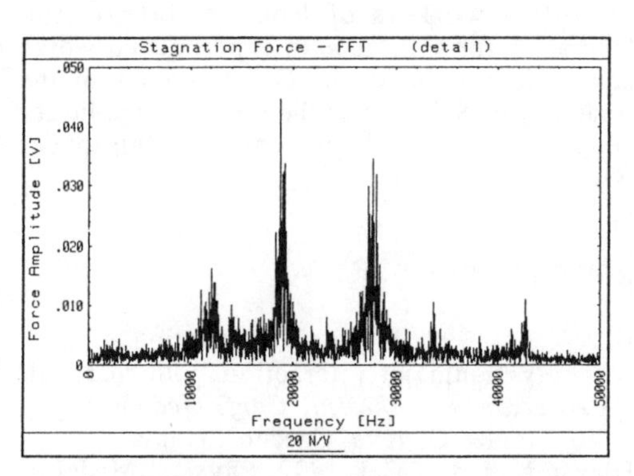

Figure 6. The record of a frequency analysis of the waveform from Fig. 4 by the FFT method.

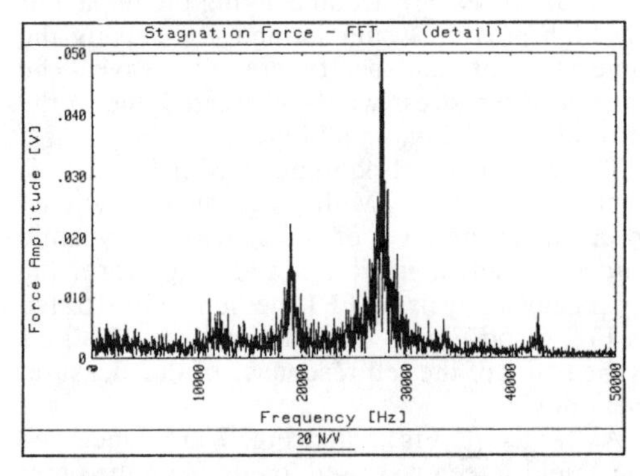

Figure 7. The record of a frequency analysis of the waveform from Fig. 5 by the FFT method.

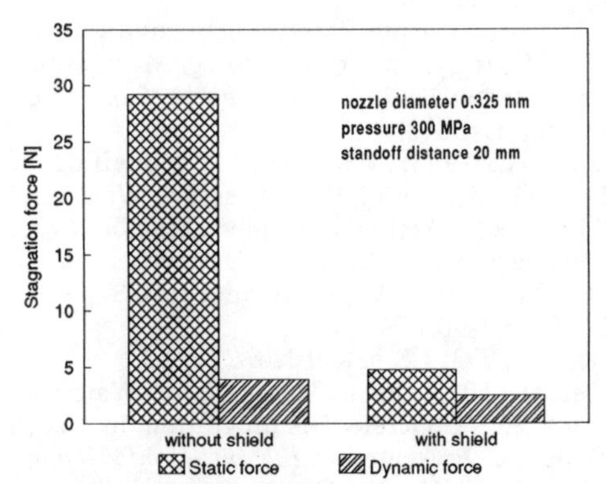

Figure 8. The level of the static and dynamic forces from Fig. 4 and Fig. 5

Figure 9. The shape of the hole in the shield screen cut by the inner zone of the jet.

2.2 *Experimental procedure*

In consequence of cutting ability of the energy-rich inner zone of the jet a hole is penetrated in the shield screen and this part of the high pressure water jet is passing throw the hole and fall down to the impingement disk where the static and dynamic stagnation forces are measured.

The effective diameter of inner zone can be determined as a value of the hole surface by a photographic technique.

3 EXPERIMENTAL RESULTS AND DISCUSSION

A short pattern of experimental results is given in Fig. 4 - 9. The experiments were carried with a nozzle exit diameter of 0.325 mm and an intensifier pressure 300 MPa. A stand-off distance of 20 mm in case when the shield screen was used was divided half-and-half.

The charts clearly demonstrate that the hydrodynamic phenomena, such as the formation of shock waves, are accompanying the impact of the high pressure water jet and are causing the generation of non-specify pressure wave. The pattern of the stress waves generated due to the jet is shown in Fig. 4 and Fig. 5.

The experimental equipment with Hewlett & Packard HP 6942 A Multiprogrammer involved to calculate the level of the static and dynamic forces as shown in Fig. 8 and to perform the frequency analysis by FFT method as illustrated in Fig. 6 and Fig. 7. The maximum at 27,8 kHz is the result of the self resonance of the measurement device.

As far as the Fig. 5 and Fig. 7 are concerned the shield screen was used to eliminate the outer zone of the jet.

The strong effect of the nozzle geometry is illustrated in Fig. 9. The observed shape was caused by the defect of the nozzle exit. It had a great influence on the rate of the forces in Fig. 8.

4 CONCLUSION

In the present paper the new method for both quantitative and qualitative study of the high pressure water jet impact phenomena has been carried. The stagnation force measuring system with the shield screen represented here is under test for the time being.

It seems that this method will be convenient not only for continuous jets but for cavitating, self-oscitating, pulsing and other new technique too. These technique which have a great potential for rock and concrete materials jet cutting applications are under investigation in the laboratory (Sitek 1993).

The first experience show that such a device like presented one is necessary for guarantee of reproduction of experimental investigation of high pressure water jet because the behavior of the high velocity jet, especially the hypervelocity impact of compressible fluid is impossible to express by main setting parameters and therefore the true values of the static and dynamic forces and the effective diameter of the jet must be obtain by the measurement.

At last, from the momentum measured when the shield screen is used the velocity of the inner active zone of the high pressure water jet can be obtained. Even though theoretical preconditions (friction-free, incompressible flow) are not satisfied, the results are sufficient for practical purposes.

ACKNOWLEDGEMENTS

The author would like to express his appreciation to fellow workers of Rock disintegration laboratory for assistance in experimental works and he is grateful to the Grant Agency of the Academy of Sciences of the Czech Republic for supporting the work presented in this paper (Grant No. 31 655).

REFERENCES

Hlaváč L. 1992. Physical description of high energy liquid jet interaction with material. *Proceedings of the Int. Conf. Geomechanics '91*, Hradec/Ostrava, Czechoslovakia

Hlaváč L. & J. Vašek 1992. Physical Model of High Energy Liquid Jet Rock Material Cutting. *International Journal of Water Jet Technology*, Vol. 3, No. 2.

Sitek, L. 1993: High Pressure Abrasive Jet Cutting Using Linear Stepping Movement. Presented at *the Int. Conf. Geomechanics '93*, Hradec-Ostrava.

Sochor, T. 1993: Mathematical Modelling of Several Aspects of High Pressure Water and Abrasive Jet Action. Presented at *the Int. Conf. Geomechanics '93*, Hradec-Ostrava.

Vala, M. 1990: Application for CS patent PV 06867-90.

Vala, M. 1993: CS patent No. 277753

Vašek, J. 1993: Basic Research of Water Jet Rock and Concrete Disintegration in Czech Republic. Presented at *the Int. Conf. Geomechanics '93*, Hradec-Ostrava.

Geomechanics 93, Rakowski (ed.) © 1994 Balkema, Rotterdam, ISBN 90 5410 354 X

Basic research of waterjet rock and concrete disintegration in Czech Republic

J.Vašek
Institute of Geonics of the CAS, Ostrava, Czech Republic

ABSTRACT: The brief survey of basic research of rock and concrete disintegration by water and abrasive jets in the area of physical of jet-abrasive interaction, mathematical modelling of several aspects of high pressure water and abrasive jet, the measurement of the non-setting parameters of the high pressure water jets, high pressure abrasive jet cutting using linear stepping movement, liquid flow from nozzle and experimental study of discharge coefficient and application of high-pressure water jet technology in a construction company s practice noted.

1 INTRODUCTION

Basic government-founded research aimed at a material waterjet disintegration, especially rocks and concretes, has been investigated in the Czech Republic in 1980 at the Institute of Geonics of the Academy of Sciences OSTRAVA (formerly the Mining Institute). In 1986 other institutions from both Czech and Slovak Republic joined the research: the University of Mining and Metallurgy of Ostrava, the Czech Technical University of Prague, the Technical University of Bratislava, the Technical University of Brno. The results of the research were published in both domestic and foreign journals and presented on international conferences as well (Foldyna 1992 a, 1992 b, Hlaváč 1992 a, 1992 b, 1992 c, 1993, Horák 1992, Martinec 1992, Vašek 1989, 1990 a, 1990 b, 1990c, 1990 d, 1991, 1992 a, 1992 b, 1992 c, Wolf 1992). In 1992 two 3-years internal grants supported by the grant agency of the Academy of Sciences entitled "Mechanism of Hard Rocks and Similar Materials Disintegration by Means of a High Energy Liquid Jet" and "Tribology of the abrasive particles contained in a liquid medium during their interaction with cut material at great velocities of their application" were acquired and their solution had started. Brief survey of research program is presented in this paper. More detailed informations in some research areas can be obtained in papers presented by Sitek (1993), Martinec (1993), Hlaváč (1993), Sochor (1993), Vala (1993), Slanec (1993), Foldyna (1993), Jeřábek (1993) (see Tab.1).

2 THE PRESENT STATE OF THE ART

The anthropogenic activity in the earth crust is put together inseparably with the process of disintegration of rocks. This process is represented from the physical point of view by a failure of the energy bonds of materials to be disintegrated by means of external power activity. Up to now that power is generated in prevailing part by means of mechanical tools or by explosive power releasing above all. The effective application of mechanical tools is limited by properties of disintegrated material and application of explosives is limited by negative incidence with neighbouring environment.

The present approaches to the solution of problem namely of hard-to-cut materials disintegration may be characterized as attempts for single improvement of existing methods of generation and transfer of power into the material to be disintegrated (e.g. research and development of new materials, geometry and mode of action of mechanical tool improvement).

It is impossible through those partial improvements only of mechanical tools effectively disintegrate the whole scale of the rock materials, encountered in earth crust. Therefore a research of some new principles of power transfer to the rock materials which make possible to overcome the problems, being extremely heavily solved by means of existing methods of disintegration, is realized. High energy liquid jet (further HELJ) is one of those perspective power transfer principles. Therefore, it is necessary to profound the

Tab.1. Basic research of waterjet rock and concrete disintegration in the Czech republic.

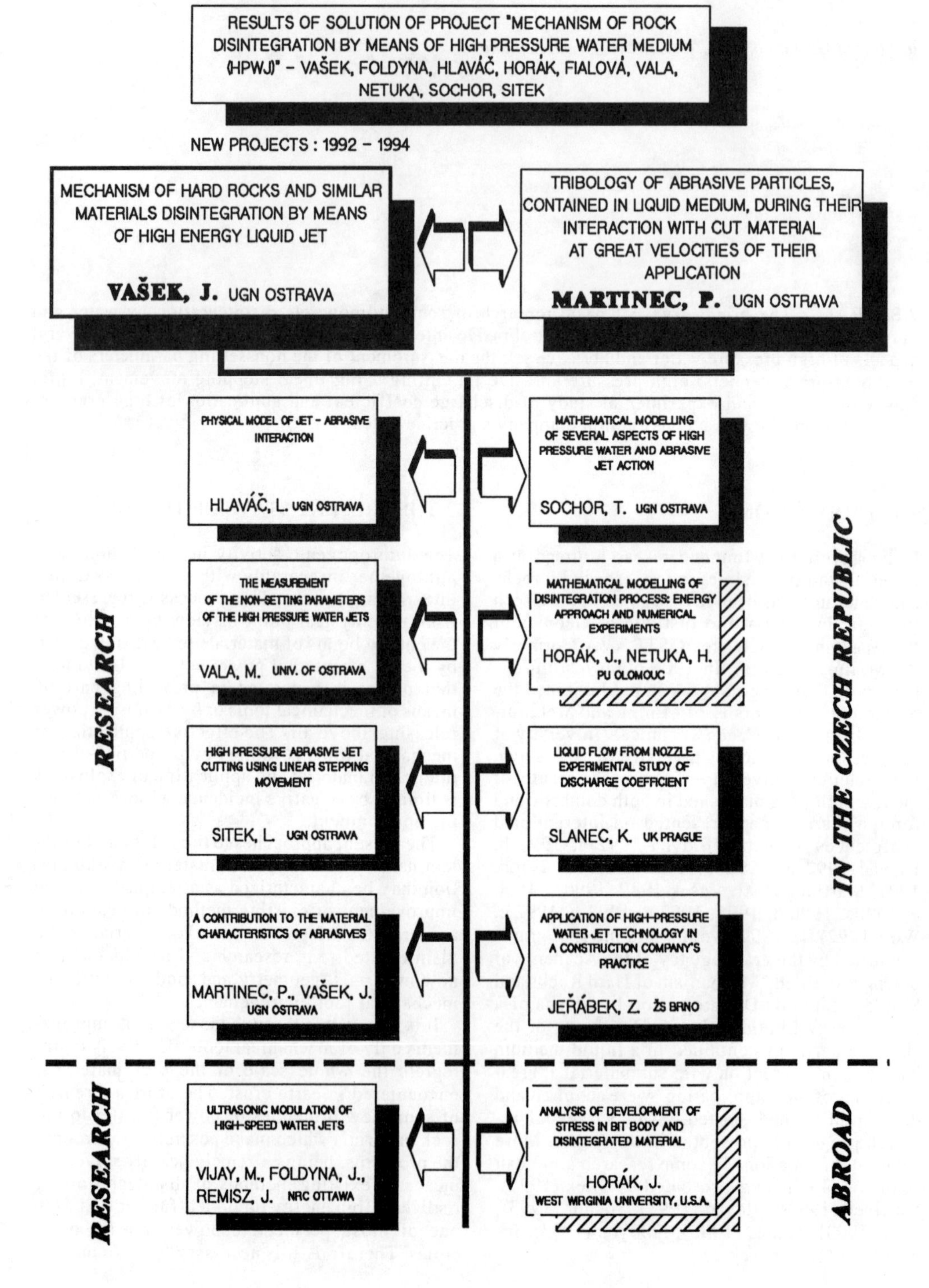
STARTING BASE : 1986 – 1991
RESULTS OF SOLUTION OF PROJECT "MECHANISM OF ROCK DISINTEGRATION BY MEANS OF HIGH PRESSURE WATER MEDIUM (HPWJ)" – VAŠEK, FOLDYNA, HLAVÁČ, HORÁK, FIALOVÁ, VALA, NETUKA, SOCHOR, SITEK
NEW PROJECTS : 1992 – 1994
MECHANISM OF HARD ROCKS AND SIMILAR MATERIALS DISINTEGRATION BY MEANS OF HIGH ENERGY LIQUID JET
VAŠEK, J. UGN OSTRAVA
TRIBOLOGY OF ABRASIVE PARTICLES, CONTAINED IN LIQUID MEDIUM, DURING THEIR INTERACTION WITH CUT MATERIAL AT GREAT VELOCITIES OF THEIR APPLICATION
MARTINEC, P. UGN OSTRAVA
PHYSICAL MODEL OF JET – ABRASIVE INTERACTION
HLAVÁČ, L. UGN OSTRAVA
MATHEMATICAL MODELLING OF SEVERAL ASPECTS OF HIGH PRESSURE WATER AND ABRASIVE JET ACTION
SOCHOR, T. UGN OSTRAVA
THE MEASUREMENT OF THE NON-SETTING PARAMETERS OF THE HIGH PRESSURE WATER JETS
VALA, M. UNIV. OF OSTRAVA
MATHEMATICAL MODELLING OF DISINTEGRATION PROCESS: ENERGY APPROACH AND NUMERICAL EXPERIMENTS
HORÁK, J., NETUKA, H.
PU OLOMOUC
HIGH PRESSURE ABRASIVE JET CUTTING USING LINEAR STEPPING MOVEMENT
SITEK, L. UGN OSTRAVA
LIQUID FLOW FROM NOZZLE. EXPERIMENTAL STUDY OF DISCHARGE COEFFICIENT
SLANEC, K. UK PRAGUE
A CONTRIBUTION TO THE MATERIAL CHARACTERISTICS OF ABRASIVES
MARTINEC, P., VAŠEK, J.
UGN OSTRAVA
APPLICATION OF HIGH-PRESSURE WATER JET TECHNOLOGY IN A CONSTRUCTION COMPANY'S PRACTICE
JEŘÁBEK, Z. ŽS BRNO
ULTRASONIC MODULATION OF HIGH-SPEED WATER JETS
VIJAY, M., FOLDYNA, J., REMISZ, J. NRC OTTAWA
ANALYSIS OF DEVELOPMENT OF STRESS IN BIT BODY AND DISINTEGRATED MATERIAL
HORÁK, J.
WEST VIRGINIA UNIVERSITY, U.S.A.
RESEARCH
IN THE CZECH REPUBLIC
RESEARCH
ABROAD

knowledge of rock materials failure mechanism and its behaviour when use HELJ.

3 PROJECT AIM DETERMINATION

The base aim of 3-years research project is to demonstrate the unique character and effectivity of application of high energy liquid jet (further HELJ) in controlled disintegration of predefined class of materials which cannot be disintegrate either effectively by mechanical instruments or without unfavorable consequences for materials itself or for neighbouring environment on a one hand and on the other hand to profound the base of knowledge concerning mechanism of disintegration of given materials when HELJ is applied. To reach this aim following manner of investigation was proposed:

- physical model formulation, defining high-concentrated energy of liquid jet by means of physical relations usable for mathematical description of HELJ interaction with rock (Fig.1a),

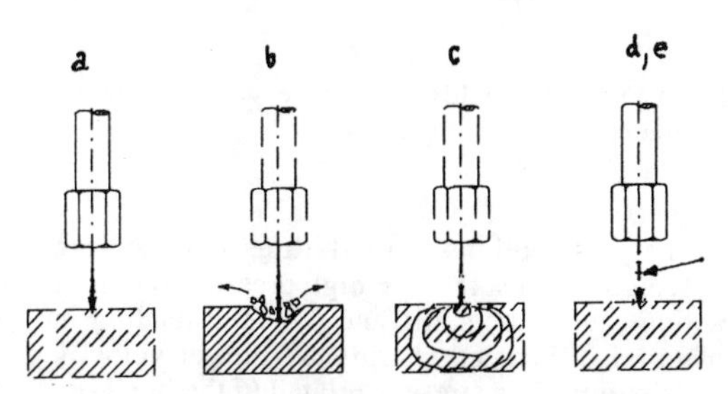

Fig.1. Schematic display of fundamental parts of proposed project solution

- physical model formulation of rock material disintegration mechanism when HELJ is applied to obtain physical relations useful for mathematical simulation of HELJ interaction with rock materials (Fig.1b),
- mathematical simulation and control of stress and deformation fields, induced in rock material, representing action of HELJ and its cooperation with mechanical tool (Fig.1c),
- physical material properties of natural and synthetic abrasives, a morphology of the abrasive grains and its quantification, structural defects in the mineral grains and their dependence on the parametres of the cutting process, an analysis of a material cutting by individual abrasive grains, possibilities of a treatment of natural abrasives for improvement of their properties, experimental research of changes of the HELJ character and rock disintegration mechanism caused by solid elements adding to the water jet (Fig.1d),
- analysis of HELJ application possibilities for supplementary energy transfer (Fig.1e),
- measurement of the non-setting parameters of the HELJ and verification of the theoretical presumptions using experimental research.

4 RESULTS AND DISCUSSION

In the area of physical modelling there was stated simple relation among fundamental values, characterizing the jet, and macroscopic effects of jet activity - depth and width of the cut, energy being converted, etc. The equation for the depth of cut evaluated from the basic jet parameters, material properties and cutting conditions was derived and utilised. In the same way there was derived relation for specific energy, transfered by the jet. With adding of micro-petrographic research there was created simplified physical image about processes, taking place during the material disintegration on material grain level. The further part of physical modelling was orientated to phenomena being connected with application of water jet for multiple cuts and for cuts with rotating and oscillating jets. Recently the physical modelling was applied for analysis of abrasive jets (Fig.2), (Hlaváč 1993 b).

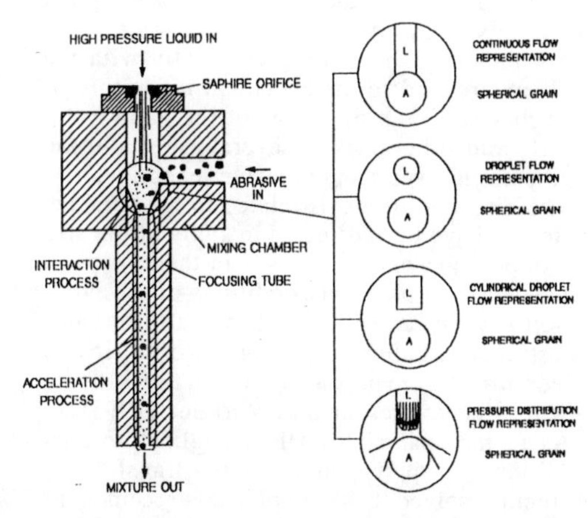

Fig.2. Scheme of liquid and abrasive grain interaction

In the area of mathematical simulation of rock disintegration process there was created an algorithm for determining of a course of material stress in interactive set composed of mechanical tool, water jet and rock (Fig.3), using the theory of variational inequalities and the finite element method.

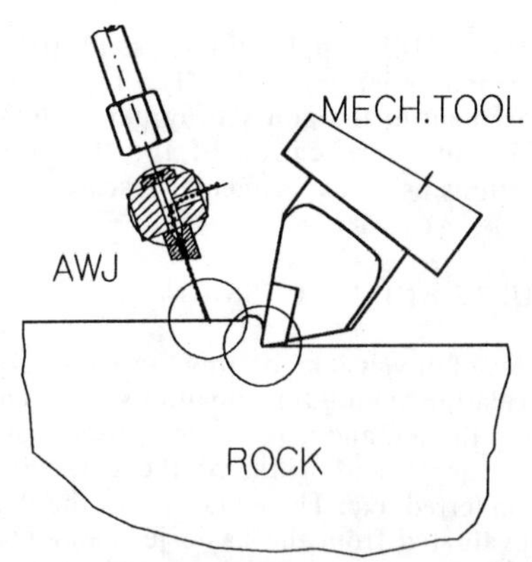

Fig.3. Interactive set of mechanical tool, water jet and rock

The interaction of mechanical tool, water jet and rock is very complicated physical process therefore an interdisciplinary access is necessary for its characterization and correct description. An external power can be generated by means of the HELJ or with assistance of mechanical tool.

Horák (1992) has been paying attention to the numerical simulation of the stress and deformation fields induced in rock and mechanical tool material. Since August 1993 he continues his research activity in this area in West Wirginia University in U.S.A.

Sochor (1993) is dealing especially with the problems related to mathematical modelling of rock disintegration by means of the plain water jet. He aimed his work at several aspects of the high pressure water and abrasive jet action.

The research problems related to the abrasive materials have been directed to study of the size of the particles and to changes in their shape in dependence on physical conditions of the interaction process as well as to physical properties and structure of abrasive particles. Results in this research area were presented several times. Some new results are presented in Martinec (1993).

As to the analysis of HELJ application possibilities for supplementary energy transfer, the problem is solved thanks to official agreement to perform collaborative research at National Research Council of Canada under Eastern Europe Fellowship Program since the end of 1992. The problems related to ultrasonic modulation of the high-speed water jet is studied and the analysis of the impact of ultrasonic modulated jet on the material surface indicate that the pulsed jet can be an effective tool for disintegration of the hard and brittle materials. The results presented by Vijay and Foldyna (1993) were obtained during joint research.

Analyses of high pressure water jet with assistance of cutting tool have lead to the conclusion that there are two different modes of use dependent on their space-time arrangement. In the first case the high pressure water jet acts in the rock at a considerable time and space distance from the cutting tool (Fig.4a) while in the second case the high pressure water jet acts immediately on the spot of the set rock-cutting tool (Fig.4b).

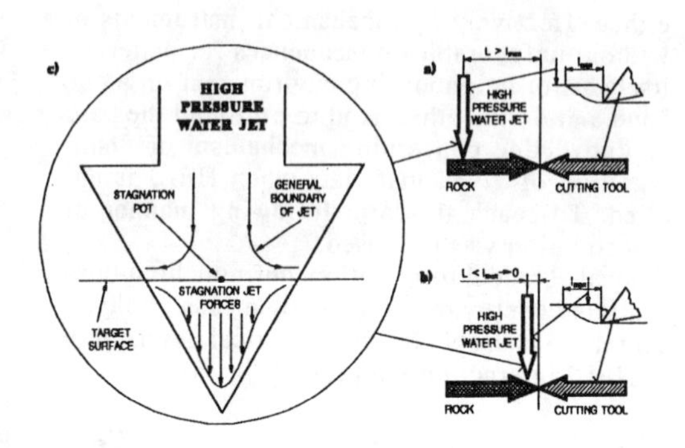

Fig.4. Scheme of high pressure water jet when interacts with target surface

The presumptions of the failure processes are based on the idea that the high pressure water jet action causes the static and dynamic loading in material influenced sharply by the jet velocity distribution. The stress is related to the break-up of the jet (Fig.4c) so much more detaile information about the interactive process are needed. Some new ideas of measurement of the so called

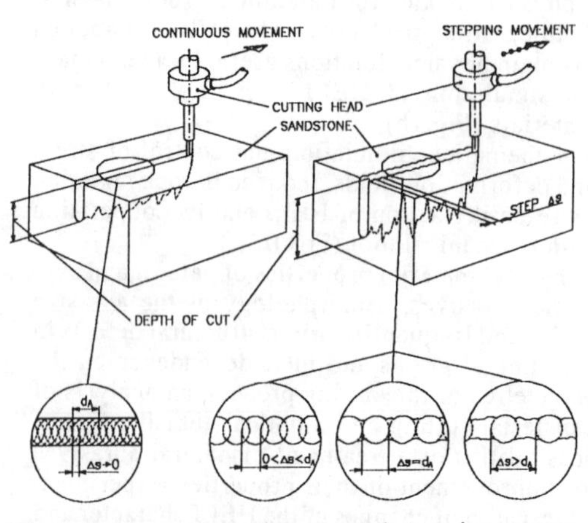

Fig.5. Scheme of continuous and stepping movement of jet

non-setting parameters of the high velocity jets are presented by Vala (1993). Some very interesting results obtained by linear stepping movement (Fig.5) of the cutting head of high pressure abrasive jet are presented by Sitek (1993).

5 CONCLUSION

A brief survey of the research activities connected to the waterjet rock and concrete disintegration in the Czech Republic since 1992 is presented in this paper to demonstrate the effort to solve very complicated waterjet technology interaction process in coordinate condition not only in the Czech Republic but also in cooperation with famous research centres and research scientists abroad. We believe that this effort will help us profound both the research knowledge related with the WATERJET TECHNOLOGY and further direct research cooperation all over the world as well.

REFERENCES

Foldyna, J. 1992 a. Experimental research of rock and similar materials cutting using high pressure water jets. *Proc. of the Int. Conf. GEO-MECHANICS 91*, p. 331-340. Rotterdam: Balkema.

Foldyna, J., Martinec, P. 1992 b. Abrasive Material in the Process of AWJ Cutting. *Proc. of the 11th Int. Conf. on Jet Cutting Technology*, St. Andrew: 135-147.

Hlaváč, L. 1992 a. Physical description of high liquid jet interaction with material. *Proc. of the Int. Conf. GEOMECHANICS 91*, p. 341-346. Rotterdam: Balkema.

Hlaváč, L., Vašek, J., 1992 b. Theoretical Model of Water Jet. *The 3rd Pacific Rim Int. Conf. on Water Jet Technology*, Taiwan.

Hlaváč, L., Foldyna, J., Vašek, J. 1992 c. Modellierung des Schneidprocesses beim Einzatz von Hochenergetischen Wasserstrahlen zur Bearbeitung von Festgesteinen. *Neue Bergbautechnik*, 22, Heft 12: 461-465.

Hlaváč, L., Foldyna, J., Momber, A. 1993 a. Das Schneiden von Gesteinen mit rotierenden Hochdruckwasserstrahlen. *Glückauf Forsch.*, 54, Nr.2: 58-62.

Hlaváč L. 1993 b. Physical model of jet-abrasive interaction. Will be presented at the Int. Conf. Geomechanics '93, Hradec n. Moravicí, Czech Republic

Horák, J. 1992. Mathematical modelling of rock cutting. *Proc. of the Int. Conf. GEOMECHANICS 91*, p. 347-352. Rotterdam: Balkema.

Martinec, P. 1992. Mineralogical properties of abrasive minerals and their role in the water jet cutting process. *Proc. of the Int. Conf. GEO-MECHANICS 91*, p. 353-362. Rotterdam: Balkema.

Martinec, P. & J. Vašek 1993: Garnets for Abrasive Water Jet Cutting. Presented at *the Int. Conf. Geomechanics '93*, Hradec-Ostrava.

Sitek, L. 1993: High Pressure Abrasive Jet Cutting Using Linear Stepping Movement. Presented at *the Int. Conf. Geomechanics '93*, Hradec-Ostrava.

Sochor, T. 1993: Mathematical Modelling of Several Aspects of High Pressure Water and Abrasive Jet Action. Presented at *the Int. Conf. Geomechanics '93*, Hradec-Ostrava.

Vala, M. 1993: The Measurement of the Non-Setting Parameters of the High Pressure Water Jet. Presented at *the Int. Conf. Geomechanics '93*, Hradec-Ostrava.

Vašek, J., Foldyna, J., Hlaváč, L., 1990 a. High pressure water jet cutting of rocks. *Proc. of 4th Int. Conf. on Machining of Nonmetallic Materials*, Rzeszow, Poland.

Vašek, J., Foldyna, J., Jeřábek, Z., Momber, A. 1990 b. Die Anwendung von Hochdruckwasserstrahlen bei der Instandhaltung von Eisenbahntunneln. *Bauplanung-Bautechnik*, Heft 3.

Vašek, J., Foldyna, J. 1990 c. Abrasive water jet cutting of hard rocks. *Proc. of the 10th Int. Symp. on Jet Cutting Tech.*, Amsterodam

Vašek, J. 1990 d. Research on High Pressure Water Jets in Mining Industry of Czechoslovakia. *Proc. of the 14th World Mining Congress*, Beijing: 449-455.

Vašek, J., Foldyna, J., Novák, J. 1991. Test on Walls of a Railway Tunnel and Samples of Concrete and Blocks of Rocks with High Pressure Water Jet Equipment for Outdoor Applications in Czechoslovakia. *Proc. of the 6th American Water Jet Conf.*, Houston: 87-101.

Vašek, J. 1992 a. Theoretical Evaluation of Pick Tips Assisted by High Pressure Water Jets. *Proc. of the 11th Int. Conf. on Jet Cutting Technology*, St. Andrew: 123-133.

Vašek, J. 1992 b. Why research water jet hard rock disintegration in Czechoslovakia. *Proc. of the Int. Conf. GEOMECHANICS 91*, p. 317-330. Rotterdam: Balkema.

Vašek, J., Horák, J., Sochor, T. 1992 c. Mathematical Modelling of Rock Cutting Process. *Proc. of the 3rd Pacific Rim Int. Conf. on Water Jet Technology*, Taiwan.

Wolf, I., Foldyna, J., Vašek, J. 1992. High pressure jets application in dam rehabilitation. *Proc. of the Int. Conf. GEOMECHANICS 91*, p. 363-366. Rotterdam: Balkema.

Water jet cutting
Technical notes

Geomechanics 93, Rakowski (ed.) © 1994 Balkema, Rotterdam, ISBN 90 5410 354 X

Mathematic simulation of combined crushing of strata by a mechanical cutter

I.K. Arkhipov, V.A. Brenner & I.M. Lavit
Technical University of Tula, Russia

ABSTRACT: The problem determining the cutting force during mining rock fracture while cutting with a simultaineous water ingress to the cutting zone through the cutter at high pressure is given in the report. Assumptions with the help of which this problem is reduced to a flat problem of linear fracture mechanics have been proved. Its solution algorithm based upon boundary element technique has been elaborated. Cutting force dependence upon water pressure, the rock's strength, depth and angle of cutting are the results of the problem solution.

1. INTRODUCTION

One of the most effective methods of cutting strong mining rock is the application of mechanical hydraulic cutters whose characteristic feature is their ability to be fed under high pressure to the cutting zone. Water under pressure gets into the rock's crack opens it and promotes its fracture. It decreases the necessary force for fracture. For a quantitative analysis of this process it is necessary to have an adequate mathematical model. The first attempt to create such a model was made by Arkhipov and Brenner (1990). In that model the process of investigating the cracks was described with the application approximation formulae giving a substantial error for long cracks. It the work presented an attempt has been made for perfecting the model. Its basic assumptions having been preserved, the process of the crack propagation undergoes simulation with the help of fracture mechanics concepts. The method of reducing the computation problem of the cutting force to a boundary integral equation has been described. Its algorithm of numerical solution has been described in details.

2. BASIC ASSUMPTIONS

The mining rock cutting scheme by the hydromechanic type cutter is given in fig.1.
The layer cutoff process of the h thickness and the t width takes place with the cutter movement. Together with the cutting force another factor of fracture is present in the process under observation: its essence is that the liquid pressure falls into the crack. Pressure in that crack brings to the growth of the said crack, thereby decreasing the cutting force. It is necessary to determine the maximum cutting force depending upon the rock strength, the t and the parametres and the water pressure. The tas formulated is a three-dimensional one. However it i possible to decrease its dimension withou substantial errors being introduced. Let us evaluat the t cutter width influence upon the cutting force. Fo that purpose let us apply the experimental data by Miller (1989) given in the table.
Divergence between R is maximum at p=0 and minimum at p=100 MPa. Assumption that the value does not depend upon equivalent assumptions of the plane scheme being applicable for solving the problem introduces errors not surpassing 25% according to experimental data, these assumptions increasing the designed the cutter factor of safety. The plane cutting scheme is given in fig.2.

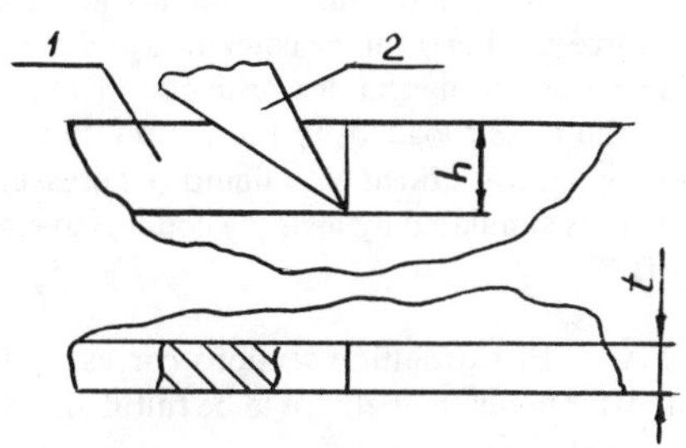

Fig.1 Mining rock cutting scheme (1 - mining rock, cutter)

Cutting force dependence upon the cutting widt (h=10mm)

p (MPa)	R (kN/mm)	
	t=30 mm	t=20 mm
0	1.58	1.93
20	1.40	1.61
40	1.16	1.40
60	1.14	1.22
100	0.78	0.85

p is the water pressure, P - cutting force, R=P/t is the specific cutting force

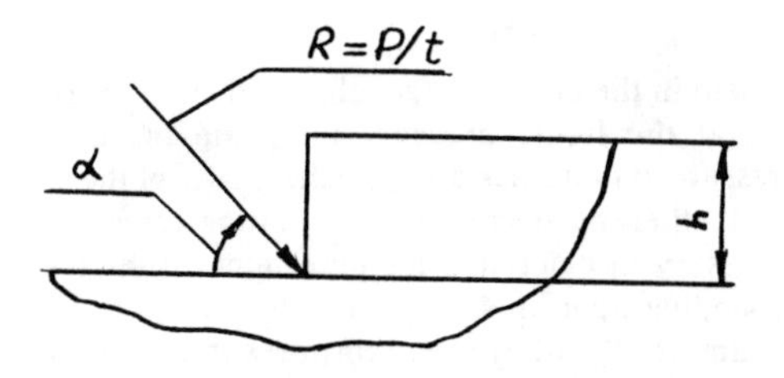

Fig.2 Plane cutting scheme

Experiments by Miller (1989) show that a angle value are bound in quite a narrow interval from 43.3° til 52.8° (at the cutter inclination angle being constant, namely 45°). Analyses embracing 11 experiments give the deviation a_m=47.1°, standart deviation s=3.78°. These date give us a possibility the assume the hypothesis cutting force inclination angle a being equivalent to the cutter inclination angle.

3. THE PROBLEM SOLUTION ALGORITHM

The specific force modulus (we shall now call it cutting force for short) can be computed as a result of the linear fracture mechanics problem having been solved. The massif loaded by the cutting force and containing the crack filled with liquid (its pressure is equal to p) is simulated by a singly connected elastic field (fig.3).

As the AF, FE, EC outline sections corresponds to infinitely far fields the H≫h are to be fulfilled.

To evaluate the massif strength it is necessary to

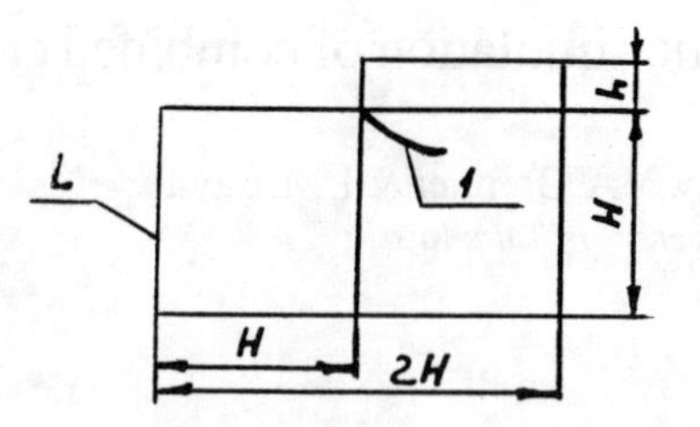

Fig.3 Design scheme (1 - crack)

know the stress intensity factors K_I and K_{II}. We shall compute their value with the application of the boundary element techniques.

The constraint reaction having been determined (see fig.3), we get the problem for a single connected finite field D restricted by the outline L where the force boundary conditions are known. Its solution is reduced to the solution of the integral Muskhelishvili equation

$$\overline{v(t)} - \frac{1}{\pi}\int_L \overline{v(\zeta)}\, d\theta - \frac{1}{\pi}\int_L v(\zeta)\, e^{-2i\theta} d\theta = A(t)$$

where v(z) iz the unknown function, regular in the D field, t, ζ are the complex coordinates of the outline point, and the A(t) function is determined by formulae

$$A(t) = \frac{f}{2} - \frac{1}{2\pi i}\int_L \frac{f\, d\zeta}{\zeta - t}; \quad f(\zeta) = \int_0^{s(\zeta)} (F_x + iF_y)\, ds$$

where complex integration in formula (4) i conducted along the L outline, beginning with certain intial point; F_x, F_y are load vecto components.

The simplest method of numerical solution o equation (1) is the following. We divide the outline into n sufficient small sections and consider the v(t function to be constant in the limits of the section. A a result we receive a system of linear equation concerning the unknown ones F_j:

$$\bar{F}_j - \frac{1}{\pi}\sum_{k=1}^{n}\left(a_{jk}\bar{F}_k + b_{jk}F_k\right) = B_j$$

The integrals (6) are easily determined analytically Equation (1) is true, if the L outline is smooth. Tha is why in calculating it it is necessary to round-off i the cornered points, for example with the help of small radius rounding-off curve. The same refers t the crack-tip (fig.4).

In the small crack-tip neighbourhood the correlation are true. Here K_I and K_{II} stress intensity factors, is crack-tip coordinate. Formula (7) is applied whe values j or k coincide with the number of the sectio of the outline surrounding the crack-tip. That is how the basic unknown problems K_I and K_{II} ar introduced into the system directly. Separatin real and imaginary components we get the 2

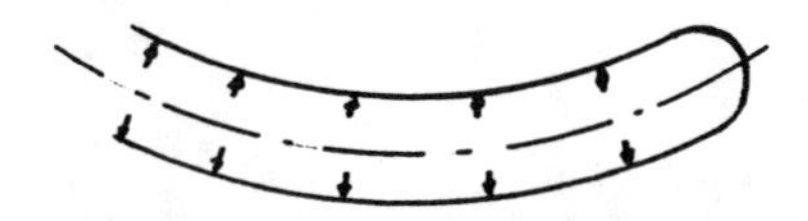

Fig.4 Crack-tip neighbourhood (p - liqui pressure, 1 - crack-tip, R - rounding-off curv radius)

system of linear equations relating to the 2 unknown ones. Having solved the equation system w get K_I and K_{II}. Applying strength criterion

$$K_I^2 + K_{II}^2 = K_{Ic}^2$$

where K_{Ic} fracture toughness gives us the possibilit to find cutting force value necessary for propagatin the cracks.

For determing the crack growth direction we assum that the crack grows along the normal to the direction of the biggest tensile stress. We further receive th equation relative to angle value q, determining th crack grown direction

$$K_I \sin q + K_{II} (3\cos q - 1) = 0$$

In equation (9) all values are determined in loca coordinate system whose abscissa axis is directed along the tangent to the crack line in its tip.

We further take the increment of the crack length, then we compute the new condition of its tip, we again determine K_I, K_{II} and R until the calculated value R begins to decrease which will mean a pass to an unstable gowth of the cracks.

4. CONCLUSION

Thus the basic equations of the mathematical model of the mining rock destruction at the mechanical hydraulic effect have been received. An algorithm of solving those equations offering a possibility to receive the dependence of the cutting force upon water pressure, strength of rock, depth and angle of cutting.

REFERENCES

Arkhipov, I.K. & Brenner, V.A. 1990. Mathematica modelling of mining rock mechanical-hydrauli destruction (in Russian). Nove poznatky vedy vyskumu a praxe v mechanike hornin. Zborni prednasok z konferencie. Vysoke Tatry: 1-5.

Brebbia, C.A., Telles, J.C.F. & Wrobel, L.C. 1984 Boundary element techniques. Theory an applications in engineering. Berlin: Springer-Verlag

Mikhlin, S.G. 1957. Integral equations. New York Pergamon.Miller, M.M. 1989. Experimental re search of the parametres of the combine designed fo destroying strong mining rocks (in Russian). PhD thesis. Polytechnical Institute of Tula.

Muskhelishvili, N.I. 1973. Some basic problems o the mathematical theory of elasticity. New York Noordhoff.Rice, J.R. 1968. Mathematical analysis i the mechanics of fracture. In Fracture, Vol.2: 191 311. New York: Academic Press.

Geomechanics 93, Rakowski (ed.) © 1994 Balkema, Rotterdam, ISBN 90 5410 354 X

Investigation into hydraulic winning with water-jet of high pressure

E. Debreczeni & I. Sümegi
Department of Equipments for Geotechnics, University of Miskolc, Hungary

ABSTRACT: For exploitation of underground mineral raw material a "bore-hole" mining method can be used. This manner has advantages especially in such cases when the traditional mining can be realized only in addition to hazard conditions of surrounding natural dangers, following the winning-operations (e.g. water inrush, gas burst, etc.)
The exploitation of mineral raw material by use of bore-hole method takes place either with mechanical winning or by means of hydraulic cutting. The transportation of won material can also be solved by the way of hydraulic transportation. At the former Department of Mining Machinery research equipment has been erected for investigation of hydraulic cuttability of certain mineral materials and for measurements of water-jet energy.
The paper offers the review of research-equipments with high pressure water-jet and measuring results as well as gives informations about the bore-hole mining planned.

1 INTRODUCTION

The present day practice of solid mineral mining is to perform the mechanized cutting of useful minerals and occasionally of partition rocks by mechanical methods.
In addition to mechanized winning, recently winning methods by other physical methods have been used. Hydraulic winning, realized by means of a high-speed water jet belongs into that group. This manner of winning has been spred recently but the most economical field of its appliance has not been satisfactory clarified yet. The hydraulic winning combined with hydraulic transportation can be consider as a mining method of novel type which is suitable to realizing the bore-hole mining. At present this mining method is primarily used for the exploitation of loosely structured deposits, e.g. phosphate deposit. Experiments have been carried out also with mineral materials of higher strength-coal, bauxite-in interest of exploitation by bore-hole method.
The ability of rocks to lend themselves to cutting has not been sufficiently clarified yet. The efficiency of hidraulic winning depends mainly on the velocity of flow and volume flow rate of water used for cutting and in addition to the properties of the rock on the relative geometrical layout of jet-rock group to each other, as well as on the direction of winning as the rocks are usually highly anisotropic. The efficiency of winning is also influenced by the macroscopic structure of rocks (stratification, fissures).

2 DEVELOPMENT OF MECHANIZED EQUIPMENT NECESSARY FOR NEW MINING METHODS

A significant amount of the Hungarian bauxite resources lies beneath the stationary water level of the rocks adjacent to the bauxite bearing rocks. In accordance with the methods introduced recently, bauxite mining is carried out by the use of active water protection. Active water protection means the lowering of the stationary water level in the adjacent, porous rock under the usable, exploitable bauxite resources.

Thus it is possible to mine bauxite without the danger of water inrush, in a dry environment. In order to facilitate active water protection, in the future it would be necessary to continuously pump quantities of water seriously damaging the environment from the porous layers. So as the avoid of environmental damage, it is necessary to develop and use new mining methods which are applicable also under stationary water level.

The coal deposits under exploitation in Hungarian black coal mining are of high gas content. It involved facing the risk of gas or rock outburst during mining. These dangers from the elements can be lessened by draining the methane content of coal deposit.

The mining tasks mentioned above, i.e. the exploitation of bauxite under the stationary water level, the mining of coal deposits rich in gas, can be performed safely by using the bore-hole mining. In order to carry out bore-hole mining, the mineral resources (bauxite, black coal) have to be explored by means of bore-holes or bored shafts and the mining and hauling equipment has to be installed through the bore-hole.

At present this mining method is primarily used

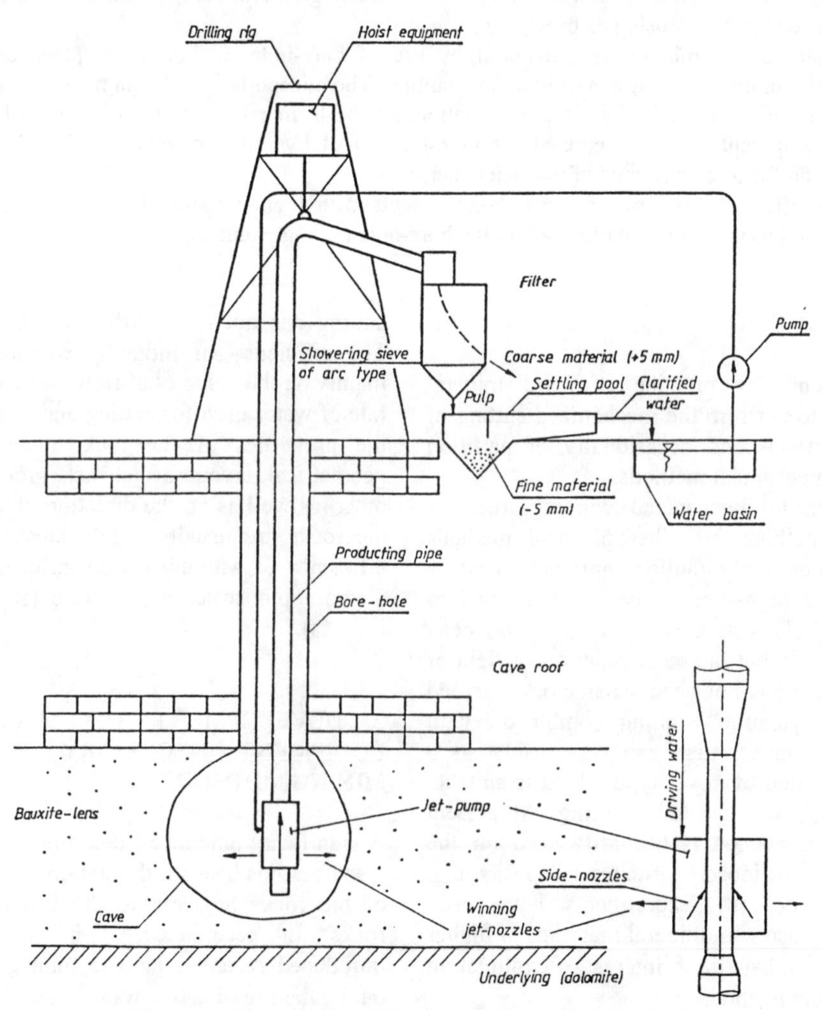

Figure 1. The scheme of the bore-hole mining method

for the exploitation of loosely structured deposits, e.g. phosphate deposits. Furthermore, it was also used experimentally for the winning of coal. Production usually took place by means of high-speed water-jet and the mixture of the exploited useful mineral and water was transported by pumps to the surface. In Hungary the method was used experimentally for the exploitation of a bauxite lens.

The first experimental exploitation of bauxite by means of bore-hole mining was realized in 1977. in Hungary by the researchers of University of Miskolc and the specialists of Bauxite Mines at Bakony. The scheme of bore-hole mining›s experimental equipment is shown in **Fig.1.** For opening up the bauxite-lens, a bore-hole of large diameter was used, excavated from the surface. The combined hydraulic winning and transportation head was built into the bore-hole. At experiments the operation of combined winning head with water-jet pump of side-nozzle type proceeded at the same pressure of feeding water. A greater part of exploited material-to be delivered by the hoist pipe-can be selected from the pulp-flow by means of showering sieve of arc type. It makes possible to recover the fine grains by the way of settling. The water, flowing over from the settler can be saved for the operation of cycle, thus it will not contaminate the environment.

At the exploitation, performed experimentally, the applied water pressure (40 bar) could not provide the effective hydraulic winning of bauxite, therefore in order of bucking and breaking the bauxite-lens, blasts were done in the bore-holes of small diameter settled around the main bore-hole of exploitation. During the operation the hydraulic winning-transportation equipment performed the exploitation and delivery of mineral material loosed by blast. The bibliography concerned also informs about the plans and conceptions regarding the bauxite mining with bore-hole method (Patvaros 1979).

As the top formation (cap rock) also was broken during the previous experiments, the exploitation of bauxite of given purity›s degree has not succeeded, therefore experiments were not continued.

It was established on the basis of the experimental experience that hydraulic winning at medium pressure is not sufficiently efficient for the production of bauxite even after loosening the deposit by blast. Therefore it is necessary either to use a remote control cutting head production unit in the vicinity of the bore-hole or to use high pressure water-jet production. In the latter case in winning under water level, the nozzle producing the high-speed water-jet has always to be taken near the bauxite.

The main objective of our research in winning technology in the future will be to develop a mechanic or hydraulic cutting head combined with hydraulic transportation which can be employed in bore-hole mining. The employment of bore-hole mining has been used also for exploitation of coal and uranian-ore as well as for the underground gasification of coal deposits (Riotoul 1981). The construction›s principle of production equipment, referred in bibliography is very similar to the applied equipment by us. The only one difference is that instead of water-jet pump of side-jet type-used by us, in the bibliography the applied water-jet pump is referred to as pump of centrical-jet type. The use of water-jet pump of side-jet type is more advantageous, as the free cross-sectional area of flow is larger therefore the size of grains to be delivered can also be larger.

3 LABORATORY INVESTIGATION OF THE WINNING PROCESS

The laboratory hydraulic winning equipment has been built out of a planing machine. The sample box, or for rocks with high strength, directly the sample embedded in concrete, was mounted on the moving table of the machine.

In the sample box, the sample was embedded in gypsum or concrete according to its characteristics and strength.

In the equipment the rock sample moves and the jet-nozzle is stationary. This solution results from the construction of the planning machine. The adjustment of the desired distance can be performed by moving the cross beam holding the jet nozzle as the winning unit in the required

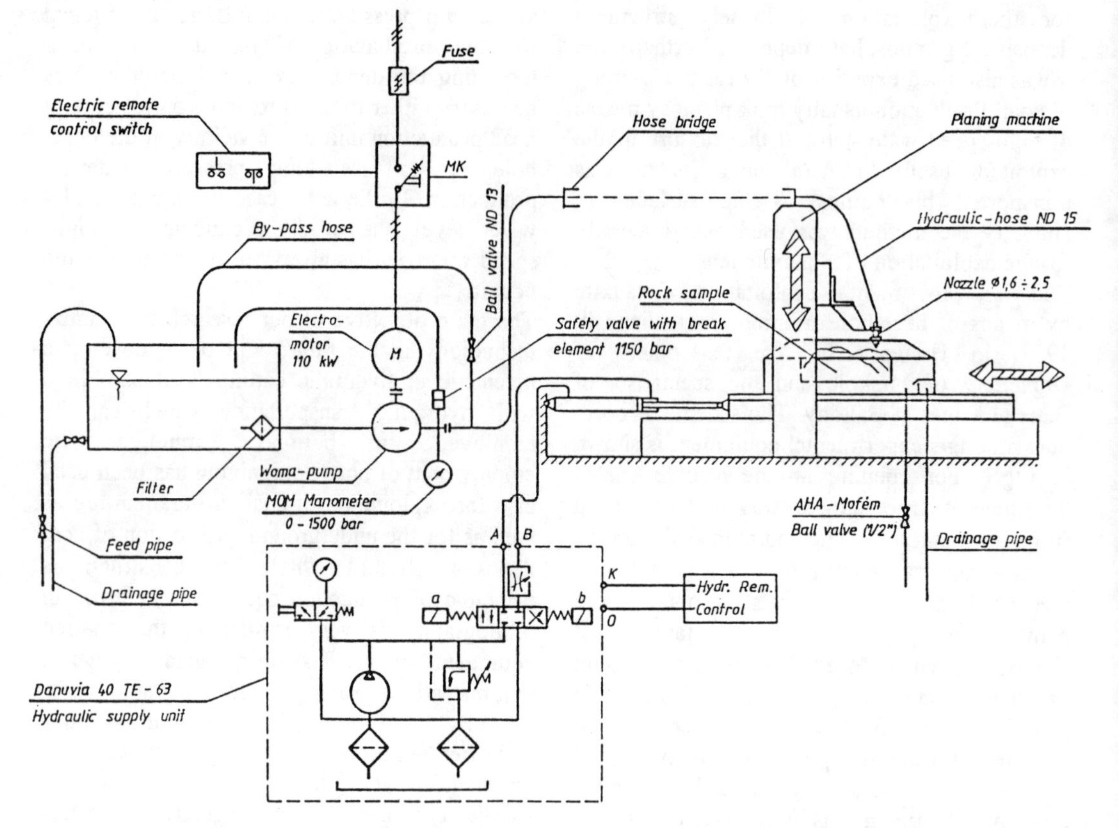

Figure 2. The layout of the hzdraulic winning equipment and the connection of table movement

direction. In hydraulic cutting the fluid has to be collected after winning. The layout of the equipment of hydraulic winning is shown in **Fig.2.** A hydraulic pump is used for producing the high pressure water. The pump available at present can produce a pressure increase up to 1100 bar before the nozzle. During the experiment, the fluid pressure before the nozzle, its volumetric flow as well as the speed of the rock sample are measured.

The high pressure system has been equipped with a fast-acting safety valve for safety reasons. The table of the planing machine is moved by a hydraulic work cylinder built specially for this purpose. The schematic representation of the hydraulic cycle can also be seen in **Fig.2.** The volumetric flow stabilizer can adjust any cutting speed. Using the stabilizer of volume-flow the moving speed of table can be set up from 0 to 0,4 m/s.

The laboratory investigations aim to determining the relationships between the mechanical characteristics of examined rocks and the parameters of hydraulic winning. This question was discussed by numerous scientific workshops, which are referred to as concise studies concerned (Hlavác 1992), (Ilias 1993).

For obtaining reliable starting data needed to designing the hydraulic winning, two methods possible are offered. In the first case the rock to be won will be tested, naturally the generalization of relationships obtained is not possible but they are reliable for the given rock. Upon the basis of another aim, in knowledge of mechanical characteristics of rock (compressive and tensile strength, modul of elasticity, Poisson number etc.) and the parameters of hydraulic winning-jet, the general relationships of hydraulic winning can also be determined. In fact, for the researchers the realizing the last

aim means the satisfactory solution of problem. Nevertheless, for obtaining the reliable starting data the concret testing of rocks to be won is needed for a long time in the future yet.
For design and manufacturing the hydraulic winning equipment a very important question is the research of jet-nozzle, operating under the conditions of most advantageous energy-utilization.
The phenomenon of energy-conversation in the jet-nozzle and the construction of fluid-jet after the nozzle were investigated by numerous researchers and also the bibliography is very rich. In this field, our endeavour is the obtaining the best energy utilization possible of high pressure fluid in the completed jet-nozzle of one or two stage types. During the experiments first of all the coefficient of efficiency of energy-conversation plays an important role and it will be examined with measurement of moment of fluid-jet. This method can be considered as a trial to selection of one of the most advantageous energetical solution among the variations possible.

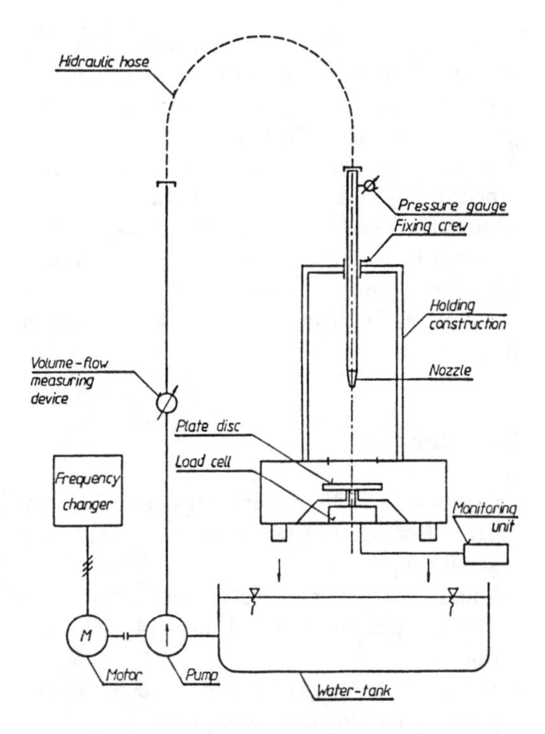

Figure 3. Laboratory investigation of winning jet-nozzles

4 LABORATORY INVESTIGATION OF WINNING JET-NOZZLES

The conversation of fluid-energy of high pressure into kinetic energy under the less losses possible on the one hand, and the forming a coherent and massive fluid-jet of high speed for a long distance on the other hand can be considered as the basic requirement, put on the winning jet-nozzles. The energy-conversation in jet-nozzles upon basis of numerous test results, with bibliography references, proceeds with investigation of jet-velocity distribution after the nozzle.
For investigation of coefficient of efficiency of energy-conversation in jet-nozzles, a measurement of moment-force acting on at right angles to plate disc is used. Varying the distance between the nozzle and plate-disc the decreasing in jet-energy can be determined. The outer dimensions, making photos or video recordings, can also be fixed exactly. The scheme of experimental equipment is shown in **Fig.3.** The feed of nozzle with water is provided by pumps of variable R.P.M. with frequency changer. In the discharge pipe-line of pump a volume flow gauge is built but in the pipe-line tract before the nozzle a pressure measuring instrument serves for determining the hydraulic power fed to nozzles. The plate disc, connected to dynamometer, is located at right angles to the jet, leaving the nozzle. The water, running off the disc flows back into the water-tank through pipe-lines of receiver. The beam construction of nozzle-holding pipe and the built in pipe-line hose-tract provide the varying the distance between the nozzle and the plate-disc, serving the measurement of moment-force.
Upon the basis of the moment-force of jet, dashing into the disc and knowing the volume-flow rate of the jet, the velocity of fluid-jet and its energy can be determined.
Using the simple measuring equipment, shown above, numerous measurements can be

performed in no time relatively and by its help the nozzle, giving coherent fluid-jet of suitable geometrically shaping, can be choosed with good coefficient of efficiency.

The investigation into the hydraulic winning proceeds by the aid of National Scientific Research Foundation, and the teachers and researchers of the Department of Mining Machinery (Dept. name changed into Dept. of Equipments for Geotechnics) are taking part in it.

REFERENCES

Patvaros,J.1979.Die Anwendungsmöglichkeiten der Borloch-Abbautechnologie bei der gewinnung der heimischen Bauxitvorhommen. Acta Geodeatica et Montanistica Acad. Sci. Hung. 14: 449-472. Budapest.

Riutoul, B.1981. Hydraulic sweep shows promise. Drilling contractor, January.

Hlavác, L. - Foldyna, J. - Vasek, J. 1992. Modellirung des Schneidprozesses beim Einsatz von hochenergetischen Wasserstrahlen zur Bearbeitung von Festgesteinen. Neue Bergbautechnik. 22. Ig. Heft 12.

Ilias, N. - Magyari, A. - Radu, S. - Achim, M. 1993. The result of high pressure jets used in rock cutting and in assisted rock cutting. Mine Mechanization and Automatization, A.A. Balhema.

Geomechanics 93, Rakowski (ed.) © 1994 Balkema, Rotterdam, ISBN 90 5410 354 X

Application of high-pressure water jet technology in a construction company's practice

Zdeněk Jeřábek
Railway Building Trade, Joint-Stock Company

1. INTRODUCTION

In our republic there are nowadays many concrete structures which are highly damaged by influence of concrete corrosion. This concerns in the first rate these structures which are exposed to weather effects, overloading and changes in using. Operating influences present another group of damage of concrete structures. When investigating the structure before iniciation of rehabilitation work we have sometimes a possibility to find out, that the users do not have basic knowledge of a harmful influence of various reagents or chemical components to a concrete structure. There is sometimes spread out an opinion that concrete is "everlasting" and can resist to everything. That results in unadecuate opinion regarding the maintenance of concrete structures. The state of many structures is a great warning.

Unter unfavorable efects we have met during our practise with - except the most prevailing, as weather efects in general, icing up, thermical oscilation, carbon dioxide, salts etc. we further have met for example damage of concrete by influence of vinegar, sugar, beer, water poor in minerals ....

It is ever necessary to realize that concrete has its own chemical composition that we exactly know and which must be respected. But in its nature, the concrete is an artificially made product which is attacked on the one hand by all influence of natural environment and of course, on the other hand by influence of weather conditions changed by civilisation. Concrete consists of cement binder, aggregates, additives and steel reinforcement. Its nature is alkaline, but it is to generally exposed acidic environment, which surrounds us. This results in corrosive procedures the purpose of which is to return all the used matters to the state of original balance in nature.

Picture Nr. 1:
Damaged concrete structure

One of the basical principal of concrete damage is so called carbonatation, that means its gradual chemical neutralisation.

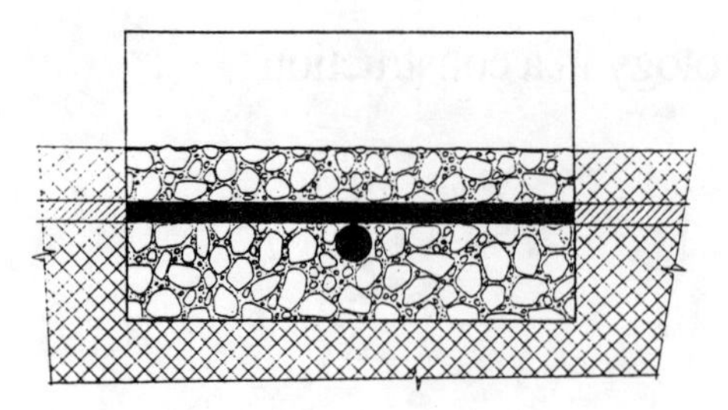

Picture Nr. 2:
New concrete structure

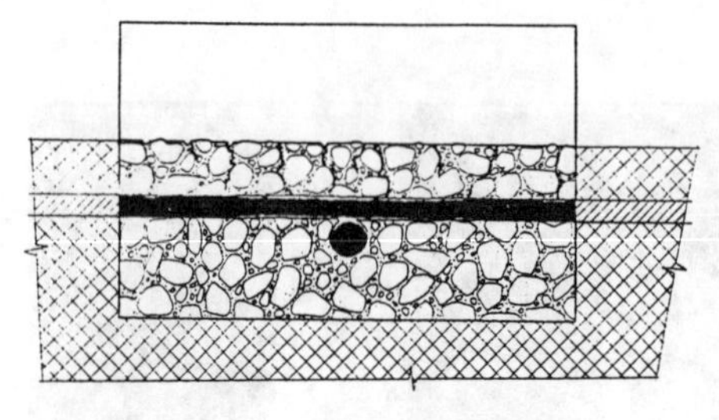

Picture Nr. 3:
Carbonatation of concrete beginning of corrosion in splitting reinforcement

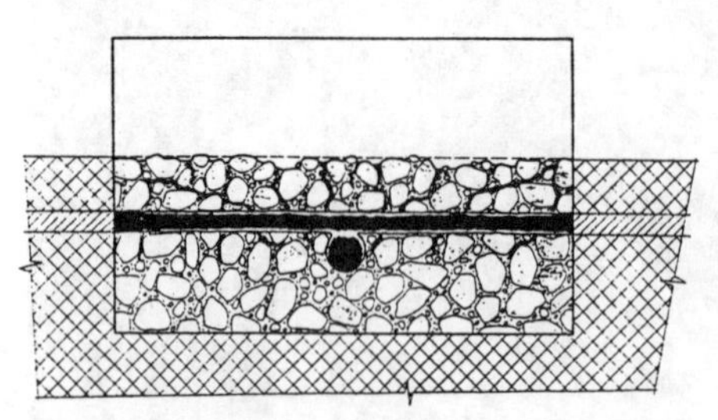

Picture Nr. 4:
Deep carbonatation of concrete, massive corrosion of reinforcement

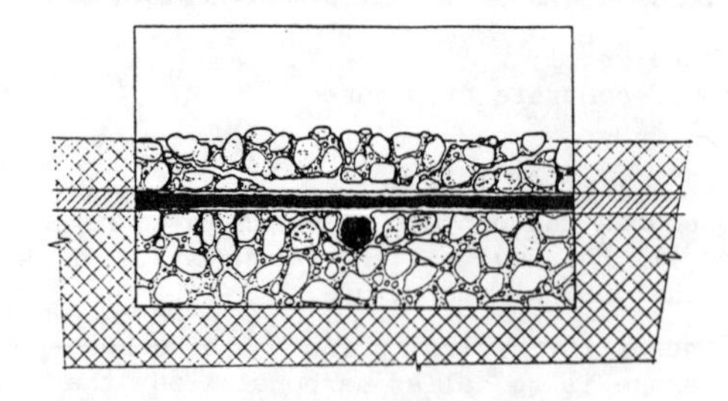

Picture Nr. 5:
Corrosive pressures, beginning of structure failure

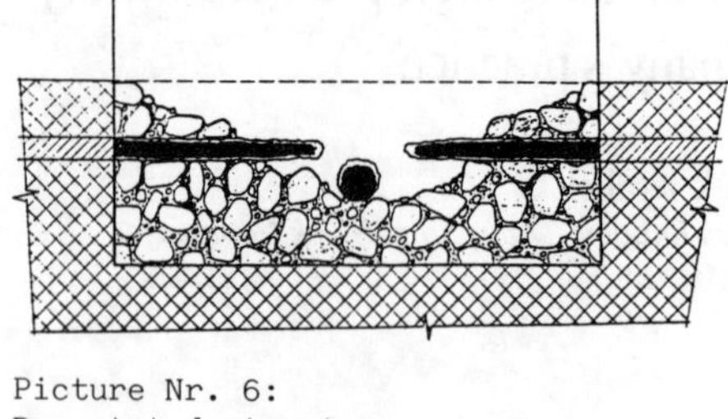

Picture Nr. 6:
Devastated structure

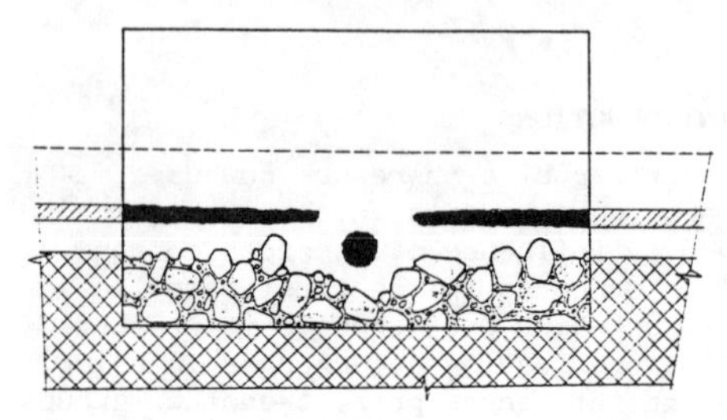

Picture Nr. 7:
Structure after removing out of failured layers with HWJ

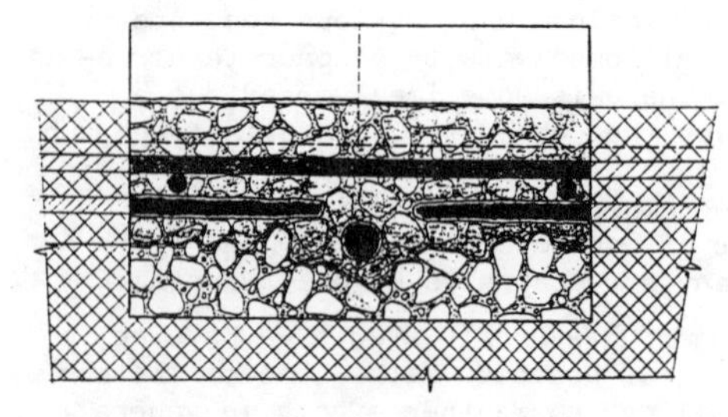

Picture Nr. 8:
Structure after rehabilitation, impregnation of concrete surface

2. PREPARATION OF THE SURFACE FOR REHABILITATION HIGH-PRESSURE WATER JET

Under surface preparation for rehabilitation procedures are to be understood:

- removing of all failured concrete zones that lost passivation capacity with steel reinforcement
- degreasing of the surface
- removing of all impurities and unsolid components of the surface
- cleaning of steel reinforcement.

As a basic rule of criterion of capability of structure surface to accept further rehabilitation technology is compression strength of 22,5 MPa and tensile strength, which is proved by so called tear-off test, (adhesion test) which amounts 1,5 MPa.

The intervention must be very considerate because there is possible to weak more the already weakened construction. The very dangerous is the vibration effect of tools and instruments by preparation are described in the chart as follow:

Procedure:
Classical, vibration instruments

Description:
It is carried out with classical vibration or rotary instruments as pneumatic hammers, cutters, grinding machines ...

Intervention suitability:
Suitable only where there is not a danger of development of micro-cracks on the construction, it is not reliable, this is not a selective operation, the surface so adapted must be worked and washed.

Procedure:
Siliceous sand blasting

Description:
It is carried out with a stream of siliceous sand carried by compressed air. As a rule it is possible to make a use of the same mechanisms as for shotcrete. For blasting of steel reinforcement it is used a special mechanisms (pressure pot). The depth of intervention depends on the fraction of siliceous sand.

Intervention suitability:
It is already a selective procedure but the effective of which is little, it envolves only a surface. It is possible to complete with it a previously mentioned procedure. Under this technology a great quantity of siliceous sand is processed which results in surplus material and transport costs. When working a huge quantity of dangerous fibroblastic powder appears, a danger of silicosis as well.

Procedure:
Steel shot blasting

Description:
It is a procedure when the rehabilitated surface is blasted with a stream of steel shot by means of rotating disk, than the shot and impurities are captured and separated, so their recirculation can follow.

Intervention suitability:
This is an effective selective procedure of work, but suitable only for horizontal structures - from above (floors).

Procedure:
Flame blasting

Description:
It a carried out with gas burners or liquid combustible burnes. By a flash heating the surface layers loose their strength and fall down. It is also possible to burn out organic matters which are situated on the surface of a structure under rehabilitation.

Intervention suitability:
Suitable only in special cases, there is a danger of development of micro-cracks on the structure.

Procedure:
High - pressure water jet (HWJ)

Description:
This is action of water jet generated under the pressure of till 250 MPa (2500 bar) by means special instruments when on the structure produces an imposed effect of direct impulse of pressure water and cavitation action.

Intervention suitability:
There is a very selective method which is very considerate to the structure (it does not develop micro-cracks). Advantageous is absolute solid and clean surface. The selectivety of this operation is provided with a continuous pressure regulation so it is possible to guarantee either taking out of bad quality material or its specified layer.

A survey of some methods of surface preparation for the rehabilitation with regard to the above mentioned survey it is obvious, that the most optimal work procedure is application of HWJ, possibly in combination with instruments for rough interventions and finishing the procedure with HWJ technology. It is possible to talk with maximum effectivness about the most considerate way which envables to minimize the possibility of development of micro--crack which could result in a great damage of the structure.

The above mentined principles must be considered even in cases of splitting of concrete structures, when it is necessary to choose such a procedure which will not cause destruction of a structure.

Now. I want to refer to HWJ application and give some points of view to this technology at a construction company. Without being equiped with this technology it is practically impossible to take over responsibility for iniciation of rehabilitation of concrete structures.

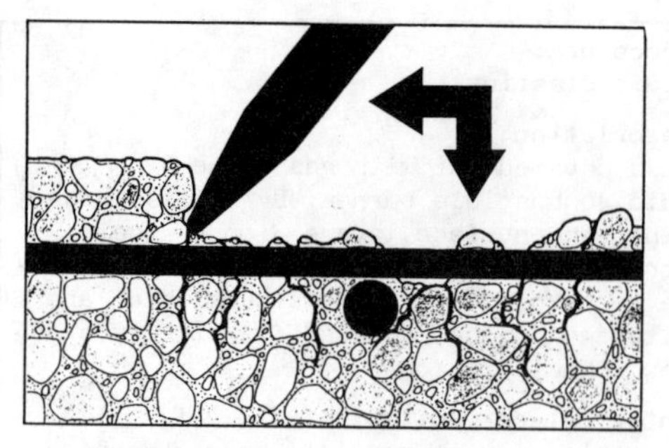

Development of micro-cracks with clasical removing out of failured layers by vibration instruments and tools.

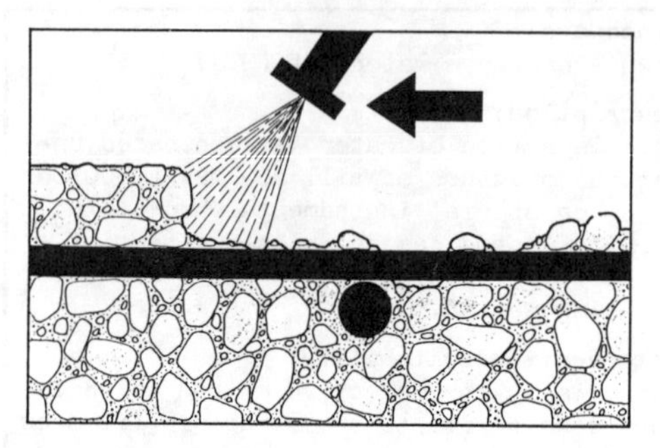

Removing out of failured layers with high-pressure water jet without development of micro-cracks.

Picture Nr. 9:
Principle of HWJ work

Picture Nr. 10 :
Preparation of ceiling construction with HWJ

3. Point of view of equipment with a type of device (pressure x water quantity)

Before purchasing of HWJ device it is nesessary to consider this investment and decide for the most suitable mechanism which enables to cope with a technological tangle - which consists in surface preparation for concrete structure rehabilitation. Generally, it is possible to choose among devices of various output categories the technical parameters of which are different especially in generated pressure and volume of water. Both parameters are connected with installed input of driving unit, which can be independent or external (e.g. diesel-agregate or electric motor).

We can distinguish among high-pressure mechanisms in categories as follow:

- low pressure,, where operation pressure amounts up to 120 MPa, volume of water can vary, in general there are used either mechanisms with lower effectivity, i.e. with volume of water up to 50 l//min. or till high effective mechanisms for removal of large deep layers of failured concrete with water volume up to 1000 m/ /min. Just with these mechanisms it is necessary to carry out the work using a manipulator because the reaction force on the tool exceeds the limit value of 250 N for a safe manual work

- medium pressure - operation pressure amounts up to 250 MPa, water volume is mostly lower, under 50 l/min., the work is done with hand tools as a rule. These mechanisms can be advantageous in using in ground constructions, which cannot be drenched so much as in the case of low pressure mechanisms. A low consumption of water is an advantage

- ultra high pressure - in reference with pressures which amount up to 600 MPa, water rate of flow are in order of litres per minute. These mechanisms are used in "under roof" applications, in industry in special technologies, especially when solving the problems related to special splitting up of materiales.

4. Point of view of incorporation in organisation

From the point of view of organisation it is necessary to incorporate the operation of HWJ agregates in order to correspond with the needs and objectives of the company. We can distinguish two principle ways of organisation of work with HWJ.

On the one sides - work on the commercial base. The advantage is building of quite special structure which is profesionally independent. It is possible to concentrate the specialists, service, maintenance, operation, manipulation.

This case can be referred only as one

working operation which is only a part of complete intervention. This principle is very spread with many companies in West Europe, but there the market is so developped to some extent which enables a representation of many other companies, who are specialised in working operations which follow HWJ.

Another type of work organisation which can be optimal for the actual situation in the Czech Republic is incorporation of HWJ into a group who can offer and carry out a complete intervention of rehabilitation. It may be referred as to some extent a dispersion of technologies - the company must be equiped with, but on the other side the company can offer to carry out a complete intervention with her own capacities.

Except of HWJ the company must be equiped with building chemistry and shotcrete technologies. The character of such a company or group is in general specialised in building and must be completed with internal or external capacities in building diagnostics and projection design.

5. Technical-developing point of view

Character of remedial work on concrete structure is almost ever new. The used technology for every structure differ as the case may be. When applicated HWJ it is necessary to dispose with many various manipulators, very often the special ways of captation, cleaning and liquidation of used water must be considered.

These operations must be precised beforehand what can have influence on the final price, but it is necessary to be equiped with all technical capacities and provide a smooth realisation of the action.

In general it can be said the special conditions of application of HWJ technologies are reflected in 40 - 60 % in costs.

6. Ecological point of view

As already previously meant the ecological point of view must be solved before application of rehabilitation and recovery technologies in general. It is connected with health protection of personnel when working.

a) Health protection

It is to be realised that tool the HWJ is applied with can be considered as a weapon. Non-observing of conditions for application of this technology can result in very heavy direct injury, but very hard damage of health with a smallest injury with HWJ can occur. Water contains various patogenous microorganisms which can penetrate the wound direct to the blood circulation. The personnel of attendance must be equiped with appropriate clothing and shoes, head and eyes protection. A great care of regular education and practice instructions led by experts must not be neglected.

b) In a broader context an attention must be paid to real influence of application of HWJ technology on the environment. Already in the phase of investigation of the specified work it is necessary to take into consideration the materials the HWJ comes into contact with and evaluate their influence on the environment. It is necessary to elaborate an ecological part of technological project where the absolute unexceptionability of applied technologies is guaranteed.

In water used for liquidation of failured materials can appear grease, chemical compounds, cruel-oil derivates ...

One of the possibilities to proceed with is an application of an effective system for sucking off used water including failured materiales.

Another possibility is to set a unit for filtration a flotation stage of cleaning, so practically cleared water is conducted away to the waste. Separated materials are liquadated with usual procedures - common methods.

7. Literature

1) Jeřábek Z.: Sanační zásahy na betonových konstrukcích za využití technologie vysokotlakého vodního paprsku, 1. symposium na téma sanace betonových konstrukcí, Brno, 23.-24. dubna 1991, ŽS Brno.

2) Jeřábek Z.: Sanace betonových konstrukcí -náplň práce SANCENTRA ŽS Brno, 1. symposium na téma sanace betonových konstrukcí, Brno, 23.-24. dubna 1991, ŽS Brno.

3) Jeřábek Z.: Rehabilitation of concrete structures using high-pressure water jet, Proceedings of the international conference GEOMECHANICS 91, Hradec n.Ostravicí, 24.-26.9.1991, A.A.Balkema, Rotterdam.

4) Wolfseher R.: Hochdruckwasserstrahl für die untergrund Vorbehandlung, Schweizer Fachverband für Hydrodynamik am Bau,1991.

5) Vanderwalle M.: Tunnelling the World, N.V.Bekaert S.A. 1990.

6) Vašek J.: Cutting of Rocks with the assistance of high pressure water jets, HOÚ ČSAV Ostrava, 1988.

7) Vašek J., Jeřábek Z., Momber A.: Die Anwendung von Hochdruckwasserstrahlen bei der Innstandhaltung von Eisenbahntunneln, Bauplanung und Bautechnik, 44 Jg., Heft 3, März 1990.

8) Sborník přednášek - I. mezinárodní symposium "Sanace betonových konstrukcí", Brno, 23.-24. dubna 1991.

9) Sborník přednášek - II.mezinárodní symposium "Sanace betonových konstrukcí a podlah", Češkovice, 26.-27.5.1992.

10) Firemní literatura firem: SIKA, ALIVA, HAMMELMANN, SIKA ROBOTICS, RENESCO, CON-JET.

11) Teichert P.: Spritzbeton, Laich SA, CH-6671 Avegno, 1980.

12) Sborník přednášek - III. mezinárodní symposium "Sanace betonových konstrukcí", Brno, rotunda pavilonu A BVV, květen 1993.

Geomechanics 93, Rakowski (ed.) © 1994 Balkema, Rotterdam, ISBN 90 5410 354 X

A contribution to the characteristics of mineral abrasives for WJC

P. Martinec
Institute of Geonics of the CAS, Ostrava, Czech Republic

ABSTRACT: In the experiments performed at the Institute of Geonics of the Academy of Sciences of The Czech Republic in Ostrava, a number of testing methods described in the following were used to determine the properties of abrasives destined for water jet cutting. A procedure for the analysis of mineral grains of the abrasives was proposed. The influence of the individual properties of abrasives on the efficiency of water jet cutting is described in papers which are published gradually.

1 GENERAL REQUIREMENTS ON PROPERTIES OF MINERAL ABRASIVES FOR AWJC

An integral part of the AWJC system are the mineral abrasives (abrasives in the following). Their properties are decisive, together with the whole technological system of the AWJC, for the cutting performance, the quality of the cut and the cost the whole of operation. The requirements for abrasives are specific: the material must by hard (Mohr hardness > 6),, specific gravity > 2600 [kg·m^3], poor cleavage and acceptable cost. The material must comply with the conditions of the Occupation Safety and Health Act (OSHA). Us. Dept. of Labor [Rankin, 1993] and other similar national occupation safety, labor and environment regulations. (Weiss, Martinec, Vítek, 1988). Under 29 CFR 1910, 1200, OSHA requires than any hazardous material in the workplace be recorded by Material Safety Data Sheet (MSDS). MSDS must be kept for empoyees to review at any time.

From the point of view, of environmental protection, mineral abrasives must be non - toxic, non - radioactive, with high stability in environmental conditions after deposition of slurries from the cutting or high pressure abrasive water blasting. The study of morphological, grain size and structural changes of abrasive grains in passing through the system of abrasive input - mixing chamber - foccusing tube - water jet - and the study of the interaction place between the jet with abrasives and the cutting material ("the system" in the following) is important for understanding desintegration processes in the system and for a better efficiency of cutting of hard material.

The main aim of our effort is be increase the efficency of the cutting process by designing optimum properties of mineral abrasive grains for individual cutting materials.

2 TESTING METHODS FOR PROPERTIES OF ABRASIVES

2.1 Mineral kind of abrasives

Many experiments have been realized with different kinds of natural and synthetic abrasive:
Natural minerals:
Oxides: quartz, rutile, ilmenite, magnetite, spinels
Silicates: staurolite, cordierite, garnets (almandine, pyrope, spessartine, grossulare, andradite and their natural mixtures), olivine, pyroxenes.
Industrial materials:
Slugs: steel - slugs, Cu - slugs
Glasses: glass (with different composition), expanded natural glasses (perlite), ballotine microspheres,
Ceramics: mullite, corundum and china ceramics, SiC.

At present, only quartz, garnets and olivine of different geological origin, deposits and technological dressing are wide by used in the practice.

2.2 Parameters of abrasives that affect the efficiency of WJC

2.2.1 Identification of mineral kind, chemical composition (crystallochemistry)

Methods and preparations

Any material used for the AWJC shall be identified mineralogically, i.e. its mineral kind shall be specified. A number of standard mineralogical and crystallographical methods are available for this purpose.

Methods of optical microscopic analysis (i.e. the identification of the mineral according to its optical properties, the size and the shape of the grains, the damage and inclusions, the areas of stress concentration and microhardness) belong to the most important ones. The following preparations are used for this purpose:

- thin sections, grain thin section,
- polished sections.

For scan electron microscopy (SCAN), specially prepared preparations are used to study the shape as well as the damage to the grains to perform local chemical analysis.

Crystals may also be investigated by means of X-ray monocrystal diffraction and X-ray powder diffraction. For such analysis, monocrystals or powdered samples are used. The identification of the given mineral is realized by means of positions of diffractions peaks (d_{hkl} values) and by means of latice parameters. This method is also suitable for the study of defects in the structure of minerals.

Infrared spectrography uses powdered samples and is suitable for mineral identification of the kind of amorphous and crystalline minerals, and for the study of structural crystallochemistry and some kinds of structural deffects.

Purity of the abrasives is determined as quantitative amount of abrasive mineral in input abrasive material. There are two ways to express purity:

- by means of computing the number of abrasive mineral grains
- by means of weighing abrasive minerals in the total sample.

2.2.2 Physical properties (specific gravity, deformation constants, mikrohardness)

2.2.2.1. Specific gravity

Specific gravity is determined as prescribed by Czech Standard (CSN 72 1154), on a powdered sample. The results are given in kg · m^{-3}.

2.2.2.2 Deformation constants

Technically, deformation constants (bulk modulus, shear constant, Poisson ratio) are very difficult to be, obtained for rock forming minerals. These constants are known for a number of minerals only. A summary of the known values for mineral abrasives was published by (Martinec, 1992).

2.2.2.3 Microhardness

Microhardness may be defined as resistance to permanent plastic deformation caused by indentation of an indentor into the studied material. Indentor is understood to be an instrument of a defined shape. Optimum indentation is by means of diamond pyramid with a square base and vertex angle of 136°.

The ratio of the mass of the weight used in the experiment and the area of the impression produced in the investigated material is the Vickers hardness number – VHN [N·mm^{-2}]. The method described here allows for the determination of microhardness on grains several times greater than the diamond indentor. Also the very important index of anisotropy of microhardness can be determined in minerals.

In Table 1, a rational classification of minerals in dependence on their microhardness and comparison with the scale is given.

Table 1 – Rational classification of the hardness of minerals (Lebedieva, 1977, adapted)

group of hardness	name	hardness VHN [N.mm^{-2}]	hardness Mohr's scale	optimum veight [grams]
I	very soft	10–588	1–2	5–10
II	soft	588–1177	2–3	15–20
III	fairly hard	1177–5394	3–5	50–70
IV	hard	5394–10787	5–7	100–200
V	very hard	> 10787	> 7	200

(Chrushchov 1949) proposed a scale of hardness in substitution of Mohr's scale. Graphite corresponds to 1, diamond to 15 in Chrushchov's scale. The formula for the calculation of hardness H_0 according to Chrushchov is the following:

$$H_0 = 0,675 \cdot (VHN)^{\frac{1}{3}}$$

NHN – Vickers hardness number

Both Vickers and Chrushchov's hardness numbers are used abundantly to describe hardness of materials and minerals.

2.2.3 Grain size, grain size distribution and shape of the grains

Grain size and grain size distribition are very important parameters. Grain size and grain size distribution may be determined using the following methods:

- measurement of characteristic dimensions of grains in a microscopic preparation, e.g. of maximum length and minimum width of the individual grains and their conversion into an equivalent circular or eliptical area. (In a narrower fraction, it is good enough to measure 100 to 150 grains approximately).

From the grains measured, the distribution function can be obtained (i.e. the number of grains of given size). For this determination, also an image analyzer with manual separation of individual mineral grains may be used.

- sieving analysis for mineral grains greater than 0.05 mm (distribution of grains according to their mass).
- for small grains, it is good to use Laser–Particle–Sizer for rapid, automatic particle size analysis in the range of 0.16 – 1250 μm.

From the cumulative distribution curve of grain size, it is possible to obtain sorting coefficients, minimum and maximum grain sizes and median of grain sizes (in accordance with the number of particles or with the mass, depending on the method chosen).

Shape of grains

The shape of the particle is substituted by the analysis of characteristic dimensions of the projection of the particle in the horizontal plane. One of the possibilities of expressing the shape of the particle is to determine the ratio of the maximum length to the minimum width after measurement of the individual particles in the microscope.

Also the automated image analyser with the corresponding software may be used with advantage (Vidas Rel. 2, Jan., 1991) especially when analyzing manually contoured grains from grain thin sections. The manual contouring of grains is necessary because the individual grains in grain thin sections may touch or overlap. For the determination of the shape of an abrasive material, we have used successfully the following parameters:

- area of individual grains and their distribution,
- D_{max} – the longest diameter of a grain obtained by selecting the largest of the Fered diameters measured in 32 different directions (i.e. at an angular resolution of 5.7 degrees),
- D_{min} – defined similarly to D_{max}, but taking the shortest among 32 Feret diameters,
- F_{circle} – circularity shape factor of a grain, defined as:

$$F_0 = 4 \cdot \Pi \cdot AREA/PERIM^2$$

The values of this parameter range between close to 0, for very elongated or rough object, and 1 for circular objects.

- Perim = object (grain) perimeter, calculated as:

$$PERIM = PERIM\ X + PERIM\ Y + PERIM\ XY \cdot 2^{0.5}$$

- F_{shape} – aspect ratio of the grain defined as:

$$F_{shape} = D_{min}/D_{max}$$

Statistical processing of the measured and calculated data serve for the evaluation of the shape. The results are presented either as the dependence of F_{circle} (F_{shape}) on the area of grains or as F_{circle} (F_{shape} ratio). It is more advantageous to make the evaluation of set of grains in a narrow grain size than to evaluate the whole sample of the material at a time.

Morphology of mineral grains

Expression of the 3-D shape of a grain is difficult. It is more simple to describe in words the surface and the shape of the grain distinguishing, at the same time, the natural effects. To obtain a certain standardization of such description, it is possible to use a comparison with typical pictures of grains.

This procedure was used, e.g., by (Vašek and Martinec 1993) for the description of garnet grains.

2.2.4 Defects in mineral grains

In natural and synthetic minerals and materials, there are found frequently defects which may be divided into the following groups:

2.2.4.1 Mechanical defects:

- inclusions of water + gas (bubbles of water and gas of a size some 10 – 1000 μm in a mineral grain), very often i.e. in quartz of hydrothermal origin
- inclusions of different types of minerals (i.e. quartz in garnet)

Their content is given as percentages, i.e. the number of grains with the individual types of defects in total number of the investigated grains (generally, some 200 to 500 grains).

- mechanical damage of grains by cracks.

This is a very frequent type of damage of natural grains. The presence of grains with such damage is given in percentages, i.e. as the number of grains with cracks out of the total number of the investigated grains (a total of some 200 to 500 grains).

In the above mentioned cases, optical microscopy and grain thin sections are used. The influence of mechanical defects on the depth of cut was studied by Vašek et al. 1993.

2.2.4.2 Structural defects

Defects in the structure of minerals can be studied by means of different methods:

- by means of polarized light microscopy, we can determine in isotropic minerals areas of stress concentration. This areas in isotropic minerals become optically anisotropic. This method is suitable e.g. for garnets. A standard type of grain thin section is used.

By means of XRD on powdered sample we can determine:

- variability of lattice parameters,
- coherently diffracting domains – linear size of crystallite and lattice strain (Stokes 1948, Wagner 1966).

Data on variability of lattice parameter, lattice strain and mean sizes of crystallites of different genetic types of quartz are given by (Weiss, Martinec 1978).

Experiments with garnets (unpublished author's data) show that, in spite of certain complication with the preparation of samples, greater structural defects were found in garnets from micaschists with tectonic deformation than in garnets from other types of rock where the tectonic deformation was not present.

2.2.5 Resistance of focussing tube to abrasion

Resistance of focussing tungsten carbide tube to abrasion is a specific criterion. Abrasive minerals and other materials can by divided into tree groups:

- very low abrasivity (less then 10 $\mu m \cdot min^{-1}$): glass, quartz, garnet from sedimentary deposits, staurolite, rutile, ilmenite, magnetite, pyroxenes,
- low to medium abrasivity (10 to 25 $\mu m \cdot min^{-1}$): garnet from primary deposits, cordierite, Cu – slugs,
- high abrasivity (greater than 25 $\mu m \cdot min^{-1}$): steel slugs, mullite ceramics, china ceramics, corundum.

For AWJC are suitable minerals and other materials from first two group only.

3 CONCLUSION

Abrasives of natural or synthetic origin have properties which were affected by natural conditions of their origin or by the technology of their production. Also the processing of the raw material to obtain the concentrate abrasive material affects their properties. During the technological processing of raw material, the aim is to standardize the composition and the purity of the abrasive, The size and the distribution of grains, their shape and sorting coefficient. In some mineral abrasives, and especially in garnets, the final processing yields a standard colour of the abrasive and, to a certain extent, some types of structural defects of the grains are eliminated. This improves the properties of the garnets.

During the passage of abrasive grains through the individual parts of the AWJC system, the morphology of the grains and their size change. The intensity of the disintegration of these mineral grains depends on technical parameters of the system and on the properties of mineral grains of the abrasive material at its input. Thus, the cutting performance is given by the properties of the grains of the abrasive material at its output from the foccusing tube. The comparison of the properties of the abrasive at the input into the AWJC system and when it has left the foccusing tube is one of the keys for the optimization of the properties of the abrasive and for the subsequent improvement of the abrasive water jet cutting.

REFERENCES

Foldyna, J., & P. Martinec 1992: Abrasive material in the process of AWJ cutting. *Procc. of the 11th Int. Symp. on the Cutting Technology*, 135 - 147, Kluwer Academic Publishers, Dordrecht, Netherlands, 1992.

Chrushchev, M.M. 1949: New scale for microhardness. *Zavodnaya Laboratoria*, 2, Moscow.

Lebedieva, S.I. 1963: *Identification methods for mineral microhardness.* Publishing House of Academy of Sciences of USSR, Moscow.

Martinec, P. 1992: Mineralogical properties of abrasive minerals and their role in the water jet cutting process. *Procc. of Inf. Conf. GEOMECHANICS 91*, Rakowski ed.,Balkema, Rotterdam: 353 - 362.

Rankin, M. 1993: On the job An Overview of Occupational safety and health act regulations, standards and requirements in the workplace. *Procc. of the 7th American Water jet Conference* (Hashish M.edit.), August 28–31., Seatle, Washington, paper 71: 685 - 689.

Stokes, R.A. 1948: A numerical Fourier-analysis method for the correction of widths and shape of lines on X-ray powder photographs. Proc. Phys. Soc. B61: 382 - 389, London.

Vašek, J., Martinec, P., J. Foldyna & J. Hlaváč 1993: Influence of properties of garnet on cutting process.- *Procc. of 7. th American Water jet Conference*, (Hashish M. edit.) August 28–31., 1993: Seatle, Washington, Paper 26, Vol. I. : 375 - 387.

Vašek, J., P. Martinec & L. Hlaváč 1993: Environmental impacts of water jet technology. *Procc. of Conf. EKOTREND 93, Mining University, Ostrava*: 29 - 33, (in Czech).

Wagner, C.N.J. 1966: Analysis of the broadening and changes in position of X-ray powder pattern peaks. In: Cohen J.B., Hilliard J.E.(edits.): *Local Atomic Arangements Studies by X - ray Diffraciton*, Chapter 6, NY.

Weiss, Z. & P. Martinec 1978: Lattice strain and linear size of quartz crystallites of different genesis.- *Čas. Min. Geol. 23*, 3: 243 - 253, Praha.

Weiss Z., P. Martinec & J. Vítek 1988: *Mine dust and dust nuisance.* SNTL Publishing House, Praha.

Geomechanics 93, Rakowski (ed.) © 1994 Balkema, Rotterdam, ISBN 90 5410 354 X

Water jet usage in German civil engineering – State of art

A. Momber
WOMA Apparatebau GmbH, Duisburg, Germany (Presently: University of Lexington, Ky., USA)

ABSTRACT: The paper gives a review of water jet application in building construction, highway engineering, foundation engineering, building sanitation, and ground decontamination. The description of selected cases contains technological details and specific aspects of high-pressure technique.

1 INTRODUCTION

In the last few years, the application of water jet technique in civil engineering has caught a lot of new fields in all parts of this industry, e.g. building sanitation, ground decontamination and demolition of constructions. In Germany some regulations in ecology (dust, noise, vibrations, deposits) and in concrete sanitation quality (crack absence, surface structure, reinforcement), but also economical aspects have supported the installation of plain and abrasive water jet tools. Fig. 1 shows the general applications of water jet technique in German civil engineering in relation to pressure and volume rate.

2 SELECTED APPLICATIONS IN FOUNDATION ENGINEERING

2.1 Water jet assisted pile driving

Sheet pilings are, as a role, driven home and pulled by means of diesel pile drivers or vibrators. According to the conditions prevailing in the surrounding area, such as soil conditions, degree of urbanisation and available cross sections, auxiliary systems to support the pile driving or vibration can or have to be employed in order to accommodate the increasing demands for an economic method of construction, sparing material and preserving the environment as much as possible. Depending on the steel sheet pile section and the soil conditions, one or more high-pressure lances are fixed along the length of the section, either on the section itself or close to the driving cap or vibrating pliers. At the start of the pile driving or vibration, the high-pressure unit is switched on. Then, depending on the soil conditions, up to 120 l/min of water at pressures between 25 and 50 MPa is conveyed through the nozzles. Two important effects assist the pile driving. At first the water flow causes the soil to loosen up at the sheet pile point, which in turn decreases the point resistance. In addition the water flowing off reduces both the skin friction and the interlock friction. Fig. 2 exhibits the general structure of a water jet assisted pile driving system.

The advantages of this system in comparison with conventional tools are (WOMA 1986, Temme 1986, Cohrs 1993):

- reduction of working time up to 75 %,
- reduction of soil vibrations up to 50 %,
- reduction of noise emission u to 30 %,
- reduction of working cost up to 50 %.

Measurements (Temme 1986) have shown that the vibrating stresses fall below decreed values for socles (≤ 2 mm/s) and for ceilings (≤ 10 mm/s), respectively. Fig. 3 contains a technical comparison between water jet assisted and conventional pile driving.

2.2 Ground decontamination using water jets

The number of decontaminated grounds and sites in Germany is estimated of about 80,000. For many of them - 26,000 - the situation is acute and their sanitation must be started without loss of time. In general one can discern in-situ and on-site

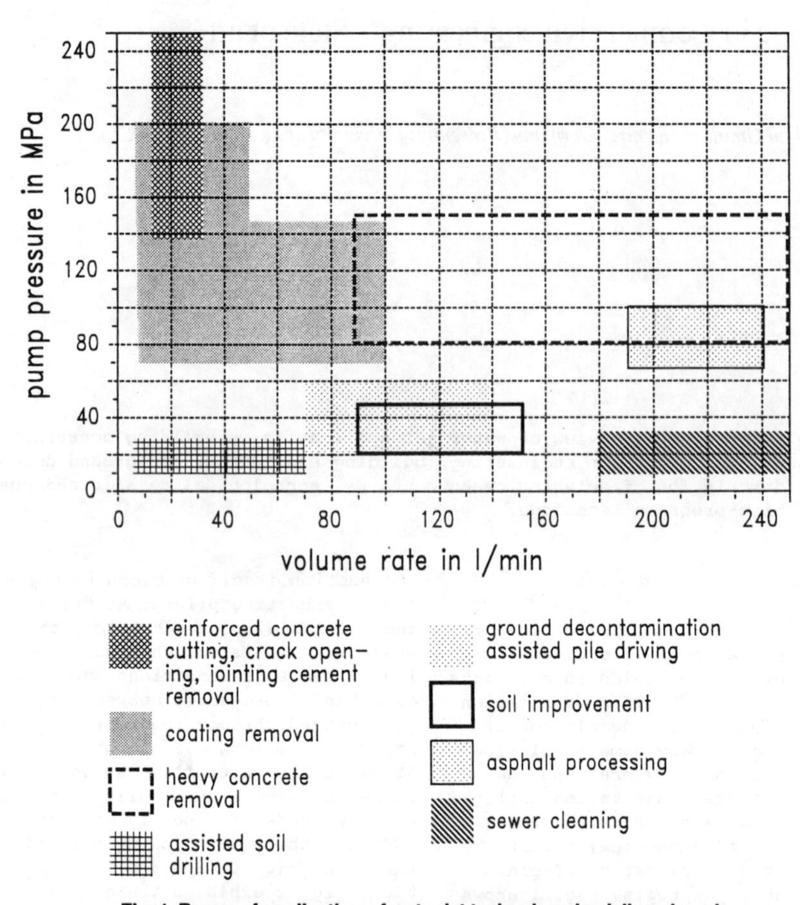

Fig. 1 Range of application of water jet technology in civil engineering

Fig. 2 General structure of a water jet assisted pile driving system (WOMA, Duisburg)

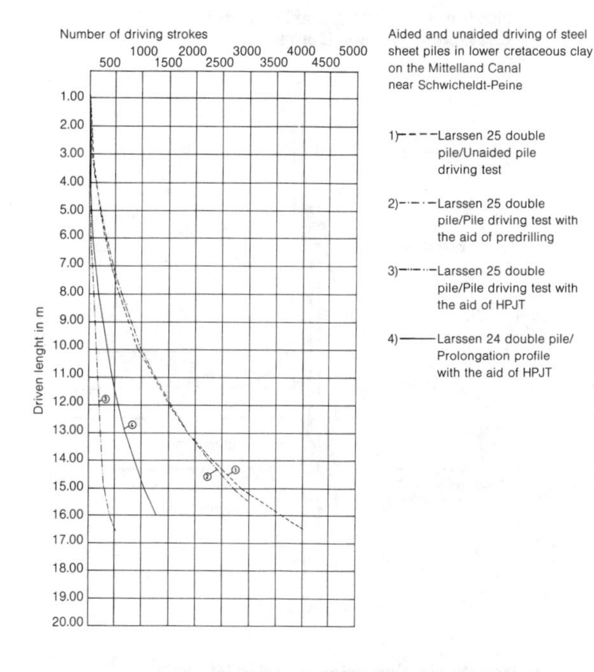

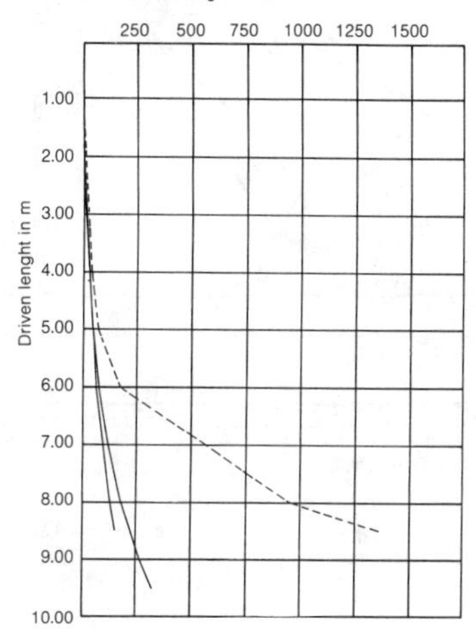

Fig. 3 Comparison between different pile driving methods (WOMA 1986)

methods. Technical and environmental demands on decontamination methods are put down in RAL (1991). Water jets are applicable for extraction and hardening methods, respectively. In Germany first experience exists in both decontamination technologies.

Decontamination using hardening method is based on the well known jet grouting principle and is described by Balthaus (1990) and Sondermann (1991). Both solutions are similar (figure 7). The process differs from jet grouting in only water being injected into the ground in the eroding stage. The eroded soil and the water are then air lifted to the surface where it enters a completely enclosed treatment system. The remainder of the material is then disposed of at a licenced site where additional decontamination, for example using biological treatment, can take place. The partial cleaning takes place in closed rooms, in jacket tubes (fig. 4). Inside this room the high-energy water jet shears the contaminants from the soil particles. The cleaning of the ground goes step by step, by an interlace pattern of several jacket tubes. The application is possible also under buildings. In general it is found that high-pressure water jet based hardening devices are excellent tools for cleaning contaminated grounds.

The advantages of this technology are:

- in situ performance,
- reusing of the cleaned soil,
- mobile devices,
- small space required,
- applicable on built-up places,
- high cleaning depth,
- applicable for high degree of contamination,
- applicable for organic and inorganic, soluble and antisoluble agents,
- combination of decontamination and stabilisation is possible.

An example for using extraction method is described by Heimhard (1987). He reports on a special jet tube. The contaminated soil and air will be drawn off through the focus point and cleaned by the water jet. The whole complex consists of distribution device, water jet cleaning tube and high-pressure pump, drainage works, waste water treatment, waste air treatment, and a device to treat and discharge the contaminants. All parts are placed in containers. The technique is applicable for removing chlorinated hydrocarbon, aromatic hydrocarbon, polycyclic hydrocarbon, alipathic hydrocarbon, heavy metal, and cyanide.

Figure 5 shows some cleaning results for contaminated soils using the water jet decontamination methods. Table 1 contains some general information on all methods. Sondermann (1991) and Balthaus (1990) have

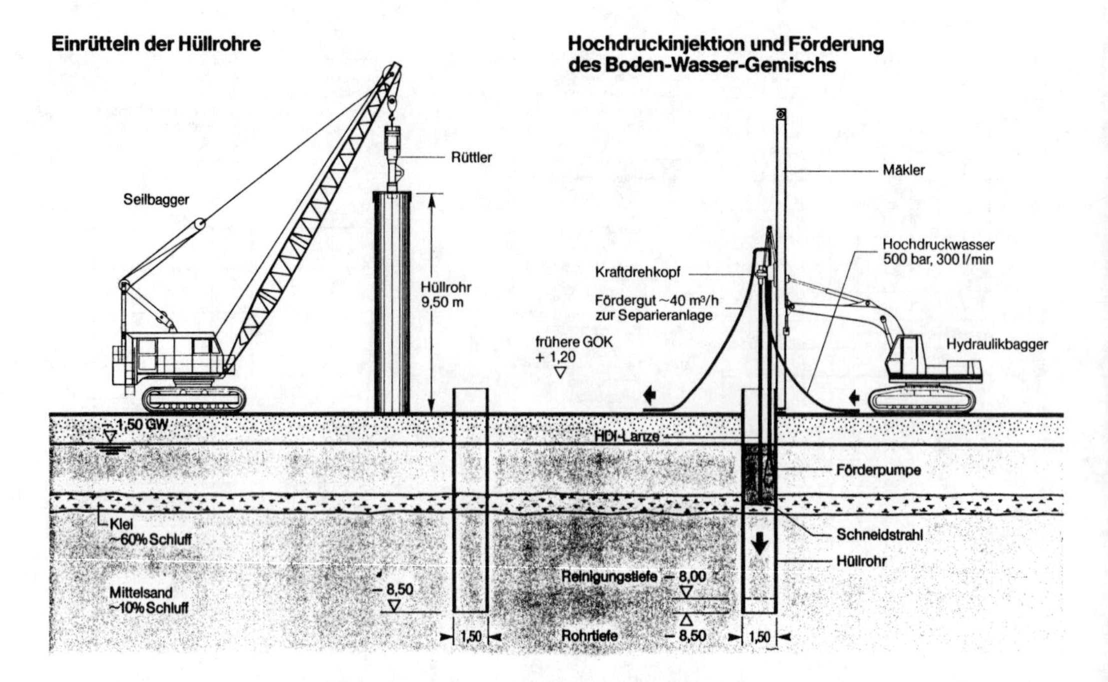

Fig. 4 General structure of ground decontamination using modified jet grouting method (Philipp Holzmann AG, Düsseldorf)

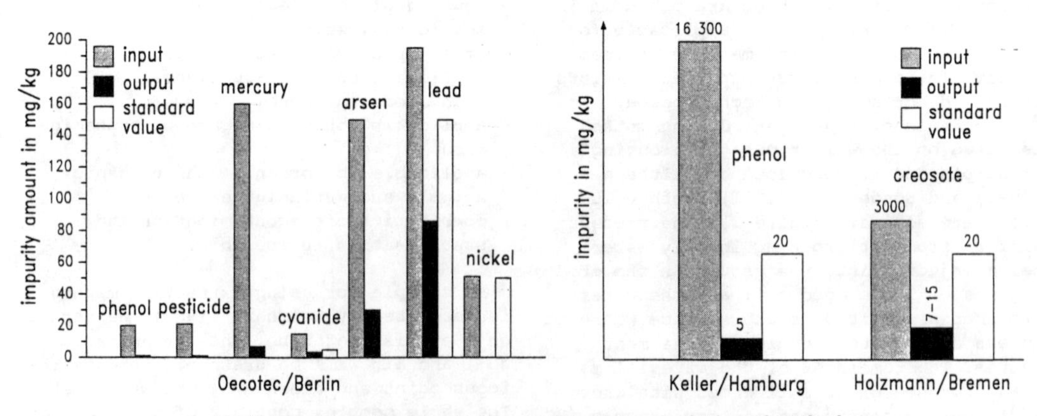

Fig. 5 Results of decontamination works using water jet based methods

discussed a combination of water jet usage and microbiological decontamination as an interesting alternative for further developments. The usage of water jets for soil cleaning is recommended by the German Society of Foundation Engineering (DGEG 1993).

2.3 Soil improvement using water jet technique

The soil improvement using water jet technique is known in Germany under the names "Soilcrete" and "HDI-Technik" and is being applied for the solution of many problems, e.g. foundation, stabilisation of collapse-endangered bridges (Brandl 1992), sealing of dams and soles (Döring 1986, Rabe and Toth 1987) and improvement of foundations of historical buildings (Stocker and Lochmann 1990). The general principle of jet grouting has been well known for many years. Fig. 6 shows the general performance. In relation to ground properties, geometry and

Table 1: Technological details of water jet based decontamination methods (Momber 1993a)

Parameter	Site		
	Bremen (Ph. Holz-mann, 1990)	Hamburg (Keller GmbH, 1991)	Berlin (Klöckner Oecotec, 1987)
Pressure	500 bar	400 bar	350 bar
Contaminated area	15000 m2	1200 m2	—
Soil volume	—	10500 m3	7000 t
Contamination depth	10 m	12 m	—
Efficiency	12 t/h	—	15-40 t/h
Costs	—	DM 1000/m3	DM 130-200/t
Occupied space	200-250 m2	—	55 containers

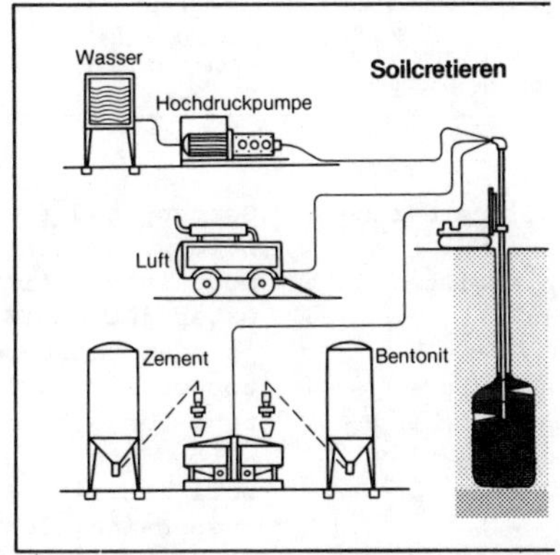

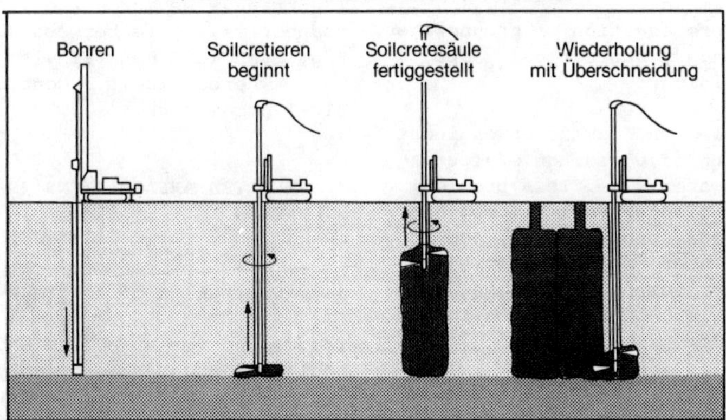

Fig. 6 General performance of water jet based soil improvement (Keller GmbH, Offenbach)

Table 2. Variations of water jet assisted soil improvement technique (Keller 1992)

name	Soilcrete S (single)	Soilcrete D (double)	Soilcrete T (triple)
principle	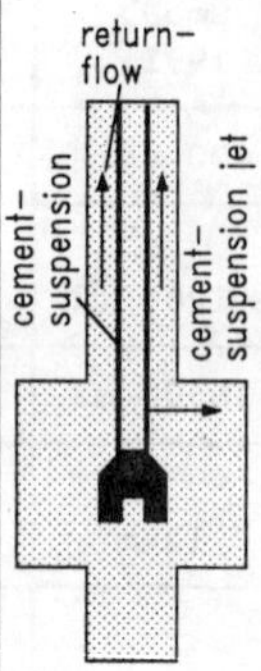 return-flow, cement-suspension, cement-suspension jet	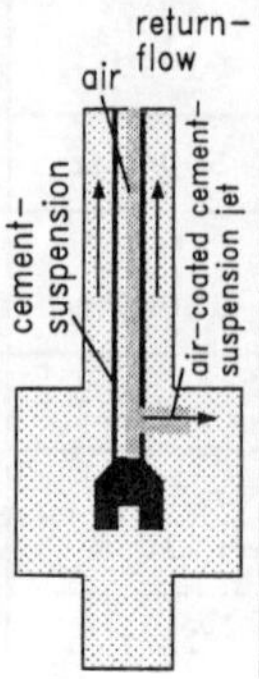 return-flow, air, cement-suspension, air-coated cement-suspension jet	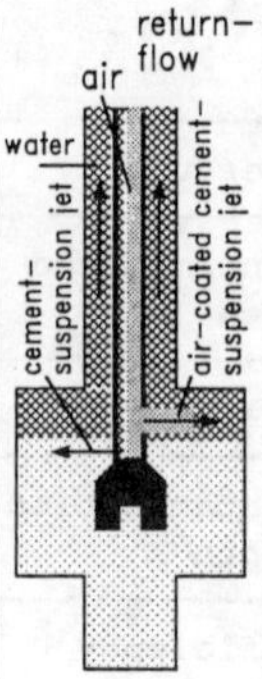 return-flow, air, water, cement-suspension jet, air-coated cement-suspension jet
pressure (MPa)	20 - 40	20 - 40	30 - 40
application	- soils with low erosion resistance - small and medium pillar diameter	- soils with medium and high erosion resistance - lamellar walls - underpin - sealings	- underpin - sealings

required quality, three variations can be applied (Table 2).

The advantages of water jet assisted soil improvement are (Keller 1992):

- independent of soil layer structure,
- applicable for rolling down soils and cohesive soils, including coarse clay,
- no ground water impairment,
- free of vibrations,
- small spatial amount,
- applicable in preservation of groundwater,
- generation of every geometrical shape also under buildings.

A similar problem in foundation engineering is the opening of cleadings of foundations for groundwater flow. This problem can be solved by placing rock-filters inside the waterproof concrete wall, see Fig. 7. The usage of water jet assisted injection for soil improvement is recommended by the German Society of Foundation Engineering (DGEG 1993).

2.4 Water jet assisted soil drilling

Considering the high traffic density in Germany and rising cost in trenching technology, trenchless methods become more and more interesting. One of these methods is water jet assisted soil drilling. Fluid jets at pressures of about 25 MPa enter the top of a rotating boring rod and work as a tool for ground loosening. The fluid evaluation depends on quality and structure of soil material (e.g. sandy ground: bentonite suspension, loamy ground: plain water). The transverse force and tensile force, respectively, are between 36 kN and 150 kN. Fig. 8 shows a general view on a complete unit, whereas Table 3 contains some technical parameters.

3 SELECTED APPLICATIONS IN BUILDING SANITATION

3.1 Sanitation of concrete constructions

The demand for effective and multipurpose tools in Germany has greatly increased within the scope of rehabilitation of buildings - especially reinforced concrete construction.

In connection with the rehabilitation

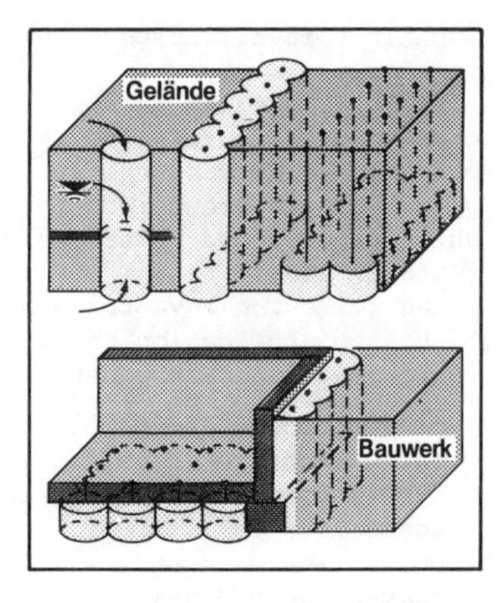

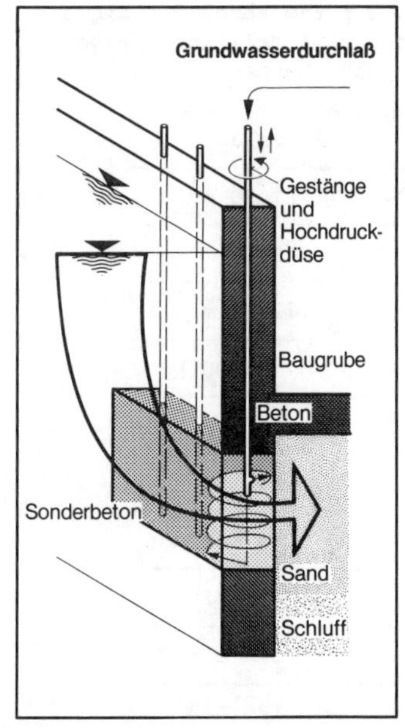

Fig. 7 Cleading of foundation opening for groundwater flow (Keller GmbH, Offenbach)

Fig. 8 General view on a water jet based soil drilling unit (Tracto-Technik GmbH, Lennestadt)

Table 3. Technical parameters of a big water jet assisted soil drilling system (Tracto-Technik GmbH, Lennestadt)

Parameter	Value
length	7.50 m
breadth	2.49 m
height	2.90 m
weight	18,000 kg
tensile force	150 kN *
traverse force	150 kN *
working rate	25 m/min *
number of revolutions	220 min *
drilling length	300 m *
power	190 PS
fluid pressure	0-25 MPa
fluid rate	0-76 l/min

* maximum values

and maintenance the following tasks usually have to be solved:

- Cleaning of concrete surfaces and building joints,
- Derusting of reinforcement steel and metal surfaces,
- Roughening of concrete surfaces,
- Removal of concrete and laying bare of rebars,
- Expanding cracks in reinforced concrete structures for press-filling.

Over the last few years the high-pressure water technique has developed from an alternative technique to a state-of-the-art tool in the construction industry and has now been adopted by all essential German regulations on concrete rehabilitation (DAfStB 1990, ZTV-SIB 1990).

For tasks in concrete sanitation the water jet technique ranges from hand-held tools for cleaning and coating removal up to heavy machines for deep concrete removal, e.g. on bridges and lock walls. On the processing of large objects like locks, bridges and barrages with water jets Rosa (1991), Knufmann und Westendarp (1992) and Schönig (1993) reported. Figure 9 shows examples for heavy concrete removal and asphalt removal, respectively, using rotating tools.

Table 4 contains technical and performance information of concrete and coating removal. In Germany the application of water jets is prescribed for derusting of chloride-corroded reinforcement bars (DAfStB 1990).

The following advantages have contributed to the application of water jets in reinforced concrete constructions:

- prevention of damage to, or loosening of, built-in parts (reinforcement, clamping sleeves) and connection areas (joint edges),
- prevention of dust,
- prevention of gases, vapours or slags,
- no heat formation,
- almost no impact noise,
- no percussions or vibrations,
- low reaction forces (good mechanisation and automation possibilities),
- wide range of applications,
- wide range of tools,
- utilisation of selective material removal (differentiating between various classes of concrete or between joint material and brickwork),
- guaranteed high pull-off strength (> 1.5 N/mm2),
- high removal rates,
- guaranteed practically crack-free cut edges.

Momber (1993) has discussed a lot of problems in relation to cleaning, roughening and removing of concrete including economical and ecological aspects.

Table 4. Removing rates for concrete and coatings (Momber 1993)

Material	Pressure in MPa	Efficiency in m2/h
Rubber	40-70	1,500
Rubber coating	95	200
Bitumen	80	25
Paint marking on roads	75	2-3 km/h
Polyurethane	90	12
Artificial resin	140	10-15
Limewash	40	60
Concrete		
76 mm deep	120	6.0
13 mm deep	120	9.0
200 mm deep	100	5.0
50 - 300 mm deep	140	3.0
50 - 80 mm deep	100	10.0

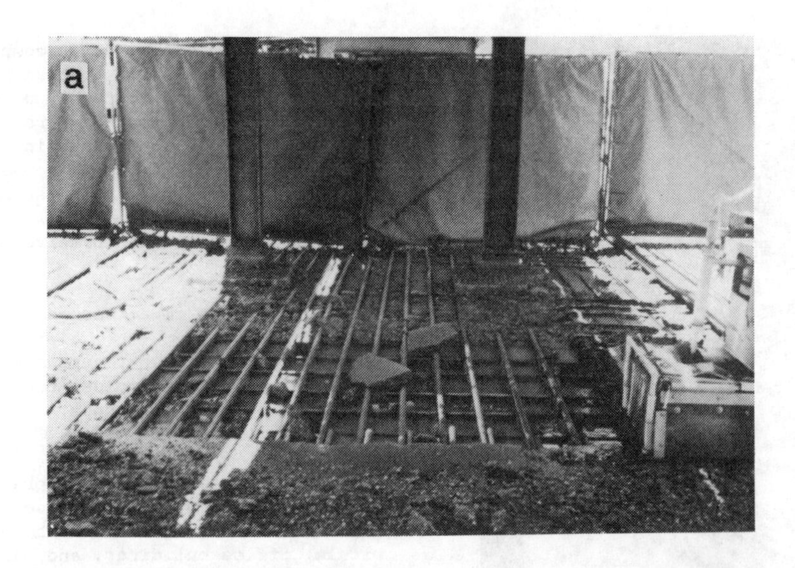

Fig. 9 Heavy concrete removing (a) and asphalt processing (b) by water jet technique (WOMA GmbH, Duisburg)

3.2 Removing of jointing filler

The removal of fillers from brick and rock joints is a general and important problem in sanitation of historical buildings. Traditional methods like hand hammers and chisel, pneumatic chipper and right angle grinder are affected with many disadvantages.

A new method is the so-called high-pressure fine-jet-cutting. The small dimensions of the tool (0.3 mm - 0.8 mm) allow a vibration-free clearing, even through relatively great depths (up to 10 cm), without damaging the surrounding original stone substance, even in the case of irregular joint courses. The efficiency of the technique is about 20 m/h for brick walls (Kauw 1993). The whole solution consists of a high-pressure pump (up to 200 MPa) and a joystick remote controlled cutting unit (Fig. 19). Water as well as removed materials can be sucked off by a disposal unit.

Fig. 10 Water-jet based joint cleaning system on site (Hamacher GmbH, Aachen)

3.3 Cutting of reinforced concrete and steel

Cutting of building materials by abrasive water jets has been a well known and used method for some years. But a comprehensive application in relation to trenching reinforced concrete and to partly demolition of construction in Germany is a result of the development of plunger pumps in the pressure range of 200 MPa. Plunger pumps have proved a success in many industrial fields under site conditions.

The main reasons for abrasive-water jet usage are:

- no significant impact sound,
- no vibrations,
- no formation of dust, vapour and slag,
- no thermal and mechanical tensions at cutting or breaking edges,
- no danger of explosion.

These operation properties predestine abrasive water jets for the use on or in cultivated buildings such as hotels, hospitals, office buildings, and buildings which are sensitive to dust and vibrations (e.g. data processing centres, locations of precision measuring instruments), explosion-hazardous surroundings and in narrow rooms. Figure 11 and Table 5 show some examples of reinforced concrete cutting.

4 ACKNOWLEDGEMENTS

The author is grateful to WOMA Apparatebau

Fig. 11 Cutting of strongly reinforced concrete beam by abrasive water jet (WOMA GmbH, Duisburg)

Table 5: Cutting rates on reinforced concrete using plunger pumps (Momber 1993)

Pressure (MPa)	Nozzle (mm)	Power (kW)	Abrasive (kg/min)	Traverse (m/h)	Kerf depth (cm)	Rein-forcement
130	1.3	80	3.6	1.0	37-40	strongly rein-forced
				2.4	25	
200	0.8	70	2.6	5.4	14	6 φ 6 mm
				1.2	40	8 φ 25 mm
			2.5	1.3	33	8 φ 20 mm
125	0.9	80	3.7	2.2		2 φ 6 mm

GmbH for preparing the paper. The author wishes to thank Philipp Holzmann AG, Düsseldorf, Keller Grundbau GmbH, Offenbach, Hamacher GmbH, Aachen, and Tracto Technik GmbH, Lennestadt, for their information and materials.

REFERENCES

Balthaus, H., 1990. In-situ-Hochdruckwäsche kontaminierter Böden. Baugrundtagung Karlsruhe, Deutsche Gesellschaft für Erd- und Grundbau e.V.: 121/133

Brandl, H., 1992. Fundamentsicherung einsturzgefährdeter Brücken. Baugrundtagung DGEG e.V, Dresden, Verlag Wehlmann Essen

Cohrs, H.H., 1993. Rammen und Pressen - Geräte, Technik und Verfahren. Hoch- und Tiefbau 3: 21/24

DAfStB, 1990. Richtlinie für Schutz und Instandsetzung von Betonbauteilen. Richtlinie Deutscher Ausschuß für Stahlbeton

DGEG, 1993. Empfehlungen des Arbeitskreises "Geotechnik der Deponien und Altlasten", Deutsche Gesellschaft für Erd- und Grundbau e.V., Verlag Ernst & Sohn, Berlin

Döring, M., 1986. Abdichtung der Sösetalsperre. Wasser und Boden 9: 444-449

Heimhard, H.-J., 1987. Die Anwendung des Hochdruck-Bodenwaschverfahrens bei der Sanierung kontaminierter Böden in Berlin. Int. Meeting of the NATO/CCMS Pilot Study Demonstration of Remedial Action Techn. for Contaminated Land and Groundwater, Washington D.C.

Kauw, V., 1993. Wasser als Werkzeug zum Ausräumen schadhafter Mörtelfugen. in: Werkstoffwissenschaften und Bausanierung, Teil 2 (ed. F.H. Wittmann), expert-Verlag, Böblingen: 1049/1057

Keller, 1992. Soilcrete - Das vielseitige Verfahren zur Baugrundverbesserung. Firmenschrift Keller Grundbau GmbH, Offenbach

Knufmann, Th., A. Westendarp, 1992. Instandsetzung am Eidersperrwerk. Beton 7: 377/382

Momber, A., 1993. Handbuch der Betonbearbeitung mit Druckwasserstrahlen. Beton-Verlag GmbH, Düsseldorf

Momber, A., 1993a. Recent developments in water jet usage in ground and traffic area rebuilding. Accepted for 7th American Water Jet Conference, Seattle

Rabe, E.W., S. Toth, 1987. Herstellung von Dichtwänden und -sohlen mit dem Soilcrete-Verfahren. Mitteilungen des Institutes für Grundbau und Bodenmechanik der TU Braunschweig, Heft 23

RAL, 1991. Aufbereitung zur Wiederverwendung von kontaminierten Böden und Bauteilen. Gütesicherung RAL-RG 501/2, Deutsches Institut für Gütesicherung und Kennzeichnung e.V.

Rosa, W., 1991. Bearbeiten von Betonflächen mit Hochdruckwasser. Berichtsband Konstruktive Instandsetzung von Stahlbetonbauwerken, Innsbruck: 17/25

Schönig, E., 1993. Betoninstandsetzung mit elastischen Schlämmen. Bautenschutz + Bausanierung 4: 12

Sondermann, W., 1991. In-situ/On-site Sanierungsverfahren zur Reinigung kontaminierter Böden. Entsorgungs-Praxis 3: 76/83

Stein, D., W. Niederehe, 1992. Instandhaltung von Kanalisationen. Verlag Ernst & Sohn, Berlin: 691/720

Stocker, M., A. Lochmann, 1990. Schonende Stabilisierung von Gründungen historischer Bauten. Vortrag Baugrundtagung DGEG e.C., Karlsruhe

Temme, W., 1986. Wasserhochdruckspülung als

Verfahrenshilfe bei der Vibrationsrammtechnik. Tiefbau-BG 2: 111/117
WOMA, 1986. WOMA-HOESCH high-pressure jetting technique (HPJT). Process documentation and photo report, WOMA Apparatebau GmbH, Duisburg
ZTV-SIB, 1990. Zusätzliche Technische Vertragsbedingungen und Richtlinien für Schutz und Instandsetzung von Betonbauteilen. Bundesministerium für Verkehr, Verkehrsblatt-Verlag

Mechanical rock disintegration
Main lectures

Geomechanics 93, Rakowski (ed.) © *1994 Balkema, Rotterdam, ISBN 90 5410 354 X*

Algorithm of control of disintegration of rocks at drilling from the point of view of costs per meter of bored hole

J. Bejda, V. Krúpa & F. Sekula
Institute of Geotechnics of the Slovak Academy of Sciences, Košice, Slovakia

ABSTRACT: The problems of the direct controlling of the disintegration of rocks from the point of view of costs are analyzed. The criterial function which has the minimum in the field of the drilling parameters that correspond to the minimum of the costs per meter was established. The experimental results obtained at drilling by means of the diamond core bits are introduced. These experimental results enable verification of a criterial function as function suitable for utilization in the development of control in-line system acceptable for criteria of the borehole economy.

INTRODUCTION

The area of problems considering the control of drilling is treated in various papers (Wijk,1991; Sekula,1983; Sitnikov, 1992). In this paper we want to concentrate on the problem of control of drilling from the point of view of minimizing of costs of drilling per drilled meter, i.e. to search the criteria that would enable to establish the algorithms of direct control of drilling that result from this requirement.

Wijk in (1991), at analysis of this problem, have used the simplified form of the well-known function of costs

$$C=\frac{R}{V}+\frac{P}{L_m} \qquad (1)$$

,where C are the costs of drilling, besides the eliminated investment, project and transport costs and similar costs, Sc/m; R are the rig costs per time unit, Sc/t; P is the price for the drill bit, Sc; v is the actual drilling rate, ms^{-1}; L_m is the lifetime of the bit measured in hole length, m. The author defines the necessary functions v and L_m by considering the function v as linearly dependent on $F^{3/2}$ and n, where F is the thrust force on the bit and n are the revolutions of the bit, and suggested the relation for the life of the bit

$$L_m=\lambda\,.\,F^{-\phi}\,.\,n^{-\upsilon} \qquad (2)$$

,where λ, ϕ, υ are the constants, $\lambda>0$, $\phi\geq0$, $\upsilon\geq0$.

Sitnikov (1992) have suggested to extend the costs function by adding the time losses at pulling out the pipe and pipe running (this is valid mainly for the geological reconnaissance drilling). Based on this extending the relation (1) can be modified to the form

$$C=\frac{R(\sum t_i+\sum t_s)}{L_m}+\frac{P}{L_m} \qquad (3)$$

,where t_i is the net time of drilling in the depth interval H_i, t_s is the time of pulling out, running in of the pipes and another time losses.

Let H_0 is the depth of the starting of the drilling with the new unworn bit and L_M is the penetration of this bit (or life of bit) up to the total wear out; let $L_M=m.l_i$, where m is the number of cycles of pulling out and running in of the pipes, l_i is the total penetration of the bit during one cycle; let v_z is the running speed, v_t is the pulling speed, v_{wl} speed of the running and pulling of the internal wire line core barrel. In respect of used symbols it is possible to derive the relations:

a/ core drilling with simple barrel

$$t_s=\frac{1}{v_z}(m.H_0+\frac{L_M}{2}.(m-1))+\frac{1}{v_t}(m.H_0+\frac{L_M}{2}(m+1)). \qquad (4)$$

In this relation is possible to take into the consideration the case of the improper filling up of the core barrel

b/ wire line core drilling

$$t_s=(2.H_0+L_M).(\frac{1}{v_{zti}}+\frac{m-1}{v_{wl}}) \qquad (5)$$

,where

$$V_{zti}=V_z=V_t$$

c/ fullface drilling

$$t_s=\frac{H_0}{V_z}+\frac{H_0+L_M}{V_t}\ . \qquad (6)$$

The relations of Wijk and Sitnikov result from the same idea that the relation (1) or (2) enables finding out of the optimal regime of drilling from the point of view of the drilling costs at partial deriving according F and n. On the first sight is possible to see the relations of the similar type as (2) and the linear function v of F and n parameters don't enable to solve this task without additional limitations that are connected with the technical parameters of the drilling rig. Sitnikov introduces the five pairs of the functions of F and n, that enable to search the extreme of the costs function, but solving of the problem in this mode is very complicated and so the suggested procedure is not suitable for creating of the algorithms of direct control of drilling from the point of view of the costs per drilled meter.

The solution of the problem seems to be the introducting of the non-conventional assumption that the life of the bit during drilling is not constant value and this value is changing in every even the infinitely small penetration ΔH. Based on this assumption is then possible to derive partially the costs function according L, where L is variable of the life of the bit. This assumption shows to be very stimulating, but it has some limitations at using.

DEFINITION OF THE CRITERIAL FUNCTION AND HYPOTHESIS

The detailed analysis of the problem under study based on analysis of the experimental stand laboratory drilling (Sitnikov,1992) so as on the knowledge achieved at realization of the monitoring systems applied on the machines for small and big diameter rotary drilling and fullface boring enables to define the set of theoretical generalizations that lead to the promising solution.

The first notion is: the every monitoring set that could be utilized as the control set (system) have to enable the measuring of the consumption of the energy of drilling. By means of this variable is possible to calculate the specific energy of the drilling, for example as the quantity of the energy consumed per volume unit (or weight unit) of the disintegrated rock. A part of this specific energy is non-reversible accumulated into the rock, but the sufficient part of this energy is transformed into the various forms. One of these forms is the part of the energy consumed to wear out the bit.

Let w is the specific energy expressed be the common formula

$$w=\frac{F.v+2.\pi.M_k.n}{S.v} \qquad (7)$$

, where M_k is the torsional moment of the bit, Nm; S is the surface of the functional face of the bit, m^2. The volume of the energy w_I non-reversibly accumulated into the rock during disintegration is the variable that can be expresses as

$$w_I=K(H).w \qquad (8)$$

,where K(H) is the variable that depends on parameters of the bit adjusted in the depth H, and K(H) also depends on the construction parameters of the bit and the abrasivity of the rock. Based on the relation (8) in the drilling interval $<H_1, H_2>$ arise the total wear of the bit 0 that can be estimated by the approximate relation

$$O\doteq K_{av}<H_1,H_2>.w_{av}.(H_2-H_1) \qquad (9)$$

,where K_{av} and w_{av} are the average values in the interval $<H_1, H_2>$.

From the relation (9) the approximate value of the wear intensity of the bit (I) can be defined as

$$I=K_{av}.w_{av}\ . \qquad (10)$$

At drilling by means of the kind of the bit in the same rock is possible to consider the value K_{av} as the constant provided that the parameters of the drilling are not very widely ranged. Based on this is possible to define the life of the bit by the expression

$$L=\frac{A}{K.w} \qquad (11)$$

,where A is the constant.

The relation (11) enables to transcribe the costs function (1) to the form

$$C=\frac{\frac{R}{w}+\frac{v.P.K}{A}}{\frac{v}{w}} \qquad (12)$$

In the relation (12) two functions of which the both are changing during drilling are in the numerator and denominator. On the basis of the analysis of the results of the monitoring of the experimental drilling so under laboratory conditions as under in situ conditions we have established the hypothesis:

THE MINIMUM COSTS OF DRILLING ARE REACHED AT THE SUCH REGIME OF THE DRILLING AT WHICH THE MAXIMUM OF THE FUNCTION V/W WILL BE SECURED.

The claim contained in the hypothesis means that at the change of the parameters of the drilling that causes the growing of the function v/w the growing of the function R/w+vPK/A is slowlier as the growing of the v/w function, i.e. it have to be valid that with growing of the v/w function the value C is going down.

EXPERIMENTAL VERIFICATION OF THE SUGGESTED HYPOTHESIS

In the Figure 1 the relations v=v(L), w=w(L) and v/w=f(L) as the functions achieved by means of the correlation of the laboratory obtained experimental data of drilling by means of the inserted diamond core bit at the constant regime of the drilling (F= 10 kN, n= 8,3 s^{-1}) of andesite of Ruskov locality are fitted. These results were correlated by means of the method of the least squares with using of the functions

$$v(L)=V_0 \cdot e^{-k_v \cdot w(L) \cdot L} , \qquad (13)$$

$$w(L)=w_M-\frac{k_w}{L} \qquad (14)$$

,where k_v, k_w, w_M, v_0 are the constants, the relation (13) is valid for L> 3 m and the relation (14) is valid for L> 10 m. These conditions were derived from the requirement so that the following conditions would be secured with increasing of the penetration L :

$$v(L_2)<v(L_1)$$

which results from the requirement formulated by Sitnikov (1992), and

$$w(L_2)>w(L_1)$$

suggested by authors of the paper.

Let us apply the obtained functions in the relation (3) at considering L_M=L; i.e. we consider the variable life of the bit. The new costs function is established :

$$C=\frac{R}{v_0 \cdot e^{-k_v(w_M-\frac{k_w}{L}) \cdot L}}+\frac{P}{L} \qquad (15)$$

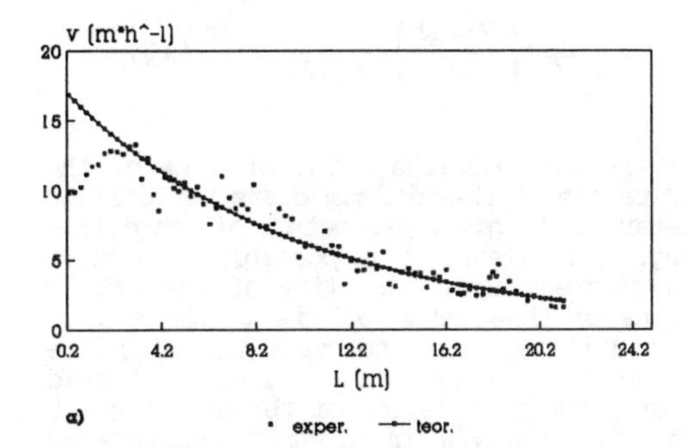

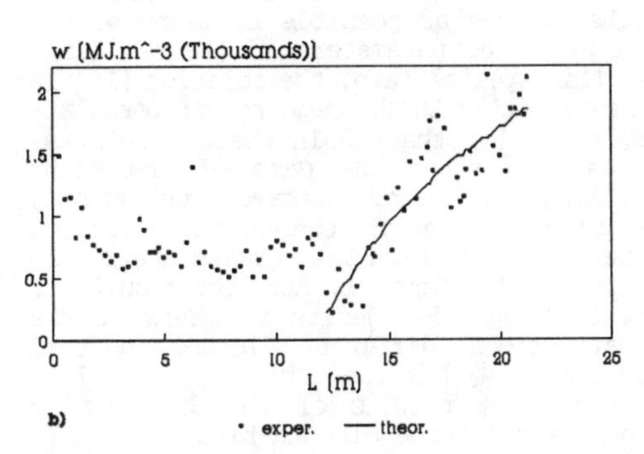

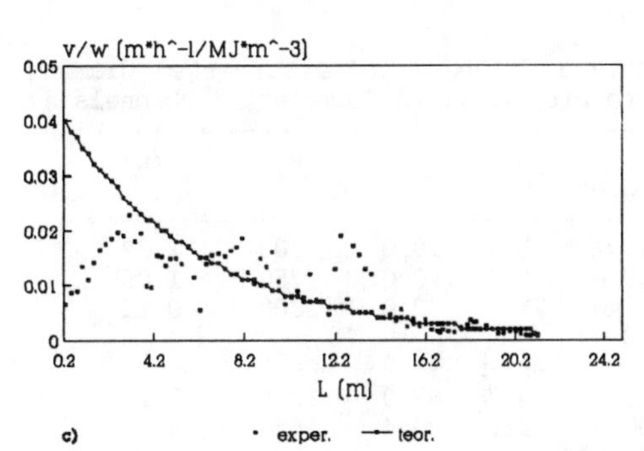

Figure 1. The results of stand research.Rock: andesite Ruskov. Drilling bit: diamond inserted, 46 mm in diameter, 2 channels, matrix hardness 400 HV, size of diamonds 100-150 pieces per gram. Regime of drilling: F= 10 kN, n= 8.3 s^{-1}.
a) penetration rates, r= 0.9549
b) specific energy, L>12 m, r=0.88517
c) v/w, r=0.88497

By means of partial deriving of the function (15) according L at using of the condition of the optimization of drilling

$$\frac{\partial C}{\partial L}=0$$

we obtain the relation (16)
, where L_M is the maximum obtainable life

$$L_M^2=\left(\frac{v(L_M)}{w_M}\right)\cdot\frac{P}{k_v.R} \qquad (16)$$

of the bit from the point of view of the criterion of the minimum costs per drilled meter at the given parameters of the drilling. It is clear, it is possible to achieve the increasing of the life of the bit by means of the change of the parameters of the drilling. In the relation (16) the value v/w is presented, i.e. we would search the such regime of the drilling for the function v/w in order to achieve as higher values as possible by means of the change of the parameters of the drilling. For this special case, the relation (16) is in agreement with the requirement formulated in the hypothesis. In the our database we have searched the data of the stand drilling by means of the same drilling bit, as in the case of the above described experiment, of which the results are illustrated in the Figure 1. Also the results of the drilling of andesite by means of the constructional different diamond inserted bit are treated in the Table 1. The Table 1 presents the results of the drilling by means of all three diamond bits:

Table 1. Stand results, inserted diamond core bit, 46 mm in diameter, 2 channels

Bit Regime	L m	v $m*h^{-1}$	w $MJ*m^{-3}$	v/w $.10^{-2}$
1.	3.8	10.0	700	2.29
400 HV	9.7	8.0	750	1.06
10 kN	21.0	2.0	1800	0.12
8.3 s^{-1}				
2.	3.8	87.0	1230	7.1
400 HV	9.7	80.4	1650	4.9
13.6 kN				
34.3 s^{-1}				
3.	1.2	7.4	2040	0.36
700 HV	2.0	5.0	2600	0.19
10 kN	3.8	3.3	2840	0.11
8.3 s^{-1}				

In the Table 1 the diamond bit No.3 differs from the bits No.1 and 2 only in the hardness of the matrix (bits No.1 and 2 had the hardness of the matrix 400 HV and bit No.3 had 700 HV).

The experimental and calculated results treated according the relations (13) and (14) are introduced in the Figure 2.
By means of the bit No. 2 only the data introduced in the Figure 2 were obtained.
According the relations (13) and (14) the values of the constants k_v, k_w, w_M and v_0 were calculated for all three bits and so it was possible to calculate the course of the function of the costs, for example according the relation (15). The Figure 3 introduces the calculated costs.

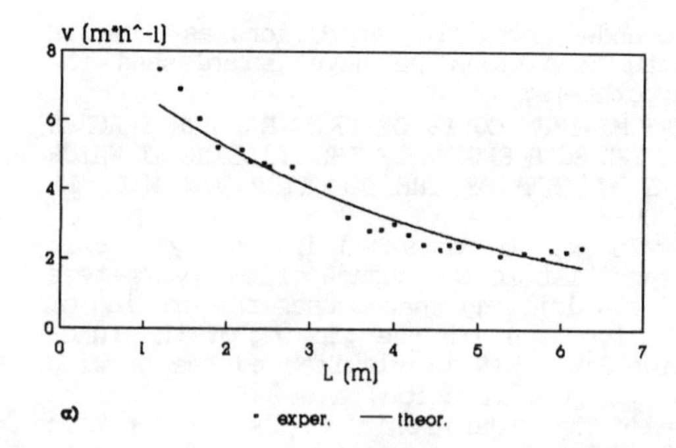

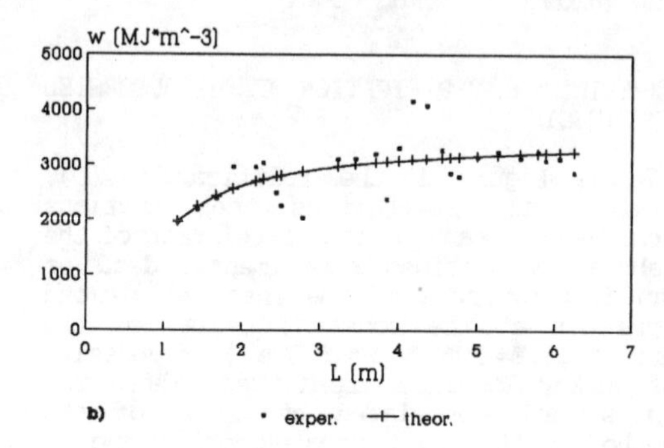

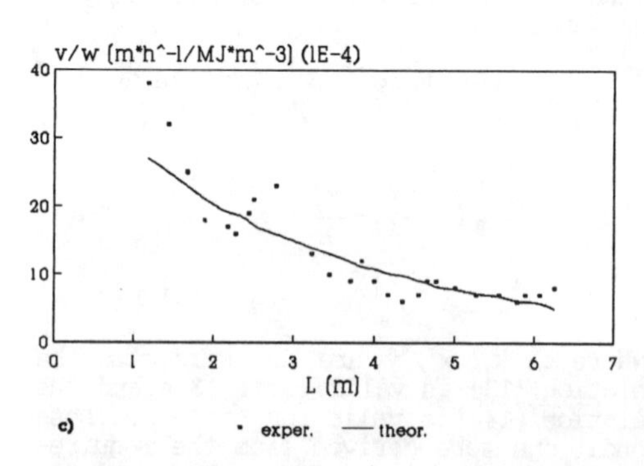

Figure 2. The results of stand research. Rock: andesite Ruskov. Drilling bit: diamond inserted, 46 mm in diameter, 2 channels, matrix hardness 700 HV, size of diamonds 100-150 pieces per gram. Regime of drilling: F= 10 kN, n= 8.3 s^{-1}.
a) penetration rates, r=
b) specific energy, r=0.61335
c) v/w, r=0.90697

We suppose the relations (13) and (14) will be satisfied for the characterization of the changes of the v and w functions for the such type of the bit of which the drillability is getting worse with the increasing of the wear at keeping the

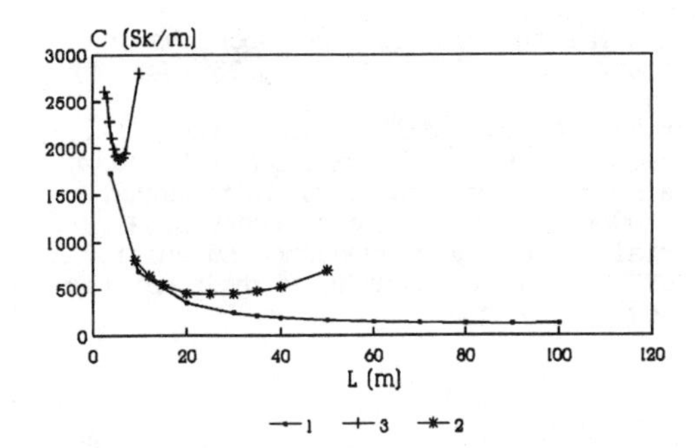

Figure 3. The costs of drilling by means of inserted diamond bits in andesite Ruskov. Bits and regimes:
No.1. 46 mm, 400 HV, F=16.6 kN, n= 34.3 s^{-1}
No.2. 46 mm, 400 HV, F=10 kN, n=8.3 s^{-1}
No.3. 46 mm, 700 HV, F=10 kN, n=8.3 s^{-1}

constant conditions of the drilling. For example, the impregnated diamond core bits, or the similar so called self sharpening bits, are characterized by drillability which is not changing during wide range of the penetration. But the life of these bits and so the costs of drilling by means of these bits have to correspond with drilling parameters. We assume the above presented hypothesis is valid equally for these impregnated bits.

Let us to document this claim on the example of one experimental result. In the Figure 4 the results of the stand trials of the drilling of **granite** by means of the impregnated diamond bit are fitted. The dividing of the interval of the penetration of 6 m to the intervals of 1,2 m we have selected intentionally so that the values of the course of v and w in the every shorter interval would be achieved similar as by means of the inserted diamond bit. The results illustrated in the Figure 4 were achieved from the drilling by means of the impregnated diamond core bit of 46 mm in diameter with using of the relatively large synthetic diamond grains (0,5 - 1,0 mm in diameter). These large diamonds caused the non-uniform impregnation of the matrix of the bit and so the impregnated diamond bit with several 'working layers' was produces. This have caused the fluctuation of the drillability of the bit between better and worse drillability. It is clear that it is possible to apply the formulated hypothesis in this case.

DISCUSSION OF RESULTS AND CONCLUSION

Based on the analysis of the experimental drilling results from the drilling by means of the diamond bits performed on the model rock the special function that enables to simplify the problem of the optimization of the drilling from the point of view of the costs per drilled meter was established. It was shown, independently on the mode of the drilling and the construction of the bit is possible to create the simply criterion enabling the construction of the electronic sets of the monitoring and controlling of the drilling by means of which is possible to approach to the minimizing of the costs of drilling.

The verification of these experimental

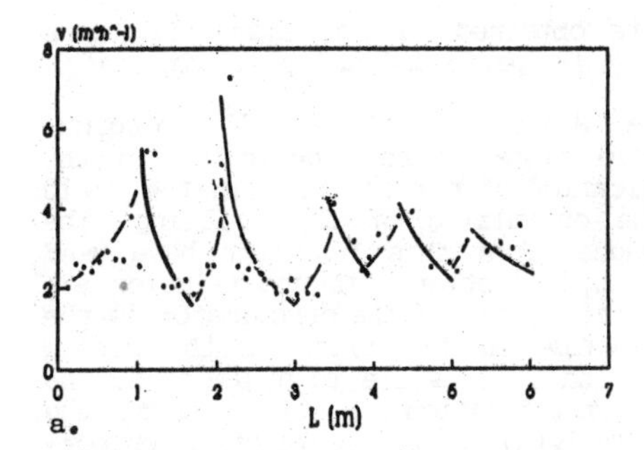

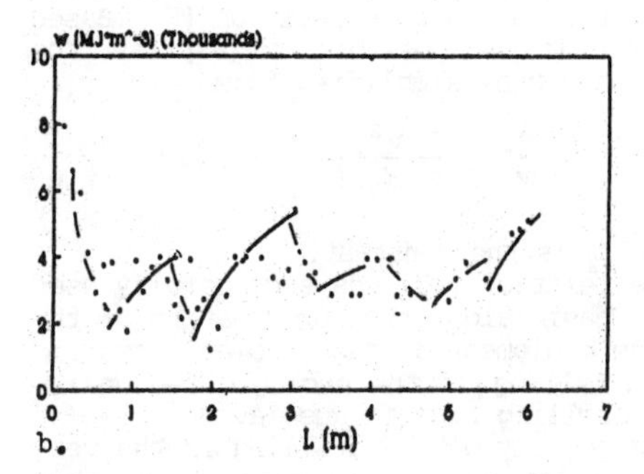

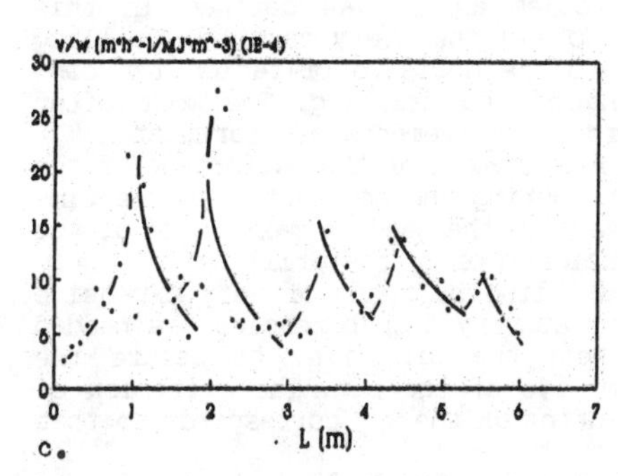

Figure 4. The results of stand research. Rock: granite Hnilec. Drilling bit: diamond impregnated, 46 mm in diameter, 4 channels, matrix hardness 400 HV, type of diamonds symthetic SDA-85 DeBeers. Regime of drilling: F= 10 kN, n= 8.3 s^{-1}.
a) penetration rates,
b) specific energy,
c) v/w,

results obtained during laboratory tests have to be performed under conditions in situ.

The Moscow Institute of Geological Reconnaissance is more dedicated to the verification of our attitude in the field of the optimizing of the drilling. The colleagues from this Institute have used some simplification : in the relation (7) the first member of the denominator is the small value in comparison to the second member and is possible to neglect it. The next simplification is replacing of the function $M_k(L)$ by linear function proportional to F in the same relation (7). Based on this it is possible to express the function v/w by simplified form

$$\frac{v}{w}=\frac{v^2}{2.\pi.K_M.F.n} \qquad (17)$$

,where K_M is the constant.

The relation (17) was successfully used as the basic algorithm for creating of the program equipment of the automatic systems of controlling of the geologic reconnaissance drilling rigs in Russia.

The economy of the drilling is the very wide problem as it was defined in this paper. Not in the every case the drilling costs are the decisive criterion for controlling of the drilling. The most often two border requirements are formulated:

- the requirement of the performing of the drilling during the shortest possible time that leads to the need of maximizing of the penetration rate of the drilling,
- at drilling by means of very expensive bits, or at very deep boreholes, is needed to protect the bit, i.e. to secure the maximum life of the bit. The criterion of the securing of the w_{min} corresponds to this requirement.

The regime of the drilling that reflects to the criterion (v/w) is the compromise between the above formulated criteria and this criterion reflects to the requirement of minimizing of the costs per drilled meter of borehole.

REFERENCES

Sekula,F.;Krupa,V.;Bejda,J.;Merva,M. 1983. Optimization of Drilling Process on the Basis of Monitoring of Some of its Param eters, Rock Mechanics in Czechoslovakia, Subcommittee for Rock Mechanics of the Czechoslovak National Committee of World Mining Congresses, p.95, Ostrava, January 1983

Sitnikov,N.B. 1992. Optimizacia Kolongko vogo Burenija Skvažin Zatupľajuščinsia Porodorazrušajuščim Instrumentom, IVUZ Gornyj Žurnal, p.59, No.12, dekabr 1992

Wijk,G. 1991. Rotary Drilling Prediction, International Journal of Rock Mechanics and Mining Sciences, and Geomechanics Abstracts, Vol.28, No.1, p.35, January 1991

Vorobjov,G.A.;Novožilov,B.A.;Vareca,S.A.; Fadejev,V.F. 1987. Upravlenie Processom Almaznogo Burenija, Sbornik naučnych trudov Sintetičeskije sverchtveryje mate rialy v geologo-razvedočnom bureniji, AN USSR, Institut sverchtvjordych materialov, Kijev, p.48

Geomechanics 93, Rakowski (ed.) © 1994 Balkema, Rotterdam, ISBN 90 5410 354 X

Coal and rock cutting

A.W. Khair & L.D. Gehi
West Virginia University, W.Va., USA

ABSTRACT: This paper describes the mechanism of coal and rock cutting from quasi-static indentation through dynamic indentation to rotary cutting. The results indicate that during quasi-static indentation, a crushed zone, and subsurface fractures develop. The crushed zone is a potential source of dust generation. Dynamic indentation produces a crushed zone due to the frictional resistance offered by the coal or rock during penetration of the bit. Confining pressure changes the mode of failure from tension to mixed tension and shear. The wider bit causes larger crushed zone and more dust generation. The rotary cutting is a combination of quasi-static and dynamic indentation which results in chip formation.

1 INTRODUCTION

The trend from conventional mining to continuous machine-cutting of coal, using more productive machinery is inexorable. Intolerable levels of dust generated at the face is an indication of inefficient coal or rock cutting by these machines. The Federal Government of the United States paid about 11.7 billion dollars workman compensation to more than 470,000 miners with Coal Workers Pneumoconiosis (Newmeyer, 1981). Bieniawski and Zipf (1985) reported that annual benefit payments were about 2 billion dollars. The need for research in improving coal cutting is therefore paramount.

The enactment of Health and Safety Act in 1969 in the United States triggered the initial investigation into coal cutting mechanism and cutting tool design. The European countries were also engaged in the same area of research, however, their emphasis was in the use of wedge shaped bits. In 1981, the United States Bureau of Mines commissioned the South West Research Institute to investigate the effect of operating parameters mainly on airborne respirable dust in order to provide a healthier and safer atmosphere for coal mine workers.

A research program by the Department of Mining Engineering, West Virginia University, USA was initiated to study the fragmentation mechanism during coal/ rock cutting. The research program was divided into three areas of research, namely quasi-static indentation, dynamic indentation and rotary cutting.

2 QUASI-STATIC INDENTATION

The experiments were conducted on two different geometrical shapes, cylindrical disc and rectangular, of specimens with three different wedge indentor angles, 20, 45 and 60° and four confining pressure equivalent to in situ overburden depth of 0, 100, 200, and 300 m. The loading conditions are illustrated in Fig. 1. Besides coal, sandstone was also subjected to the test in order to study the mode of failure. The mechanical properties of rock and coal are summarized in Table 1.

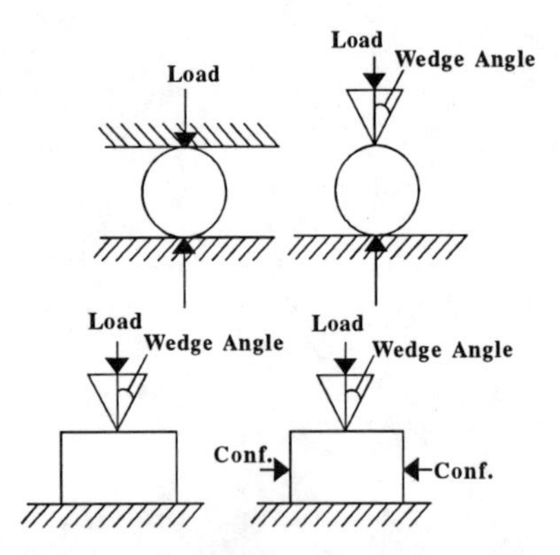

Figure 1. Loading condition in quasi-static indentation

Table 1. The geo-mechanical properties of coal and rock

Material	Compressive strength (kg/cm^2)	Young's Modulus (10^3 kg/cm^2)	Poisson's Ratio
Coal	521	35	0.47
Rock	668	157	0.15

2.1 *Shape of specimen*

The pattern of failure suggests that the cylindrical disc fails due to propagation of tensile fracture created in the center of the specimen and extends to the points of load contact. The failure of the rectangular block specimen is caused by the crushing of the specimen around the loading point, penetration of the indentor and eventually splitting of the specimen along the loading line.

2.2 *Wedge Indentor Angle*

The wedge angle of an indentor plays an important role in fracture mechanism. When the angle is 90° i.e under flat plate, the failure takes place due to tensile crack developed in the center along the line of loading. The mode of fracture gradually changes to mixed tension and extension with a decrease in wedge angle. In rectangular specimen, the crushed zone of the specimen around the loading point increases with the increase in the indentor angle from 20 to 60°. The crushing is followed by the penetration of the indentor. An indentor with a lesser wedge angle is able to penetrate deeper than the wider one. However, the failure for all angles was in extension during the final stage of splitting. The coal specimen offers less resistance when force is applied in the direction of the butt cleat plane than when in face cleat plane direction. In general, the force per length of the rectangular block increased for wedge angle from 20 to 45°.

2.3 *Confining Pressure*

When the coal specimen is confined along the butt cleat plane, it requires more force before failure occurs. However, confining the specimen along the face cleat exhibits no significant effect on failure load. Under low confining pressure, the specimen splits and fails in extension as the load increases to the level of failure load. High confining pressure causes the failure along a plane perpendicular to the direction of the wedge indentor (Fig. 2) and during such failure more chipping and higher debris of fragmentation develop (Khair, 1984a).

3 DYNAMIC INDENTATION

The dynamic indentation was simulated in the laboratory by a free falling bit from a height. Force due to impact was calibrated using a specially designed spring box as shown in Fig 3.

The spring was displaced by a free falling load from a fixed height and equivalent static load and kinetic energy were calculated, result of which are summarized in Table 2.

3.1 *Fragmentation Process*

The free falling bit indents and compresses the coal block. Initially, a tensile crack is developed by the tip of the bit. The crack

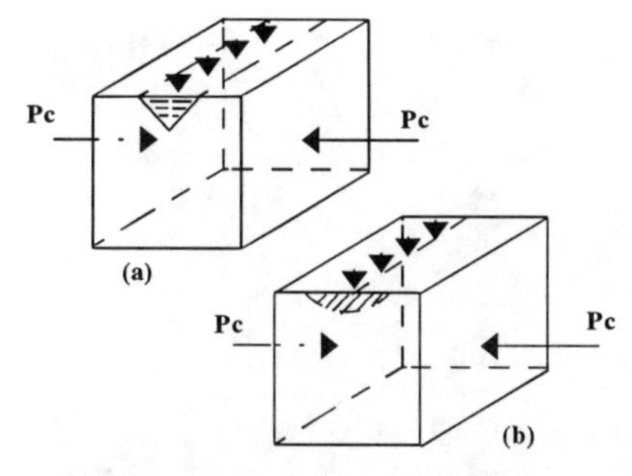

Figure 2. Direction of failure plane (a) parallel to line of action at lower confining pressure, (b) perpendicular to line of action at higher confining pressure

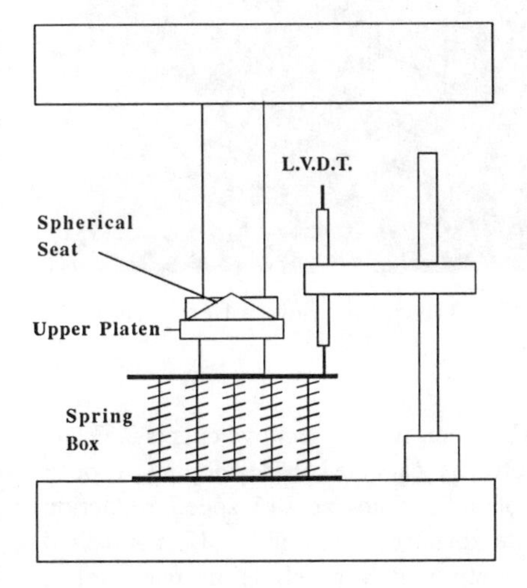

Figure 3. Spring box for calibration of impact force

Table 2. Calibrated force and energy

Drop Height (m)	Equivalent Static Load (kg)	Kinetic Energy (joules)
0.15	225	61
0.30	272	122
0.45	317	183
0.60	369	244

propagates until the kinetic energy is consumed. The bit body works against the friction offered by coal block while penetrating in it. This forms a crushed zone which is, in fact, a source of dust generation. The damaged area is in proportion to the applied kinetic energy, however, face cleat orientation produces a larger area as compared to the butt cleat. The crater formed due to dynamic indentation is classified into three zones as shown in Figure 4.

As a result of indentation the upper most surface is chipped off up to a depth of about 2/3 to 3/4 of the penetration depth. Zone 1 is the crushed area immediately in contact with the bit body. Below this zone, the coal or rock is fragmented due to propagation of radial tensile cracks up to about 19 mm and is termed as Zone 2. Thereafter, the cracks further extend, but no fragmentation occurs in Zone 3.

The subsurface damage was studied using OYO's Sonic viewer. The dynamic Young' modulus decreases due to sub-surface damage. The difference in dynamic Young's modulii before and after indentation increases with the confining pressure, indicating that higher confining pressure results in larger damage area.

The particle size is generally less than 400 mesh size and also depends upon the type of bit. Four types of bits were used (Fig. 5).

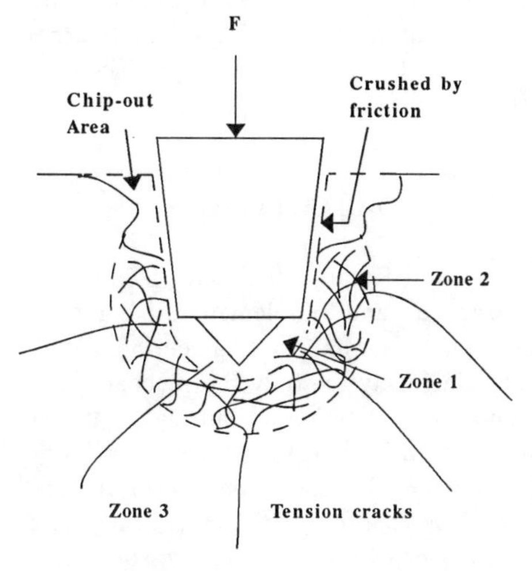

Figure 4. Schematic view of a crater formed due to dynamic indentation

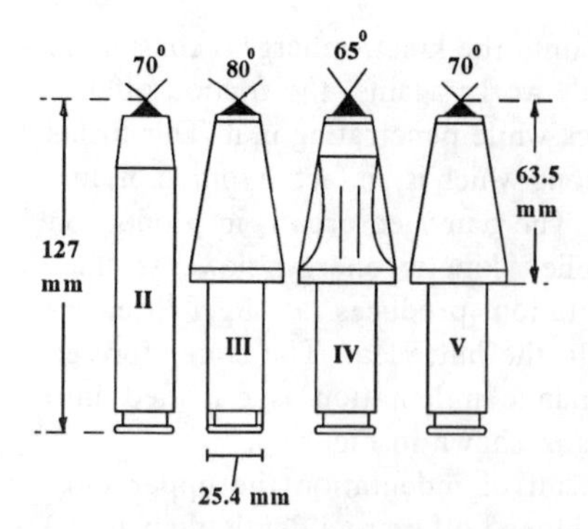

Figure 5. Different types of bits

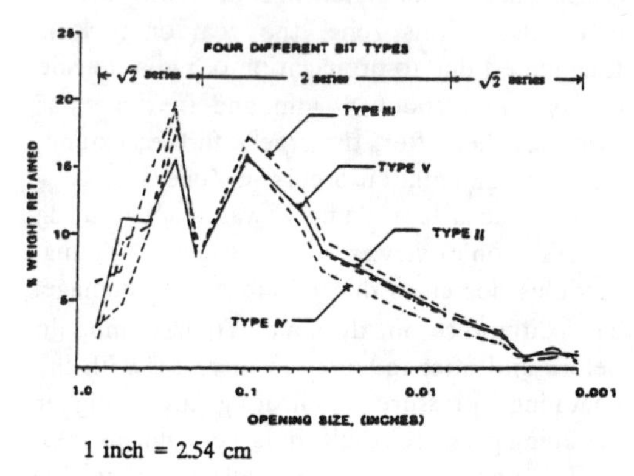

Figure 6. Size analysis of fragments for different types of bits

The bit type III with wider tip produced less larger fragments and more fines. The Sharper the bit the less fines were produced as illustrated in Fig. 6.

4 ROTARY COAL CUTTING

An Automated Rotary Coal Cutting Simulator was designed and implemented at the Department of Mining Engineering, West Virginia University to study the mechanism of fragmentation during rotary cutting. A typical view of experimental facilities is shown in Fig. 7. The equipment basically consists of a steel chamber. The coal or rock block is firmly fixed in this chamber with a confining pressure which can be regulated by the help of hydraulic cylinders. The cutting drum is fixed with 22 bit-blocks placed on its periphery in such a way that

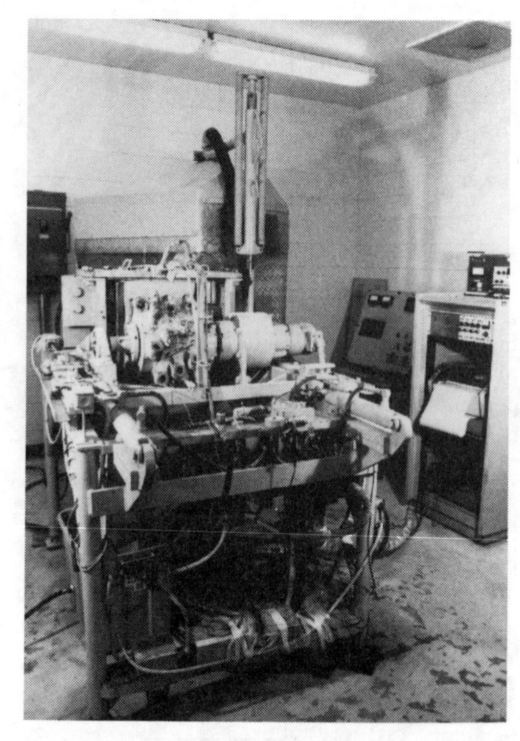

Figure 7. A typical view of experimental facilities

spacing between two consecutive cuts is 0.63 cm. The motor is capable of developing a peak torque of 3051 joules at 20:1 speed reduction ratio. The advance of cutting head is controlled by a remote switch which is prefixed for a specific distance of advance. The cutting pressure is automatically recorded on the chart recorder. However, the detail description of the machine is given elsewhere (Khair, 1984b).

4.1 *Mechanism of Fragmentation*

Fragmentation of coal during cutting with continuous miner is not a continuous process. Observation during rotary coal cutting in the laboratory indicated that it is similar to linear cutting, comprising of two phases, crushing and chipping. The mechanism of fragmentation during rotary cutting is illustrated in Fig. 8.

The bit first induces elastic deformation followed by inelastic subsurface cracking. This results in crushing of all the surface

irregularities. As the bit advances, the subsurface cracking continues. In contrast to the rock, coal has inherent weak planes in the forms of cleats. The length of subsurface cracks is influenced by these weak planes as they provide free face and helps propogating cracks to break prematurely, producing fine fragments and dust particles. As the bit advances, the subsurface cracking makes the uncrushed material steeper and steeper inducing a high cutting resistance. When the cutting resistance reaches high enough to overcome the strength of the coal, a major crack develops under the tip of bit.
This crack propagates all the way up to the free surface and forms chips. Some fine fragments and dust particles which escaped from clearing off get deposited on the surface of the new coal face due to electrostatic attraction. Due to heat, these particles are plastered onto the rough surface of the coal surface as a thin layer of film (Reddy, 1988). The thickness of plastered material ranges from 0.8 to 1.6 mm (Khair and DeVilder, 1986). This layer acts as a major source of dust generation during subsequent cut.

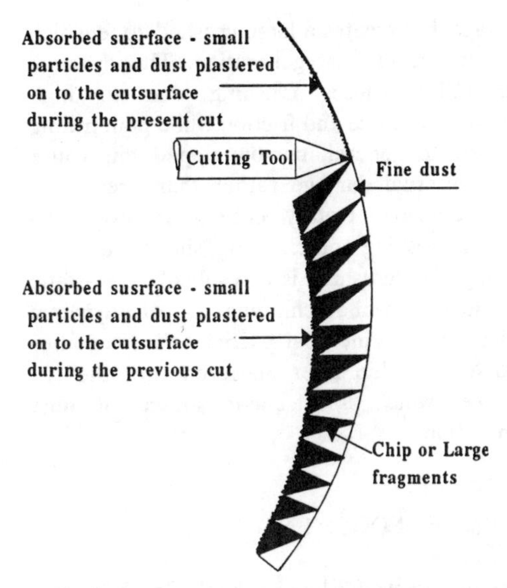

Figure 8. Fragmentation mechanism during rotary cutting

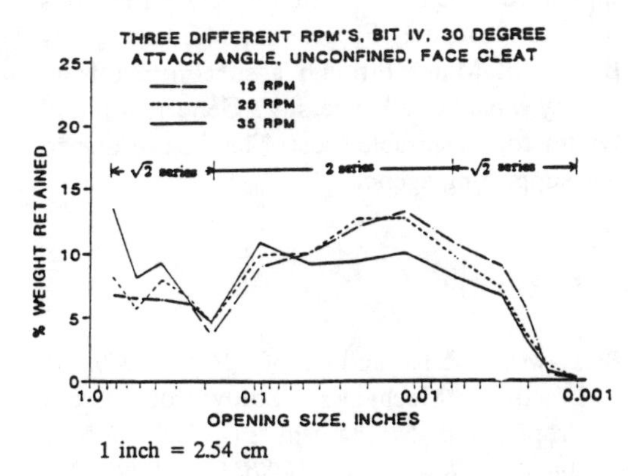

1 inch = 2.54 cm

Figure 9. Size distribution at different drum speed

4.2 *Effect of operating parameters on fragmentation*

During rotary cutting the bit produces a larger damage area and more fines. The frictional resistance increases with the increase in spacing between the two cuts, resulting in more fines.

The higher cutting velocity makes the crack propagate up to a greater depth, producing large fragments. As the velocity increases, the production of fines decreases and large fragments increases (Fig. 9). However high drum velocity may cause a high fanning action and may give rise to more airborne respirable dust.

Due to confining pressure, the coal or rock specimen offers high resistance in the cutting path resulting in more rough fracture surface. The higher confining pressure produces less fines in case of butt cleats orientation. However, there was hardly any difference in face cleat orientation.

The bit attack angle has apparently shown no difference in size distribution. The face cleats produce more large fragments and less fines than butt cleats at 6mm depth of cut. However, no difference is observed at a depth of cut of 0.8mm.

5 CONCLUSIONS

The mode of rock failure under quasi-static indentation changes from pure tension to mixed tension and shear as the confining pressure is increased. The confining pressure restricts the propagation of fractures thus restricting internal damage, but creates a large crushed zone which is a source of dust generation. The dynamic indentation causes crushing due to high compressive force and friction while penetrating deeper. In general, narrow tipped bits cause more internal damage rather than creating a crushed zone. Rotary cutting involves two phases, crushing and chipping, similar to linear cutting. The crushing is a result of quasi-static indentation while chipping is of dynamic indentation. Fine coal particles fuse together and form a thin layer plastered over the cut surface, which is a major source of dust generation.

ACKNOWLEDGEMENT

The authors would like to thank Dr. D Xu and Mr. M. Ahmad for their assistance in the research study. The research study has been supported by U.S. Department of the Interior's Mineral Institute Program administered by Bureau of Mines through a subcontract from Pennsylvania State University's Generic Mineral Center for Respirable Dust. Thanks are due to this supporting agency.

REFERENCES

Bieniawski, Z.T. and Zipf, R.K. 1985. A Fracture Mechanics Study of Crack Propagation Mechanism in Coal. *Second Annual Report on USBM Grant No. G1135142, No. 4201 (PSUI)*, Oct., 116 pp.

Khair, A.W. 1984a. Study of Fracture Mechanism in Coal Subjected to Various types of Surface Traction Using Holographic Interferometry. *Proc. of the 25th U.S. Symp. on Rock Mechanics*, Northwestern University, Evanston, IL, pp 103-114.

Khair, A.W. 1984b. Design and Fabrication of a Rotary Coal Cutting Simulator. *Proc. of the Coal Mine Dust Conference*, West Virginia University, Morgantown, WV, pp 190-197.

Khair, A.W. and Devilder, W.M. 1986. Correlation of Fragment Size Distribution and Fracture Surface in Coal Cutting Under Various Conditions. *Proc. of the Int. Symp. on Respirable Dust in Mineral Industry*, The Pennsylvania State University, University Park, PA, 46 pp.

Newmeyer, G.E. 1981. Cost of Black Lung Program. *Mining Congress Journal* Vol. 67, No. 11, Nov., pp 74-75.

Reddy, N.P. 1988. Characterization of Coal Breakage by Continuous Miners. *Ph.D Dissertation*, West Virginia University, Morgantown, WV, 189 pp.

Geomechanics 93, Rakowski (ed.) © 1994 Balkema, Rotterdam, ISBN 90 5410 354 X

Development of a non-explosive mining system for underground hard rock mining

F.L.Wilke & F.Spachtholz
Technical University of Berlin, Germany

ABSTRACT: The common mining method in hard rock mines is drilling and blasting. Blasting vibrations, the resultant surface or mine damages and the time for smoke clearing are some reasons to substitute a mechanical - breaking or cutting - mining system for heading by blasting. Therefore the application of an impact ripper is proposed because of the small tool wear and the possibility to utilize rock properties like jointing and layering. In compact hard rock the breaking efficiency will be significantly enhanced by the introduction of additional free faces to which the tool can break. To produce such free faces the practical and economical machines are diamond wire saws and circular saws. The paper describes the theoretical backround and first results obtained from underground tests in a base metal mine.

1 REASONS FOR A NON-EXPLOSIVE MINING SYSTEM

In the past it was habitual to use explosives wherever rock had to be freed or broken and until today a great deal of effort has been put into search for greater productivity in rock excavation. Drill and blast methods have been speeded up mostly by better drilling techniques, but the progress is still slowed down (Horton, 1993, 20-21). The use of explosives in underground hard rock mines is one of the major potential sources of accidents, and explosives dictate a discontinuous, cyclic mining system on the mine which does not make the most efficient use of haulage and processing machinery (Shaw, 1988). In addition, there is lost waiting time after the blast for smoke clearing. Unless the blast is very tightly controlled, and the rock mass is completely cosistent, time and money must be spent on rock support. Besides, mine and quarry operators are coming under increasing public pressure to minimize blasting to control shock waves and to reduce noise, dust and general nuisance to their neighbours. Also, the physical risks, regulations and safety routines involved in the handling, storage, transport and use of chemical explosives have always been, if not a problem, certainly an unproductive expense (Horton, 1993, 22-26). And in special cases where the site is adjacent to vibration-sensitive structures like dams or close to the surface below nature reserves, the operators are anyway compelled to use quite different techniques for rock breaking because of the resulting surface or mine damages as well as the mixture of ore and surrounding rock.

2 NON-EXPLOSIVE BREAKING SYSTEMS

For the reasons described above, there is an incentive to develop systems which do not require explosives for breaking. Breaking methods not including explosives may be summarized as follows in table 1.

To date, roadheaders have been the main alternative to drilling and blasting techniques for development. Their economic application in harder, more abrasive ground than that encountered in coal mining is limited. Cutter wear increases rapidly when the rock becomes harder and/or more abrasive (Chadwick, 1993). Generally, a rock with a compressive strength up to 120 MPa is cuttable with roadheaders and the maximum wearing factor for rock to be cut economicly is 0.5 N/mm (Warnecke, 1993). The Tunnel Boring Machine (TBM) has seen little application by the mining industry, largely because they are restricted to circular profiles which are not ideal for haulage applications. A considerable amount of research and development time and money has been put into mobile mining machines which will be capable of excavating rock with an unconfined compressive strength of 250 MPa (Chadwick, 1993). But the mobile miner can not utilize rock properties like jointing and layering just as roadheaders and TBMs.

Tab. 1 Breaking methods not including explosives

Breaking System	Methods of use	Results
Steady pressure & shear		
1. Diamond bits	Saws, corers	Cut kerf, core
2. Rockcutter	Drag bit cut	Cut kerf
3. Raise borer	Bores hole	Full ore removal
4. Large dia. drill	Bores hole	Full ore removal
5. Mobile Miner	Disc cutters	Full ore removal
6. Roadheaders	Pick cutters	Full ore removal
Impact		
1. Rock drill	Small holes	Cut kerf
2. Swing hammer	Hammers face	Two pass, full cut
3. Impact Ripper	Hammers face	Full face cut
4. Projectile	Hammers face	Full face cut
Thermal		
1. Plasma torch	Drills hole	Kerf cutting
2. Flame jet	Drills hole	Kerf cutting
3. Lasers	Fractures rock	Kerf cutting
4. Electron beam	Fractures rock	Kerf cutting
5. Microwaves	?	?
Chemical		
1. Bristar-1000	Reaction in hole	Pre-fracture
2. Additives	Cutting water	Assist cut
Electric		
1. Electro-hydrodynamic	In hole fractrue	Pre-fracture
Water-jet		
1. abrasive waterjet	Cuts rock	Kerf cutting

There are numerous other methods of breaking. Most of them have not been tested for anything other than their rock-breaking capability, no attempt has been made yet in many cases to try them out in a mine (Shaw, 1988).

Contrary to that, impact rippers have been tested in several ways and it seems that the most common alternative to conventional drilling and blasting techniques is the hydraulic hammer mounted on a movable base (hydraulic excavator or wheel loader).

Hydraulic hammers can be pedestal mounted at crusher stations for clearing blockages and they can even be used underwater if they are suitably modified. In the mining industry the impact ripper is used to break rock directly off the full face (Bartels, 1992). Especially in German hard coal mines gate roads kept on the line are frequently headed with impact rippers.

The hydraulic hammer is fast, efficient and flexible. The tool wear is very small and the impact ripper can utilize rock properties like jointing and layering. Provided that the rock is prefractured by pre-existing fracture patterns by mechanical fracture methods (diamond wire saw or circular rock saw) the impact ripper could be an economic competitor in stoping in the near future.

3 THE EC RESEARCH PROJECT

The primary objective of the research project "Non-Explosive Mining Systems For Hard Rock Mines" was to develop a hard rock breaking system with the use of diamond saws and impact ripper. The work at the Royal School of Mines, London was concentrated on the development of the theories behind the rock cutting and the optimization of the use of diamond tools. Laboratory sawing tests were carried out at Diamant Boart's pilot station in Brussels. Underground trials of the proposed rock breaking system were carried out in the test mine at Laporte Minerals near Sheffield. The research work at Technical University of Berlin was to investigate and evaluate the performance of the impact ripper and to investigate the application of the impact ripper in vein ore deposits.

4 INFLUENCES ON THE PERFORMANCE OF FREEING

The main influences on the performance of freeing (Fig. 1) achieved by the impact ripper were determined qualitatively (Korf, 1992; N.N., 1986; N.N., 1992, Firmeninformationen). They are in the fields of equipment, mode of operation, stope layout, characteristics of free faces, rocks and formation as well as mining method.

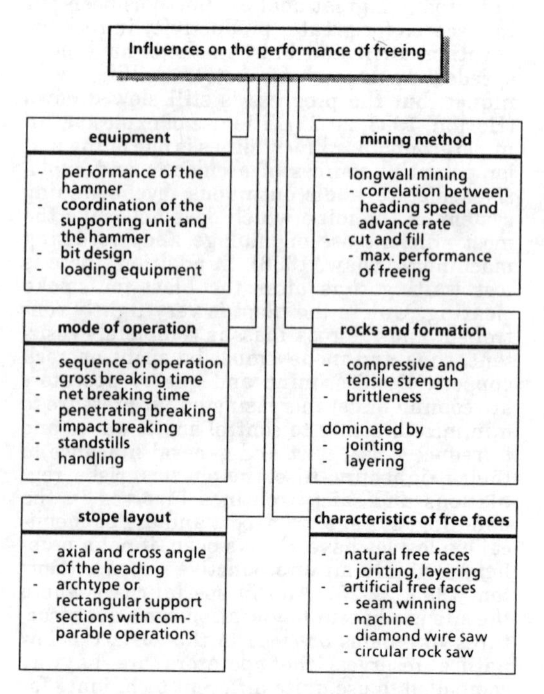

Fig. 1: Influences on the performance of freeing

4.1 Equipment

The performance of freeing increases with the performance of the hydraulic hammer. The performance of the hammer P can be influenced by variation of the single impact energy E (140 to 12,000 J) and the impact frequency z (1,500 to 300 1/min) depending on the properties of rocks. In hard rocks it is better to work with a low impact frequency and a high single energy.

With increasing single impact energy the service weight of the hydraulic hammer and the supporting unit increases in a linear way. The best weight relation of the supporting unit and the hydraulic hammer was determined in the range of 13 or 18 to 1.

In hard rocks it is better to use a low bit junction surface like a moil point or flat chisel than spades or blunt bits while penetrating breaking. When impact breaking (boulder breaking) a blunt bit is necessary.

Because impact rippers can only bring up the rock a further loading device is needed. This gives the impact ripper a sufficient mobility and the operator a good view to the rock face. In German hard coal mines usually sidetipping loaders or face-conveyors or combinations of both are used.

4.2 Mode of operation

The operation in headings can be divided in breaking, loading, timbering and standstills. In German hard coal mines the gross breaking time is about 7 % of the total cycle time. Compared to that the time for timbering takes about 70 %.

The gross breaking time can be divided into the net breaking time, when only loosening work is done, and the time needed for the correct positioning of the boom and the bit. In German hard coal mines the net breaking time is in the range of 75 to 90 %. The absolute time for the positioning of the boom respectively the bit is in the range of 5 to 10 seconds.

In the course of penetrating breaking only the apex or the blade of the bit is penetrating the rock. This causes a cleavage procedure towards a free face. In the course of impact breaking the blunt bit introduces a strong mechanical pressure into the rock surface. Due to this mode of operation multiple free faces are needed. The compression waves will be reflected at the free faces. The result are tensile waves which will destroy the rock.

The handling of the impact ripper is a very important point. Education, experience and motivation of the operator influences up to 50 % of the performance of freeing.

4.3 Stope layout

The axial angle of inclination of gate roads should not exceed 15 gon and the cross angle of inclination 10 gon. But there seems to be no other limit for the application of impact rippers in headings with rectangular or archetype support.

Due to the different bridging effects at the rock face the efficiency of an impact ripper depends on the point of the bit attack. Therefore in German hard coal mines gate roads kept on the line the road face is subdivided into sections with comparable operations (Fig. 2); these are the basement rock under the seam, the roof above the seam and the surrounding rock of the roadway.

An internal study of the German hard coal mine Monopol shows that the average numbers of bit attack for profilizing the surrounding rock are nearly twice as much for working at the roof above the seam to receive the same rock quantity.

4.4 Characteristics of free faces

As mentioned about the breaking mechanisms using an impact ripper there is a need for free faces. Heading gate roads kept on the line the

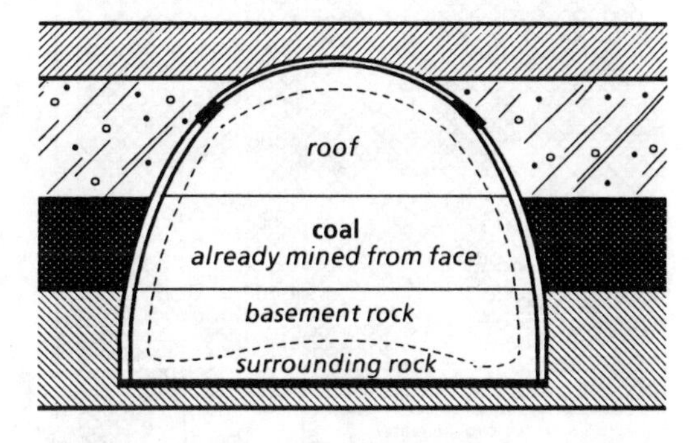

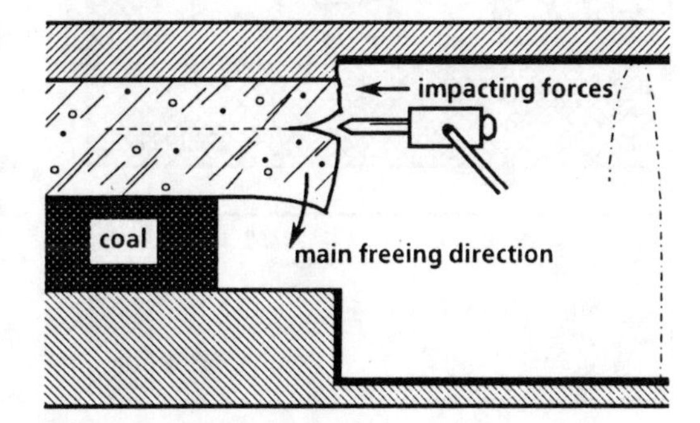

Fig. 2: Driving a roadway by impact ripping

free face is determined by the seam thickness and cutting depth of the seam winning machine. Artificial free faces have to be created, if natural free faces (jointing and layering) or free faces as described above do not exist. The amount of free faces needed depends on the degree of weakness of the rock and the field of effect of the applied breaking mechanism.

Creating these free faces the application of both diamond wire saw and circular rock saw as secondary freeing machines supporting the impact ripper in vein ore deposits is possible.

A diamond wire saw was tested in underground at Milldam Mine, Laporte Minerals. During these tests Pelligrini's Televar TVD 55 diamond wire saw was used to cut slots of the side walls of a rectangular raise. The sawing layout for cutting the walls parallel to the drive where the saw machine was installed is shown in Fig. 3.

The results showed that an average cutting rate of 8.5 m2/hr can be achieved for sawing this kind of grey limestone with an average

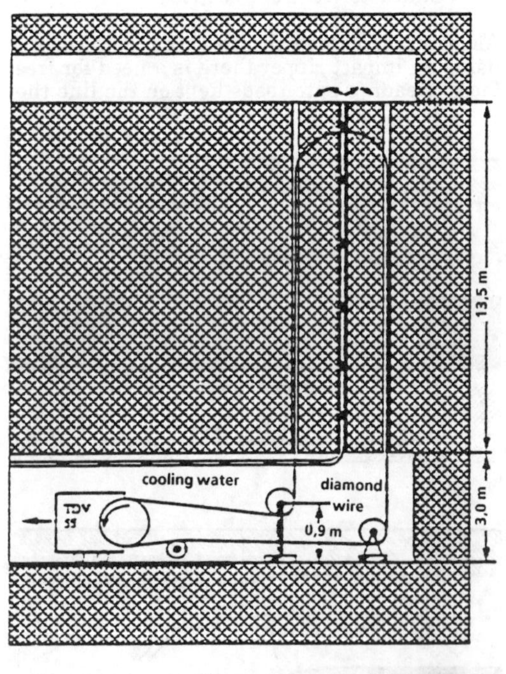

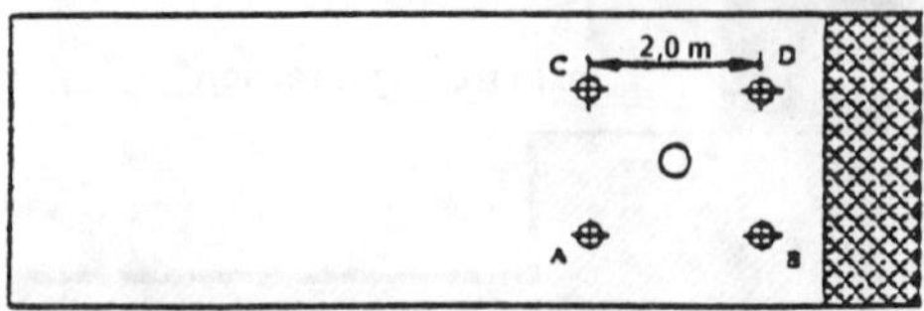

Fig. 3: Sawing layout for parallel cuts (A-B, C-D)

compressive strength of 150 Mpa (Shaw, Wilke, Bramley, Thoreau, 1992). At present, experimental and theoretical work will be carried out on diamond wear mechanisms to determine the optimum tool life.

To support the impact ripper as a secondary freeing machine the diamond wire saw will be economical while producing horizontal and/or vertical slots along the whole length of the proposed cross cut headings (25 m).

Contrary to that, the position of the free faces had to be reconsidered from advance to advance while working with a circular rock saw. This might be of great interest because geometry as well as rock and formation properties of vein ore deposits can change significantly on short distances. Different studies on the kinematic of the circular rock saw (Singewald, 1992; N.N. 1992, Sägelafette) proved that with minor construction changes the proposed supporting unit (Webster Miner 2000) suit as the basic tool for the circular rock saw and the hydraulic hammer.

4.5 Rocks and formation

Basically, the performance of freeing decreases with an increasing compressive strength (Hermann, 1987) and an increasing ductility. Working with a Krupp hydraulic hammer HM 800 in compact sandstone (130 MPa) the achieved performance of freeing is in the range of 8 m3/hr and working in weak shale (20 MPa) the performance of freeing ranges up to 13 m3/hr.

But different applications of impact rippers demonstrated that the rock properties were dominated by the formation parameters jointing and layering. In spite of high compressive strength and high ductility the work of freeing becomes easier with an increasing degree of jointing and layering of the formation.

4.6 Mining method

In impact ripper headings in German hard coal mines there is a close correlation between the heading speed and the rate of advance of the corresponding face. To ensure that the heading advance is at least as high as the advance of the face, the possible maximum performance of the heading is not used. With 3 shifts the average heading performance is about 3.5 m/d in headings with an average internal cross-section of 22 m^2 and an average rock cross-section of 12 m^2.

In hard rock mines with cut and fill mining methods such a correlation is unknown. So, the maximum performance of an impact ripper will be reached.

5 IN SITU TESTS AT MILLDAM MINE

The performance of freeing achieved by an impact ripper in cross- cutting headings in a fluorspar mine had to be examinated depending on different test situations. The test goal was the evidence of suitability of an impact ripper in vein ore mining. Especially, it must be tested whether the performance of freeing of the impact ripper is a function of direction, number and thickness of the free faces. In addition, this enables a quantification of the in chapter 4 qualitatively described influences on the performance of freeing.

5.1 Geology

Milldam Mine, one of Laporte Minerals' Glebe Mines is located near Stoney Middleton in Derbyshire (England), within the Peak District National Park. Fluorspar mineralization is found mostly on the eastern side of the Derbyshire dome. The deposits are of fissure vein type and are restricted stratigraphically to Visean limestones of the Carboniferous System. The workable depth of vein is usually about 100 m. Veins can vary in width between 1.5 and 12 m over relatively short strike length. They are composed of varying proportions of fluorspar, barytes, calcite and galena (Bramley, 1990).

5.2 Planned mining method

The in-vein sub-level stoping mining methods had become unsafe, due to the incompetence of the vein. Furthermore, the overlying weak shale strata had collapsed in a number of places with two deleterious effects: mixed with the ore it had rendered the mill flotation process inoperable, and the collapse had caused subsidence holes in the sensitive landscape.

To avoid subsidences in the nature reserve (Peak District) an underhand mining operation protected by an artificial roof is planned.

Starting from two sublevels different drifts will be cross cut through the vein for 3 meters into the limestone to obtain an abutment for a reinforced concrete level. This work will be done by the impact ripper protected by the diamond wire saw or a circular rock saw. In addition, up to 4 m before the vein (footwall) drilling and blasting is possible as the limestone is a very competent rock. Then vertical boreholes have to be produced to join the floor of the upper drift to the roof of the lower ones with a diamond wire cut. After creating the artificial level the drift will be filled and after being consolidated on one side of the filled drift a new drift can be flanked off. The vein ore between the sublevels will be triggered with a mobile front loader because of the minor hardness (Mohs hardness=4) protected by the artificial roof.

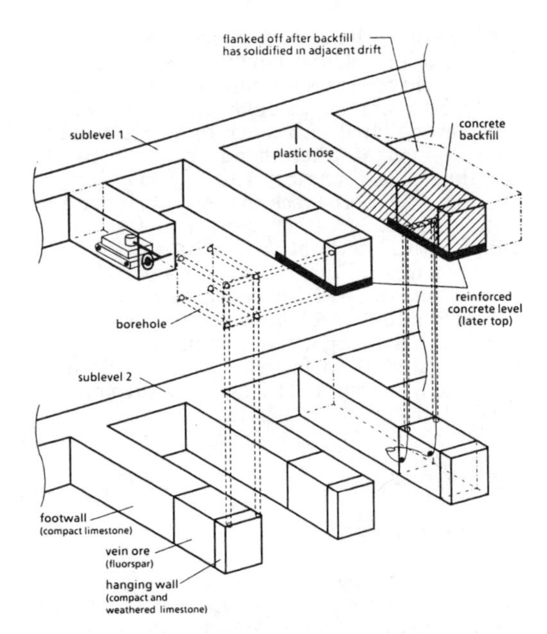

Fig. 4: Underhand cross-cut with filling

Having finished mining between these two sublevels mining progresses into the depth (Fig. 4).

5.3 Short description of the tests

The in situ tests were realized with an impact ripper which combines the Webster Miner 2000 and the Krupp Hammer HM 560 CS and several different bits (pyramidal moil point, flat chisel, profile chisel, blunt bit). The Webster machine was powered by a 112 KW, 1100/550 V motor. The hydraulic pump has a capacity of 350 ltr./min. The maximum oil pressure in the hydraulic circular is 170 bar.The weight without hammer is 15.500 kg, the length is 8.8 m, the width is 1.6 m and the height is 2.2 m. The hammer has a weight of 825 kg. The working oil flow is 110 ltr./min and the working pressure 170 bar. The impact frequency of 850 blows/min is adjusted to the hydraulic system of the Webster Miner to get the maximal single impact energy of 1,700 J. The necessary hydraulic feed force of the hammer is about 10 KN.

In order to create extra free faces Pelligrini's Televar TVD 55 diamond wire saw (kerf) and the PTT drilling rig (borehole perforation) were used. The application of a circular rock saw was not possible. Instead, the needed free faces were established by closely drilled boreholes. This perforation should be comparable to the sawing cuts concerning the support of the impact ripper. The muck was conveyed by the Torro 250 BD mobile front loader.

Altogether, the impact ripper was tested in four headings. Three headings with an excavated archtype cross-section of 13 m^2 had already been mined through the vein, which were consolidated with concrete. There was the scope to elongate these headings for 3 m into the limestone (hanging wall). Depending on the properties of the rock the impact ripper had to work to faces without artificial free faces, with vertical perforation, with vertical (distance = 1 m) and horizontal (distance = 0.7 m) perforation and with blowed spaces (Pentaflex 100 detonating cord) between the boreholes. The fourth heading with a rectangular support was prepared in the footwall with a horizontal diamond wire saw cut (1.2 m above the bottom) to which the impact ripper could break.

5.4 First results

5.4.1 Servicing

The personnel operating the impact ripper was not as experienced as those typically found in a German hard coal mine, where ripping is routine and a considerable operator experience is gained. Missing experience was particularly evident in terms of slewing and positioning the hammer at an appropriate position and attitude. It is likely that with more experienced operators the single slewing time could be three times quicker (not 20 sec but 7 sec).Thus, improved operator experience will double output from the ripping system.

In addition, during the tests it became obvious that the productivity could be improved by modifications of the impact ripper. This can be achieved with a better position for the driver seat to give the operator the possibility to see the actual point of bit attack. Also an additional swivel axis only for the hammer at the front of the boom would be of great interest just for profilizing the side walls.

5.4.2 Interruptions

Almost 33 % of the test time was spent on interruptions. In table 2 the percental time and the reasons for the interruptions are shown.

5.4.3 Performance of freeing

The performance of freeing for each particular advance is shown in figure 5.

The achieved performance of freeing in the main cross section (3I) was 2.8 times larger than for profilizing the side walls (3IIA, B, D). During the work at the main cross section the size of material broken down generally ranged up to about 30 to 50 cm but profilizing it is in

Tab. 2: Percental time and reasons for interruptions

time [%]	reason	activity
49,8	discussion	
28,6	mechanical	- maintaining - reparing - adjusting
12,5	loading	pulling back the muck from the face with the impact ripper for better positioning of the bit
4,2	mining	greater rock pieces were pinched at the roof which had to be plucked
3,5	measurements	pictures, videos, ...
1,5	driving	driving the supporting unit for a better positioning of the boom and the bit

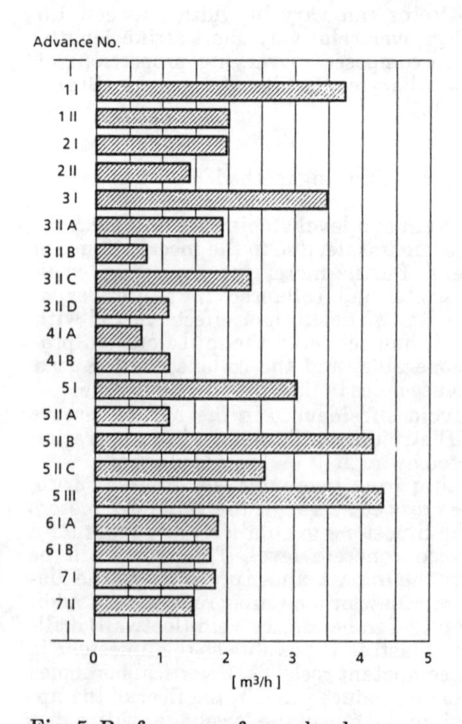

Fig. 5: Performance for each advance

the range of 5 to 10 cm. In the hard limestone of the main cross section the distance between the points of bit attack should not exceed 20 cm but while profilizing it is in the range of 5 to 10 cm.

In addition, on an average for working at the main cross section is a need for 20 to 35 points of bit attack to get 1 compact m^3 and for profilizing 65 to 95 points.

Working in the main cross section the achieved performance of freeing was 1.5 times larger than for the roof (3IIC). That means that the performance of freeing for the roof was 1.8 times larger than for profilizing the side walls. This possibly results from the fact that it was not necessary to drive the machine in another position while working on the roof. Furthermore, positioning the hammer in the roof section was much easier.

Contrary to the hypothesis which was made about the borehole perforation there was no remarkable difference between headings with (6IA, B) or without (2I, II) a perforation in the hard limestone. But after blasting the spaces between the boreholes (5I) the performance of freeing was 4.4 times larger when impact breaking with the blunt bit and 6.2 times larger when penetrating breaking with the flat chisel (5IIB, III) which was the chisel with the best performance. The moil point (1I, II) and the profile chisel (5IIA) penetrated very fast into the rock that they were pinched. In addition, the temperature of the top of these chisels increased to such a point that plastic deformations ocurred while ripping more than 15 seconds on one point.

Also it was found that the performance of freeing using a diamond wire saw slot (7I, II) with a thickness of 1 cm was 1.8 times higher than without such a slot (4IA, B) but with the same chisel in the same hard limestone.

With the comparisons before the following hypothesis could be made provided that the slot is open for the whole working time:

$$P' = (x + 1) * P \quad [1]$$

where P' = performance of freeing with slots; x = number of slots; and P = performance of freeing without slots.

For the underground mining operations at Milldam Mine there are only 390 min/shift effective working time in a 8-hour shift. The average hammering time for the impact ripper would be 350 min/shift and for loading 40 min/shift. In the limestone a drifting rate of nearly 13 compact m^3/shift (1 m/shift) could be reached with an average performance of freeing of 2.1 compact m^3/hr. The average drifting rate achieved by a roadheader (based on the Webster Miner) in weathered limestone with jointings (nearly the vein) is also 13 compact m^3/shift and by drilling and blasting operations in the limestone 26 compact m^3/shift.

6 FURTHER ACTIVITIES

To make more profound statements for the achievable performance of freeing as a function of the position of the free faces, especially, those built by diamond wire saw and circular rock saw further in situ tests will be necessary. Horizontal and vertical free faces will have to be produced (Fig. 4) for a better profilizing work and to reduce the amount of the necessary mechanical energy of the impact ripper

In addition, a modified impact ripper will be tested in another base metal mine where the cut and fill mining method is used. This method (Fig. 6) offers an additional, procedual involved free face comparable to German hard coal mines while upwards stoping.

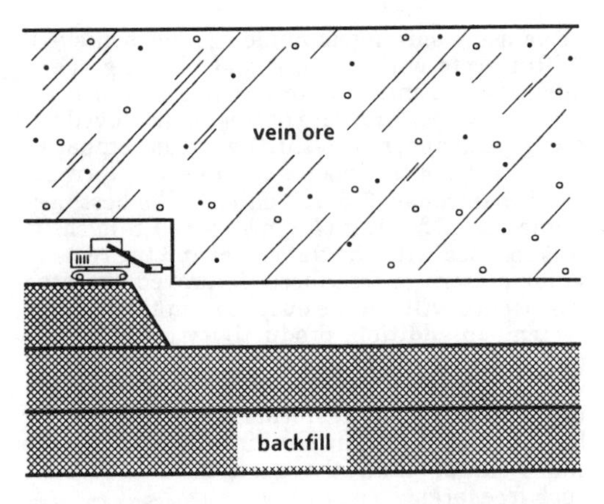

Fig. 6: Cut and fill with an impact ripper

Besides, the interdependence between the achievable performance of freeing and the abrasiveness of rock has to be examined.

7 CONCLUSIONS

The aim of the research activities is to develop an underground mechanical mining method for hard rocks with comparable costs to drilling and blasting operations. One of the most common alternative to conventional drilling and blasting techniques is the impact ripper supported by a secondary freeing machine. In German hard coal mines the observed influences on the performance of freeing achieved by an impact ripper are in the fields of equipment, mode of operation, stope layout, characteristics of free faces, rocks and formation as well as mining method.

To avoid subsidences in the Peak District National Park an underhand mining operation protected by an artificial roof with a primary (impact ripper) and secondary (diamond wire saw, circular rock saw) freeing machine is planned at Milldam Mine. The paper describes the first results of the impact ripper tests at Milldam Mine.

The most important statement which could be made is that the suitability of an impact ripper in vein ore mining is evident. The achieved drifting rate with drilling and blasting operations is 2 m/shift and only twice as much as the average performance of freeing achieved by the impact ripper (1 m/shift) what is comparable with impact ripper headings in German hard coal mines (3.5 m/3 shifts). The best performance of freeing (4.3 m^3/hr ≈ 1,8 m/shift) was reached with the flat chisel and the rectangular pattern of free faces. Improved operator experience will double output from the ripping system. In addition, productivity could be improved by modifications of the impact ripper. A very interesting hypothesis could also be set up: The achieved performance of freeing in headings with x additional free faces is (x + 1) times larger than the performance without such free faces.

The impact ripper could be an economic competitor for stoping in the near future. Therefore, the impact ripper must be tested in other base metal mines where the cut and fill mining method is used.

REFERENCES

Bartels, R. 1992. Wer richtig hämmert, hämmert am längsten. Sonderdruck bd-baumaschinendienst. 2.

Bramley, J.V. 1990. Fluorspar Mining in Derbyshire, reprinted from Mining Magazine.

Chadwick, J. 1993. Hard rock mobile mining. Mining Magazine.1: 27-29.

Hermann, A. 1987. Hydraulik-Hämmer Entwicklung, Einsatz, Leistungen. Bergbau. 3. 109-113.

Horton, N. 1993. Plasma blasting - a breakthrough for rock breaking? World Mining. 2: 20-21.

Horton, N. 1993. No blast, no damnation. World Mining Equipment. 2: 22-26.

Korf, P. 1992. Datenauswertung aus der ZPD der Ruhrkohle AG, Herne, unveröffentlicht.

N.N. 1986. Betriebsstudien von Ripperörtern der RAG, unveröffentlicht.

N.N. 1992. Firmeninformationen von Krüsser Stahlhammerwerk, Krupp, Montabert, Hausherr, Stanley, Teledyne, Webster, Rammer.

N.N. 1992. Sägelafette auf mobilem Trägergerät und Sägewagen für die Blockgewinnung im Schiefer, Schellinformationsprospekt, Höpfingen.

Shaw, C.T. 1988. Future develoments in stoping methods with special reference to non-explosive methods of rock breaking. Bergtechnik des Untertagebaus: 409-414.

Shaw, C.T., Wilke, F.L., Bramley, J.V. Thoreau, B. 1992. Third report on progress of research activity for the EC Raw Material Research Programme: Non-explosive Mining Systems for Hard Rock Mines.

Singewald, C. 1992. Naturwerkstein - Exploration und Gewinnung. Steintechnisches Institut Mayen, Rudolf Müller Verlag, Köln.

Warnecke, G. 1993. Application of roadheaders for mining the Eagle Point Uranium Deposit. Diplomarbeit TU Berlin. 8-11.

Geomechanics 93, Rakowski (ed.) © 1994 Balkema, Rotterdam, ISBN 90 5410 354 X

New technology for excavating of vertical workings in rock mass

F. Sekula, T. Lazar & M. Labaš
Institute of Geotechnics of the Slovak Academy of Sciences, Košice, Slovakia

ABSTRACT: The paper analysis the new technological principle of sinking of vertical workings by means of the thermal melting of rocks. The theoretic geomechanic tasks of the radial fracturing of the rockmass for a possibility of the transport of the molten rock into the arisen cracks are analyzed. Also the theoretical tasks of the controlling of the process from the point of view of the securing of the continual optimal advance of the flame injector as the transferrer of the thermal energy into the rockmass are analyzed.

Contemporary state of the technics and technology of sinking of vertical workings is limited at the conventional mechanic systems of disintegration by some border of the maximum obtainable productivity which is given by the possibility of the speed of the transport of disintegrated material from face of the working to the surface. The same concerns the thermal techniques of the disintegration where under the influence of the thermal energy the mechanical disintegration by means of the thermal stresses occurs again. Also in this case the debris of the disintegrated rock has to be transported under the speed corresponding to the speed of sinking of the working. The speed of the transport of debris is limited due to the confined space between the working tools and diameter of the working. Due to this fact the disintegration is strongly dampened and limited, and so the penetration rate of the particular specialized disintegrating system at these technologies is also strongly dampened and limited. The above presented facts are documented in the research reports (Sekula, 1980; Krupa, 1984) considering the mathematical modelling of mechanical drilling, of the Institute of Geotechnics where it was shown that due to increasing of the actual penetration rate of the drilling, respectively with decreasing of the drilling resistance $R=R_1$, the parameter of the dampening C and C_1 (see Fig.1) is rising. The Figure 1 documents the fact the process of the dampening at the small drilling resistance becomes to be infinitely large.

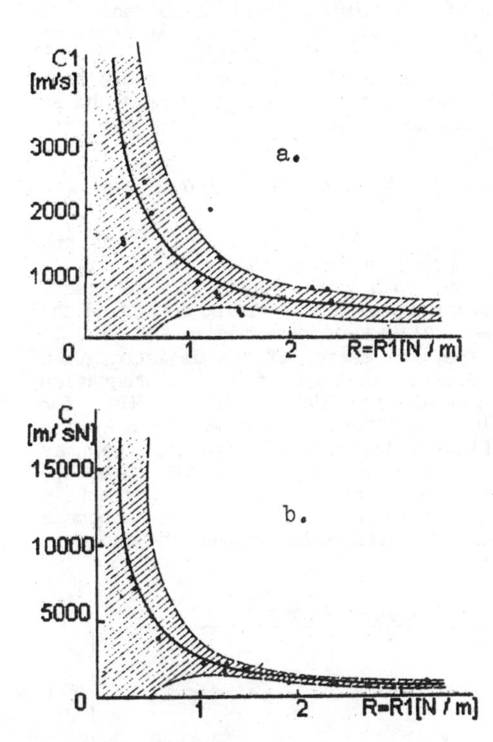

Figure 1. Parameters of dampening.
a/ constant C_1 vs drilling resistance $R=R_1$
b/ constant C vs drilling resistance $R=R_1$

The overcoming of this limiting range, respectively the move of this technology into the new technical dimensions, is possible to achieve in present time by removing of the transport of the material created by the disintegration of the volume of the working. The removing of this complex transport is possible to reach by the fracturing of the rockmass by means of the highpressure mediums and the melting of the rock. In this case the molten rock would penetrate the radial cracks in the rockmass under the high pressure of the combusted mediums and a part of this molten rock would be transported into the incrust lining replacing the conventional lining. This system would improve the productivity and effectiveness of the technology.

For the working with the circular cross-section is necessary to create the tangential tension stress in the wall of the working in order to fracture the rockmass. The value of this stress for the rocks of the highest strength can be expressed by the relation

$$\sigma_t = \sigma_H + (p_0 + \sigma_H) \cdot \left(\frac{a}{r}\right)^2 \qquad (1)$$

, where σ_t is the tangential fracturing stress [MPa], σ_H is the tangential component of the stress caused by the pressure of the overburden [MPa], p_0 is the internal pressure of the mediums [MPa], a is radius of the working [m], r is the radial distance from the center of the working [m].

The tensile strength for the rocks with the highest strength is about 10 MPa and sc the fracturing pressure is then approximately

$$p_0 = 10\ MPa + 0.5\ \rho . g . h \qquad (2)$$

, where ρ is the specific weight of the rocks [kgm^{-3}] and h is the depth [m]. With the depth of the working the necessary fracturing pressure is rising according the linear relation illustrated in the Fig.2. The specific energy of the disintegration at the mechanical systems of disintegration is ranged from 100 to 10.000 MJm^{-3} for various diameters of the working. At the melting of the rock the specific heat is about 5.000 MJm^{-3}, besides the carbonate rocks where this value is approximately twice as high as at the non-carbonate rocks. The difference between the mechanical disintegration and the melting of the rocks from the point of view of the energy demands is very high. This unfavorable effect is possible to replace by increasing of the productivity of the sinking relatively by one order.

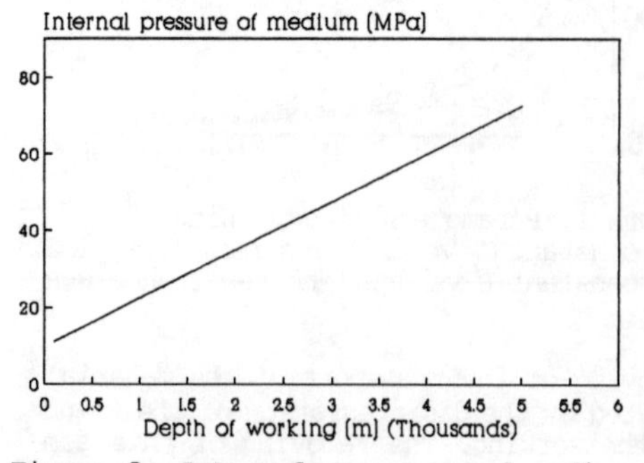

Figure 2. Internal pressure of medium vs depth of working

The necessary energy of the combustion is possible to reach by the special transferrers of the thermal energy into the rock that are called the flame injectors. The function of the injectors is to transport the combustion mediums into the space below the injector by means of the special nozzles. In this case it will be secured the thermal energy is not going directly from the flame injector into the rock as in the case of the mechanical disintegration the mechanical energy is going. The result is that the intensive thermal erosion typical for mechanical disintegration (abrasion) doesn't occur.

The surface of these injectors have to be protected by the special ceramic coating material for the case of the incidental interaction between the injector and molten rock (Foppe,1975;Lazar,1993). The necessary pressures of the combustion fumes for the fracturing of the rockmass and creating of the radial cracks is possible to attain by the special forming of the head of the injector. This forming will secure the sealing of the space in the calibrating zone of the whole equipment for creating the necessary pressure of the fumes under the flame injector. For illustration the values of the necessary thermal outputs, consumption of the oxygen, and for example hydrogen, per time unit for the various diameters of the working and the various rates of the penetration of the melting system are presented in the Table 1. At the same time the values of the quantity of the molten rock per time unit, which is necessary to transport under the high pressure into the radial cracks and the incrust zone is stated. At the same time the quantity of the arisen water steam as the product of the combustion is stated.

The fundamental condition of keeping of the flame injector as functional under the complicated conditions of the sinking is the achievement of the information about the physical state of the construction of the injector, the advance of sinking, and the back influence of the molten rock and the function of the whole system that have to be continuously controlled. The selected parameters can be: the pressure in front of the head of the flame injector, the rate of advance of sinking, the temperature of the molten rock, respectively of the flame.

Let us to indicate the method of the obtaining of the information about the object of which the rate of the indefiniteness is given by the conditions of existing of the functional object.

Let us observe the free object S_1 (see

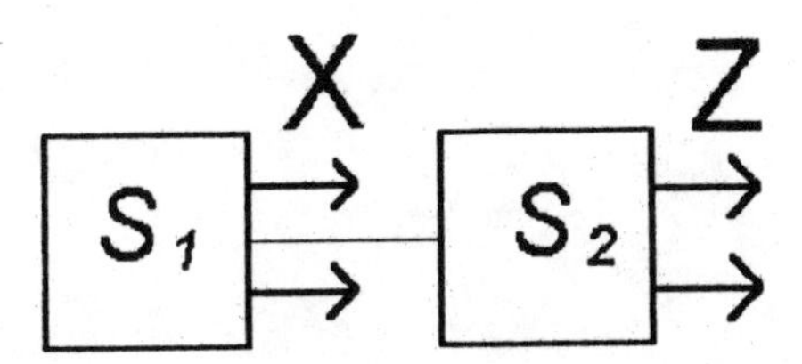

Figure 3. Relation of objects S_1 and S_2

Fig.3) of which the input signal is zero. By linking of the output of the object S_1 to the input of the next object S_2, from behavior of the object S_2 we can ascertain the behavior of the object S_1. In this case the object S_2 will observe the object S_1 and we can call the object S_2 as the observer. The presented verbal description of the performance of the objects S_1,S_2 can be formulated mathematically as follows:

Let S_1 is the independent object that can be described by the state equation

$$\dot{\mathbf{x}}(t) = \mathbf{A}.\mathbf{x}(t) \qquad (3)$$

which is acting as "the disturber" on the object S_2 described by the state equation

$$\dot{\mathbf{z}}(t) = \mathbf{F}.\mathbf{z}(t) + \mathbf{H}.\mathbf{x}(t) \qquad (4)$$

, where $\dot{x}$, x, $\dot{z}$, z are the derivations and the state variables of the objects S_1, S_2; **A**, **F** are the constants of the objects S_1, S_2; **H** is the input matrix of the object S_2; t is the time.

Let us suppose the existing of the transformation matrix **T** that complies the equation:

$$\mathbf{T}.\mathbf{A} - \mathbf{F}.\mathbf{T} = \mathbf{H} \qquad (5)$$

Let us consider the zero initial conditions

$$\mathbf{z}(0) = \mathbf{T}.\mathbf{x}(0), \; \textit{for } t \geq 0 \quad \mathbf{z}(t) = \mathbf{T}.\mathbf{x}(t)$$

On the basis of this we can expect for the case when **z**(0)≠**T.x**(0) will be valid also **z**(t)=**T.x**(t).

By the calculation and the following arranging of the equations (3),(4),(5) we obtain the equation of the dynamic work of the system S (associated objects S_1, S_2), of which the performance can be expressed by Laplace picture:

$$(p-\mathbf{F})(\mathbf{Z}(p) - \mathbf{T}.\mathbf{x}(p)) = \mathbf{z}(0) - \mathbf{T}.\mathbf{x}(0) \qquad (6)$$

By mans of the transfer of this equation into the time domain we obtain the complete idea about the dynamics of the system S:

$$\mathbf{z}(t) = \mathbf{T}.\mathbf{x}(t) + e^{\mathbf{F}t}[\mathbf{z}(o) - \mathbf{T}.\mathbf{x}(0)] \qquad (7)$$

From the relation (7) results that at the negative own values of the matrix **F** the vector of the state **z**(t) will approach the vector **T.x**(t). After introducing the requirement of the obtaining of the unit matrix **T** the equation (7) then provides the complete compendium of the behavior of the state variables **x** in the time.

Similarly we can derive the observer S_2 for the case of the non-zero control inputs **u**(t) of the object S_1. But in this case we have to link the same inputs to the observer S_2. The object S_1 is in this case described by the equation

$$\dot{\mathbf{x}}(t) = \mathbf{A}.\mathbf{x}(t) + \mathbf{B}.\mathbf{u}(t) \qquad (8)$$

and the equation of the observer S_2 with connected input will be

$$\dot{\mathbf{z}}(t) = \mathbf{F}.\mathbf{z}(t) + \mathbf{H}.\mathbf{x}(t) + \mathbf{T}.\mathbf{B}.\mathbf{u}(t) \qquad (9)$$

, where **B** is the matrix of the constants of the controlling.

It was shown that by means of the introducing of the unit matrix **T** we create the possibility of the identification of the states of the object S_1. This requirement

Tab.I. Technological data of the working of the flame injektor

diameter [mm]	penetr. rate [mm/s]	thermal output [MJ]	consumption of oxygen [kg/s]	consumption of hydrogen [kg/s]	volume of molt. rock [l/s]	quantity of water steam [kg/s]
100	3	0.118	0.0086	0.001	0.0235	0.0096
	5	0.196	0.013	0.0015	0.0393	0.015
	10	0.393	0.026	0.003	0.0785	0.029
300	3	1.06	0.07	0.009	0.212	0.079
	5	1.766	0.116	0.015	0.353	0.1316
	10	3.532	0.232	0.03	0.706	0.263

we can comply by the selecting of the parameters of the observer S_2, i.e. the elements of the matrix **F** and **H**.

Let us keep the premise **T=1**. In this case the equation (5) will have the form: **E=A-H**. From this results the object S_1 has in this case the same order as the observed object S_2. It is clear the properties of the observer depend on the matrix of the inputs **H**, concretely on the choice of its elements.

The presented paper results from the results of the laboratory research performed by Mr. W. Foppe. This is possible to documents by videotape. The following research is necessary to direct on the problems of the burning of the hydrogen at the various pressure conditions in the special pressure chamber, the forming of the flame, the focussing of the flame, respectively the spacial group arranging of several flames interacting with rock.

After the experimental and theoretical simulation is possible to start the research under in situ conditions.

REFERENCES

Sekula,F.; Krupa,V.; Bejda,J. 1980. Partial differential equation of drilling rate / in slovak /, Folia Montana, Proceedings of Mining Institute of SAS, Extraorinary issue, Veda SAV Bratislava, p.43

Krupa,V.; Sekula,F.; Bejda,J.; Koci,M.1984. Another evolution of mathematic modelling of diamond drilling / in slovak /, Folia Montana, Proceedings of Mining Institute SAS, Veda SAV Bratislava, p.95

Foppe,W. 1975. Thermal melting drilling equ ipment / in german /, Germany Patent DE 2554101, Germany Patent Office

Sekula,F.;Lazar,T.;Foppe,W.; Sabol,A. 1993. Equipment for hole sinking by flame /in slovak/, PV 0443-93, Office of Industry Property of Slovak Republic, Bratislava

Lazar,T.;Sekula,F.;Foppe,W.;Sabol,A. 1993. Equipment for hole sinking by flame with combined control /in slovak/, PV 0444-93, Office of Industry Property of Slovak Republic, Bratislava

Lazar,T.;Sekula,F.;Foppe,W.;Sabol,A. 1993. Equipment for hole sinking by flame with flow control /in slovak /, PV 0445-93, Office of Industry Property of Slovak Republic, Bratislava

Mechanical rock disintegration
Lectures

Geomechanics 93, Rakowski (ed.) © 1994 Balkema, Rotterdam, ISBN 90 5410 354 X

Possibilities of construction of underground openings of large diameters by means of soils extruding

J. Bejda, F. Sekula & O. Krajecová
Institute of Geotechnics of the Slovak Academy of Sciences, Košice, Slovakia

ABSTRACT: The theoretical and experimental research in the field of pushing of cylindric indenters to the cohesive and non cohesive soils was carried out. Suggestion of the extruding equipment for construction of underground openings of a large diameters without mucking out was the main purpose of the research. The verifying of the mathematic model was performed on the special stand with maximum diameter of pushed indenter 300 mm with possibility of the soil compaction and creating of the geostatic pressure up to 1 MPa.

INTRODUCTION

The extruding equipment for microtunnelling in soils without mucking out has been realized by prominent companies only in small diameters. For tunnelling of larger diameters it seems to be more proper using of the technologies of tunnelling with mucking out of soil. The reason of this is the necessity of disproportionate powerful force units for such tunnelling above 1000 mm in diameter. The such powerful force units (range 500 000 kN and more) are not common in the technical practice.

But at possibilities that results from combined acting of the several power aggregates the idea of construction of the large diameter extruding machines for soils without mucking out becomes to be real.

THE THEORETICAL BACKGROUND OF THE PROBLEM

The solution of the problem of pushing of the indenters into the soils was concentrated on the task connected with the piloting and indenting of the indenters of the small diameters for purpose of the penetrometry in the compacted zone up to c. 0.5 MPa, i.e. to the depth of about 25 m (Proceedings...,1975;Ferronskij,1979).

For proper solution we have to extend the area of the solution into the bigger depth and so to introduce the tensor of the stresses of the soil into the solution at least in a limited range. The another important requirement was to take into the consideration the influence of all input parameters at the pushing of the model indenters of rotary type into the soil, i.e. diameter of the indenter, length of the indenter, roughness of the surface of the model indenter, the properties of the soil etc.

In the cooperation of the Institute of Geotechnics SAS, the Mining Faculty of Technical University Kosice and the AT Consult company (Germany) the complex of the problems connected with this task was solved. The solving was based on the patent No. 4209695 of AT Consult Alternativ Technology Zukunftsforschung (Mr. W. Foppe the author of the patent). This patent was the fundamental idea for creating of the laboratory testing equipment on the Institute of Geotechnics SAS in 1991-1992. By means of this equipment it was possible to evaluate the thrust force on the indenter, the stress-strain state of the soil in surroundings of the pushed indenter and, in limited range, the displacements of the soil.

Equally with this testing the establishing of the mathematic models of the determination of value of the thrust force and the stress-strain state in surroundings of the indenter was done at the Technical University Kosice on the basis of known values of the geostatic pressure and the parameters of the mechanical characteristic of the soil (Ferronskij,1979;Hatala,1992).

THE CALCULATION OF THE THRUST FORCE

The calculation is based on the premise that we have the solid cone indenter of which the deformations during pushing can be neglected. The deformation properties of the cohesive and non-cohesive soils are evaluated by the following parameters:

E_0 - modulus of deformation of intact soil [MPa]

μ - Poisson's ratio of intact rock

At the attainment of the state of limiting stresses the soils behave as the contractant or dilatant soils. The influence of the change of the volume of the soil on the forces distribution at the construction of the underground opening will be transferred to the mathematic model as the coefficient of the dilatation D.

Let the primary state of the stresses in surroundings of the underground opening is characterized by the hydrostatic stress field c_i. This state could be reached with some accuracy for extruding of soils in the bigger depth and for extruding of the vertical workings.

Let us suppose that in the smaller depth is valid $\sigma_1=\sigma_2=\sigma_3.\mu/(1-\mu)$ and the construction of the underground opening is realized by means of the cylindric rotary body consisting of the cylinder of the length L and radius A ended by the cone with the apical angle 2α. For calculation of the thrust force necessary to extrude the soil by means of the cone body Hatala (Sekula, 1992;Hatala,1992) have established the relation for non-cohesive soils

$$P_k=\pi.A^2.\cos\alpha.\tan(\alpha+\varphi_k).q \quad (1)$$

where φ is the angle of internal friction of the soil, φ_k is the angle of the surface friction of the soil with the surface of the cone body, φ_r is the residual angle of the internal friction of the soil; and the pressure acting on the surface of the cone body can be expressed by the relation

$$q=\frac{2.\sigma_i}{(k+1)\left[1-\frac{1}{2-D}.\left(1-\frac{\sigma_i.(1-k)}{2.G.(1+k)}\right)^2\right]^{1-\frac{k_r}{2}}} \quad (2)$$

,where

$$G=\frac{E_0}{2(1+\mu)}\ ,\ k=\tan^2(45°-\frac{\varphi}{2})\ ,\ k_r=\tan^2(45°-\frac{\varphi_k}{2})\ .$$

For the calculation of the displacements in the non-cohesive soils (Hatala,1992;Sekula,1992) the radius of the zone of the non-elastic deformations R was calculated according the relation

$$R=A.\left[\frac{1}{1-\frac{1}{2-D}.\left(1-\frac{\sigma_i.(1-k)}{2.G.(1+k)}\right)^2}\right]^{\frac{1}{2}} \quad (3)$$

and for r>R is valid

$$U_r=\frac{R^2.\sigma_i.(1-k)}{2.G.r.(k+1)} \quad (4)$$

, where r is polar coordinate (the distance from the axis of the extruded tunnel) and U_r is the displacement of the soil.

For the calculation of the thrust force acting on the cone body in the cohesive soils the author (Sekula,1992;Hatala,1992) have established the relations

$$P_k=\pi.A^2.q.\cos\alpha.\tan(\alpha+\varphi_k+\beta) \quad (5)$$

,where q is the pressure acting on the surface of the cone body given as the relation (6) (see Appendix) and the angle of the friction β that characterizes the adhesiveness between the soil and the surface of the cone body is expressed as the relation (7) (see Appendix), where C is the apical cohesion and C_r is the residual cohesion.

A part of the resistance of the soil at extruding is caused by the friction between the cylindrical part of the indenter and the soil. Let us mark this force as P_t. For calculation of this force is known the relation that is used for the calculation of the friction force of the shield in the soils

$$P_t=f.[4.(\sigma_1+\sigma_3).A.L+G_f] \quad (8)$$

, where f is the coefficient of the friction between the cylindrical part of the shield and the soil, G_f is the gravity of the extruding equipment, A is the radius of the indenter, σ_1 is the horizontal component of the stress in the soil and σ_2 is the vertical component of the stress in the soil.

In the relation (6), as it was stated above, the additional pressure on the cylindric part of the equipment caused by the cone compacted soil was neglected. In order to take this pressure into the consideration the relation (6) is necessary to re-arrange to the form of the relation (9) (see Appendix), where j=1 is valid for the non-cohesive soils, see the relation (2), j=2 is valid for cohesive soils, see relation (6), and the meaning of the symbols is the same as in the relation (6).

By means of the summation of the forces expressed by the relations (1) and (9) we obtain the relation (10) (see Appendix) , where P_{c1} is the total force necessary for extruding of the non-cohesive soil by means of the cone indenter. Similarly, by summation of the relations (5) and (9) we obtain the relation (11) (see Appendix), where P_{c2} is the force necessary for extruding the cohesive soil by means of the cone indenter.

THE DEVELOPMENT AND REALIZATION OF THE STAND

The stand for experimental verification of the mathematic model is illustrated in the Figure 1. In the effort to reach the maximum possible diameter of the pushed indenter the 1/4 circle sector of the cylindric vessel was selected for the experiments. Similarly the extruding indenters were created as 1/4 circle sectors of the cone-cylindric body.

The equipment enables to imitate the geostatic pressure on the cylindric part of the vessel (imitation of the σ_3 component) by means of the rubber wall pushed by compressed air and the lid pushed by the hydraulic jack (imitation of the $\sigma_1=\sigma_2$ components) to the soil. Into the soil the sensors for evaluation of the pressures and the displacements was possible to arrange according the need of the task under study.

On this stand equipment several experiments resulting from the requirement of the verification of the relations (1,4,5 and 9) was realized. The trials were performed on two types of the soils of following parameters:

1. fine grinded clay from the locality Borovany (Czech republic) dampened to the moisture 21 % resulting from the requirement of the maximum compacting of this clay. The specific weight of the clay was under these conditions 1743 kgm^{-3}, the angle of internal friction was $\varphi=26°45'$ and the cohesiveness was 0.005 MPa.
2. the mixture (fine grinded clay of Borovany and fine magnesite sand of Bankov locality) with the characteristics: the moisture at the maximum compacting 17,345 %, the specific weight of the material at this moisture 2244 kgm^{-3}, the angle of the internal friction 35°, cohesiveness 0.0116 MPa.
3. fine magnesite sand of Bankov locality, moisture at the maximum compacting 6.67 %.

The additional parameters were used according the literature sources.

For the best imitation of the natural conditions of the occurrence the soil was placed into the vessel gradually as 10 cm layers compacted by electric vibrator. After filling up the vessel, due to compaction of the soil to the imitated geostatic pressure, the soil was gradually loaded and unloaded by means of above presented two systems that imitate $\sigma_1=\sigma_2$ in the range 0.3 - 1 MPa and σ_3 in the range 0.3 - 0.7 MPa. Before the measuring the pressure in the experimental vessel was set to the value corresponding to the depth of the occurrence from 25 to 50 m.

The pushing of the experimental cones of radii 0.05 and 0.15 m was realized by means of the hydraulic equipment instrumented with the logging of the axial thrust force measured from the oil pressure by the electric sensor.

The measured and calculated values of the thrust force in dependence on the diameter of the indenter are introduced in Figures 2,3 and 4. The data considering the pushing of the indenters of radii 0.01 and 0.02 m were obtained earlier on the experimental stand of smaller diameter.

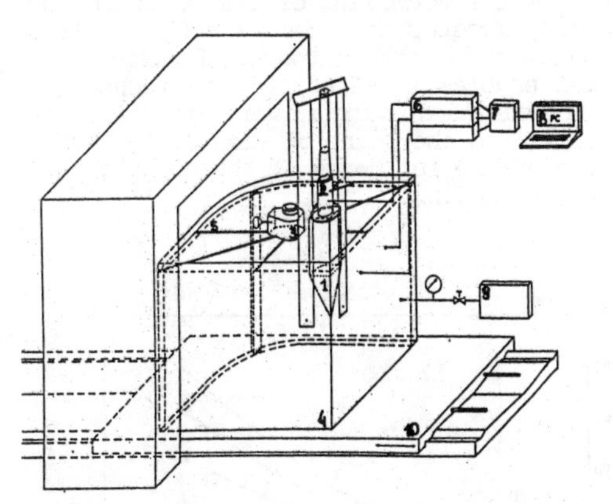

Figure 1. The outline of the experimental stand for the research of extruding of soils.
1-indenter, 2-hydraulic equipment of the indenter, 3-hydraulic equipment of the vessel, 4-vessel, 5-lid of the vessel, 6-measuring system Mikrotechna, 7-A/D transducer, 8-personal computer, 9-pumping equipment, 10-movable table

THE DISCUSSION OF THE RESULTS AND CONCLUSION

As it can be seen from the Fig.2 for the experimental course of the thrust force in dependence on the radius of the indenter at pushing into the clay the most suitable is the theoretic course (fitting) correspond-

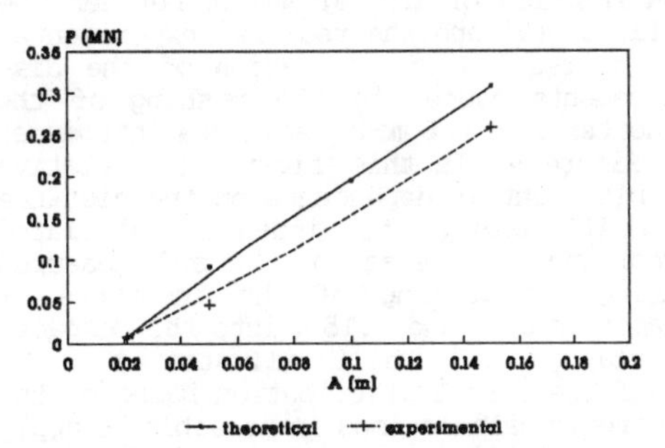

Figure 2. The theoretic and experimental courses of the thrust force in dependence on the radius of the indenter A. Clay, f=0.4, L=0.5 m

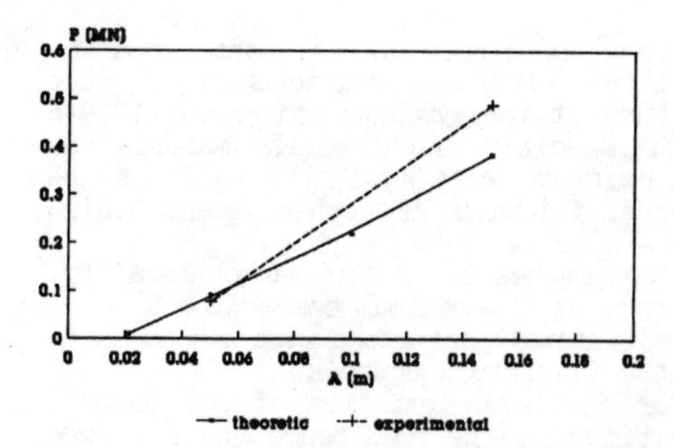

Figure 3. The theoretic and experimental courses of the thrust force in dependence on the radius of the indenter A. Mixture, f=0.4, L=0.5 m

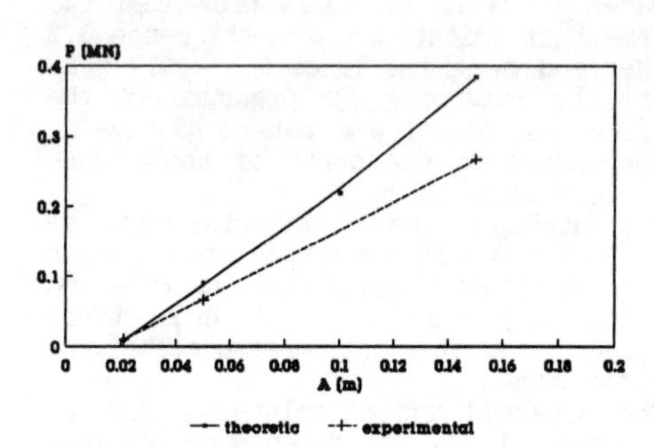

Figure 4. The theoretic and experimental courses of the thrust force in dependence on the radius of the indenter A. Sand, f=0.4, L=0.5 m

ing to the depth of the occurrence about 15 m.
In the case of pushing into the mixture (clay and sand) and sand for the experimental fitting of the presented dependence the theoretic course for the depth of the occurrence about 25 m is more suitable.

The courses of the thrust force illustrated in Figures 2,3 and 4 present the confirmation of the agreement between the relation (9) and the realized experiments.

The results of evaluation of the displacements caused by the pushing of the indenter into the model soil are introduced in Figure 5. In this figure the relative displacement in dependence on the distance from the axis of the indenter calculated according the relation (4) and measured during the pushing of the indenters of diameters 0.05 and 0.15 m into the compacted clay (σ= 0.7 MPa) is illustrated.

If the forces distribution based on the theoretic calculations is possible to apply directly in the construction of the openings the expecting values of the displacements are always higher as the observed real displacements. This fact was observed

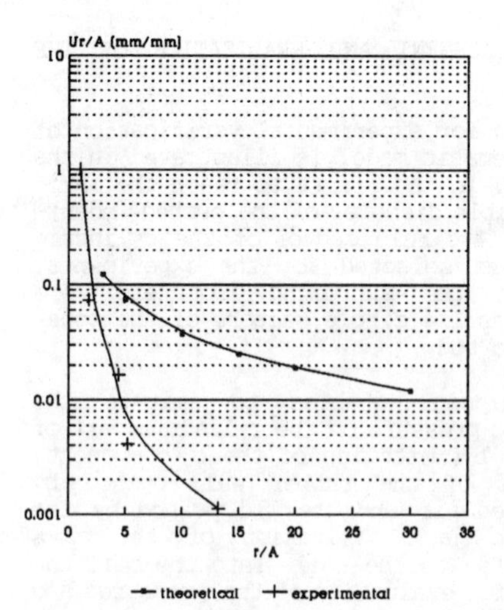

Figure 5. The relative displacement of the soil in dependence on the distance from the axis of the indenter. Clay, 2A=0.3 m and 2A=0.1 m

at the study of the displacements of the soil during shield driving (Bakoš,1992).

The established theoretic relations served for calculation of the expected forces distribution and the hydraulic output necessary for extruding. The Figures 6,7 and 8 illustrate the examples of a such calculations. This relations present the theoretic background necessary for the construction of the extruding equipment of large diameters. The experimental stand enables the modelling of the extruding up to the diameter of indenter 300 mm. After some not very complicated modifications the stand enables the study of the technology of the shield driving in very effective way mainly from the point of view of the evalu ation of the influence of the opening on its surroundings.

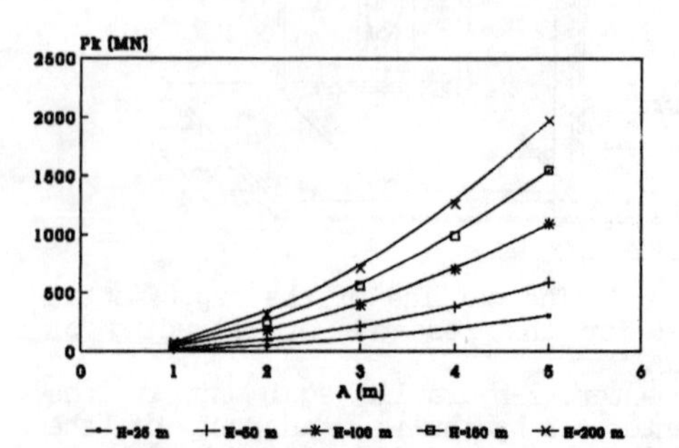

Figure 6. The calculated relation between the thrust force on the indenter and radius of the indenter in non-cohesive soils

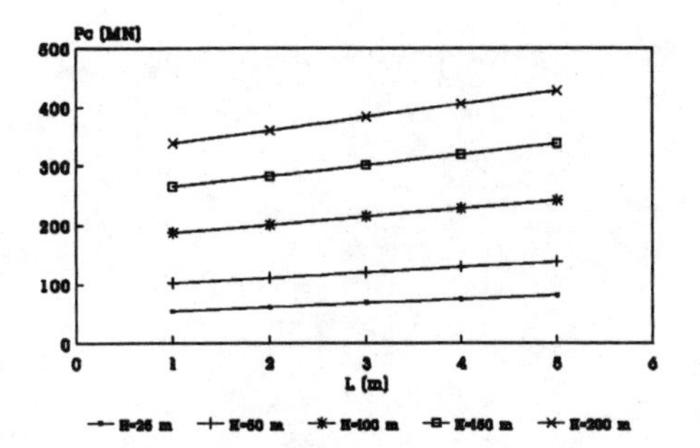

Figure 7. The calculated relation between the total extruding force and length of the cylinder L for A=2 m in non-cohesive soils

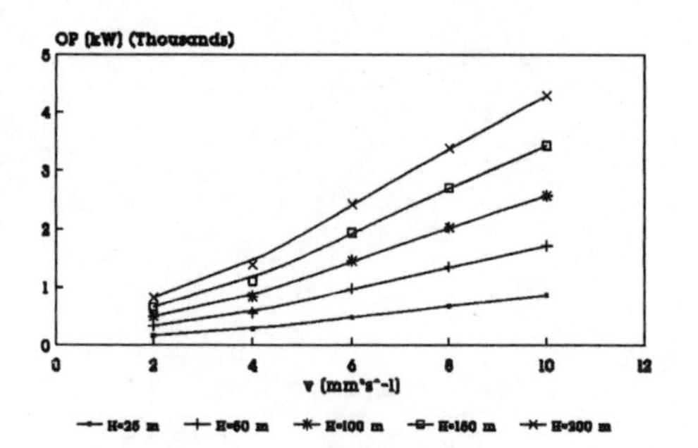

Figure 8. The relation between the total hydraulic output and the rate of advance of extruding in non-cohesive soils, A=2 m,L=5m

REFERENCES

Bakoš,M.;Klepsatel,F. 1992. Unfavorable influence of construction of sewers on residential area, measuring and prediction of deformations of roof in Bratislava and its environs /in slovak.,In: Proceedings of 8th Mining Conference, section New Achievements of Science, Research and Practice in Rock Mechanics, Kosice

Hatala,J.;Grejtak,A. 1992. Calculation of thrust force at creating of underground openings by compaction of cohesive and non-cohesive soils /in slovak/, In: Proceedings of 8th Mining Conference, section New Achievements of Science, Research and Practice in Rock Mechanics, Kosice

Ferronskij,V.I.; Grjaznov,T.A. 1979. Penetration logging /in russian/, Nedra, Moscow Proceedings of the European Symposium on Penetration Testing, 1975, Stockholm, Vol.1, Vol.2

Sekula,F. and col. 1992. Method and equipment for continual fullface extruding of tunnels in soils /in slovak/, Research report, Institute of Geotechnics SAS, Kosice

APPENDIX

$$q=\frac{(2.\sigma_i+2.C.k^{\frac{1}{2}}).(1-k)+2.C_r.k_r^{\frac{1}{2}}.(1+k)}{(1-k_r).(1+k).\left[1-\frac{1}{2-D}.\left(1-\frac{\sigma_i.(1-k)+2.C.k^{\frac{1}{2}}}{2.G.(1+k)}\right)^2\right]^{\frac{1-k_r}{2}}}-\frac{2.C_r.k_r^{\frac{1}{2}}}{1-k_r} \quad (6)$$

$$\beta=arccot\left[\frac{q.\cos\alpha}{C_k.\cos\varphi_k.\cos(\alpha+\varphi_k)}\right]+\tan(\alpha+\varphi_k) \quad (7)$$

$$P_{tm}=f.[(4.(\sigma_1+q_j.\cos\alpha)+(\sigma_3+q_j.\cos\alpha)).A.L+G_f] \quad (9)$$

$$P_{c1}=\pi.A^2.q_1.\cos\alpha.\tan(\alpha+\varphi_k)+f.A.L.[4.(\sigma_1+\sigma_3)+8.q_1.\cos\alpha+\frac{G_f}{L}] \quad (10)$$

$$p_{c2}=\pi.A^2.q_2.\cot(\alpha+\varphi_k+\beta)+f.A.L.[4.(\sigma_1+\sigma_3)+8.q_2.\cos\alpha+\frac{G_f}{L}] \quad (11)$$

Geomechanics 93, Rakowski (ed.) © *1994 Balkema, Rotterdam, ISBN 90 5410 354 X*

Application of diamond wire for cutting various materials

Bożena Ciałkowska
Institute of Production Engineering and Automation, Technical University, Wroclaw, Poland

ABSTRACT: The paper presents the concept of wire cutting. The method is compared with other techniques used in cutting difficult-to-machine materials. The prospects for applying the method to contour cutting of unconventional materials, especially composites,are discussed.

1. INTRODUCTION

Wire cutting may be readily compared to the well known technique of cutting stone with a cable drenched with abrasive slurry. This ancient method has been recently modernized by employing modern control systems enabling the most intricate shapes to be fabricated. By contrast, the wire cutting technology has been originated and developed during the last twenty years but has recently undergone the same changes connected with improved control means as other manufacturing processes [1].

Similarities between a cutting cable and wire are largely superficial and on gaining a deeper insight into the physics of the processes involved they become less evident.

A simple classification will help to outline the area of our interest. Cables coated with diamond grit have from 7 to 12 mm in diameter and are used for rough cutting of large size parts. Wire tools are much finer with diameters from 0.02 to 1 mm. The finest sizes (0.02 - 0.1 mm) are used for precision slicing operations in the semiconductor industry. The thicker wires (0.1 mm and above) are preferred for straight-line and contour cutting of various materials.

There is a variety of wire cutting machines available on the market from simple hand-controlled devices to advanced models featuring computer control systems [6,8,10,12]. Various aspects of the wire cutting technology are presented in block diagram form in Fig.1.

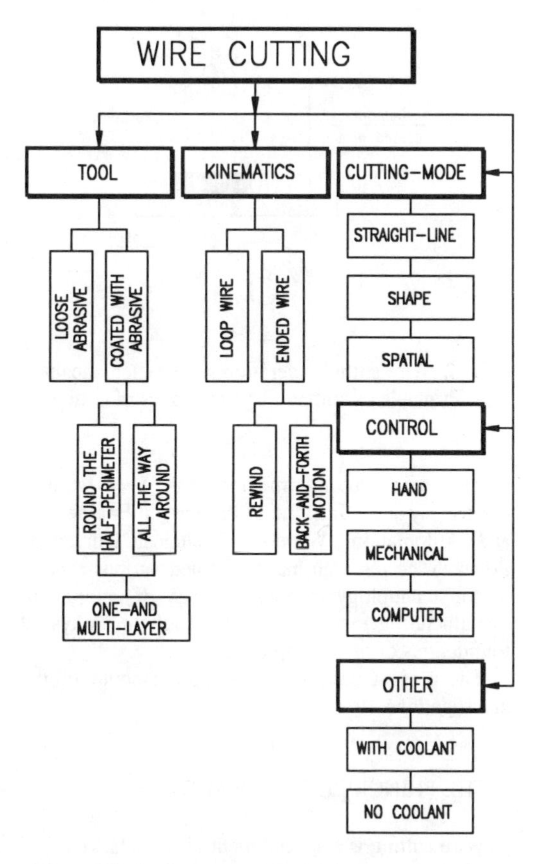

Fig.1. Block diagram presenting various aspects of the wire cutting technology.

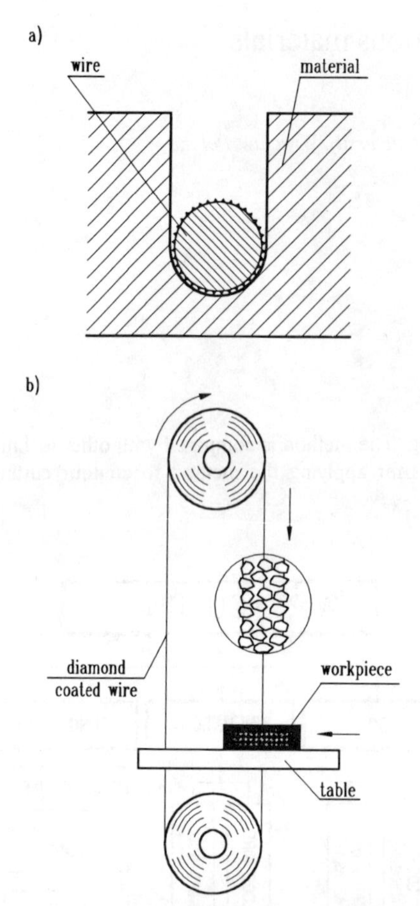

Fig.2. Wire cutting operation show schematically
a) Principle of cutting b) Kinematics of cutting

The wire cutting method has been studied extensively at the Institute of Production Engineering and Automation, Wroclaw Technical University, Poland. The research has concerned various aspects of the technology including design of equipment, manufacturing methods for wire tools and physical foundations of the cutting process.

The present paper gives a short account of the investigations.

2. THE PRINCIPLE OF THE METHOD

Wire cutting is a special method of metal removal in which a pulled up wire coated with diamond powder or charged with loose abrasive makes a cut [Fig.2][2,3]. The most important features of the method are shown graphically in Fig.3.

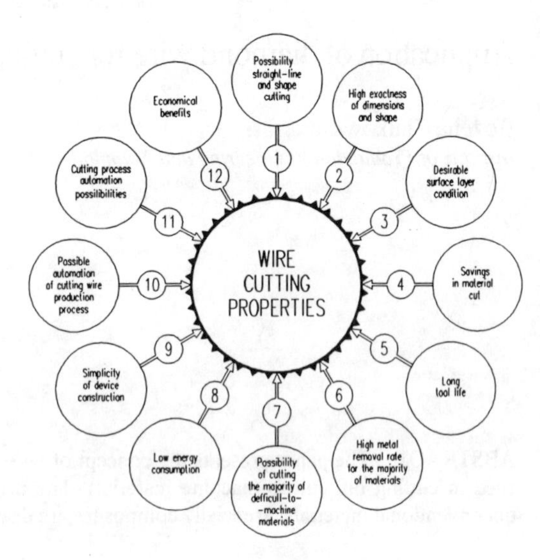

Fig.3. Wire cutting characterization.

3. POSSIBILITIES OF THE METHOD

Both our own experimental evidence and the literature data [4,5,6,8,9,10,12] show that the method is unique because it can be applied to any conceivable material. The material can be either extremely tough or brittle, hard or soft or may be a combination of widely different constituents. Expensive materials can be cut with kerf losses much lower than those required by other methods. Fragile crystalline materials such as oxide or semiconductor monocrystals can be cut into slices as thin as 100 μm without damaging their crystal structure or causing edge chipping. The small cutting forces (of the order of a few newtons) result in negligible heat generated within the cutting zone - an interesting feature for the semiconductor industry!

The wire tool is elastic so its cutting ability never becomes impaired by the removed material particles and the process remains stable for hours.

High accuracy of the cut and a fine finish are readily attained with this method if only a few basic requirements are met. It is not the aim of this paper to discuss such points in detail, so we will mention only the most vital ones: the wire must be driven very precisely throughout the process as no misalignment is allowed, it must be also kept under uniform tension at all times. The wire can travel at 50 m/s or more, the speed being easily adjusted to a particular application.

The wire saw compares favourably with other methods when it comes to power consumption. A motor under 1 kW will successfully drive a saw capable of cutting even large blocks of hard material.

It is to be noted that there are other methods that can be used with difficult-to-machine materials such as water jet, plasma arc or laser beam cutting and conventional diamond disks. Neither of them however can compete with wire cutting when it comes to cost and flexibility. Thermal methods such as plasma or laser cutting affect unfavourably the surface layer of the workpiece and additional finishing operations may be required. The laser devices do not accept specimens thicker than 20 mm. The same applies to plasma cutters which also yield rough surfaces of the cut.

The water jet method has been widely accepted by many industries owing to its wide range of application. It is however generally acknowledged that with such materials as monocrystals, ceramics, glass or some composites the method is either physically inapplicable or cost prohibitive. Water jet cutting is frequently accompanied by delamination of material being cut and other undesirable effects. Limited thickness of the work and high energy consumption are yet other weaknesses of this otherwise excellent method.

Conventional tools such as band and disk saws have been long found of little use for cutting composite materials or fabricating intricate shapes.

A wire tool is free of all these disadvantages. There is no other tool that can accept practically any material, of any size ranging from microtome dimensions to massive blocks. Limited durability of the wire itself may be at present regarded a little unsatisfactory but recent advances promise longer wire lives.

4. EXPERIMENTAL RESULTS

The investigations were conducted on a custom-made set-up designed by the author. The wire used had from 0.3 to 0.8 mm in diameter and 2.5 m in length and was also fabricated at the Institute. Fig.3 shows some SEM images of the wire.

A variety of difficult-to-machine composite materials were tested including graphite and glass laminates and honeycomb blocks. Fine slices of honeycomb material (thinner than 1 mm) could be obtained which is unconceivable with other methods.

Similar excellent results were obtained for various ceramic materials, sintered carbides and such exotic materials as wax or foam. Steel strings fabricated by the push-out method were also successfully sectioned.Some shapes obtained from the experiments are shown in Fig.5 [6,9,11].

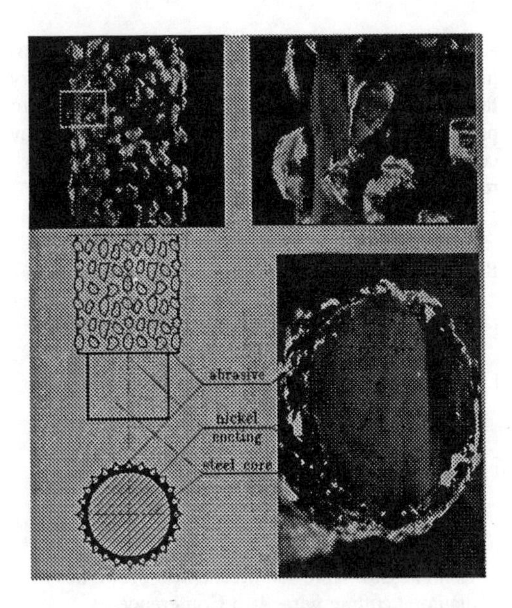

Fig.4. The wire coated with diamond powder.

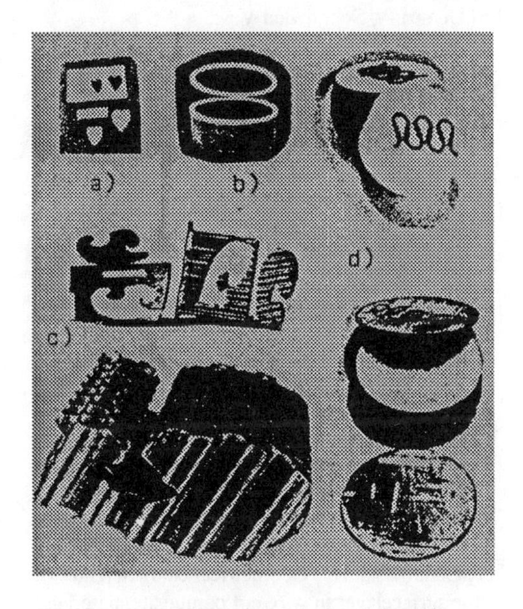

Fig.5. Example shapes obtained from the experiments.
a) ceramics b) bronze c) honeycomb
d) multicomponent material with steel reinforcement

5. CONCLUDING REMARKS

The experimental investigations have fully proved applicability of the wire cutting method to sectioning and contour cutting a wide class of difficult-to-machine materials. The method compares very favourably with other techniques when it comes to productivity, accuracy, quality of finish and energy consumption. For composite materials it is the most recommended method.

REFERENCES

[1].Diamantseilsäge für das computergesteuerte Konturensägen von Naturstein; IDR, 4,1991.

[2].Ciałkowska B., Kubik K, Presz R. A cutting tool in the form of endless wire. Polish patent No 155 908 1992

[3].Ciałkowska B., Kubik K, Presz R. A manufacturing method for the wire cutting tool. Polish patent No 155 909 1992

[4].Ciałkowska B., Presz R. Effect of technological factor on mechanical properties of diamond cutting wire, 4-th Conference "Advances in theory and practice of material machining", Kraków, 1990, Poland.

[5].Dennis P., Schmieden V. Präzisionsbearbeitung mit vorgespannten Diamantwerkzeugen; Werkstattstechnik, 79,1989.

[6].Vogel M. - Diamantdrahtsägen -- ein neuartiges Trennverfahren, seine Möglichkeiten; Werkstattstechnik, 79,1989.

[7] Ciałkowska B. Presz R. Manufacture of wire for contour cutting of composite materials, NM'90 Conference "Mechaning of non-metalic materials", Rzeszów 1990, Poland

[8].Cialkowska B., Presz R. - Contour cutting of difficult-to-machine materials using a diamond wire; Mechanik, 5/6,1991,Poland.

[9].Ciałkowska B. - Diamond wire tools: the future in cutting composite materials; 4-th Conference on Machining Engineering, Łódź, 1991, Poland.

[10].Diamantdrahtsäge für Problemwerkstoffe; Metallbearbeitung, Juni,1990.

[11].Ciałkowska B. , Geometrical characteristics of the surfacelayer in wire cut composite materials, 2-nd Conference "Effect of manufacturing methods on the surface layer characteristics", Gorzów Wielkopolski 1993, Poland.

[12]. Laser Technology Inc., North Hollywood, California., USA- Promotional material

Geomechanics 93, Rakowski (ed.) © 1994 Balkema, Rotterdam, ISBN 90 5410 354 X

Mathematical model for determination of pressure forces

J. Hatala & A. Grejták
Mining Faculty, Technical University, Košice, Slovakia

ABSTRACT: The contribution describes a mathematical model for the determination of the pressure force magnitude at the creation of underground objects by the jacking of the cone shape indentor in cohesive and cohesionless soils respectively. The magnitude of the pressure force is a function of the magnitude and geometrical shape of the indentor being pushed, strength and deformation properties of soils.

1 INTRODUCTION

With the increase of the number of constructed underground engineering structures and with the rapid technique development worldwide, projects of new technological procedures of their excavation appear.

One of new, so far very little theoretically and technically worked out procedures, is the creation of underground objects by the soil compacting in the way of the indentor pushing, shape and dimensions of which correspond with the required profile, without the necessity of the soil mucking from the excavated profile.

From the great number of so far unambiguously unclarified problems of theoretical, technical and technological character respectively as the prior one is the problem of the magnitude of the pressure force needed for the pushing of indentor which creates the underground structure of the required cross-section.

In 1991 and 1992 the Mining Institute of the Slovak Academy of Sciences completed laboratory testing equipment enabling the determination of the pressure force magnitude as well as the measurement of the stress-strain state of the soil in the surrounding of the indentor being pushed. Simultaneously at the Department of Mining and Geotechnics, Mining Faculty, Technical University of Košice was created the mathematical model for the determination of the pressure force and stress-strain state in the indentor surrounding from known values of the geostatic pressure and parameters of the soil mechanical characteristic.

In this contribution there are described only geotechnical assumptions of the theoretical solution and final equations for the determination of the pressure force of the conical indentor for the creation of undergroung circular structures.

2 GEOTECHNICAL ASSUMPTIONS OF THE SOLUTION

The mathematical model for the calculation of the cone pressure force at the creation of the underground circular structure is based on the following assumptions.

2.1 The creation of underground structures by the cone pushing is realized in compressible soils wich from the point of view of their properties could be divided into two basic groups:

2.2 *Cohesionless soils* - sands, gravels (vithout cohesion)

a) compacted ones, which at the shear failure behave as dilatant (they increase their volume). Their properties are stipulated by Figure 1:

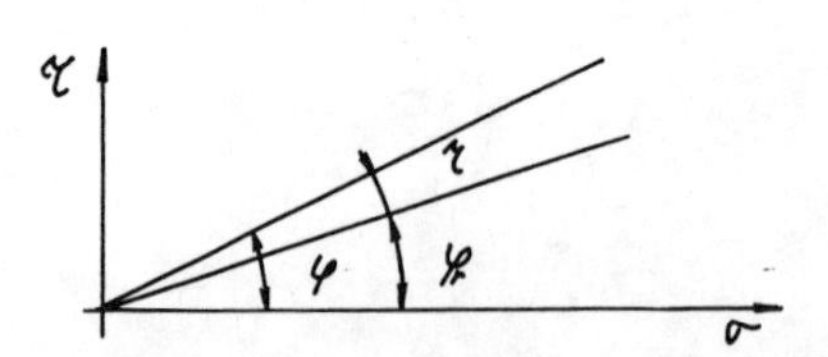

Fig.1 φ_r - residual internal friction angle, φ - vertex internal friction angle, τ - angle taking into accout the dilatancy influence

b) loose cohesionless soils with the porisity $e < e_{kr}$ ($e_{kr} \equiv CVR$ - critical void ratio), Figure 2:

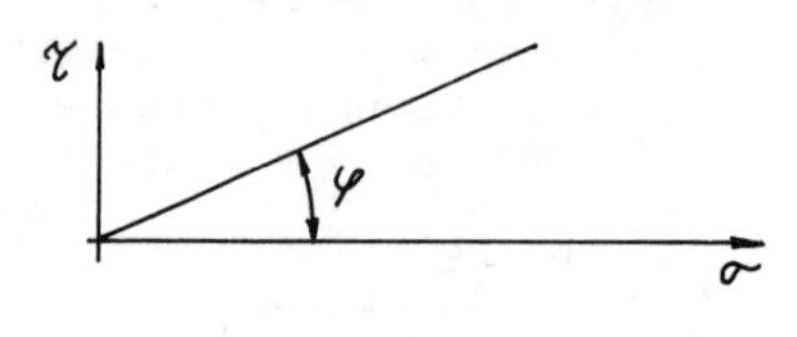

Fig.2

Their properties are determined by the φ- internal friction angle.

2.3 *Cohesive soils* which properties are given by parameters according to Figure 3:

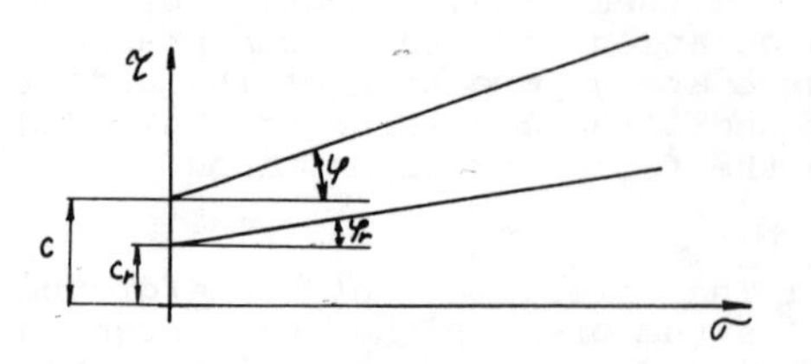

Fig.3 φ - vertex internal friction angle, c - vertex cohesion, φ_r - residual internal friction angle, c_r - residual cohesion

Deformation properties of cohesive and cohesionless soils are giver by the following parameters: E_o - deformation modulus of the unbroken soil, μ - Poisson's number of the unbroken soil.

2.4 After the achieving of the limit state of stresses the soils behave as the contractant or dilatant. The influence of the soil volume change on forces at the creation of the underground space into the mathematical model is projected by the dilatancy coeficient D.

$$D = 1 + \frac{\triangle V}{V} \qquad (1)$$

where $\triangle V$ is the volume change at the failure.

a) at contractant soils $D < 1$
b) at dilatant soils $D > 1$

2.5 Virgin stresses at the place of the cteation of the underground structure are characterized by the hydrostatic stress field with the value of σ. This assumption holds true for the rotary cone pushing in greater depths by the force P with the vertex angle 2α, with the maximal radius r.

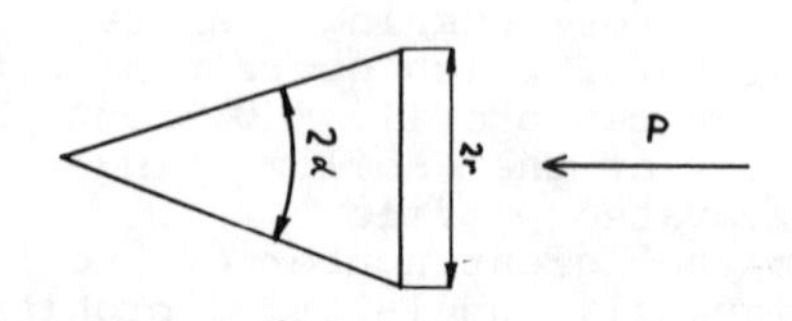

Fig.4

2.6 The model takes into account the influence of further parameters: φ_k - friction angle between the soil and the cone surface, c_k - adherence (binding power) of cohesive soils on the cone surface.

3 PRESSURE FORCE CALCULATION

For the calculation of the pressure force P the following equations were derived.

3.1 Cohesionless soils

For the calculation of the pressure force P in cohesionless soils the folowing parameters are given: σ - geostatic pressure at the depth of the constructed object, φ_r - residual internal friction angle of soils, τ - angle taking into account the dilatancy influence, φ-φ_r +τ - vertex friction angle (it is given at loose soils only), E_o- deformation modulus of the soil, μ - Poisson's number of the soil, D - the dilatancy coefficient of the soil, φ_k - friction angle between the soil and the indentor surface, α - half vertex angle of the cone indentor, r - maximal cone radius.

Calculation of auxiliary parameters:

$$G = \frac{E}{2(1+\mu)} \tag{2}$$

$$k = tg^2(45° - \frac{\varphi}{2}) \tag{3}$$

$$k_r = tg^2(45° - \frac{\varphi_r}{2}) \tag{4}$$

$$n = (1-k_r)/2 \tag{5}$$

$$q = \frac{2\sigma}{(k+1)\left[1 - \frac{1}{2-D}\left[1 - \frac{\sigma(1-k)}{2G(1+k)}\right]^2\right]^n} \tag{6}$$

Pressure force:

$$P = \pi r^2 q \; cotg\alpha \; tg(\alpha+\varphi_k) \tag{7}$$

At loose, unconsolidated soils at which $\tau = 0$ into the equation (6) we substitute $k_r \equiv k$.

The minimal value of the pressure force P we obtain at the cone with the vertex angle $2\alpha = 90° - \varphi_k$. If $2\alpha > 90° - \varphi_r$, then in the equation for the calculation of the pressure force (7) we substitute

$$\alpha = 45° - \frac{\varphi_r}{2} \tag{8}$$

$$\varphi_k = \varphi_r \tag{9}$$

From the given follows that at $2\alpha > 90° - \varphi_r$ the magnitude of the pressure force does not depend on the magnitude of the vertex angle of the pressure cone and in given soil is constant. In such a case the equation (7) changes to the shape

$$P = \pi r^2 q \; cotg(45°-\varphi_r/2)\, tg(45°+\varphi_r/2) \tag{10}$$

3.2 Cohesive soils

For the calculation of the pressure force in cohesive soils the following input data are given: σ - geostatic pressure at the depth of the constructed object, φ_r - residual internal friction angle of soils, φ- vertex friction angle of the soil, E_o - deformation modulus of the soil, μ - Poisson's number of the soil, D - the dilatancy coefficient of the soil, φ_k- friction angle between the soil and the indentor surface, α - half vertex angle of the cone indentor, r - maximal cone radius, c_r - residual cohesion, c_k - cohesion between the soil and the cone surface.

Calculation of auxiliar parameters: Calculation of G,k,k_r and n according to equations (2),(3),(4) and (5).

$$B = 1 - \frac{\sigma(1-k)+2c\sqrt{k}}{2G(1+k)}$$

$$q = \frac{(2\sigma+2c\sqrt{k})(1-k_r)+2c_r\sqrt{k_r}(1+k)}{(1-k_r)(k+1)\left[1 - \frac{1}{2-D}B^2\right]^n} - \frac{2c_r\sqrt{k_r}}{1-k_r} \tag{11}$$

$$\beta = arccotg\left[\frac{q \; cos\alpha}{c_k cos\varphi_k cos(\alpha+\varphi_k)} + tg(\alpha+\varphi_k)\right] \tag{12}$$

$$P = \pi r^2 q \; cotg\alpha \; tg(\alpha+\varphi_k+\beta) \quad (13)$$

Also in this case the minimal value of the pressure force P we obtain at the cone with the vertex angle $2\alpha > 90° - \varphi_k - \beta$. If $2\alpha > 90° - \varphi_r$ then in equation (12) for the calculation of the angle β we substitute $c_k = c_r$ and $\varphi_k = \varphi_r$ and in equation (11) for the calculation of the P we substitute $\alpha = 45° - \varphi_r / 2$. At $\alpha > 45° - \varphi_r / 2$ the magnitude of the pressure force does not depend on on the magnitude of the vertex angle 2α and for the given soil is constant. In such a case equations (12) and (13) will change to the shape

$$\beta = arccotg\left[\frac{q \; cos(45°-\varphi_r/2)}{c_r cos\varphi_r cos(45°+\varphi_r/2)} + tg(45°-\varphi_r/2)\right] \quad (14)$$

$$P = \pi r^2 q cotg(45°-\varphi_r/2) tg(45°+\varphi_r/2+\beta) \quad (15)$$

In cases where the indentor being pushed is ended by the cylindrical part (Figure 5) the magnitude of the pressure force is increased by the friction force P_1 value on the cylindrical part.

$$P_1 = 2\pi r l_c q$$

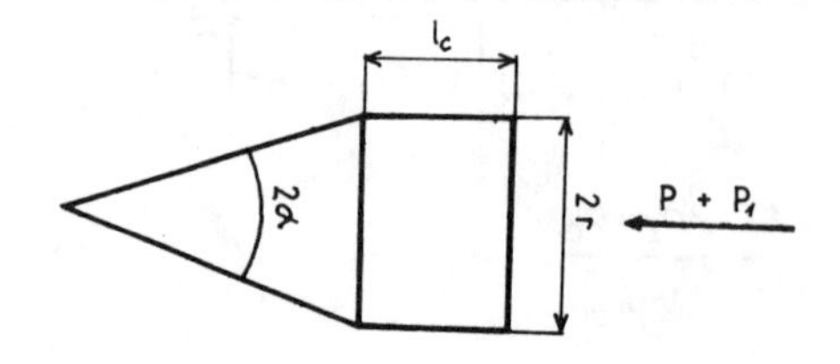

Fig.5

The correctness of the mathematical model was verified by the laboratory experiments at the Institute of Geotechnics of the Slovak Academy of Sciences and at the Mining Faculty, Technical University of Košice. Laboratory experiments on indentors with diameters from 2 to 30 cm have proved a good agreement between the experimental and mathematical modelling results. Some of them are described in the article of authors: Bejda, Sekula, Krajesová, Hatala, which is published in these proceedings as well.

In the conclusion of this contribution we would like to say that the mathematical model also enables the determination of the distribution of stresses and deformations in the surrounding of the cone being pushed as well as the shape and the zone area of the intensive shear deformations.

REFERENCES

Hatala, J. and Grejták, A. 1992. Výpočet tlačnej sily pri vytváraní podzemných diel stlačením nesúdržných a súdržných zemín. 8. Banícka vedecko - technická konferencia, Košice.

Hatala, J. and Trančík, P. 1987. Mechanika hornín a masívu. ALFA Bratislava.

Jaeger, J.C. and Cook, N.G.W. 1976. Fundamentals of Rock Mechanics. Chapmen and Haal, London.

Sekula, F. and kol. 1992. Metóda a zariadenie pre kontinuálne plnoprofilové pretláčanie kolektorov a tunelov v zeminách. Banícky ústav Slovenskej akadémie vied Košice.

Geomechanics 93, Rakowski (ed.) © 1994 Balkema, Rotterdam, ISBN 90 5410 354 X

Problems of correlation of the breaking characteristics of rocks

J.Jurman & J.Vrbický
VSB, Technical University of Ostrava, Czech Republic

ABSTRACT: The breaking characteristic of rocks is expressed for individual excavation technologies, and, with regard to differences of individual processes, there is no universal criterion of such breaking evaluation. A demand of an information's quality and quantity increase in a sense of the breaking characteristic with surface mining, led to a consideration of using drilling process monitoring results that could be applied easily on given terms. However, it means the expression of a correlation between breaking when drilling = drillability, and between breaking when drilling with a rotary excavator = digging resistences. On the basis of operational measurements, the problems analysis was performed and possibilities of a solution were outlined.

1 INTRODUCTION

Resistance which excavated rocks put up during an excavation process on a working machine's cutting tool is called a breaking characteristic of rocks. The term breaking characteristic of rock is used in the broadest sense of meaning to express the interaction between a rock and a cutting tool. By this the position of a breaking characteristic within a set of rock characteristics is determined as a technological characteristic.

Due to the great number of various aspects effecting excavation it is practically impossible to formulate the universal and generally valid definition of a breaking characteristic. For this reason we usually define breaking characteristic for individual excavating technologies (e.g. for drilling - drillability, for mining - cutting resistance, digging resistance, plough ability, etc.). In all cases these are processes with a different degree of dispergation reached by a different technological process, thus to express correlative relations between the individual sorts of breaking characteristic is very difficult.

2 ANALYSIS OF A BREAKING CHARACTERISTIC OF ROCKS

Complexity of an excavation process can be best explained by the thorough analysis of all factors entering the process which can be divided into four groups:

- geological (expressed by geological characteristics)
- technical (expressed by features of a cutting tool)
- technological (expressed by regime parametres of disintegration)
- other effects.

Rock properties effecting a breaking characteristic are represented by a file of technical and petrographical properties. Strength and deformable characteristics are the most important properties of rocks. As to petrographical property it is mineral composition, especially a content of hard minerals, grain size and grain size distribution.

Some of the above mentioned characteristics have a very important and even decisive effect upon a breaking characteristic of individual rocks, the other can be of only of secondary importance or they become evident only at certain excavating methods. Effects of various properties can overlap, reciprocally strengthen or even cancel.

Though some properties can increase or decrease a breaking characteristic in a decisive manner, none of them by itself can characterise a breaking characteristic of rock enough objectively. Therefor it is obvious that a breaking characteristic is always given by a group of various and co-operating properties.

A change of any of them sometimes seeming neglect able can cause a considerable change of a breaking characteristic.

Rock properties are unchanging.

The type of a cutting tool is determined by a respective excavating process and by a rock. Features of a tool are created by its geometry and a material. A cutting tool is not usually changed when rock properties alter, its selection is carried out before setting into operation.

Regime parameters determine the shape and size of the chip of broken rock and they are obviously formed by two parameters (for instance during drilling there are revolutions and velocity of feed), that can be partially or entirely regulated. Here enters the human factor to the breaking process (in case, that the process is not fully automated) with subject evaluation.

Outer conditions are created by stress state of massive, chemise, temperature and they are largely unaffectable.

It is clear on the base of performed analysis that objective evaluation of the breaking characteristic is not possible without taking into account all the factors mentioned above.

Every technology of breaking and crushing of rocks needs methodology for determination of breaking characteristic. We can find out by detailed analysis that these methodologies differ from each other both in substance and in result expression. Some of them find the breaking characteristic in laboratories on rock samples, other directly in situ, but not with full objectivity always. The most appropriate criterion for evaluation of breaking characteristic of rocks is specific volume energy, the substance of which is evaluation of consumption of energy that is needful for breaking of unit volume of rock.

3 BREAKING CHARACTERISTIC AT SURFACE MINING

Managing of mining process on surface mines is conditioned by sufficient number of information about geologic conditions, excavation resistance, e.g. about breaking characteristic of excavated cuts and benches. The information about geologic conditions of deposit and its bedrock and top wall are obtained from deep holes. Excavation resistance is largely investigated by measurements on wheel excavator, that represents one substantial disadvantage: After evaluation in situ measurements we know the excavation resistance in location that has been excavated already and the value of excavation resistance in rock that has not been broken yet we can estimate only. This lack of information about excavation resistance before the excavation process leads to looking for other methods of investigating.

The great volume of drilling works on surface mines offers solving: To find correlation between drillability and excavation resistance.

It was developed the measuring device for measurement on the drilling rig that evaluates from monitored values the specific volume energy minimised by regulation of regime parameters (revolutions, thrust). This quantity is the criterion for evaluation of drillability. As the example we present in the fig.1 the record of operation measurements that shows drillability course along the length of the hole. Drillability value is not in absolute units.

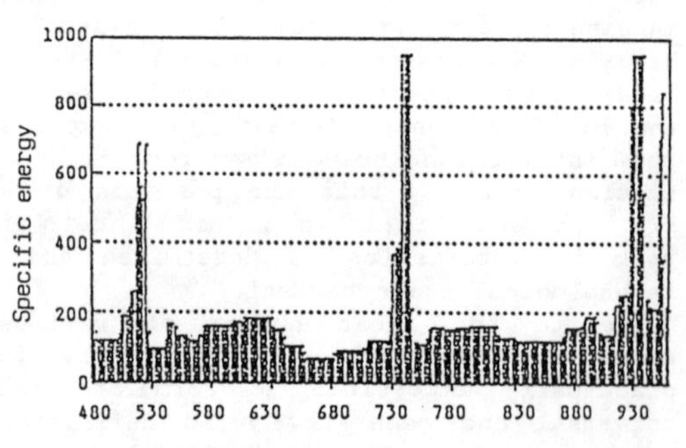

Fig.1 Drillability course along the length of the hole

Measuring of specific excavation resistance on wheel excavator by indirect method. That means that loading of wheel motor drive in dependence on excavated chip is measured. According to standard methodology the excavation resistance are determined in kN/m, however, to express correlation relation was necessary to evaluate the energy consumption in unit capacity of broken rock, e.g. in J/m^3

During the drilling process the

instantaneous value of specific energy is being measured but during the excavation by excavator the average value of specific energy in excavated chip is evaluated. Taking into account to difference of the two breaking processes it is necessary to accept these basic re quirements for keeping objecti-vity:

- minimised specific energy is considered as criterion of breaking characteristic
- operating measurements have to be performed in the same conditions always.

4 DESIGN OF CORRELATION BETWEEN DRILLABILITY AND EXCAVATION RESISTANCE

For the first design of correlation were used the results of in situ measurements on Vršany surface mine. From 22 holes were chosen courses of specific energy in hole No.3 and 12 (Fig.2) and from measurements on the excavator wheel were chosen the average values of specific energy in individual benches (Fig.3). These two factors enter to the correlation relation:

- construction design of the excavator cutting medium
- breaking characteristic during drilling process
- effect of thickness of hard partings
- effect of location of hard partings regarding the excavator wheel.

Regarding the different shapes of buckets and different number of them on the excavator wheel it is necessary to

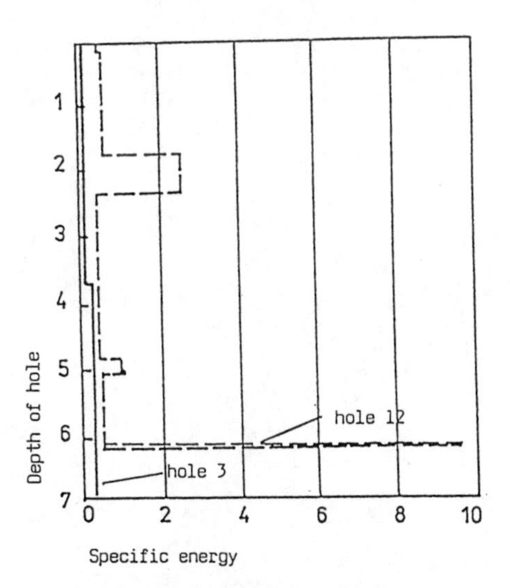

Fig.2 Course of specific energy in the holes No.3 and 12

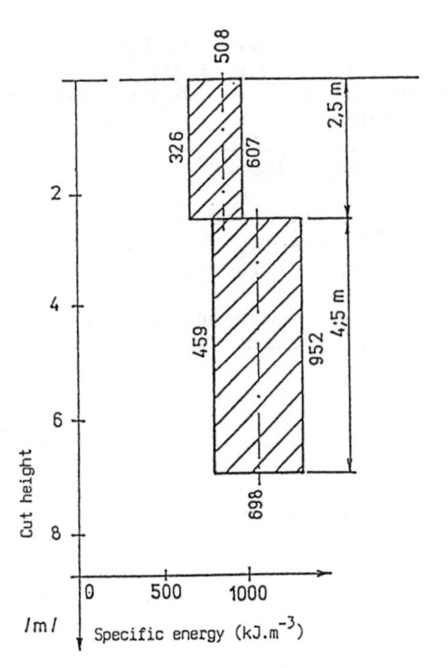

Fig.3 Course of specific energy on the excavator wheel

perform comparative measurement for each excavator and to establish the constant expressing the effect of construction of cutting medium. The occurrence of hard partings in excavated rocks evokes the growth of breaking characteristic values and at determined thickness of parting the rock becomes unsuitable for excavating by specific type of machine. The effect of hard partings location regarding the excavator wheel is evident of Fig.4.

If the parting is located in upper part of the bench the peripheral strength to

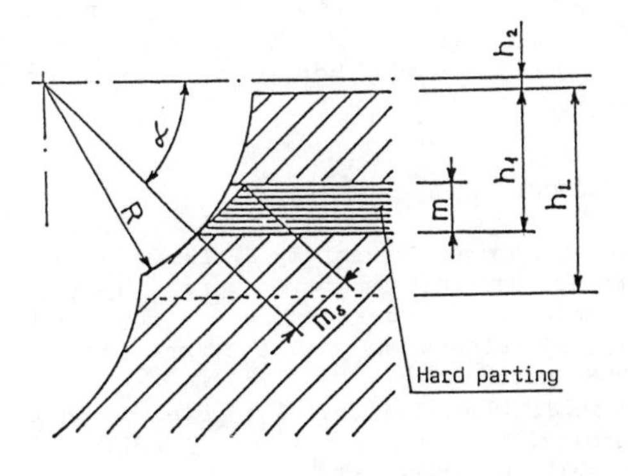

Fig.4 Location of hard partings regarding to the excavator wheel

breaking it is many times lower then if it is located in lower part of the bench.
That is why we reduce the thickness of parting by relation:

$$m_s = m/\sqrt{(1 + (h_1 + h_2)^2/R^2)} \quad (1)$$

This reduced thickness of parting we introduce to the relations (2) resp. (3), but only at layers with expressive higher excavation value.

From operating measurements analysis and the ideas mentioned above ensues design of two alternative correlation relations that will be tested by another measurements in situ.

Breaking characteristic of rock bench:

Alt.1

$$R = K_1 . K_2 . \sqrt{\Sigma(R_i^2 . m_i)/\Sigma m_i} \quad (2)$$

Alt.2

$$R = K_1 . K_2 . \ln(\Sigma(e^{R_i} . m_i)/\Sigma m_i) \quad (3),$$

where

R_i - breaking characteristic of the rock layer during drilling

m_i - thickness of rock layer

K_1 - constant expressing effect of the construction design of excavator wheel

K_2 - constant expressing the rate of breaking characteristics during excavating by excavator and during drilling.

5 CONCLUSION

This design of correlation between drillability and excavation resistances is the first approximation that takes out greater sample of measured values in various rock conditions. Further it takes out to establish for every excavator R_{KRIT} on the base of in situ measurements that would be the limit border of extraction ability.

REFERENCES

Bašta,L.,Jurman,J.,Vrbický,J. 1993. Experimentální mčrení pro vyuzití vrtných souprav k hodnocení rozpojitelnosti masivu. Výzkumná zpráva. VÚHU Most.

Jurman,J. 1986. Zjištování tvrdých poloh v nadlozí hnčdouhelých lomu. Kandidátská disertacní práce. VSB Ostrava.

Jurman,J. 1988. Rypné odpory jako kritérium rozpojitelnosti pri povrchovém dobývání. UHLÍ 5: 222 -226.

Rabia,H. 1982. Specific energy as a criterion for drill performance prediction. Journal rock mechanics and mining scienes and geomechanics abstract 19: 39-42.

Sekula,F. 1979. Teoretické a technologické aspekty rozpojova-nia hornín. Doktorská dizertacná práca. SAV Banícký ústav, Košice.

Geomechanics 93, Rakowski (ed.) © 1994 Balkema, Rotterdam, ISBN 90 5410 354 X

The cutting constant of the rock and the influence of the secondary fracturing of the rock on its value

V. Miklúšová, E. Lazarová & M. Labaš
Institute of Geotechnics of the Slovak Academy of Sciences, Košice, Slovakia

ABSTRACT: The ratio of the depth of the indentation of the indenter into the rock and the depth of the chipping out of the crater under the indenter at the static loading of the indenter was experimentally verified. The aim of the experiments was to determine the character of this ratio and to verify the influence of the secondary fracturing of the rock on its value, i.e. this ratio was verified in the domain of the intact rock and in the domain of the rock fractured by the previous pushing of the indenter.

The fundamental assumption of the mechanical disintegration of the rock is the mutual contact force acting of the indenter and rock. The force acting of the indenter on the rock causes the stresses and deformations in the indenter and the rock under the indenter and its contact surrounding. At the sufficiently high values of the stresses under the indenter that reach the critical values for the rock disintegration the disintegration of the rock and the fracturing of the contact area occurs. In some distance from the acting of the indenter is possible to observe the various phenomena of the rock fracturing (cracks, microcracks, dislocations) also in the area out of the volume of the disintegration.

With regard to the fact the mechanism of the disintegration at the technology of the boring by means of the large diameter fullface tunnel boring machines fitted with roller disc cutters is identical with the mechanism of the disintegration at the static pushing of the indenter into the rock the laboratory press as the experimental device for the determination of the cutting constant of the rock and the influence of the rock fracturing on its value was selected.

On this press the indentation strength σ_{vt} according the Srejner's method was being determined. The indenters were shaped as the cones with the flat end of the surface 1,2, and 3 mm^2. In the moment of the brittle failure the maximum depth of the indentation of the indenter into the rock h_z was determined. The depth of the chipping out of the crater h_v, after removing of the compacted core, was determined by means of the device for the length measuring SOMET 643.

For the experimental research we have chosen two types of the rock, andesite and limestone. Several samples of the particular rocks was used. the reproducibility of the results is good in spite of the fact the particular samples show some variance of the values of the reduced indentation strength, as it is shown in the Fig.1.

The aim was to work in the kvasi intact domain and, on the other hand, in the domain of the rock fractured by the previous pushing of the indenter. At the realization of the experiment we have utilized the results of the research of the russian authors (Baron,1977) considering the achievement that the particular domains of the pushing of the indenters in the distance larger then 7 mm are not mutually influenced. On the basis of this research the first indentation was performed into the

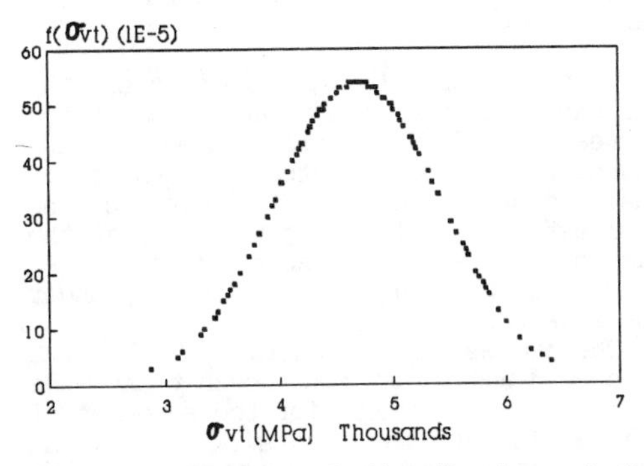

Figure 1. The normal distribution of the indentation strength measured in the andesite.

kvasi intact domain of the rock and the second pushing was in the distance 4 mm from the first one. By means of the described procedure we have obtained the statistic set of the data. With regard to the fact the depth of the indentation h_z is directly proportional to the indentation force F, and the depth of the chipping out h_v is also directly proportional to the indentation force, as it is known from the literature (Baron,1969), it have to be valid the ration h_z/h_v doesn't depend on the indentation force F (see Fig.2).

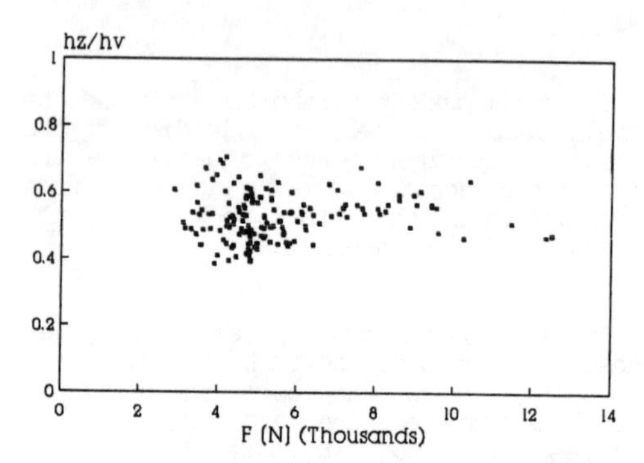

Figure 2. The dependence of the ratio of the indentation and the depth of chipping out on the pushing force.

The value of this ratio is given as the slope of the linear dependence between h_z and h_v (see Fig.3).

From the set of the measured data the average values of the maximum pushing force in the moment of the brittle failure of the rock under the indenter was calculated. Also the ratio h_z/h_v was calculated separately for the intact and fractured zone of andesite and limestone. Both calculations were performed for various surfaces of the indenters. The results are stated in the Table 1.

From the Table 1 is possible to see the influence of the volume factor. With the increasing of the contact area of the indenter the rock shows the lower values of the indentation strength, what is the manifestation of the larger number of the discontinuities or the ranges of the larger discontinuities in the rock directly under the indenter.

The values of the ratio h_z/h_v for the fractured and intact zone doesn't show the sufficient differences for the same set of the samples at the given size of the indenter. It is possible to assume that the fracturing of the rock doesn't influence the value of this ratio. Also from the Fig.2 can be seen the bigness of the varia-

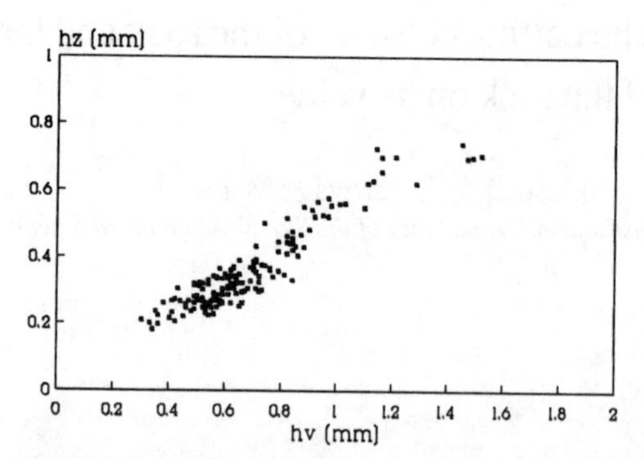

Figure 3. The dependence between the depth of the indentation and the depth of chipping out, regression r=0.873408.

tion of the values is comparable for both the fractured and intact zone. Equally the values of the ratio h_z/h_v for different surfaces of the flat end of the indenters don't show any trend and the difference of its values is probably caused by using of the another set of the samples (variation of strength properties).

For the whole set of the following characteristics were determined:

arithmetic average h_z/h_v = 0.5268
standard deviation S = 0.0639
coefficient of variation v_k = 12,13 %.

Table 1. The results of the experimental measurements.

Indenter surface [mm²]		Intact zone	Fract. zone
		Andesite	
1	F[N] σ_{vt}[MPa] h_z/h_v	5 122 5 122 0.5297	4 596 4 596 0.5144
2	F[N] σ_{vt}[MPa] h_z/h_v	8 254 4 127 0.5581	7 203 3 601 0.5522
3	F[N] σ_{vt}[MPa] h_z/h_v	12 110 4 037 0.4839	
		Limestone	
2	F[N] σ_{vt}[MPa] h_z/h_v	3 391 1 695 0.4828	2 941 1 470 0.4978

We can state the deviations from the average are probably caused only by variation of the rock properties.

At the experiments on the limestone the less numerous set of the results was obtained. From this set similar conclusion as at the andesite can be drawn. The value of the arithmetic average for limestone is : $h_z/h_v = 0.4903$.

On the basis of the experimental results and the literature sources (Krupa,1993) is possible to state the ratio h_z/h_v is the cutting constant of the rock.

This enable us to determine the contact surface of the disc cutter with the rock and so the critical stress under the disc cutter at the disintegration can be evaluated.

By means of this procedure it would be possible to determine the values of the strength of the disintegrated rock in the face of the working directly under in situ conditions.

REFERENCES

Baron,L.I.; Glatman,L.B.; Kozlov,J.N.; Melnikov,I.I. 1977. Disintegration of the rocks by means of the tunnel boring machines - Disintegration by the grouped bits /in russian/, Nauka, Moscow, p.54

Baron,L.I.; Glatman,L.B., Zagorskij,S.L. 1969. Disintegration of the rocks by means of the tunnel boring machines - Disintegration by roller bits /in russian/, Nauka,Moscow,p.27

Krupa,V.; Krepelka,F.; Bejda,J.; Imrich,P. 1993. The Cutting Constant of the Rock does not Depend on Scale Effect and Rock mass Jointing, 2nd International Workshop Scale Effect in Rock Masses 93, Balkema, p.63

Geomechanics 93, Rakowski (ed.) © 1994 Balkema, Rotterdam, ISBN 90 5410 354 X

Fuzzy decision-making of wear intensity of small diameter diamond core drilling

P. Imrich & V. Krúpa
Institute of Geotechnics of Slovak Academy of Sciences, Košice, Slovakia

ABSTRACT: Processes of wear intensity at diamond drilling are complex, integrated and relatively ill-known. Conventional experimental knowledge records (usually tables, graphs and/or equations) are not suitable for an efficient uncertainty reasoning. A fuzzy knowledge base is a suitable framework for acquisition of vague, sparse and inconsistent knowledge. A revitalization of Wear Intensity records and re-used literature knowledge items is a retrospective application of knowledge engineering algorithms. This is an ad hoc process and a general theoretical background does not exist. Therefore this paper presents a detailed description of the technique of the knowledge base establishing. No a priori knowledge of fuzzy mathematics is needed.

INTRODUCTION

At solving of any new technical problem we are forced to model huge systems, in which the man plays the main role. For description of such huge systems, it is necessary to handle a large amount of information and so accurate models of complicated systems becomes to be mathematically unsolvable, as we work with systems of which the accurate mathematic description doesn't exist. Fast evolution of theory of fuzzy sets indicates the fuzzy modelling is the part of mathematics that enable to integrate linguistic and mathematic description of real complicated systems and introduces the vagueness into the solution.

The domain of fuzzy applications, qualitative and semiqualitative modelling is chemical industry and food industry, some applications are utilized for diagnostics in engineering industry (Dohnal,1983; Babinec,1989).

Non-standard apprehension of modelling in the field of geotech nics is becoming to be relevant only in this period. This modelling leads to the utilization of various types of expert systems, techniques of neural networks, dynamic fuzzy expert systems and techniques that enable integration of deep and shallow knowledge.

In past, conventional modelling of primary models of disintegration of rocks and minerals we had utilized. This modelling was solved by means of "black box" method, that enables at large amount of experiments, by using of conventional statistic data handling, to find adequate model of proper solution of some theoretic problems of modelling of rock disintegration and optimization of disintegration. This attitude doesn't enable to solve a lot of problems in rock disintegration. Another attitude in non-standard modelling represents the utilization of fractal properties of objects under study. From the mathematic point of view such solutions lead to the models with real numbers derivations. As we define systems described by means of whole numbers derivations as standard, so the non-standardness, non-whole numberness expresses level of chaos in real systems, in which the real system differs from model.

At present the great attention is devoted to this description of uncertainty in various systems and area of its application is rapidly growing. First results are known in the field of rock disintegration as by means of this technique as succeeded to recognize properties of stochastic self-similarity on rocks (LiGongbo,1991).

OPTIMIZATION OF DRILLING

The effectiveness of drilling depends on many technical parameters (type of drilling rig, organization of drilling operations, technology of drilling,etc). If the technology of drilling is chosen the main parameter that influences the drilling performance is the selection of the bit and application of the optimal regime parameters of drilling on this bit. Improper choice of the bit for supposed rock and unsuitable regime parameters guarantee the drilling costs are rapidly rising.

The research of the drilling by means of the diamond core bits on the Institute of Geotechnics SAS was focussed from inserted bits with natural diamonds, through impregnated bits with natural or synthetic diamonds, to the bits with using of the synthetic polycrystallic diamond 'buttons' of various shapes. From the point of view of the wear and the construction of the bit the last mentioned type of bits substantially differs from conventional inserted and impregnated bits.

From the point of view of the construction the inserted diamond core bits are characterized by using of the relatively large natural diamonds inserted to the surface of the matrix of the bit. This construction indicates the mode of the wear of the bit that is wear out mainly on the diamond grains. The grains are ground down and their functional surfaces are diminished. This deterministic process is intruded by stochastic phenomena of the falling out of the bad inserted diamonds.

It is possible to pursue the gradual wear out of the bit explicitly by means of the measuring of the height of the bit and the decrease of the weight of the diamonds. This implicitly results in the decrease of the penetration rate at the constant regime parameters. The decrease of the penetration rate is possible to pursue as function of the time or penetration and the decrease is defined by exponential curve (Krupa,1981). The wear is the integral quantity the parameter of which is the wear intensity that is defined as the decrease of the height of the bit per drilled meter. Our research prove that there is a relation between the specific energy of the disintegration or the wear intensity and the decrease of the penetration rate of the drilling. This relation can be expressed by formula

$$v(F,n) = v_0(F,n) \, . \, e^{-kw(F,n) \, . \, L} \qquad (1)$$

where kw(F,n) is the wear intensity, which is indirectly determined by monitoring of the specific energy w(F,n), k is pro-rata coefficient that depends on the properties of the rock and bit, L is penetration, $v_0(F,n)$ is the initial penetration rate of drilling, F is thrust, n are revolutions.

From the point of view of the construction the impregnated diamond core bits are characterized by grains of the diamonds of small size whose are dispersed in the working edge of the matrix of the bit. When the one layer of the dispersed diamonds is worn out, the next layer becomes to be the working surface with the new diamonds. During drilling by means of the such bit the cycles of the grinding and polishing of the diamonds and matrix are repeated and such performance results in the accidental fluctuation of the achieved penetration rate at constant regime parameters of drilling. The analysis of the character of fluctuations shows the each layer of the diamonds of the impregnated bit is worn out in the same way as the inserted diamond bit, so the decrease of the penetration rate can be described by similar exponential curve (Bejda,1993). The difference is that the intervals of the penetration are shorter and the whole process is repeated according the number of such fictive layers of the bit.

The performance of the diamond bits inserted with polycrystallic 'buttons' is similar to the hard-metal 'button' drilling bits. The disadvantage of these bits is the affection of the 'buttons' to be destroyed or fallen out from the matrix.

The paper is focussed to the area of the impregnated diamond core bits because of the ambiguity of the performance of these bits. This ambiguity causes problems in selection of the proper bits and so hamper the optimization of the drilling. This paper indicates on the example of the evaluation of the wear intensity how is it possible to overcome such a problem by using of the fuzzy mathematics.

BASICS OF THE FUZZY REASONING

The theory of fuzzy sets allows the existence of a type of uncertainty due to vagueness or fuzziness rather than due to randomness alone. In its most basic sense, a fuzzy set is a set where objects have gradual rather than abrupt transition from membership to non membership.

A fuzzy set theory is based on the premise that the key elements in human thinking are not numbers but words. The most important feature of human thinking is not yet well-known ability how to extract from a collection of masses of data only such items of knowledge which are relevant to the task at hand.

Use of this theory is suitable when the subjective estimates are necessary or rather often, aspects under study cannot be taken into the consideration because of the vary nature of the objective methods. It is very difficulty e.g. to analyze engineering experience or human ability to draw conclusions which are based on analogy, using conventional statistics.

So the fuzzy theory is a generalization of conventional set theory. Because of this generalization, fuzzy set theory has a wider scope of applicability than conventional theory in solving problems which involve, to some degree, subjective evaluations.A linguistic value is a "value" that is given by words, e.g. high, low. Let us take as an example of a verbal variable DIA (Diameter of grains of diamonds). To quantify expert knowledge a set of verbal values is needed. Such an DIA "dictionary" could be: low, medium, high, very high. The DIA dictionary is rather subjective. Therefore, another expert can use a different dictionary. The reason why such quantifications of DIA are chosen by the expert is that he/she is used to quantify his/her knowledge by the entries in the dictionary. A purely verbal quantification is too vague (Dohnal,1983).

The fuzzy set represents an optimal trade-off between the absolutely (pseudo) precise number and the vague verbal quantification. The linguistic value is transformed into a fuzzy set by the specification of a grade of membership, e.g. the verbal value "medium" DIA is transferred into a fuzzy set by the grade of membership function given in Fig.1. Any fuzzy definition is a formalization of a rather subjective expert feeling. So by means of the functions of a grade of membership it is possible to define intersections of functions that determine similarity of linguistic values. The similarities indicate the feeling of the expert, for example, how similar is the low and medium wear intensity.

If the fuzzy values have a non-empty fuzzy intersection this intersection is a "bridge" through which information about one value can support reasoning about the next (similar) value and vice versa.

A fuzzy model is a set of conditional statements:

if $A_{1,1}$ and and $A_{1,n}$ then B_1 or
if $A_{2,1}$ and and $A_{2,n}$ then B_2 or
. . . .
. . .
if $A_{m,1}$ and ...and $A_{m,n}$ then B_m

where fuzzy sets

$A_{i,j}$, B_i; i=1,2,... m, j=1,2,...n

are one-dimensional fuzzy sets and can be easily specified or/and modified using points a, b, c, d (see Fig. 1), $A_{i,j}$ is one-dimensional fuzzy set quantification of the jth variable in the ith conditional statement, B_i is one-dimensional fuzzy set, quantification of a dependent variable in the ith conditional statement, m is number of conditional statements, n is number of variables (dimensionality of knowledge bases). However, flexibility of man-machine dialogue requires flexibility in the definition of dependent variables. Any statement represents a certain point of view. Therefore, the same fact can be presented in many ways.

An example of a "Wear Intensity" conditional statement is:
"If DIA is Very High then Wear Intensity is roughly 0.8 mm/m".

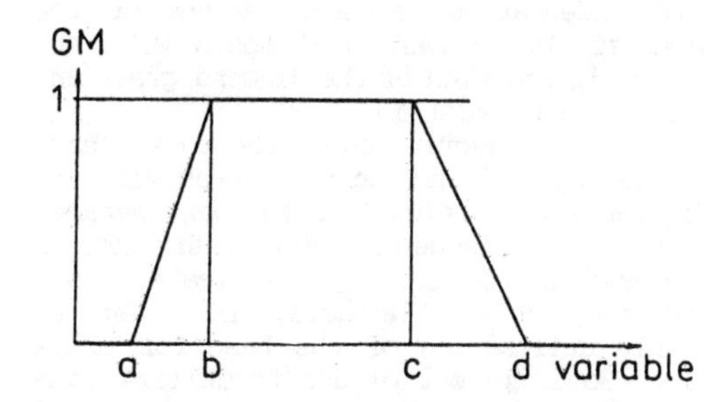

Fig.1 Function of a grade of membership, a, b, c, d characterize GM (grade of membrership) from non-membership (a, d) to full membership (b, c).

TECHNIQUE OF FUZZY ATTITUDE IN DRILLING

For evaluation of the problem under study by means of the fuzzy technique it is necessary to establish a knowledge base. The knowledge base contains statements whose represent experiments or data from literature sources or experiences in heuristic form. The most important item in establishing of the knowledge base is selection of the only but all possible parameters that influences, in our case, the wear intensity of the impregnated diamond core bits. So these parameters as independent variables result in the dependent variable of the Wear Intensity.

The independent variables consist of the parameters of the diamond drilling bit, rock and applied regime of drilling. The main parameters of the diamond bit are the quality of the diamonds, hardness of the matrix of the bit and size of the diamonds. For establishing of the knowledge base only experiments with 8-channel bits of the same shape were used so no additional construction parameters of the bits are needed.

Another group of the parameters is the regime parameters. These parameters were evaluated continuously during drilling on the experimental drilling stand. Besides the thrust and revolutions of the bit, it was possible to neglect the hydraulic regime of the experimental drilling so as the drilling was carried out with water mud without additives at constant throughflow.

The last group of the independent parameters is the rock characteristic. This characteristic is rather subjective and so the fuzzy technique was fully utilized at evaluation of the rock characteristic.

As the main parameter of the rock from the point of view of the diamond drilling the Srejner's cone indenter hardness of the rock was chosen. The main reason of using of this kind of the point load tests is that indentation hardness is one of the most similar conventional tests with the actual interaction of the diamond grain and rock (McFeat-Smith,1977).

For full description of the rock behavior during drilling the rock abrasivity and drillability were used as the next parameters. The knowledge base was established on the basis of the not very complicated tests and easy obtainable data, in order to enable utilization of the base for practice. So it is out of use to utilize some of the complicated conventional tests of the abrasivity or drillability. Here the experiences of the drilling experts were used. Another reason of this assessment is that at using of some extensive testing of the drillability or abrasivity with output that is distributed into the 10-20 sorts the dimensionality of the base is rapidly rising and the base becomes to be out of practical use (Howarth,1986).

Another item of the creation of the base is establishing of the dictionaries of the particular parameters and functions of a grade of membership that represents expert's feeling of similarity between items of the dictionary. The number of the linguistic values of the particular dictionaries gives the dimensionality of the base. The more values the higher dimensionality, the higher accuracy of the model, but the more conditional statements is necessary to describe the model. So it is the task for expert to recognize how many values to choose for proper description of the problem under study. For example it is non-sense to use 20 items of the dictionary of the rock hardness, if the variance of the distribution of the hardness tests of the same rock covers 4-5 such sorts and although the Gauss's distribution gives some typical value this value cannot represents the whole rock (Bejda,1988). The same rock according such testing could be sorted to the different sorts as non-similar and it haven't to be true. On the other hand too coarse dictionary causes diminishing of the accuracy of the model. So, for example the wear intensity dictionary uses the six values as the most relevant output and the regime parameters, that are less relevant from the point of view of the selection of the bit have three (revolutions) and four (thrust) linguistic values. Relatively coarse are the dictionaries of the rock parameters, but the combination of three rock parameters gives 36 possibilities of rock sorting.

Another attitude of the expert is applied in creating of the functions of a grade of membership. Based on experiences it is possible to define how similar are the items of the dictionary. The higher is the similarity of the item the higher level of the generalization of the output can be achieved, but improper similarity causes confusion of the output (Babinec,1992).

So based on above presented attitude the knowledge base for evaluation of the wear intensity of the impregnated diamond core bits was established.

POSSIBILITIES OF APPLICATION OF THE KNOWLEDGE BASE

The fuzzy technique presents a generalization of conventional statistic methods. So the established knowledge base presents the generalization of the selection of the optimal impregnated diamond bit from the point of view of the life of the bit (so as the wear intensity is the criterion of life of cutters). The conventional statistic methods don't enable multi-parametric evaluation of the influence of the particular parameters because of various types of rocks and bits. So the fuzzy technique represents the higher level of the generalization of the selection of the optimal bit.Another advantage of the fuzzy technique is the input of the base is optional. So the query to the base can be generalized with ALL specification (any value is possible) of at some unknown parameters, or another variable can be chosen as dependent output, for instance in order to find the rock on which the given bit will have the minimum wear intensity.

The knowledge base is handled by the fuzzy expert system with using of personal computer. The input of the base is the specific query and the output is a recommendation of the expert system, which is presented by the list of all possible values of the dependent variable, resulting from the query, with their grades of membership to the query (Babinec,1989). As a grade of membership can be from the interval <0;1> the most relevant "answer" is the value that posses the grade of membership from the interval ,for example, <0,8;1>. It depends on expert's feeling which value of the output is the threshold between relevant and non-relevant.

Another possibility of the utilization of the base is vector optimization which enables, in transparent form ,to present the multicriterial optimization of any parameter of the base. The output is then a list of solutions from which the expert can choose the most appropriate solution in given conditions.

So the above presented techniques give the possibility for utilization in any standpoint of the optimization of the drilling or boring and, after some upgrade of the base, there is a possibility of optimization of drilling from the point of view of costs per drilled meter or penetration rate of drilling. So the technique covers the main three criteria of optimization:

- the minimum wear or the maximum life of the cutters
- the minimum costs per drilled meter
- the maximum penetration rate of drilling.

In order to achieve the most appropriate output it is possible to utilize another techniques of data handling. For example for the evaluation of the level of chaos of the knowledge base the fractal evaluation of chaos can be used (Dohnal,1992).

CONCLUSION

Evolution of new non-conventional methods of data handling enables more generalized decision-making without extensive lost of information (in fuzzy technique it is possible to utilize incomplete inputs or failed experiments,e.g). These methods are based on highly efficient utilization of computers and enable quick solving of the tasks based not only on statistic data, but on experiences and feelings of the experts.

Output of these methods is possible to obtain in conventional form so the output can be used for another conventional data handling.

For fully utilization of these techniques there is a demand of cooperation of drilling or rock mechanics expert with experts in knowledge engineering in order to achieve the best transition of the achievements of the knowledge engineering in applied rock mechanics.

REFERENCES

Dohnal,M. 1983. Linguistic and fuzzy models, Computers in Industry, No.4, pp.341-345.

Dohnal,M. 1983. Fuzzy simulation ofindustrial problems, Computers in Industry, No.4, pp.347-352.

Babinec,F.;Dohnal,M. 1989. Transfer of knowledge in Chemical Equipment Reliability,In:Collect.Czech.Chem.Commun., Vol.54, pp.2692-2710.

Li Gongbo,Tang Chunan,Xu Xiaohe 1991.Description of G-S particle-size distribu tion of rock communition with fractal geometry, Transaction of Nonferrous Metal Society of China, Vol.1, No.1, pp.35-38

Krupa,V. 1981. Mathematic modelling of diamond drilling on the basis of the stand research. /in slovak/ PhD thesis. Mining Institute SAS, Kosice.

Bejda,J.;Krupa,V.;Sekula,F. 1993. Algorithm of control of disintegration of rocks at drilling from the point of view of costs per meter of bored hole, Geomechanics'93, Balkema

McFeat-Smith,I;Fowell,R.J. 1977. Corre lation of rock properties and the cutting performance of tunnelling machines, Proceedings of a Conference on Rock Engi neering, Newcastle upon Tyne, England, pp.581

Howarth,D.F. 1986. Review of rock drillability and borability assessment methods, Trans. Inst. Min. Metall. (Sect.A: Min. Industry), 95, October 1986.

Bejda,J. & col. 1988. Statistical evaluation of the reduced indentation strength of the model rocks for testing of the diamond tools. /in slovak/ Folia montana, Proceedings of the Mining Institute SAS, Kosice.

Babinec,F. 1992. Cognitive analysis of fuzzy knowledge base, Advances in model ling & analysis, ASME Press, France, Vol.23, No.1, pp.27

Babinec,F. 1989. Diagnostic and Reliability Expert System, Proceedings of the 3rd Intern. Symposium "Operational reli ability in Chemical Industry", Bratislava, pp.284

Dohnal,M. 1992. Reliability Knowledge and Fractal Evaluation of Chaos, Microelectronics and Reliability, Vol.32,No.6, pp.867-874

Geomechanics 93, Rakowski (ed.) © 1994 Balkema, Rotterdam, ISBN 90 5410 354 X

Prediction of weak zones and rock boundaries on the basis of monitored parameters at full-face boring

F. Krepelka, V. Krúpa, F. Sekula & O. Krajecová
Institute of Geotechnics of the Slovak Academy of Sciences, Košice, Slovakia

ABSTRACT: Paper presents a method that enables continuous evaluation of change of rockmass on the face of bored working by means of indirect methods in real time. Case study presents examples of evaluation of monitored parameters on raise borer Bespadrill P-1 and tunnel boring machines Wirth TB-II-33H and RS 24-27H. The sensibility and utilization of results for prediction of weak zones a rocks boundaries is introduced.

INTRODUCTION

One of the achievements of the research of rock disintegration by means of fullface tunnel boring machines (TBM) is the notion that it is possible to express quantitative and qualitative state of disintegrated rockmass by means of monitoring of TBM performance. As change of quality we comprehend the rock type change which is accompanied by change of all mechanical characteristics. We characterize the quantitative change by means of changes of some properties of the same rock, for example the change of state of rock weakness, the change of hardness, the change of strength, etc. The identification of the rock boundary is then the identification of qualitative change of state of disintegrated rockmass, the identification of weak zone (or fault) is, in most cases, the identification of quantitative changes of rockmass. In the process of boring the cases when simultaneous changes of qualitative and quantitative characteristics of rockmass arise or these changes are concentrated in spatially and timely short interval of the TBM advance may occur. The areas of existence of weak zones in rockmass are, as our results show, always in advance accompanied with changes of quality or quantity. This enables the prediction of weak zones. Based on present results we can state the identification of quantitative changes is easier to be done.

DESCRIPTION OF THE METHOD

From the monitored variables of fullface boring we can calculate additional variables that more complex describe the studied phenomena in the zone of interaction disc cutter - rock or cutterhead of TBM - rockmass. One of these variables is the specific energy of disintegration defined as the quantity of the energy consumed per unit of the disintegrated volume of the rock. For technical purposes the following relation is used for its calculation

$$w = \frac{2\pi . n . M_k}{S . v} \qquad (1)$$

,where n are revolutions of cutterhead of TBM, M_k is torsional moment of cutterhead, S is surface of disintegrated face of the tunnel and v is penetration rate of boring.

The torsional moment and penetration rate of boring are variables that are influenced by construction of cutterhead of TBM, construction of disc cutters, state of functional surfaces of disc cutters (surfaces of direct contact with rock), parameters of regime of boring (thrust by means of which the TBM acts on rock, and revolutions of cutterhead) and parameters of the rock which is just disintegrated.

If we want to determine the rock parameters for purpose of prediction of weak zones and rock boundaries it is necessary to stabilize the influence of the complex of other parameters, i.e. to secure their kvasi constant character. At the same time it is needed to give plausible answer to the question, by means of which precautions is possible to secure it. These precautions will be discussed later in this paper.

The second variable which is possible to calculate from monitored data is the disintegrating stress under disc cutter σ. We determine σ by calculation from the relation

$$\sigma = \frac{F_d}{S_k} \qquad (2)$$

where F_d is the force by means of which the disc cutter acts on rock at disintegration, S_k is the contact surface of the disc cutter with the rock in the moment of failure. The force F_d is determined as the vector sum of axial and tangential force that is acting on the disc cutter at combination of the rotation of the cutterhead of the TBM and the disc cutter. The contact surface S_k is calculated by means of the relation

$$S_k = \frac{\pi . h . D}{\cos\frac{\alpha}{2}} \cdot \frac{\arcsin\frac{2\sqrt{D.h - h^2}}{D}}{180} \qquad (3)$$

,where h is the depth of chipping of the rock under disc cutter, d is diameter of the disc cutter, α is the angle of the edge of functional surface of the disc cutter. At using of the relation (3) for calculation of S_k we consciously admit some inaccuracy which influences the accuracy of σ calculation determined by the relation (2). This inaccuracy results from the calculation of the relation (3), where it would be more exact to use the value of the depth of indentation of the disc cutter into the rock in the moment of chipping out p, as it is valid $p<h$. But it is impossible to determine this p value from the monitoring of boring. The results published in (Krupa,1993;Miklusova,1993) show for the same rock weaken in various degrees the ratio of the depth of indentation of the bit and the depth of chipping out of the rock is constant and so this mistake in the calculation of the relation (3) is only formal.

The classic theory of elasticity defines the quadratic law of growing of the specific energy of elastic deformations in dependence on the stress as

$$w = \frac{1}{2} . \frac{\sigma^2}{E} \qquad (4)$$

which is valid for domain of elastic deformations and purely elastic material (Mellor,1972;Novikov,1988). In the modified form in respect with defined proportionate deformation ϵ is possible to interpret the quadratic law (4) as linear growing of specific energy, respectively deformation work, on stress and strain

$$w = \frac{1}{2} . \sigma . \varepsilon \qquad (5)$$

where the domain of validity is domain of purely reversible deformations (Jaeger, 1979). But in rock disintegration the loading of the rock that leads to the destruction of the rock occurs. At destruction, it is necessary to give rise the plastic non-reversible deformations that cannot cause the growing of the deformation energy and so the quadratic law of growing of specific energy on stress is changed to the linear law in the domain of prevailing plastic deformations (Kostak,1982) which can be expressed by common statement

$$w = k . \sigma \qquad (6)$$

Some authors affirm (Teale,1965; Hustrulid,1971) at rock disintegration by mechanical modes exists such parameters of regime of disintegration at which the specific energy of disintegration has minimum. In the same breath they add that at this regime of disintegration of rocks functional bonds between the specific energy of disintegration and the mechanical properties of rocks is possible to reveal. Our measurements (Krupa,1981) support this claim. The situation becomes to be a little confused if we are not able to secure this conditions during rock disintegration. The approximate solution, sufficient for technical interpretation of mutual bonds of specific energy of disintegration and disintegrating stress, is the case of securing of approximately constant conditions of loading of rockmass. In the process of boring the constant regime that is characterized by kvasi constant penetration rate of boring is approaching to this state. Respectively, the information powerfulness of the specific energy of disintegration determined by calculation considering disintegrating stress is not fully disappearing even at control of the process so that the kvasi constant thrust and revolutions would be achieved. By means of this method the data illustrated in the paper were obtained.

Also it is needed to exclude the cases of abrupt change of quality of functional surfaces of disc cutters because this case acts on the value of specific energy so as the change of qualitative or quantitative characteristics of rockmass.

From comparison of the values of the specific energy of disintegration calculated from the data monitored on various TBMs is clear the important role in above mentioned process has, besides the construction of cutterhead and various disc cutters (range of tracks of discs and parameters α and d of discs), the diameter or size of bored tunnel. Explicitly, it is possible to document in the relation (1), see parameter S, but implicitly, in our opinion, this phenomenon results from the scale effect of rocks. As the size of the tunnel is growing so the surface of the face and 'size of rock sample' is also growing. Therefore the stresses necessary to disintegrate face are going down in sense of the statistic theory of strength (Kraev,1989).

In reverse order is possible to give answer on the problem of the scale effect by statement: if functional bonds between specific energy of disintegration and disintegrating stress are valid specific energy of disintegration have to reflect the scale effect of rock.

From the above presented discussion results that with the simple mathematic apparatus used in our technique of the assessment of state of rockmass is impossible to generalize the technique for various TBMs, as we are not able to explain exactly the influence of construction differences of cutterheads of particular types of TBMs. The technique enables to solve the given aim, defined in the title of the paper, only for given TBM at discussed limitations.

The example of evaluation of the monitoring of input and output parameters of boring according above presented technique at two tectonic faults that were later identified physically in bored intervals 50 - 50.5 m and 84 - 79.5 m is demonstrated in Figure 1. The measurements were carried out at the uphill boring of the enlargement borehole of 1.8 m in diameter in porhyroide (uniaxial compressive strength 114 MPa) by means of the raise borer Bespadrill P-1, locality Slovinky, Slovakia. The regression and correlation parameters are stated in Table 1. From the data that belong to the raise borer Bespadrill P-1 in Table 1 imply the linear relation σ=f(w) for both measured intervals possess the consistent parameters of the toleration of the errors of the measurements.

The measurements illustrated in Figure 1 confirm the linear relation between specific energy of disintegration and disintegrating stress at which the absolute values of specific energies and calculated stresses enable still to identify the influence of the depth of occurrence. It is clear, the strength of porphyroide in the rockmass is higher as the laboratory defined uniaxial compressive strength. This strength is growing with the depth of occurrence. The influence of the depth of occurrence is indicated in Figure 1. by higher value of the specific energy of disintegration of the same rock in the bigger depth at the boring of the enlargement borehole. The values of specific energy of disintegration and disintegrating stress measured in the zone of tectonic faults are lower as in the zone of intact rock. This fact is good visible in Figure 2 which illustrates the values of logged parameters in dependence on penetration of boring. At the same time, it is necessary to state the absolute values of disintegrating stresses are lower as the stresses of corresponding uniaxial compressive strength and so it is impossible to compare absolutely these methods of evaluation of the rock.

We suppose these deviation are caused by the inaccuracy of the determination of the depth of indentation of the disc cutter as it was discussed at the relation (3).

In Figure 3 the results of the monitoring of boring of the new drainage gallery at Hodrusa from interval of boring 4540-5020 m, bored by means of the TBM Wirth TB-II-330H, are treated. The TBM was driven

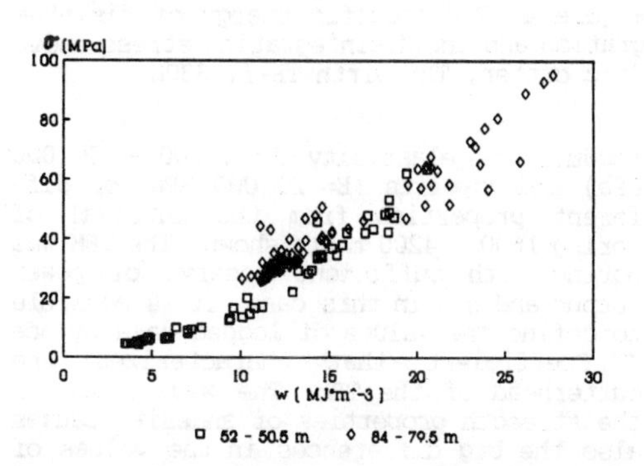

Figure 1. Specific energy of disintegration and disintegrating stress under disc cutter at boring by means of the raise borer Bespadrill P-1.

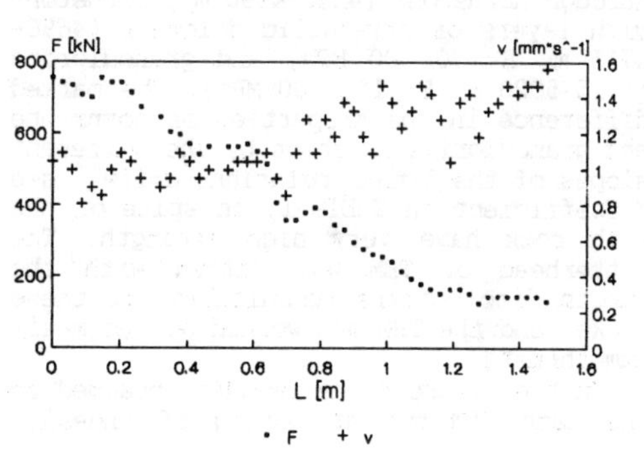

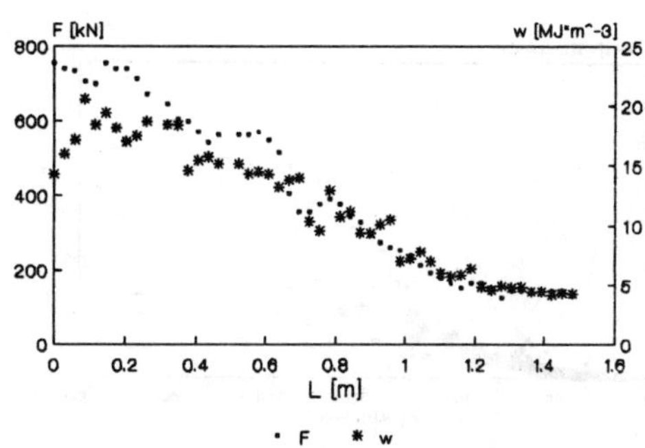

Figure 2. The monitored data from the enlargement borehole, bored by means of the raise borer Bespadrill P-1

Tab. 1 Regression Output: $\sigma = f(w)$

	1	2	3	4	5	6
Constant	-11,3966	-9,61561	-2810,84	-799,183	-870,000	-44,6959
Std Err of Y Est	4,323992	5,179993	1128,733	227,4781	355,7268	14,75578
R Squared	0,941367	0,906218	0,915921	0,957081	0,876399	0,913204
No. of Observations	54	88	50	14	85	259
X Coefficient (s)	3,196145	3,409807	66,62251	18,11241	26,09598	5,365622
Std Err of Coef	0,110615	0,118283	2,913487	1,107221	1,075706	0,103185

1 - Bespadrill P-1, porphyroide, 52-50,5 m
2 - Bespadrill P-1, porphyroide, 84-79,5 m
3 - Wirth TB-II-330H, hornstone, 4540-4690 m
4 - Wirth TB-II-330H, hornstone with layers of crystallic dolomite, 4690-4715 m
5 - Wirth TB-II-330H, granodiorite, 4715-5020 m
6 - Wirth TB-II-330H, andesite, rhyolite, 1600-4200 m

through hornstone (4540-4690 m), hornstone with layers of crystallic dolomite (4690-4715 m, E= 70 000 MPa) and granodiorite (4715-5020 m, E= 150 000 MPa). The marked difference in the properties of hornstone and granodiorite is shown by the different slopes of the linear relations $\sigma=f(w)$, see X Coefficient in Table 1, in spite of the both rock have very high strength. The cutterhead of TBM was fitted with the smooth disc cutters (unsuitable for these rocks) and the TBM had worked at the maximum thrust.

In the Figure 4 the results obtained on the same TBM but at boring of andesite

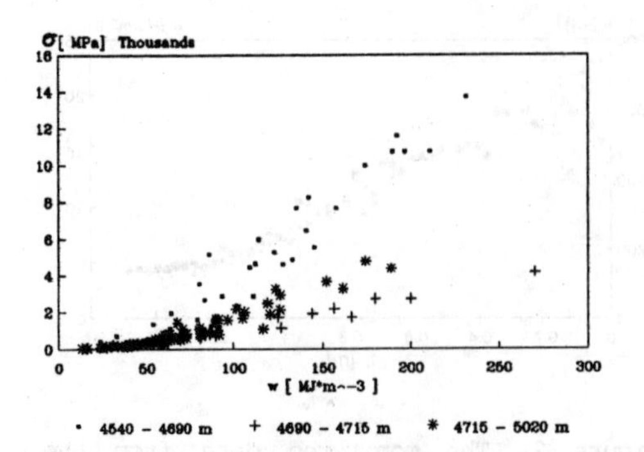

Figure 3. The specific energy of disintegration and the disintegrating stress under disc cutter, TBM Wirth TB-II-330H

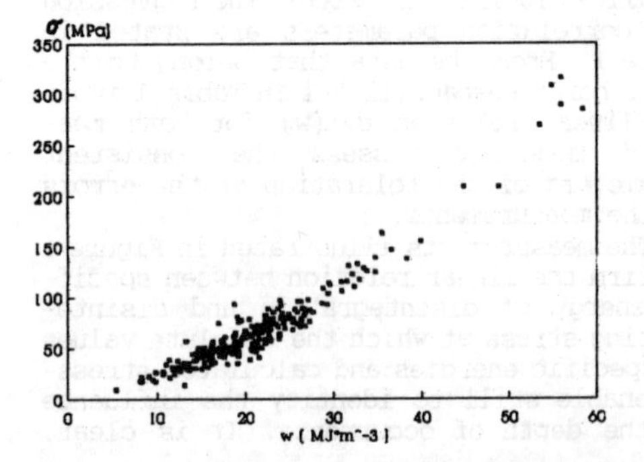

Figure 4. The specific energy of disintegration and the disintegrating stress under disc cutter, TBM Wirth TB-II-330H

(modulus of elasticity E= 6 000 - 26 000 MPa) and rhyolite (E= 20 000 MPa) of different properties from the interval of boring 1600 - 4200 m are shown. The TBM was boring with sufficient reserve of power output and so, in this case, it is possible to define the values of logged data by one X Coefficient that characterizes the cutterhead of the TBM. The wide range of the strength properties of andesite causes also the big differences in the values of the specific energy of disintegration.

CONCLUSION

By the monitoring of selected parameters of the process of boring under in situ conditions is possible to obtain information about relative stresses needed to disintegrate the rock on the face of the tunnel. Under precisely given conditions of boring these stresses of disintegration enable to determine in relative proportionate values approximate properties of disintegrated rock and so continuously monitoring of the changes of rocks. By means of the presented mathematic procedure is possible to determine for given type of TBM the mutual domain of values of monitored and calculated parameters corresponding to the stable and unstable parts of the tunnel. This method enable to identify the weak zones and faults in the rockmass during boring, and because these weaknesses are in most cases accompanied by the changes of the rock properties in its surroundings, enables to predict the occurrence of weaknesses. The more difficult task is precise determination of properties of rocks and rock boundaries. This task inevitably needs to obtain more data from boring by means of TBMs for further successful solving. In spite of this, the present knowledge show the task is solvable.

REFERENCES

Hustrulid,W.A. & Fairhurst,C. 1971. Int.J. Rock Mech.Min.Sci.Vol.8,p.335

Jaeger,J.C.and Cook,N.G.W. 1979. Fundamentals of Rock Mechanics, Chapman and Hall, London

Kostak,B. 1982. Propabilistic Strength Prediction In a Rock Massif, In: Proceedings of Institute of Geology and Geotechnics, Acta Montana, No.58, Praha

Kraev,Ju.K. 1989. IVUZ Gornyj zurnal. Vol.6, p.14, (in Russian)

Krupa,V. 1981. Mathematic modelling of diamond drilling on the basis of the stand research./in slovak/ PhD thesis. Mining Institute SAS, Kosice.

Krupa,V.;Krepelka,F.;Bejda,J.;Imrich,P. 1993. The cutting constant of the rock does not depend on scale effect and rock mass jointing, 2nd International Workshop Scale Effects In Rock Masses 93, Balkema, Roterdam, pp.63

Mellor,M. 1972. Int.J.Rock Mech.Min.Sci. Vol.9,p.661

Miklusova,V et al; 1993. The characteristic coefficient of rock and influence of the secondary failure on its value. Geomechanics'93

Novikov,G.Ja. 1988. IVUZ Gornyj zurnal. Vol.4,p.8, (in Russian)

Teale,R. 1965. Int.J.Rock Mech.Min.Sci. Vol.2,p.57

Geomechanics 93, Rakowski (ed.) © 1994 Balkema, Rotterdam, ISBN 90 5410 354 X

Secondary rock breaking by use of impactors

Š. Krištín & M. Maras
Technical University, Košice, Slovak Republic

ABSTRACT: The influence of the secondary explosiveless breaking of hard rocks. The results of the experimental research, theoretical considerations. The selection of a suitable impactor. The possibilities of the impactor gripping in mining conditions.

1 INTRODUCTION

Even at the optimal technology of the primary rock breaking by blasting a certain share of oversize lumps is developed. This share is influenced by:

* the technical level and the technological discipline of drilling and blasting,
* the tectonic state, physical-mechanical and technological properties of the broken rocks,
* the parameters of the technological line, where further manipulation and processing of primary broken ore is carried out (LHD's shovel volume, dimensions of primary crusher input opening, belt conveyors' width etc.).

In the Slovak ore and magnesite mines the share of oversize lumps in the ore after the primary breaking is 5 - 20 %. Lower share is in the magnesite mines were $d_{max} \leq 0,8$ m while in ore mines $d_{max} \leq 0,4$ m. Relatively small d_{max} value in ore mines is influenced by the dimensions of the ore pass discharge opening.

In the Slovak ore and magnesite mines the secondary breaking is carried out:

* directly in the stope in the process of loading (especially after blastings),
* and in last five years by impactors as well.

The operation of the secondary breaking in the stope by blasting indirectly influences the continuity of the loading. At the loading of oversize ore lumps the LHD performance decreases.

The mechanical secondary breaking with airpicks either at the loading place or on the ore pass grate is of low efficiency and very strenuous.

Experiments of the secondary breaking in abroad (Byzov (1975), Duyse (1978), Lobanov (1983)) have shown that at the utilization of heavy-duty loading and hauling technique the most suitable seems to be the mechanical breaking by heavy hammers - *impactors*.

That is why the investigation of methods of mechanical secondary breaking was carried out by many authors and in a complete way by Krištín (1981 - 86). This investigation was oriented on these problems:

a) the development of the optimal impact energy methodology determination for the secondary breaking of ores and magnesite,

b) the selection of suitable impactors for conditions of underground mining of ores and magnesite by the use of the railless mechanization (drilling jumbo, drilling rig, LHD),

c) the kinematic solution of the manipulation beam for the impactor gripping with the reach over the total area of the grate and the ore pass with the possibility of its tilting out of the breaking area after the interruption of the breaking.

2 DETERMINATION OF THE IMPACT ENERGY IN THE CONNECTION WITH THE OPTIMAL SECONDARY BREAKING OF GIVEN ROCK TYPE

In the case of the secondary breaking the tool is in the contact with the rock. A part of this energy is consumed for the crushing of a rock part under the tool, a part is changed into the heat and penetration or wedging of the boulder into the platen and only the remnant of the energy is used for the breaking. At the working out of the fundamentals of the theory of the boulders' mechanical breaking we came out from the theoretical works of Rittinger, Kick-Kirpitchev, Bond, Rebinder, Levinson, Kuklin and others.

Kuklin (1969) was searching the rock breakability by the impact energy. In the empirical formula determining the total impact energy raquired for the breaking he accepted: punch strength, Poisson's number, modulus of elasticity, the area of newly created surfaces and scale (dimension) factor. At the same time he used also some coefficients which include: elastic weves' energy transmission and structure heterogeneity. According to the breakability by the impact energy he divided rocks into three groups:

1. easy breakable rocks (c_f=0.84, s=0.25),
2. medium-breakable rocks (c_f=1.81, s=0.27),
3. hardly breakable rocks (c_f=1.81, s=0.23).

Coefficient

$$c_f = \frac{\sigma_c^{1.5} . \sigma_p^{0.5} (1-\nu)^{2.5} . (1+\nu)}{\nu^{1.5} . E}, \quad (1)$$

where σ_c is the uniaxial compressive strength, σ_p is the punch strength, ν is the Poisson's number, E is the modulus of elasticity and s is the coefficient depending on the rock structure heterogeneity.

The total energy required for the secondary breaking of a rocky boulder is expressed by the following equation:

$$W_c = c_f \; . \; c_w \; . \; S_p^{1.5} \; . \; n^5, \quad (2)$$

where c_w is the coefficient expressing the elastic wave transmission energy and the process of changes of the impact load into the forces which break the rock, S_p is the area of newly created surfaces and n is the number of the hammer's strokes needed for the breaking to the required fragmentation.

The total energy calculated according to the equation (2) is influenced by the great extent of input data and coefficients. The secondary breaking by the impact energy is also influenced by the position of the impact place (point) against the discontinuities, thrust on the tool, quality of the platen on which the boulder is placed, tool geometry, extent of fractures developed at the primary breaking. The acquired results in this way have a great dispersion. According to Kuklin (1969) the optimal impact energy for individual rock types is 1 - 3.5 kJ.

Weustenfeld (1980) carried out experiments on the breaking by the HM 600 impactor. His experiments showed that for the optimal rock breaking by the impact energy it is necessary to know the rock structure, tectonics and thereby also those places in which the cohesion is the lowest. He says that at the impact energy of 7 kJ most rocks could be broken. This datum holds true more for rocks with the boulder volume over 2 - 3 m^3.

Henry V. Duyse (1987) divided rocks from the point of view of the mechanical secondary breaking into 5 groups in the dependence on their uniaxial compressive strength and achievable breaking performance at various impact energies of the impactor (Figure 1).

1st class - rocks with the uniaxial compressive strength over 100 MPa,

2nd class - rocks with the uniaxial compressive strength 70 - 100 MPa,

3rd class - rocks with the uniaxial compressive strength 35 - 70 MPa,

4th class - rocks with the uniaxial compressive strength 7 - 35 MPa,

5th class - rocks with the uniaxial compressive strength below 7 MPa.

But Figure 1 does not take into

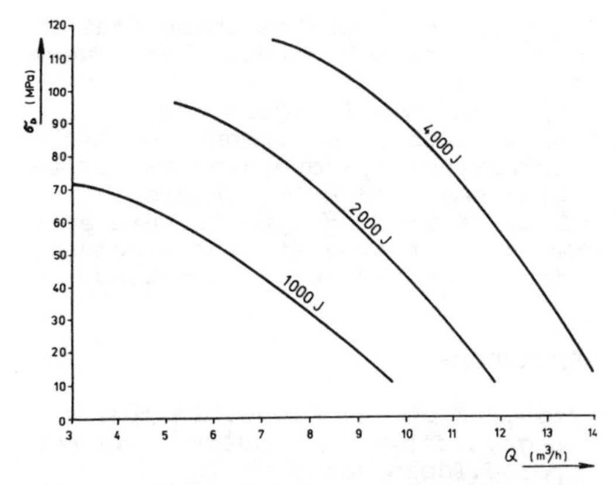

Fig. 1 Dependence of the breaking performance on the strength σ_D at various impact energies of the impactor

consideration the rock with the σ_c > 120 MPa and breaking performances over 14 $m^3.h^{-1}$.

Ores and magnesite have often uniaxial compressive strength higher than 100 MPa and according to Figure 1 the performance 12 - 14 $m^3.h^{-1}$ would not suffice. NORDSTAHL (Germany), one of the greatest impactor producers in Europe, recommends for higher breaking performances impactors with the impact energy 3 - 10 kJ.

Generally, at the determination of the impact energy value required for the secondary rock breaking of given boulder type it is possible to come out from the failure criterium of the energy quantity which the boulder absorbs till the moment of its failure, i.e.

$$W_{sp} \geq W$$

where W is the maximal energy quantity accumulated in the boulder being broken till the moment of its failure.

The limit energy W_{sp} is a very complicated function and comprises the following factors:

* shape characteristic of the boulder,
* rock's strength,
* dimension factor,
* moduluses of elasticity and deformation,
* boulder anisotropy and microtectonics,
* strength and state of the platen on which the boulder is put,
* impact energy, impact frequency,
* thrust on the impactor tool,
* orientation of the boulder being broken against the tool.

That is why we concentrated ourselves on these factors which substentially influence the breaking process. For this condition the most suitable are experiments of breaking with adjusted samples and at the end the direct boulder breaking on a special stand. Besides the stable gravitation stand also a mobil one was developed. This stand can be used for experiments directly in the operation.

3 RESULTS OF THE EXPERIMENTAL INVESTIGATION AT THE DETERMINATION OF THE OPTIMAL IMPACT ENERGY

The author suggested and constructed the gravitation stand which made possible to adjust the impact energy within the extent from 100 J to 12 kJ with the possibility of change after each 100 J. The impact velocity was 1.4 - 6.26 $m.s^{-1}$. Dimensions of samples were maximally 2 x 1.5 x 0.7 m, i.e. the volume was maximally 2.1 m^3.

For the breaking experiment of unadjusted samples the following rocks were used: graphitic phyllite, porphyroid, metamorfic siderite and magnesite. The strength of the samples σ_c was 80 - 202 MPa. At samples with the σ_c ≤ 120 MPa the impact energy W_i = 1500 - 2000 J was sufficient for the breaking. At samples with the σ_c ≥ 120 MPa W_i was 2500 - 3000 J. For this required impact energy for example H-4x - H-7x NORDSTAHL impactors are suitable and for the impact energy till 2000 J domestic HK-2000 hammer is suitable.

Experiments in mining conditions were carried out by the H-3x impactors installed on adjusted PN-1500 LHD manipulation beam (Figure 2).

On the basis of the experience from experiments in Banské stavby, š.p., Prievidza were developed and produced 2 prototypes of mobil carriers on the basis of the PN-1700 LHD with a special manipulation beam for the NORDSTAHL -255 impactor, type IRV-250.

The achieved performances at the magnesite breaking with the boulder volume V = 0.9 - 6.8 m^3 were Q =

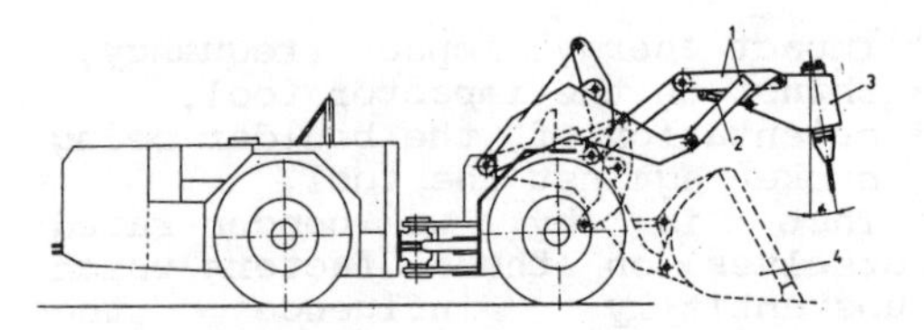

Fig. 2 Impactor on the LHD manipulation beam. 1 - manipulation beam, 2 - hydraulic jack of the beam, 3 - H-3x impactor.

12.58 - 17.7 $m^3.h^{-1}$ to the fragmentation of $d_{max} \leq 0.8$ m.

4 SELECTION OF THE MANIPULATION BEAM FOR THE IMPACTOR INSTALLED AT THE ORE PASS

At the selection of the manipulation beam it is necessary to fulfil the following conditions:

* separately solve the ore pass grate construction with taking into account its load during the boulder breaking in order to enable the boulder stabilization during the breaking,
* to chose a suitable place at the ore pass grate for the impactor instalation in order to have a reach over the whole grate area,
* the operator site has to be safe.

Limit positions of the manipulation beam with the impactor with the weigth of G_1 = 1000 kg, hammer length with the tool - 1700 mm over the 3 x 3 grate are shown in Figure 3.

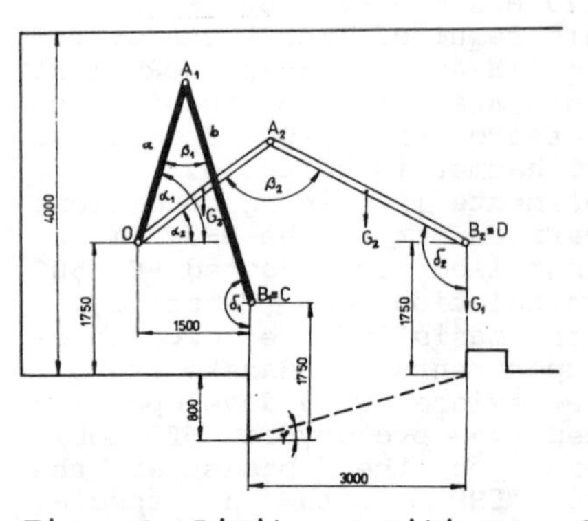

Fig. 3 Limit positions of the manipulation beam for the impactor installed at the ore pass grate in the stope. a - OA_1 boom, b - A_1B_1 beam.

As a result of the investigation the following prototypes were developed:

* IRV-250 mobile equipment,
* SN-1 stable equipment with the hudraulically operated manipulation beam and H-3x impactor.

Both of the equipments were successfully tested at the secondary siderite and magnesite breaking.

REFERENCES

Byzov,V.F et al. 1986. *Razrushenie negabaritnykh kuskov gornykh porod*.Technika: Kiyev.

Grantmyre,I. & Hawkes,I. 1975. High-Energy Impact Rockbreaking. *CIM Bulletin* 8:63-70.

Krištín,Š. 1983. Mechanické sekundárne rozpojovanie skalných hornín v banských podmienkach. *Dizertačná práca*. Košice: Banícka fakulta.

Krištín,Š. 1986. Niektoré aspekty sekundárneho beztrhavinového rozpojovania hornín v banských podmienkach. *Rudy* 10:291-296.

Krištín,Š. 1987. Príspevok k rozmerovému riešeniu manipulačného ramena ťažkého zbíjacieho kladiva. *Rudy* 6:159-161.

Krištín,Š. 1992. Wtórne rozdrabnianie bryl skalnych za pomoca mlotków udarowych bez stosowania strzelania. *Seminar*, Gliwice:1-4.

Kuklin,I.S. et al. 1969. Rozrusheniye negabarita gidropnevmatitcheskim butoboyem. *Gorniy zhurnal* 2:34-37.

Van Duyse,H. 1978. Utilisation des brise-roches dans les carrierés. *Annalen der Mijnen van Belgie*:558-606.

Lobanov,D.P. 1980. *Mashiny udarnovo deystviya dlya raurusheniya gornykh porod*. Moskva:Nedra.

Weustenfeld,H. 1980. Experimentelle und theotretische Untersuchung über das Eintreiben eines keilformigen Meissels ins Gestein. *Dizert. Arbeit*,p.10-20, 28-41. Aachen:Technische Hochschule.

Mechanical rock disintegration
Technical notes

Geomechanics 93, Rakowski (ed.) © 1994 Balkema, Rotterdam, ISBN 90 5410 354 X

Electromagnetic preventer

S. Jakabský, M. Lovas, T. Lazar, F. Sekula & S. Hredzák
Institute of Geotechnics of the Slovak Academy of Sciences, Košice, Slovakia

ABSTRACT: The article describes a non-conventional system of the killing a well. The killing a well is solved by the preventer based on an electromagnetic principle. Control system of the preventer equipped with pressure unit and data processing one is automatic.

INTRODUCTION

The preventers serve for mouths killing during pressure effects, i.e. the seepage and blow-out of oil and gas wells. Their construction has to enable the performing of necessary operations during the drilling, or during the pressure effects recovering. The present constructions of preventers enable to take up the pressures up to 17.5 MPa, at the diameter of casing strings up to 4 1/16". In connection with the catastrophic situation in Kuwait oil fields in 1991, it was solved the task of non-destructive mouth killing, having kept the possibility of their further utilization without the costly recovering work. It was suggested the magnetic way of mouth killing, which enables to take up the pressures up to 22 MPa. It was considered the application of special magnetic circuits for strings in safe depth under the ground level.

SYSTEM SPECIFICATION

At the indication of pressure start in the mouth, the annular space is closed by the preventer jaws, then in accordance with the pressure on the stand and in the annular space, it is controlled the process of pressure effect deadening. After the preventer shut-off it arises the immediate increase of pressure as a result of the so called hydraulic impact, in dependence on the flowing medium velocity. According to Zukovski the increase of pressure is

$$\Delta p = (10 \div 14)\, v$$

where
Δp – the difference of the original pressure and the pressure after shut-off
v – the fluid medium flowing velocity

It follows from the above mentioned equation, that it is required the gradual killing to eliminate the hydraulic impact, that is why it is necessary to take into consideration this fact in the design of preventer. The suggested magnetic closing consists of a magnetic circuit, which surrounds the casing and makes a magnetic field inside of it, where it is caught the ferromagnetic material dosed from outside. The amount, grain size and magnetic properties of the dosed material depend on the casing diameter, magnetic point length, intensity of generated magnetic field and also on the maximum value of seepage pressure.

For the calculation of magnetic field induction necessary for magnetic plug fixing, or the ferromagnetic material in annular space, it was used the following relation :

$$B \geq \sqrt{\frac{\mu_o . P . S}{S'}} \;,$$

where
μ_o – vacuum permeability
P – pressure in the well
S – annular space area
S' – contact surface of the ferromagnetic material with casing and the drill string

It is required the magnetic field

with the induction above 1 T to make the magnetic stuffing-box, which withstands the pressures about 20 MPa. The length of such magnetic stuffing-box, created by the ferromagnetic material catching, achieves 2 m at least. The ferromagnetic particles are structurized and kept by cohesive magnetic tensile in the annular space, which is induced by the field, whereby the intensity is perpendicular to the well axis. The required intensity of magnetic field can be reached using the solenoid, or the multipole magnetic circuit.

MAGNETIC CIRCUIT DESIGN

There were calculated the parameters of the disk shaped coils with turns, with non-uniformly distributed current density. The disks are radially cut up and in the cut point they are connected to series on the outside circumference.

The disposable sources of electric power and the required value of magnetic field intensity in the cavity of solenoid $H_o = 10^6$ $A.m^{-1}$ are the basic entering data for calculation. The relation between the coil power input P and the magnetic field intensity H is determined from the power output $P = R.I^2$. The coil resistance R was expressed at the calculation by means of the fictive resistance of coil R_f with one disk. For the roll system, which contains N disk we will calculate the value of magnetic tension:

$$U_m = \frac{P.\xi.h.(r_2-r_1)}{\pi.\rho.(r_2+r_1)}$$

ξ - coefficient of coil feeding
h - disk height
ρ - specific resistance of the used mate rial, of which the disk is made
r_1, r_2 - outer and inner radius

The coil efficiency is given by Fabry's coefficient G (α,β), the values of which for various shapes of coils are accessible in appertaining literature. By introduction of Fabry 's coefficient and the notation $\alpha = r_2/r_1$, $\beta = h/2r_1$, we get the following relation after arrangement for the intensity of magnetic field in cavity of assumed coil:

$$H_o = G(\alpha,\beta)\sqrt{\frac{P.\xi}{r_1\rho}}$$

It was performed the calculation for the coil, made of the following materials : aluminum bronze, tin bronze, carbon steel and cast steel. The common entering data for the calculation of coils of the above mentioned materials was the following : the magnetic field intensity (or the magnetic induction) in the cavity of coil, the solenoid height, the inner radius r_1, the unexceedable output of power source and the amount of flowing cooling medium (water).

As an example it can be shown a hypothetic proposal of the coil using the disks made of aluminum bronze.

- outer diameter D_2 = 1 000 mm
- inner diameter D_1 = 180 mm
- specific resistance $\rho = 13 . 10^{-8}$ Ωm
- disk section $S = 0.328 . 10^{-3}$ m^2
- disk ohmic resistance R = 0.735 mΩ
- number of disks in solenoid N = 1 124
- insulation layer thickness d = 0.25 mm (laminated glass)
- coil height h = 1 200 mm
- coil ohmic resistance R' = 0.82 Ω
- volume of active part of coil V = 0.684 m^3
- mass of active part od coil m = 5 200 kg
- power input of coil p = 131.1 kW
- current density in disk j = 1.22 $A.mm^{-2}$
- rate of flow of cooling liquid Q = 15 $l.s^{-1}$
- factor G = 0.132.K (K-regulation coefficient, it was taken into consideration K=3)

Designing the cylindrically arranged magnetic circuit, which creates a harmonic magnetic field, it was considered the relation for magnetic field intensity :

$$\bar{H}=H_o\left[e_r\left(\frac{r}{r_o}\right)^{n+1}\cos n\theta+\bar{e}_\theta\left(\frac{r}{r_o}\right)^{n+1}\sin\theta\right]$$

where
e_r, e_o - unit vectors
H_o - magnetic field intensity for $r = r_o$
n - half of poles number

EXPERIMENTAL RESULTS

To confirm the theoretical considerations there were performed the experiments with the dipole, quadrupole and solenoid type of magnetic circuit. The achieved results, i.e. the maximum pressure taking up in annular space, are in annular space, are in harmony with the calculations.

In Fig.1, there is shown the arrangement of experiment, which simulates the situation in the well during pressure effect. There were used the casings of 7" and 3" diameter, the length of

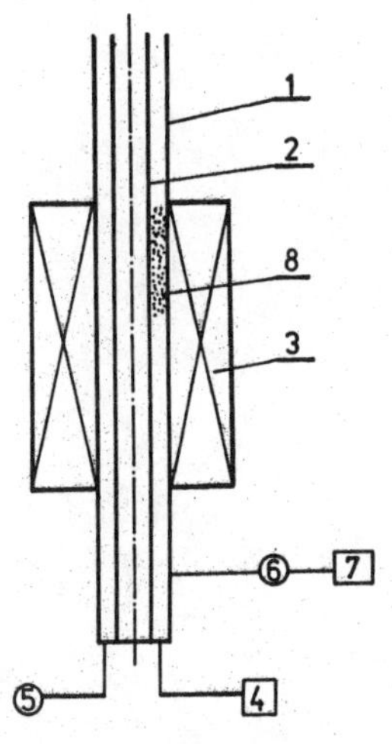

1,2 – casing pipe
3 – electromagnetic circuit
4 – high-pressure pump
5 – pump for suspension
6 – pressure gauge
7 – computer
8 – ferromagnetic material

Fig.1. The arrangement of experiment

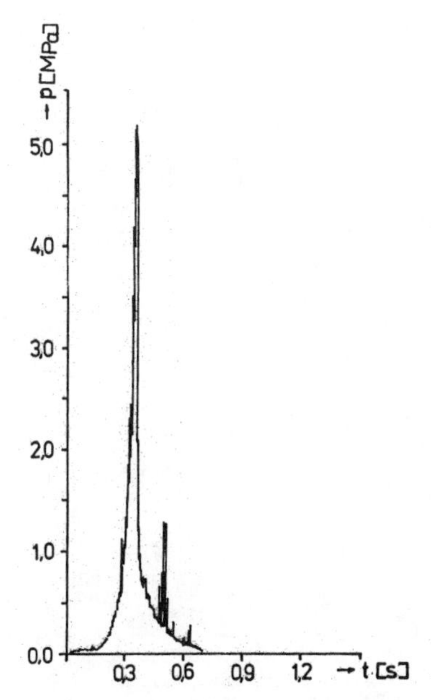

Fig.2. The pressure scanning

pole circuit was 0.45 m, it was used the powder iron like a ferromagnetic material. The pressure scanning was performed by the pressure gauge the electric output signal was analyzed by computer (Fig.2). The magnetic stuffing-box remained undisturbed at the generated pressure of 6.0 MPa, which was reachable as maximum by means of the used pump.

CONCLUSION

The magnetic plug can be used for stopping the fluid media flow in piping, whereby it does not occur their destruction. The magnetic system can work automatically in various modes after its fitting with a pressure gauge and a control power unit.

REFERENCES

Parkinson, D. H. & Mulhall, B. E. 1971. Inducement of strong magnetic field. Moscow. (in Russian)

Kulda, J. 1974. Magnetic field in heavy-current electrical engineering. Prague. (in Czech)

Geomechanics 93, Rakowski (ed.) © 1994 Balkema, Rotterdam, ISBN 90 5410 354 X

Processes of disintegration in rocks and minerals when blasting

K.L. Kolev & P.Y. Petrov
University of Mining & Geology, Sofia, Bulgaria

ABSTRACT: The results from theoretical and practical research are presented concerning the state of the stresses induced in heterogeneous Rock mass under the impact of the blast. The effects predicted, such as destruction and weakening of rocks and minerals, have been confirmed by experiments in laboratory and in situ conditions. A technology for a following mechanical treatment and floatation of the extracted rocks and ores has been proposed, based upon a proper determination of parameters of the drilling and blasting works, the effectiveness of which has been proven by its industrial application.

1 INTRODUCTION

The uncovering of the profitable components from the rocks yielded is related to the processes of disintegration at different levels: in mining where the rocks are extracted from the bowels, in the different stages of processing (grinding, milling, etc.), and during the floatation. In the sequence mentioned more and more smaller volumes, compared with the initial one, are subjected to disintegration and as a result of that the corresponding losses of specific energy for destruction respectively are growing up (Shemyakin, 1981). For initial fracture of rocks energy of the blast is mostly applied. Logically arises the question: is it possible to use this energy for alleviation and facilitation of the next processes of disintegration, i.e. the extracted minerals to be prepared for the following mechanical treatment and floatation yet in the stage of winning? For the clarification of this question theoretical and practical research has been conducted.

2 RESULTS FROM THEORETICAL RESEARCH

The main purpose of the theoretical research was to determine the magnitude and character (in elastic approximation) of the stresses in heterogeneous rock mass originated under the impact of the blast or, what is the same, under the influence of the waves of stresses propagating through the medium investigated. The research was conducted on the basis of fundamental principles of the Theory of elasticity and the Mechanics of continuum.

In the frames of the dynamic task of the Theory of elasticity the mechanism of emergence of stresses when dynamic impulse of pressure propagates through the polycristalline rock medium was studied. As a model an elastic bedded semi-space consisted of definite number of layers with different physico-mechanical properties was regarded. The examined model is very similar to that used by Kochetkov (1985), but we have considered more complicated situation when the dynamic load is applied at different angles to the surface of the semi-space.

Mathematical description of the task may be presented as follows: it is necessary to solve the wave equation:

$$(\lambda + 2G + \lambda tg^2\alpha)\frac{d^2u}{dx^2} = \rho\frac{d^2u}{dt^2} \tag{1}$$

under the respective initial, border and contact conditions, where: x is distance from the surface of semi-space; u is displacement of particles of the medium; t is the time; ρ is the density of material; α is the angle of application of dynamic load; λ and G are the Lamy's coefficients.

The solution was drawn in accordance to the laws for internal movement of continuous media and the requirements for dynamic equilibrium. The results obtained permit determination of magnitude and character of the stresses both into the layers and the bonds between them, depending on the parameters of a dynamic load, the conditions of its application and physico-mechanical characteristics of the rocks. A relation for displacement of the particles is derived, which is of importance in assessing the possible ruptures in the medium when defects of a different type exist (interstices, fissures, etc.)

The numeral verification reveals that the layers and surfaces between them are subjected to the influence of fast oscillating, sign-reversing loadings which in some moments exceed the outer load by amplitude. If an assumption were made that defects in medium do not exist the stresses tend to concentrate along the planes between layers of which the semi-space is built up.

3 RESULTS FROM EXPERIMENTAL WORK

The aim of experiments performed under especially designed methodology was to ascertain the validity of theoretical results. According to this methodology three experimental blasts were carried out by using explosives of different specific energy, respectively 1000, 1400 and 1750 kJm^{-3}. The experiments were performed on a stand, represented by a dead-end metal pipe, with diameter of 1200mm, height 600mm and thickness of wall 8mm. The dynamic impulse was transmitted from the charge towards the specimens by means of cement-sand mortar in ratio 4:1 in which the specimens of mineralized granodiorite in a form of regular cube (20 ±2mm) were arranged according to a scheme (on three levels and along three concentric circles) so as to avoid screening effects. After more than 28 days, needed for solidification of the mortar the blastings were done. During the experiments an representative number of non-loaded specimens was left aside for check tests. These specimens along with those gathered after each blasting were subjected to laboratory tests for determination of the compressive strength, owing to which is possible to judge of the effect of weakening due to the dynamic load. In the blasting conducted with a 1000 kJm^{-3} specific energy charge an additional number of 27 pyramidal samples (20x20x100mm) were put in a pipe for examination of some acoustic parameters of the rocks.

The analysis of the results from the strength tests shows that with increase of intensity of dynamic loading the uniaxial compressive strength of specimens fired is significantly less than the strength of those unfired. Reduction of strength of the samples fired with the lowest energy blast (1000 kJm^{-3}) is about 10 % compared to those unfired, whereas for more intensive loadings this difference is more than evident. So, in the charging with explosive of 1400kJm^{-3} specific energy most sensitive is diminution of the strength of samples closest to the chamber (1st circle) - about 40%, while for the rest of them (respectively 2nd and 3rd circles) it is lesser - 20-30%. The picture is some more distinct for the last case of charging with the blast of 1750 kJm^{-3} energy, where the reduction of the strength is of a scale of 40% and is typical for all of the specimens. Thus was drawn the conclusion that the strength of rocks could be significantly reduced for greater volumes of rock mass, by means of proper increasing of intensity of dynamic loading.

The considerable number of specimens tested and comparatively low values of the coefficient of variations of data assure fairly high reliability of the results obtained. The diminution of strength of the rocks is related, most probably, with the emergence of different defects, non-reversible deformations, etc. For examination of the structural changes, due to the dynamic charging, the velocities of longitudinal (v_p) and transverse (v_s) elastic waves were measured onto pyramidal samples cited above, by using ultrasonic unit Sonic Viewer, model 5217A, of a Oyo firm. With the data obtained the values of dynamic Poisson's coefficient (μ_d) were calculated, by which the eventual structural changes could be discussed. The velocities of elastic waves respond differently to the structural changes of the rocks, but, in any case, they tend to diminish when micro- and macrodefects occur. According to the studies of Margues (1974), Chen Rong (1979), etc., the ratio v_s/v_p used for assessing the fissuration, tends to decrease, because of the higher sensitivity of v_s. The analysis of the data reveals a steady tendency of decreasing of v_s for the majority of samples after blasting, while for the v_p, although not distinctively, the opposite tendency was observed. The ratio between velocities for most of the specimens before charging was 1:2 (on the average), while after the loading the same was reduced with about of 5%, what is evidence for some structural changes into the samples. Another conformation of this conclusion was the relative growth of the Poisson's coefficient μ_d, but it is known that with the increase of μ_d, for all real materials, the increase of plasticity is observed, as well as reduction of strength and hardness.

4 CONCLUSIONS

The results quoted along with the regularities determined persuasively prove the expedience of application of the impact of the blast not only for demolition of rocks and ores in extraction, but also for modification of their technological properties by means of increasing of the coefficient of efficiency of the blast. It is important to note that for achieving the purposes set it is essential to determine proper parameters of the explosive impulse, what is often neglected. A proven fact is that the character of dynamic load must strictly correspond to the character of deformation of the rocks.

On the basis of the results presented a set of practical decisions has been proposed for the optimization of extraction in congruence to the needs and economic interests of the mining firms. An illustration of the mentioned is the technology for blast preparation of rocks and minerals for next mechanical treatment and floatation introduced in practice. For specific conditions an optimal level of energy saturation of the rock mass was determined, whereupon the total expenses for all of the processes before floatation, i.e. getting, charging, transport and mechanical treatment were minimum. From practical point of view that means realization of significant economical effect.

REFERENCES

Bieniawski, Z.T. 1967. Mechanism of brittle fracture of rock. *CSIR Report MEG 580.* p. 226. Pretoria.

Brace, W.F., B.W. Paulding & C. Scholts 1966. Dilatancy in the fracture in cristalline rocks. J. Geophys. Res., vol. 71 No16.

Kochetkov, P.A. 1986. Mechanism of weakening of rock mass under the impact of explosion waves.*Gornoi Jurnal*. 6: 63-66.

Lomtadze, V.D. 1990. Physico-mechanical properties of rocks. *Nedra.* 216-219. Peterburg.

Margues, J., S. Derlich 1974. Etude de quatre roches sous tres hautes pressions.*Comph. 3rd Congr. ISRM.* vol. 2A: 493-498, Denver.

Rong, Ch., Y. Xin 1979. Studies of the fracture of gabbro: *Int. J. of Rock Mech. Min. Sci.& Geomech.*, Abstrs., vol. 16: 187-193.

Rummel, F. 1974. Changes in the P-wave velocity with increasing inelastic deformation in rock specimens under compression. Proc. 3rd Congr. ISRM. vol 2A: 517-523. Denver.

Shea, V.R., D.R. Hanson 1988. Elastic wave velocity and attenuation as used to define phases of loading failure in cool. *Int. J. of Rock Mech. Min. Sci.& Geomech.* Abstrs. vol. 25: 431-437.

Shemyakin, E.I., V.I. Revnivtsev 1981. About one approach of assessing the energy outgoings of ore disintegration. *Obogoshtenie rud*. 6: 10-12.

Geomechanics 93, Rakowski (ed.) © 1994 Balkema, Rotterdam, ISBN 90 5410 354 X

Experiments in thermal spallation of various rocks

R. E. Williams & R. M. Potter
Petroleum Engineering Department, New Mexico Institute of Mining and Technology, Socorro, N. Mex., USA

Stefan Miska
Petroleum Engineering Department, The University of Tulsa, Okla., USA

ABSTRACT: The greatest limitation of the spallation process is its inability to spall (or to consistently spall) many rocks encountered in petroleum drilling and mining operations. The New Mexico Institute of Mining and Technology has conducted a series of experiments to investigate the possibility of expanding the use of the spallation process to the penetration of rocks generally considered not to be spallable. The methods used during this work were (1) spalling at temperatures below that produced by the stoichiometric burning of fuel oil and air, and (2) spalling by alternately heating and quenching the rock surfaces. No success was experienced in spalling at the lower temperatures, but initial tests showed the alternate heating and chilling system to be successful, particularly in penetrating travertine limestone. However, continued testing indicated that, unless the rocks are extremely uniform in composition, spalling will result in highly irregular holes or holes that cannot be directionally controlled.

1. INTRODUCTION

Many rocks flake - or spall - when they are heated. Because rock conducts heat poorly, rapid application of heat produces a thin, very hot layer on a rock surface. The rest of the rock, still cool, constrains the thermal expansion of the surface layer. Rapidly, stresses within the layer become so high that it breaks away from the cooler rock and flies or falls off as thin flakes or slabs, called spalls. In commercial applications the rock is usually heated with a supersonic flame jet. Reportedly, thermal spallation was known and used as a way to break rock in ancient times. Previous studies have shown that it works best in hard, compact, abrasive rocks where conventional drilling tools penetrate slowly and wear out rapidly. Spallation methods have been developed primarily for drilling shallow blast holes in open-pit mining of very hard rock or for quarrying operations. More extensive application, however, requires thermal spallation drills that penetrate softer sedimentary formations found in seeking oil or gas. Possible application could also be found in demolition and rescue work if this process could be used to rapidly penetrate concrete.

Starting in 1988, the New Mexico Institute of Mining and Technology at Socorro, New Mexico, conducted a series of tests to investigate the possibility of expanding this process for rocks considered to be unspallable. This work was an extension of work previously performed by the Linde Division of Union Carbide Corporation, which developed a system used for many years in mining taconite, the Browning Engineering Corporation of Hanover, New Hampshire, which developed a system and manufactured tools for quarrying granite, and the Los Alamos National Laboratory, which investigated this process as a possible method of drilling deep holes for its Hot Dry Rock Thermal Energy Extraction Program. The Los Alamos work showed that spallation could be used for making large cavities in granite.

2. OBJECTIVES

The goal of this project was to evaluate new techniques for spalling concrete and the rocks commonly encountered in the mining and petroleum industry and to renew interest in thermal spallation. If successful, this process could greatly reduce drilling costs.

In order to achieve these goals, the following objectives were set forth:

1. To test the spallability of rocks encountered in oil and gas drilling and in producing blast holes for mining. This work would be done in an outdoor laboratory using rock specimens and in-situ.
2. To improve the design of thermal spallation drilling equipment, particularly as it pertains to the drilling head itself.
3. To demonstrate thermal spallation to prospective users with the intention of establishing a collaborative industry/university research initiative.
4. To provide well documented drilling experiences which can be analyzed theoretically.
5. To find commercial applications for this work that would demonstrate the feasibility and the economic advantages of spallation.

3. EQUIPMENT

To achieve these objectives, a small burner or *Lance* and ancillary equipment was designed and fabricated. It was constructed to allow the burning of fuel oil (or diesel fuel) in an air stream and provided for electric ignition assisted by oxygen during start up. Also, the lance provided for the injection of water into the exhaust stream downstream from the combustion zone, thus allowing control of the exiting gas temperature. This allowed the investigation of the effect of lower temperatures than those provided by the stoichiometric combustion of air and fuel only.

The operational and control set up for field testing includes a small winch with a variable speed control supported by a tripod. This tripod system was later replaced by a small drill rig. Air pressure for the tests were limited to no more than 760 KPa by the output of the commercial compressors available within a very limited budget. Bottled oxygen was used during ignition and the ignition was provided by an automotive type spark plug fired by a 5 to 7 KV spark provided by the 110 V commercial electric power elevated in voltage by a 100:1 transformer and controlled by a variac.

Flow of the fuel, air, oxygen and water streams was measured by float type variable area flow meters. Provisions were provided in the lance for the insertion of a thermocouple into the combustion chamber and into the exhaust jet. However, malfunctions of the equipment and difficulties in positioning the thermocouple resulted in temperature data that are not considered reliable.

4. EXPERIMENTS AND RESULTS

Earlier experimentation by others showed that lowering the temperature of the flame jet improved the spallability of some sedimentary rocks. Therefore, in order to verify this prediction, the lance was designed so that water could be injected into the exhaust gases downstream from the combustion chamber to produce a lower temperature than is obtainable by the stoichiometric burning of air and fuel oil or diesel fuel of approximately 2100°C. It is estimated that the flame temperatures in these experiments did not exceed 1800°C.

An examination of video records of spallation tests showed that accidental dripping of exiting cooling water on previously heated surfaces allows continuation of the spalling process. In the absence of this cooling process, spallation ceased due to overheating of the rock. As a result of this observation, some laboratory tests were made in which the rock samples were alternately heated and quenched by rotating them on a turntable so that the rock would be subjected alternately to the heat from the exhaust jet of the lance and a cooling stream of water. The jets were placed about 7 cm from the center of rotation so that a hole of approximately 20 cm diameter would be produced. Rotational speeds were varied, but the most favorable results were obtained at a rotational speed of 2 rpm, so that the rock was subjected to a cold and hot blast every 30 seconds. Rocks tested were largely those of the wide variety that could be found in the vicinity of Socorro, NM. The most interesting tests are discussed here and more complete information on these tests is included in an appendix which, due to space limitations, is not included in this report, but is available from the author upon request. Also the figures showing equipment details and photographic information of some tests are available upon request.

4.1 *Limestone*

Travertine limestone from a quarry west of Belen, NM was tested using a variety of temperatures of exhaust gases obtained by varying the flow of the tempering water. Initially none of these experiments proved successful. However, later tests using the alternative heat and quench procedure produced a uniform hole 15 cm in diameter and 15 cm deep (the thickness of the rock sample). The advance rate was about 60 cm per hr with the limited flows available. The rocks that were successfully spalled were very compact. Later tests were made with less compact rocks from the same quarry. Portions of these rocks spalled poorly or not at all, producing irregular holes.

A possible explanation for these surprisingly successful results has been provided by Dr. Thomas Dey (1988) of the Los Alamos National Laboratory. He informed us of an earlier comprehensive study of the effect of temperature on the brittle to ductile transition of Solenhofen limestone by H. Heard (1960) which shows this transition temperature at zero confining stress to be approximately 480°C.

Rauenzahn (1986) and Wilkinson (1989) observed a transition temperature in their spallation tests. They define "global spallation onset temperature" as the temperature of the heated surface at which the first spall is observed. For several spallable rocks this transition temperature ranges from approximately 350°C to 600°C over a wide range of heat flux. Further heating usually results in increased surface temperatures with increasing heat flux.

We may surmise that attempts to spall limestone without alternate quenching will result in a few onset spalls, but then the effective near surface temperature will rise past the brittle to ductile transition temperature and spalling will cease.

4.2 *Granites*

Local granites from Jordan and Anchor canyons in the Magdalena Mountains were tested and we were able to produce satisfactory holes without resorting to the heat and quench method. Some tests were made with the temperature reduced with tempering water, but the best results were obtained when operating the lance at a maximum temperature. The spalls were small and readily blown from the surface. These rock samples were all procured on the surface and had experienced some weathering. Because the samples were small they frequently would crack or split during the test.

4.3 *Rhyolites*

Local rhyolite proved to be spallable, though not as rapidly as the granite rocks. Here again the spalls were small and readily blown from the holes. The tests of the rocks brought to the field laboratory tested quite well. However, when we went to a mountain location, and tested the rocks in-situ, we found that some of the formation spalled much more readily than other parts, thus producing irregular holes. The holes varied greatly in diameter or drifted from the center line of the tool. We believe this was at least partially due to weathering of the rock. A fact that confirms this viewpoint is that we tested the same formation in an experiment at the base of a vertical cliff by making a horizontal hole in the vertical face of the rock. This hole, though slightly irregular, proceeded quite well for 3 meters when a malfunction in our equipment caused the test to be discontinued.

4.4 *Sandstones*

Several sandstones were tested and most samples spalled reasonably well, producing very small spalls. A sandstone that was cemented with calcite did not spall either when tested in-situ at the Fite Ranch or later when subjected to the hot and cold treatment on the turntable. Those sandstones that did spall produced very small spalls. A sample of sandstone was spalled successfully using the heat and quench method producing a hole 12 cm in diameter and 15 cm deep.

4.5 *Concretes*

Finished concrete spalls very rapidly on the surface where a thin layer of sand and cement is exposed. The continuation of the spallation process to any significant depth depends largely on the characteristics of the coarse aggregate used in the mix. We found some concretes made with rhyolite aggregate would spall without using the heat and quench method, though generally the heat and quench system was required to continue to any significant depth. In no test did the spallation exceed a rate of 30 cm per hr. The concrete samples were rotated and subjected alternately to heating and quenching as were the limestone samples, but in the concrete an annular groove was formed and the center did not break free.

4.6 *Amphibolite and Basalts*

These rocks spalled on the surface producing large and very irregular spalls, as large as 5 cm x 10 cm x .5 mm thick, and action stopped when the hole had advanced no more than a few centimeters into the rock.

4.7 *Quartzite*

Holes 12 cm diameter were produced at a rate of 3.5 m per hr in a quartzite sample from a mine located in northern Quebec and in two samples of quartzite from a mine located in Newfoundland. These holes were spalled with a direct flame. A very dense rock from the Quebec mine, described as pyrite, bornite and quartzite, spalled very poorly when subjected to the direct flame and also to the alternate heating and quenching method. One sample from the Newfoundland mine spalled while glowing red, a phenomenon witnessed only when testing this rock and one sandstone.

4.8 *Dolomite*

Tests on locally available dolomite produced only a few small spalls when maximum temperatures were used and no action was observed at lower temperatures.

5. CONCLUSIONS

The spallation process has rather limited applications. It has been proven to be quite capable in spalling some granite and has been used in the quarrying of granite. The ability to produce cavities in granitic rock could find numerous applications and it is probable that with further development this system could be used in drilling deep holes in granite for use in the Los Alamos Hot Dry Rock Program. With further development work it could be used in producing cavities in granite for numerous purposes. The testing that we have conducted in spalling various other rocks at New Mexico Tech indicates that its use seems to be very limited. However, as this work was largely done on weathered samples of rock it is possible that some of the rocks, notably rhyolites and sandstones, could be penetrated satisfactorily in uniform rocks found at depth. The testing required to prove this would only be applicable to rocks found in a local area. The successes in drilling granite and the formation of cavities reported by the Los Alamos National Laboratory should be investigated further to insure that this promising drilling technique is applicable to more than one granitic formation.

6. ACKNOWLEDGEMENTS

Credit is due to F. Eugene Beck for his many contributions to this work while he served as a Principle Investigator for this project at New Mexico Tech and to Robert M. Colpitts for his work in determining the rock descriptions included in the appendix (not included in this paper). Credit is also due to the American Society of Mechanical Engineers who originally sponsored publication of this information.

7. REFERENCES

Dey, T. Los Alamos National Laboratory, Internal Memorandum to Hugh Murphy, Symbol ESS-5:88-310, 6/21/88.

Heard, H., "Rock Deformation," GSA Memoir #79, pp. 193-266, 1960.

Rauenzahn, R. M., "Analysis of the Rock Mechanics and Gas Dynamics of Flame-Jet Thermal Spallation Drilling," Ph.D. Dissertation, Massachusetts Institute of Technology, 1986.

Wilkinson, M. A., "Computational Modeling of the Gas-Phase Transportation Phenomena and Experimental Investigation of Surface Temperatures During Flame-Jet Thermal Spallation," Ph.D. Dissertation, Massachusetts Institute of Technology, 1989.

Author index